Gas Turbine Powerhouse

The Development of the Power Generation
Gas Turbine at BBC - ABB - Alstom

Dietrich Eckardt

Oldenbourg Verlag München

Editor: Dr. Gerhard Pappert
Production editor: Tina Bonertz
Cover picture: Irina Apetrei
Cover design: hauser lacour

Library of Congress Cataloging-in-Publication Data
A CIP catalog record for this book has been applied for at the Library of Congress.

Bibliographic information published by the German National Library
The German National Library lists this publication in the Deutsche Nationalbibliografie; detailed bibliographic data are available on the Internet at http://dnb.dnb.de.

© 2014 Oldenbourg Wissenschaftsverlag GmbH
Rosenheimer Straße 143, 81671 München, Germany
www.degruyter.com/oldenbourg
Part of De Gruyter

Printed in Germany

This paper is resistant to aging (DIN/ISO 9706).

ISBN   978-3-486-73571-0
eISBN 978-3-486-76968-5

Foreword

Gas turbines have become one of the most important engine types for electric power generation on a global scale – either as simple cycle installations or to a rapidly increasing extent as part of combined cycle power plants. This development is the result of many inherent advantages of this technique in comparison to alternatives – not the least its superiority in the category of cost of electricity.

During more than a century engineers of the Swiss development centre of A.-G. BBC Brown Boveri & Cie., from 1988 onwards ABB Asea Brown Boveri Ltd. and since 2000 Alstom Power Ltd. in Baden, Switzerland have significantly contributed to the achievement of today's advanced gas turbine concept and its successful integration in combined cycle applications. The present book provides a comprehensive and detailed insight on both technical and business aspects of the long-term genesis of this unique high technology product. This rather rare description of the development history of a thermal turbomachine puts also special attention on the human touch – a characterisation of leading engineering personalities of the time and the corresponding teamwork; triumphs and drawbacks accompany this fascinating professional account over decades – from a Swiss company nucleus of less than thousand to a global industrial company with several ten-thousands of employees. Numerous historical 'firsts' in gas turbine technology for power generation are highlighted – as summarized in Section 7.1 – ranging from the first realisation of the industrial, heavy-duty gas turbine in the 1930s to today's high technology gas turbine products, which combine excellent performance, extraordinary low environmental impact with commercial attractiveness.

Twenty years after Ernst Jenny's commendable book on 'The BBC Turbocharger' follows now the 'Gas Turbine Powerhouse' with a comprehensive description of gas turbine developments for power generation at BBC–ABB–Alstom. The book outlines not only the corresponding activities in gas turbine design and related disciplines, but covers also the historic development milestones of the major components – axial compressor, combustor, turbine and turbine cooling; the latter area with surprising, so far unknown revelations. The author Dietrich Eckardt provides a rather rare combination of technical and historic-economic insight, using engineering experiences of more than 40 professional years in turbomachinery research and gas turbine development, both for aero and power generation applications. The inclusion and thorough assessment of a broad variety of new sources and archive materials unveiled many interesting details of the gas turbine history, not the least surprising connections during the early parallel development of industrial and aero gas turbines.

On Friday 7 July 1939, 10.10 h started at Baden, Switzerland the full load certification test of the first power generation gas turbine for the utility plant at Neuchâtel, Switzerland – the first step of what has become a global success story and as such, a well-documented cause to celebrate this key event 75 years ago. This plant remained in service for 62 years – not bad for the first such order ever. The book provides the details to understand this engineering 'miracle', at best for the pride of present and the encouragement of future generations of engineers in this fascinating field of advanced technologies.

Baden, June 2013

Juerg Schmidli

Table of Contents

The writing of history, historiography,
is the 'doing of history', and engineers
should make sure that the historiography
of engineering is not left entirely to historians.
Recognition of this activity by the profession is very important.[1]

1   Introduction

This book tells the story of the power generation gas turbine (GT) from the perspective of one of the leading companies in the field over a period of more than 100 years and written by an engineer. With a global economic crisis imminent – triggered and accompanied by virtual stock market bonanzas and a few years after the illusions of a 'new economy' visibly failed – the time has come to reflect on real economic values based on engineering ingenuity and enduring management of technological leadership. For more than 120 years, engineers of the Swiss development centre of A.-G. Brown Boveri & Cie., ABB Asea Brown Boveri Ltd. from 1988 and Alstom Power Ltd. in Baden, Switzerland (CH) since 2000 have significantly contributed to the art of turbomachinery design in general.

In the history of energy conversion, the gas turbine is relatively new. The first utility gas turbine to generate electricity at Neuchâtel[2], Switzerland ran at full power at Baden, CH on 7 July 1939[3] and was developed by Brown Boveri. The scope of this book starts somewhat earlier to explain influencing ideas and consequently foregoing developments contributing to this invention. It then describes the steady and impressive engine growth from the first local 4 MW emergency power unit until present-day advanced configurations that generate power for large metropolitan areas with unit sizes above 300 MW. This was no success story throughout, especially not in view of the actual business fortunes achieved. There were changes in the company affiliations of the various branches, changes in product portfolio and also some development dead ends. But a few of the original BBC product lines managed to stay in the frontline of advanced superior technology even from present global perspective.

---

[1] Presumably, this quotation has several fathers; the author, Werner Albring, put his encouragement in these thoughts, see also Footnote 7 for the whole context.

[2] Neuchâtel is the capital of the equally named Swiss canton in the French-speaking part of Switzerland, some 150 km south-west of Baden. The city of 30,000+ inhabitants is sometimes referred to by the German name Neuenburg, since it originally belonged to the Holy Roman Empire, and later Prussia ruled the area until 1848.

[3] Interestingly, the first gas turbine flight also took place in 1939 – only 51 days later. A Heinkel He 178 aircraft, jet-powered by a HeS3B engine of (only) 600 kN thrust, developed by Hans-Joachim Pabst von Ohain (1911–1988) took off on Sunday, 27 August from Rostock-Marienehe, D. In England, the 1930's invention and development of the aircraft gas turbine by the British Royal Air Force (RAF) engineer officer Sir Frank Whittle (1907–1996) resulted in a similar British flight in 1941. As will be outlined in a separate documentation, BBC was indirectly involved in both sides' developments of jet engines with advanced all-axial engine configurations, as they contributed their superior compressor design know-how.

In the meantime, the sheer duration of certain engineering developments over several decades allows interesting historic observations and deductions on inherent business mechanisms, the effects of technology preparations and organisational consequences. A look into the past bears revelations on the impact of far-reaching business decisions. The positive influence of strong, courageous and visionary personalities becomes visible to the same extent as the negative consequences of hesitation and idle waiting. These prospects of an in-depth review of its own historic background have led a number of companies to launch similar assessments. BBC started from a modest nucleus to become one of the largest engineering companies worldwide with operations in around 100 countries. In 1990, it had more than 220,000 employees under the label of ABB. The engineering conglomerate was and still is amazingly innovative and successful in a broad variety of engineering product ranges: DC and AC (direct and alternating current) motors, turbo-, diesel and water generators, transformers, high-voltage transmission and grid equipment, switches and relays, steam turbines for power and ship propulsion including related gearboxes, steam-generating Velox boilers and power generation gas turbines, electric and gas turbine drives up to and including complete railway and streetcar systems, turbocompressors for the iron and steel industry, but also for aero engine applications, turbochargers, wind turbines, electro boilers and furnaces, nuclear power plant equipment, high-frequency radio and telecommunication equipment, liquid-crystal displays, vacuum tubes and semiconductors, power plant controls – to name just the most important. These product ranges comprise development activities as well as manufacturing and service operations in general.

Only one of these areas found comprehensive documentation in a stand-alone book so far: The BBC Turbocharger[4]. In 2002 an ASME paper[5] with a review of 'ABB/BBC Historical Firsts' in advanced gas turbine technology got some attention. It was then indeed Ernst Jenny, the author of the successful turbocharger book who became a strong proponent of the idea to write a follow-on history of gas turbine development[6]. Industry history, especially that of Germany and Switzerland in the context of the 2nd World War, received the special attention of professional young historians in recent years. We owe them many fundamental clarifications and also sometimes rare technical 'golden nuggets' as a result of their unrelenting, in-depth ploughing through archive materials. When it comes to the technical interpretation of their findings, pure historians sometimes reach their limits, however. In line with this thinking Werner Albring[7], the 'nestor' of German fluid mechanics at TU Dresden for years provided strong encouragement to this project.

---

[4]   See Jenny, The BBC Turbocharger
[5]   See Eckardt and Rufli, ABB/BBC Historical Firsts
[6]   Actually, the launching date of this book project can be exactly reconstructed. We met the late Ernst Jenny on 5 April, 2003 on the occasion of Georg Gyarmathy's 70th birthday (who passed away too soon on 24 October, 2009 at the age of 77). Gyarmathy, himself a former BBC director and successor to Traupel's chair for Fluid Machinery at ETH Zurich (1983–1998), was interested in technical history where he especially promoted the role of his Hungarian fellow countryman, the inventor G. Jendrassik, who built a small all-axial gas turbine and made early suggestions to apply the gas turbine for aero propulsion.
[7]   Prof. Dr. Werner Albring (1914–2007), was head of the Institute of Applied Fluid Mechanics at TU Dresden from 1952 until 1979 and author of one of the most intelligible textbooks on fluid mechanics. After his retirement he wrote several outstanding papers on the history of engineering and science (Helmholtz, Hagen, etc.) and in this context is known for his credo that qualified engineers should return to a responsible leadership role in view of the complex environment in industry and society, http://www.albring.info/

In fact, this is not the first approach to the subject. Our files contain a collection of material for a 'Swiss History of the Turbomachinery Industry' which was obviously planned in 1978/79, but the idea was dropped with the disruption of the BST industrial venture[8] at that time.

As had already been reported in the addressed ASME paper[5], one of the most intriguing aspects of the early BBC gas turbine history is the frequent in-depth involvement in parallel aero engine developments. Correspondingly, astonishing findings were made in the meantime as a consequence of further investigations and they would disrupt in full breadth the general scope of this stationary GT company history. A full roll-out has to wait for a follow-on publication.

	Book Survey					BBC/ABB/ Alstom
Year	Company Section 2	Headings	+ GT Key Components	++ Other Contents	Section	Historical Firsts
1900	02.10.1891 ↑ WW I BBC Brown, Boveri & Cie.	GT Forerunners		- Turbomach. & Turbocharg. - Early GT Attempts - Holzwarth etc. GTs	3	1893 1st AC Th. Powerplant 1895 1'000th Dynamo 1900 BBC Mhm. 1905 BBC centr. compr. for Armengaud-L. GT 1923 2-st. Turbo-Charger 1931 All-axial VELOX turboset
1950	WW II	→ The 1st Power Gen GT 1927-1945	I Axial Flow Compressor	- Turbom. Dev. - Prom. Engs. - Early BBC GTs	4	1936 Houdry 'GT' 1939 4 MW Utlty. GT Plt. Neuchâtel 1948 40 MW GT Plant Beznau
	31.12.1987	Gas Turbine Technology Developmt. 1945-1988	II Combustor	- GT Dev.mt. - Prom. Engs. - Compet. Des. - Mech.Design - Prod. Sites - Special Projects	5	1957 4x27 MW All-GT PP(record) 1970 BBC #2 in GT PP sales 1980 12-st. trans. compress., PR=16 1984 GT with premix. comb. 1994 165 MW high-eff. GT24
2000	↓ ABB Power Generation Alstom Power	CCPP GT Techn. Breakthroughs since the 1990s	III Cooled Turbine	- GT24/GT26 - Prom. Engs. - Comb.Cycle - Palmarès	6	1997 365 MW CC GT26, 58+%

Figure 1-1   *Survey of the book structure*

Figure 1-1 illustrates the structure of this book in graphic form. The left-hand scale covers the period from the formation of BBC until today, with the various Company names shown

---

[8]   BST Brown Boveri-Sulzer Turbomaschinen AG was founded in 1969 as a joint venture between BBC Brown Boveri & Cie., Baden CH and Gebrüder Sulzer AG, Winterthur CH (after Sulzer had already decided to cooperate with Escher Wyss AG before) as part of a necessary concentration process to become more competitive e.g. with a common, standardized product portfolio (Section 5.1.3). The effort failed and BST was already resolved again on 1 July, 1974. It appears that the planning for a joint Swiss turbomachinery history was a relict of foregoing BST times. Existing materials and correspondence – see BBC, Geschichte des Schweizer. Turbomaschinenbaus, 1982 – between 07/1978 and 09/1979 foresaw BBC contributions from Cl. Seippel, L.S. Dzung and H. Pfenninger, but the parties obviously agreed to stop the effort after Prof. W. Traupel's excuse that the 3rd edition of his own book had higher priority.

over time in the next column. The history of this succession of companies from BBC Brown, Boveri & Cie. via ABB Power Generation Ltd. to the present ALSTOM will be told in Section 2. Sections 3 to 6 in principle follow the chronology, with a few exceptions. Section 3 outlines in short the centuries of collecting experience in turbomachinery, a description that normally starts with the introduction of the reaction principle by Heron of Alexandria. I tried not to follow the trodden path and looked for some lesser known examples with reference to the gas turbine and the Swiss location.

Edward Constant[9], the author of one of the most comprehensive and well-researched books on GT history, differentiates between a first, aborted gas turbine revolution (1900-1920, Stolze, Armengaud) and a second, successful attempt, mainly led by BBC in the 1930s.

I have maintained this structure in Section 3, where all 'early attempts with the GT principle' belong to the first category. Section 4 describes the path to the 1st power generation gas turbine at Brown Boveri, Baden, Switzerland in the timeframe from approx. 1927 until 1945. Besides a description of the actual development activities, the text focuses on the decisive component for the GT development success: the axial flow compressor. This principle of a combined chronological and subject-oriented order has been carried through in the following sections. In Section 5, the GT's 'middle component' – the combustion chamber – has been linked to the development period for the BBC gas turbines between 1945 and 1988, the end of BBC as an independent, stand-alone company after 97 years. The narration about the 3rd GT component along the flow path – the cooled turbine – then follows in Section 6 in the context of the most recent technology breakthrough – the success of the combined-cycle power plants after 1990. This presentation has a certain benefit, since the individual GT component histories are kept together, letting the inherent development rationale become more transparent. Moreover, with a few exceptions, like e.g. the first introduction of a transonic compressor design in the 1980s, this deliberately chosen structure fits the development highlights touched on surprisingly well:

- The successful realisation of the axial compressor was *the* precondition for the BBC success towards the 1st utility gas turbine in the 1930s. Vice versa, the lack of an efficient compressor unit was in most of the foregoing efforts the reason of failure.
- The intermediate phase from 1945–1988 saw at its backend the breakthrough of BBC's unique low-emission combustion technology.
- Finally, the highly demanding, integrated turbine designs with the combination of advanced aerothermodynamics and sophisticated production technology only materialised after 1990; but they also triggered reflections on the recently rediscovered, early beginnings of BBC's turbine blade cooling technology in the 1930s.

Besides this repetitive link of section headings/contents and the 3 key GT components in the core Sections 4 to 6, each of these chapters with varying emphasis contains a treatise of

- the developments during that period,
- the relevant organisational changes,
- the most prominent, dominant engineering personalities,
- the relevant market observations and – where applicable
- the competitive developments.

---

[9]   See Constant, The Origins of the Turbojet Revolution

In a short final Section 7 'Les Palmarès', the historical 'firsts' in power generation technology by BBC/ABB and Alstom are listed in chronological order, together with a list of the responsible GT Development Directors in Baden, CH during the covered period of nearly 90 years and of the dedicated members of the GT Development 'Hall of Fame' which since 1995 is awarded annually for individual, outstanding contributions to the gas turbine development activities.

This book is written to be read from start to finish as a continuous story, once in a while interspersed by summarising description and analysis. Details and lengthier excursions have been shifted to the Footnotes on the same page, where the patient reader may find a few 'nuggets'. I hope that the interaction of the various elements of the story as described above will not confuse but rather enlighten the readers together with the presumed advantages as the narrative proceeds. References to the used literature have been collected in the comprehensive Section 8 'Bibliography'.

On the other hand, anyone who prefers to use the book as a kind of reference is recommended to turn to the Index listings – of Names, Section 9 and of Subjects, Section 10. Section 11 shall assist understanding with a comprehensive list of used 'Nomenclature and Abbreviations', followed finally by Section 12 – a short portrait of 'The Author'.

At the end of these introductory remarks, special thanks go to Juerg Schmidli and Peter Rufli from Alstom Power, Gas Turbine Development in Baden, Switzerland for defining frame and pace of this project in a generous manner. This work was considerably facilitated by a thorough preparation of the relevant, notwithstanding huge literature body for this task. The Alstom-internal database 'GT History References' in the meantime covers nearly 1'200 objects (papers, journal articles, books) that have been collected, digitised and put into a searchable form by Robert Marmilic, who herewith prepared the reliable foundation of this project. The numeration of this database is also given in the attached Bibliography in brackets [...], as an extra-benefit for Alstom-internal readers. Several colleagues contributed extensively from their own broad development experiences and by carrying out a careful proof reading of the manuscript. Mrs. Joanna Stone helped to smooth the English text and so considerably alleviated the 'readability' of engineering explanations; the endeavour to produce English technical diction was followed in the tradition of former, internationally established house publications such as 'Brown Boveri Review'. Special thanks are owed to Claude Seippel's son Olivier (1926-2012), also employed in various functions at BBC, who helped to revive personal memories of his father, especially by providing insight into the BBC part of Cl. Seippel's diary notes.

Invaluably, the great resources of the ABB Historic Archive, Baden-Daettwil, CH (Docuteam Tobias Wildi, Mrs. Raffaela Luetolf and Norbert Lang) and of the ABB ZX Test Dept. Archive (Bernhard Schoenung, Hueseyin Coskun) have been made available for these studies – with thanks to ABB HR Management (Renato Merz, Volker Stephan). Mrs. Cornelia Bodmer maintained contact to the ETHZ Library, squeezing rare information sources out of NEBIS[10]. What could not be made available in Switzerland was still in reach of the ever-ready specialists of the MTU Aero Engines, Information Services team in Munich, Germany (Helmut Schubert, Reinhard Glander, Mrs. Sabine Hechtl), in the meantime probably the

---

[10] ETHZ on-line catalogue: NEtzwerk von Bibliotheken und Informationsstellen in der Schweiz (Swiss libraries network).

best-assorted library for gas turbine and turbomachinery issues in Europe. The powerhouse graphic[11] on the book cover is by, and courtesy of Mark Welsh, Jersey City, NJ, USA. Finally, the author wants to thank a number of present and former colleagues for significant help; they spent considerable time and energy digging deeply into their memories, archives and files to reconstruct the past in as much detail and colour as possible: Jan-Erik Bertilsson, Franz Farkas, Fredy Haeusermann, Jaan Hellat, Wolfgang Keppel, Hanns-Juergen Lichtfuss, Uwe Schmidt-Eisenlohr, Martin Schnieder and Konrad Vogeler.

No one can be an expert on such a long period and on so many different technical subjects as addressed in this book. Notwithstanding, the broad external support and a thorough study of the available sources, deficits and drawbacks in the presentation of the comprehensive materials cannot be ruled out. The overall responsibility for this book, the selection of contents and the picking of individual aspects, technically and otherwise, its pros and cons, inherent correctness and hopefully not too many flaws remains with the author – and his necessarily subjective view on this fascinating profession. Clearly believing in the 'wisdom of the many' I look forward to receiving comments and proposals for improvements. In this respect the book may find many generous readers who enjoy the intended broad, nevertheless concise approach to nearly one hundred years of unique, technical company history.

Baden / Munich, October 2013

Dietrich Eckardt

eckardt@bluewin.ch

---

[11] © Mark Welsh. The illustration stands here to visualise **the** powerhouse in generic form; actually, it represents the Hudson & Manhattan Railroad Powerhouse designed as a 'technical cathedral' by John Oakman and erected 1906–08, at the time of first gas turbine trials and 30 years in advance of the first practical introduction of GT power generation. http://jclandmarks.org/campaign-powerhouse.shtml

2   Survey

The following survey covers 120 years of uninterrupted company history; may the next three figures suffice to provide an initial idea thereof. Figure 2-1 shows the historic development from the early beginnings in the late 19th century until the present Alstom in a condensed graphic form as a 'company tree'. The ancestry unites famous names from all over Europe that have left traces in one form or another. The relevant links and intersections will be discussed in the following chapters of the individual company histories as they evolve along the mainstream of GT development.

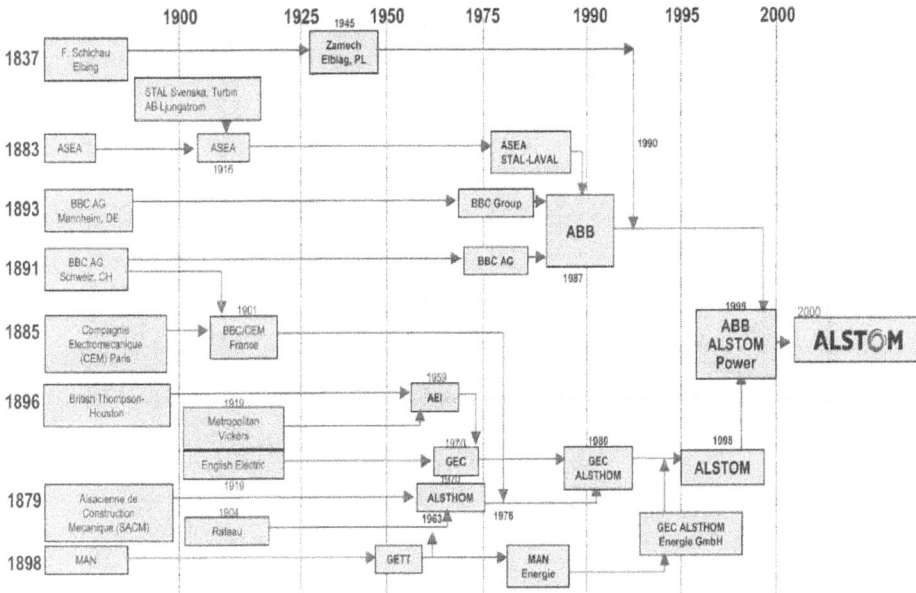

Figure 2-1   *Alstom company tree*

The whole industrial structuring process of course did not follow a straight and consequent course. There were drawbacks and detours. Economic and political crises[1] and management disappointments alternated with unforeseen successes over longer periods, all of this interrupted by the two world wars. The personnel numbers over time reflect these unpredictable events, nevertheless with a surprisingly stable growth trend over several decades.

---

[1]   The example of the economic figures for BBC Mannheim during World War II, Figure 2-12, shows that even a global political crisis could develop considerable economic momentum for BBC.

Figure 2-2 illustrates this trend on a global scale. While the BBC personnel grew rather steadily to a total of 50,000+ until the mid 1950s, the subsequent variations are much more pronounced and dynamic. At first, the number of employees nearly doubled until 1980. This was followed by stagnation and decline until BBC's official end in 1987. Next, the newly founded ABB Asea Brown Boveri in the 'Era Barnevik'[2] initiated another duplication boost with mergers and acquisitions to reach a top value of 220,000 ABB employees worldwide before 1995, followed by a steep decline of 27 percent in the following five years which saw substantial divestments, also including the transfer of the power generation business (and the addressed gas turbines) to Alstom S.A. in the year 2000. Thereafter, the reference employee numbers became somewhat uncertain, but to simplify matters, the present Alstom and ABB figures combined show an apparent recovery of the personnel numbers to approx. 200,000 in total for the year 2010.

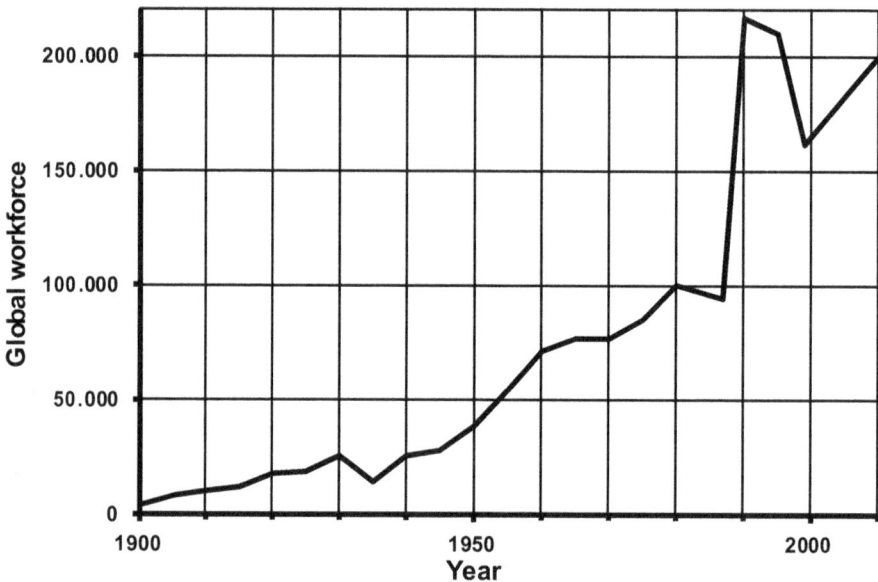

Figure 2-2   *BBC–ABB–Alstom global development of personnel numbers*

Figure 2-3 shows the corresponding employment figures for Switzerland alone, where the Brown and Boveri workforce may have comprised approx. 800 people in 1900. The Swiss employment followed the global trend in general, so that up to the 1960s the employment

---

[2]   Percy N. Barnevik, born 1941, CEO of Asea (1980–87) and of ABB (1988–1996), ABB Chairman (1997–2001). *'During his 8 years as CEO of Asea followed by the 9 years as CEO of ABB, the company achieved an increase of stock value of 87 times or 30% average per year over the 17 years. The net profit increased 60 times and sales 30 times. Based upon these extraordinary results, Barnevik received a one-off payment of 148 million Swiss Francs when he retired as CEO in 1996. In 2002, six years later under a second succeeding CEO, ABB's stock market value plummeted from 54.50 Francs in 2000 to just under 15 Francs. Barnevik recognized his involvement on this matter, both agreeing to repay half the amount of the payment and resigning from the advisory board.... Since 1999,... he has received special attention in the media for his work with entrepreneurship, Self Help Groups, and microfinance among women as a way out of poverty. By the end of 2007, 272,000 women have been organized and trained and have started 106,000 small enterprises.'* (Quote Wikipedia)

at the Swiss mother company consistently represented 25 percent of the total. Thereafter, the Swiss content decreased – slowly at first – and is now at a level of close to 5 percent only.

The companies BBC–ABB–Alstom have always been at the forefront of technological development over time. Proof of this statement over a company history of more than 100 years is not easy, though. An indicator may be the number of patents granted annually in the name of the various companies over time, Figure 2-4.[3] Since there is always a time shift of several years between patent filing and approval, there is no complete correlation with the development of the global workforce, Figure 2-2, but the coincidence is striking. A first innovation peak occurs for BBC between 1930–1934 with the axial compressor and the corresponding Velox boiler developments, Sections 4.2 and 4.3.1. During WW II and the early 1950s, innovation progress appears to stagnate, and a somewhat delayed recovery back to the 1930s level can be only realised by the mid-1960s. The best BBC values occurred between 1975–1980. Thereafter, quite strangely, there is a decrease in patenting activities, almost as though a certain innovative exhaustion anticipates the coming end of the company. Possibly, this trend was also caused by a change in the company's patenting policy for cost reasons.

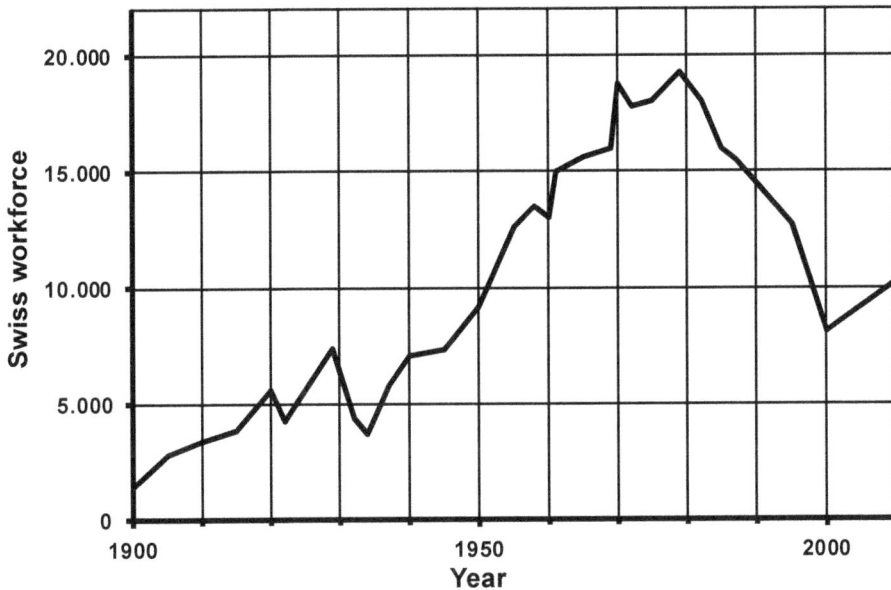

Figure 2-3   *BBC–ABB–Alstom personnel development in Switzerland*

[3]   The data result from mid-2010 <espacenet> searches in the database of the European Patent Office; they cover the complete industrial product portfolio of the investigated companies. Not shown is the first BBC decade in the 1890s, during which the young company already received 31 patents, presumably most were for inventions of C.E.L. Brown, equally distributed in the years 1893 and 1899. Up until the mid-1930s, BBC filed patents solely in the name of the company. Thereafter the name of the inventor is also noted in most cases.

The drawback of this decision could still be felt 15 to 20 years later, when the company realised that e.g. the IPR Intellectual Property Rights protection of Alstom's GT 'base fleet' and of corresponding spare parts showed painful gaps in its defence against attacking product 'pirates'. The last decade of the 20th century saw an unparalleled IPR growth trend for both ABB and Alstom[4], followed by a similarly steep downturn by a total of forty percent in the number of issued annual patent grants until 2005/2006, before rising numbers can be observed again.

In a combination of Figure 2-2 and Figure 2-4, Figure 2-5 illustrates the development of innovation power over time and thus eliminates the sheer workforce impact on patent output. The peak of engineering creativity between 1929 and 1934 becomes clearly visible with nearly 20 patents per 1,000 employees annually. The subsequent steep decline is only interrupted by a short, war-related spike in 1942. The slow post-war recovery has already been mentioned, culminating in a flat, nevertheless solid ridge over 20 years at a level of minus forty percent below the peak values of BBC's high time some 45 years before. Again, the steady decrease covers nearly the complete decade of the 1980s until a pronounced low, before another peak accelerated by energetic engineering activities in power generation marks the success of sustainable research and development efforts.

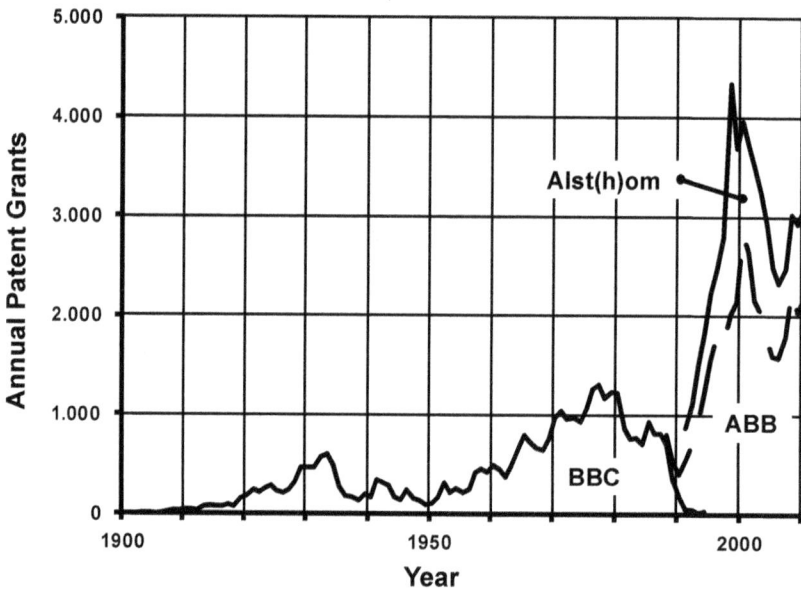

Figure 2-4   *Technology growth: annual patent grants for BBC–ABB–Alstom*

---

[4]   Though Alstom only bought ABB's power business in 2000, for comparison purposes it is worthwhile extending the Alstom view over the foregoing decade as well. This confirms that the growth in the number of patents granted to ABB was no standalone phenomenon. Consequently, the Alstom figures also comprise the corresponding Alsthom results before the name change in 2007.

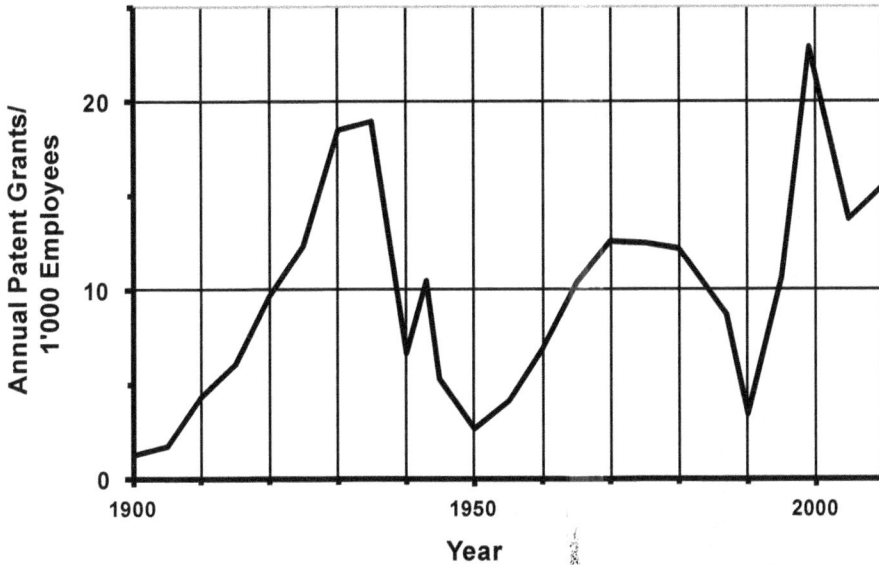

Figure 2-5   *Power innovation over time: annual patent grants per 1,000 employees of the global*
BBC–ABB–Alstom workforce

2.1   Short Company History – Power Generation

BBC and its successor companies are part of the history of technology that gave rise to the
greatest inventions of the past century. Gas turbines and more generally 'turbomachinery'
emerged in the wake of early electrification. Interestingly, the names of those pioneering
companies are today still amongst the leading players:

In 1867, the German Werner von Siemens presented the first 'dynamo' after having discovered the principle of electrodynamics. In 1879, Thomas A. Edison invented the light bulb,
establishing the basis for the creation of the powerful General Electric (1895). Already before 1891, the year of BBC's foundation, Charles E. Brown had succeeded in two historic
strides to demonstrate the possibility of transmitting AC power over longer distances:

–   First in Switzerland, between Kriegstetten and Solothurn[5] (8 km, 25–40 kW) in the
    year 1888
–   and then in 1891 by transmitting 220 kW over the record distance of 175 km in Ger-
    many from a hydraulic power station on the Neckar River at Lauffen, 80 km upstream
    of Heidelberg, to an electric exhibition site in Frankfurt-on-Main via a 15 kV power
    line and thus, unheard of minimal losses.

After Ferraris at Turin (approx. 1885) and Tesla in the USA (1887) had discovered the principle of the rotating electric field, propagating the usage of alternating current, several de-

---

[5]   The end user Sphinx Werke AG was a turning shop working for the local watch industries and preferred the
      expected, low-vibration electric drives for precision manufacturing over the established roughly running
      steam engines. See Lang, Brown and Boveri, p. 24.

signers focussed on the three-phase AC motor. For the Frankfurt electricity exhibition Charles E. Brown built the 40 pole 200-kVA-generator for Lauffen at MFO MaschinenFabrik Oerlikon, Section 2.1.1, while the corresponding motor at Frankfurt was developed by Dolivo-Dobrowolsky, AEG Allgemeine Electricitäts-Gesellschaft. The three-phase Alternating Current (AC) generator contributed considerably to the success of the new company, when this ingenious design became part of a power generation turboset with steam turbine (ST) drive from 1900 onward.[6] Instead of producing the required magnetic field in the stator or in rotors with salient poles, Brown's design for the first time generated it in a solid, fast rotating smooth rotor, on which the field coils were arranged in milled grooves.

Similarly, Brown's two unique oil-cooled transformers used to step up the voltage to 15 kV for transmission were an essential contribution to the rapid spread of electricity applications. This configuration reduced losses to an absolute minimum – and the young pioneer and his new works gained an international reputation, Figure 2-6.

Figure 2-6   *BBC AC generator and 'control centre' at Lauffen for power transmission to Frankfurt, approx. 1891   © ABB*

From this moment on, driving power no longer had to be generated and consumed at the same site. An electric cable could now link the source of energy with the place at which it was utilised; centuries of restricting mechanical transmission equipment and consequently, of only slow industrialisation were over. And also, the master of all electric trades had been proven wrong:

> *"Fooling around with alternating currents is just a waste of time. Nobody will use it."*
> – Thomas A. Edison, ca. 1880

---

[6]   For the early BBC history and the pioneering works see Lang, Brown and Boveri. For the decisive business-generating combination of Brown's high-speed AC generator with a steam turbine drive, see Footnote 13 and Figures 2-8 and 2-9.

2.1.1　BBC Aktiengesellschaft Brown Boveri & Cie., 1891–1988

On 2 October 1891 the following entry was made in the commercial register of the Swiss Canton of Aargau: *'Charles E. L. Brown of Brighton, England and Walter Boveri of Bamberg, Germany, both residing in Baden, have established a limited partnership under the company name Brown, Boveri & Cie., Baden. The nature of the business: Fabrication of electrical machines'*. The two founders complemented each other perfectly: Brown was the technical wizard and Boveri, the dynamic businessman, Figure 2-7.

Figure 2-7　*The pioneers Charles E. L. Brown (1863–1924), J. Walter D. Boveri (1865–1924) in 1891, the year of BBC's foundation　© ABB*

The colourful personal history of both pioneers and their families over the years found technically competent biographers on the occasion of various milestone anniversaries.[7][8]

Charles Eugen Lancelot **Brown** was born in Winterthur, CH in 1863, the son of Eugénie Pfau and Charles Brown sr. His mother was the daughter of a tiled-stove builder, and his father was a steam engine designer who had come to Switzerland from Uxbridge, today the north-western end station of London's underground Piccadilly Line. Brown sr. played a major part in building up the Winterthur operations of the SLM Schweizerischen Lokomotiv- und Maschinenfabrik (1871) and years later was involved in setting up the Electrical Engineering Department at the Oerlikon[9] Machinery Works (MFO). In 1885, Brown sr.[10] handed over this MFO-post to his 22-year-old son. It soon became apparent

---

[7]　See Lang, Brown and Boveri

[8]　See Dietler and Lang, Tradition and Innovation

[9]　Oerlikon, a district in the north of Zurich since 1934; MFO Maschinenfabrik Oerlikon, founded 1876, became part of BBC in 1967, and today is where ABB is headquartered. The Oerlikon 20mm-AA-gun came from the Oerlikon-Buehrle trust company WO Werkzeugmaschinenfabrik Oerlikon, an MFO spin-off since 1906.

[10]　Actually Brown sr. had decisive influence on the early success of the coming power generation company: dealing with high-speed steam engines brought him in contact with Charles A. Parsons, the inventor of the multi-stage axial-flow steam turbine, a proven concept also for the future gas turbine. Brown senior's many

that Charles Brown jr. had inherited not only a position from his father, but also extraordinary gifts as well. Prior to his debut in Oerlikon, Brown studied mechanical engineering at the Technikum Winterthur. It was at MFO that Brown met his subsequent partner, Walter Boveri, who soon recognised the talents of the mechanical engineer two years his senior and promoted Brown to head of the MFO 'Assembly Division'. Finally in 1891, after the short joint MFO intermezzo, the two BBC founders moved to their own company in Baden. Soon thereafter, when BBC was transformed from a limited partnership to a joint stock company of the same name in 1900, Brown was named chairman of the board. As is typical for his profession, the creative and impulsive engineer became unhappy with this management post and the changes due between 1909–1911, including the adoption of mass production. Consequently, he withdrew into private life as a disappointed man. In his 20 years of designing, Brown had dozens of inventions patented; technology was his world, but he cared little for the commercial aspects of BBC. In this respect the end of the 'over-engineered company' BBC (with an anti-bookkeeping mentality) sounded like a last echo of the spirit of its original founder. Along the same lines, the match with Walter Boveri was a stroke of good fortune.

Walter **Boveri** was born in the provincial town of Bamberg in northern Bavaria in 1865[11]. His father Theodor Boveri sr. was a physician, and his mother Antonie ('Toni') Elssner was the daughter of a lawyer. Walter Boveri attended the Royal School of Mechanical Engineering in Nuremberg, where he completed his studies within three years. He was just 26 years old when BBC was founded. Company visitors, asking him to direct them to the 'boss' or 'your father' he replied unshaken *"The father, that's me"*. Apart from his technical qualifications, Boveri was a tough and smart businessman. When Brown had left the company, Walter Boveri stepped in as chairman of the board; his son Walter E. Boveri (1894–1972) held the same position from 1938 until 1966 and was the last family member to lead the company. Aside from the founders, key company positions were also held by a brother of Charles jr., Sidney William Brown, who was technical manager from 1891 to 1900 and thereafter the board's delegate in upper management. One of Walter Boveri's cousins, Fritz Funk, headed up the administrative staff from the outset. Funk also served as board chairman for ten years, following Boveri's early death after a car accident in Belgium in 1924. Walter Boveri's brother Robert Boveri (1873–1934) was one of the directors of BBC Mannheim from the

---

11   years of correspondence with Parsons led to a license agreement in 1900 that enabled BBC to build the first steam turbine on the 'continent', Figure 2-9. In combination with Charles Brown's AC generator this resulted in a successful head start into the new century.

Boveri's family originally came from Savoy [it. 'i poveri' – poor people] and settled near Iphofen, lower Franconia at the beginning of the 17th century. They moved to Bamberg in 1835. Besides Walter, *1865, who was the 3rd of four brothers, the 2nd eldest, Theodor Heinrich Boveri (1862–1915), became a renowned German cell biologist, whose ideas are still followed today [see Neumann, Theodor Boveri]. He reasoned as early as 1902 that a cancerous tumour begins with a single cell in which the make-up of its chromosomes becomes scrambled, causing the cells to divide uncontrollably. He proposed the existence of cell cycle check points, tumour suppressor genes, and that uncontrollable growth might be caused by radiation, physical or chemical insults or by microscopic pathogens. It was only later that researchers demonstrated that Th. Boveri was correct with these predictions. He was married to the American biologist Marcella O'Grady (1863–1950), who was the first woman to graduate from the MIT Massachusetts Institute of Technology. Their daughter Margret Boveri (1900–1975) became one of the best-known post-war German journalists. Actually, Margret's accounts somewhat surprisingly also give insight into the early commercial BBC history. After Theodor Heinrich Boveri's death the family decided to split and sell the joint family property at Hoefen/Stegaurach near Bamberg, which Margret and her mother tried to keep together, but *'this hindered uncle Walter, in whose company BBC all our money was invested'*, so that they were only able to buy the still marvellously maintained 'Seehaus' ('Boveri-Schlösschen').

foundation of the German company in 1900 until his death in 1934; Robert's son Dipl.-Ing. William Boveri was the director at Mannheim until the 1970s.

The decision to locate the firm in Baden was made only after a very careful assessment of all the alternatives. Once made, everything proceeded very rapidly: Five months after the company was founded, 124 BBC employees moved into the offices and factory halls 200 m north of the Baden train station. Baden ultimately won out due to the Limmat River hydro-electric power station being planned there and other advantages. The power station meant a reliable supply of electricity and more importantly the first major order. In the months after its foundation, the generators and switchgear for the Kappelerhof hydroelectric power station filled the factory buildings. Brown and Boveri were pursuing a strategy that has since become a guiding principle: Manufacture products in the markets in which you sell them.

Charles Brown and Walter Boveri recognised electricity as the key form of energy for a new age. Their insight was shared by few at the time, making it difficult for Brown and Boveri to procure money. After a long search, Walter Boveri finally found a willing inves-tor for the required seed capital of 500,000 Swiss Francs (SFr)[12] in late 1890: Conrad Bau-mann, a silk magnate from Zurich, who would later also become his father-in-law.

Switzerland offered the BBC founders what they needed to get off to an excellent start, for the Alps had huge, as yet untapped, resources for hydroelectric power. The orders of the young company included equipment for power stations and subsequently electric railroads. The Jungfraubahn, a railroad electrified by BBC, carried its first tourists to the base of the Eiger Glacier in 1898 and by 1912 had been extended all the way up to the Jungfraujoch, 3454 m above sea level. The locomotives were considered technological wonders of the world at the time.

In addition to supplying the flourishing domestic market, BBC was export-oriented even in its first years of business. But high tariffs in other countries were an obstacle to expansive exportation. With the aim to penetrate new markets and avoid long-distance shipping, the company had to grant licenses or even establish local subsidiaries. The transformation into a joint stock company in 1900 was both an occasion for and a consequence of the interna-tionalisation already underway and helped to procure capital for the cost-intensive manu-facture of steam turbines. By the outbreak of the 1st World War, BBC had established a foothold in the key industrialised countries of Europe. After 1945, it proceeded to do the same on the American continent. Despite a number of setbacks and sell-offs, the BBC Group grew steadily, as already illustrated in reference to Figure 2-2.

Only one year after its foundation, BBC nominated their engineer Jean Jacques Heilmann to become the company's representative in France. The inventor of the Heilmann locomo-tive[13] was in close contact with Weyher & Richmond, steam engine manufacturers at

---

12 See Lang, Brown and Boveri

13 The Heilmann locomotives were the first steam-electric designs, using a reciprocating steam engine to drive DC generators, which in turn powered electric motors mounted directly on the axles. The initiative to improve the Heilmann locomotive brought Charles Brown sr. in contact with Charles A. Parsons (1854–1931), the inven-tor of the multi-stage axial-flow steam turbine. It was Brown sr. who saw the ideal qualification of the compact, high-speed steam turbine as a power generation drive. After corresponding over several years, a license agree-ment was finally signed in 1900; it permitted the young Swiss enterprise to prosper by building Parsons steam turbines in combination with its own high-speed generator exclusively for the European continent and several

Pantin near Paris. This enterprise laid the ground for C.E.M. Compagnie Électro-Mécanique to build and operate electric power plants as early as 1885, Figure 2-1. Since 1894 C.E.M. via Heilmann took the right to produce and distribute BBC engines in France and became BBC's full subsidiary BBC France in 1901.

The internationalisation of BBC was essentially shaped by one outstanding product:

In 1900 the company made the courageous and momentous decision to include steam turbines in its range of products, Figure 2-8. Watt's steam piston engines had triggered the first industrial revolution in the early 19th century; a hundred years later steam turbines coupled with generators were to play a role of similar importance. Rotating turbo-engines subject to constant impingement by jets of steam replaced the venerable piston steam engine.

Consequently, the production of turbine generators soon became a major line of business at BBC.[14] The fast-rotating, alternating current generator, a stroke of genius on the part of Charles Brown, led to the breakthrough of turbine generators at the turn of the century and to an influx of orders for BBC from around the world. By 1902, BBC had delivered 17 steam turbines, one of them with an output of three MW. By 1905 the product was already accounting for half of the total company sales, Figure 2-9.

Figure 2-8   *Steam turbine assembly, Baden ca. 1902*   © *ABB*

In retrospect, the early years must not be viewed solely as a time of technical innovation and success; they were also rife with intense labour and social disputes. Shareholders for their part suffered major disappointments in the 1920s and 1930s; economic difficulties

---

other countries. See Lang, Brown and Boveri, p. 21. According to Traupel, Marksteine – first steam turbine tests as part of BBC's coming series production were carried out in 1901, but the first Parsons steam turbine on the European continent was ordered directly and built at C.A. Parsons and Co., Newcastle, UK in 1899/1900 for the public utility services of the city of Elberfeld, Germany.

[14]   A special company was founded, the 'AG für Dampfturbinen System Brown-Boveri-Parsons' (public stock company for steam turbines Brown-Boveri-Parsons) for the acquisition of additionally required capital to erect the necessary, separate manufacturing plant with corresponding machinery in 1900, Figure 2-8. The license agreement with Parsons ended in 1912.

have been as much a part of BBC's history as grand triumphs. From 1903 to 1914 the German AEG Allgemeine Elektricitäts-Gesellschaft held a large part of the BBC shares. After WW I, BBC entered into a licensing agreement with the British manufacturing firm Vickers Ltd. giving the British firm the right for to produce and sell BBC products throughout the British Empire and in some European areas. The agreement gave BBC the promise of considerable future annual revenues, especially at times when protectionist international policies inhibited further expansion. Nevertheless, the BBC Group had no choice but to join forces with the powerful British Vickers Ltd. for a short time. Paying out a dividend was out of the question from 1921 to 1924 and from 1931 to 1938, but – in parallel – the technical, innovative progress in the newly founded company appears to have continued nearly unaffected. In Switzerland and abroad many BBC employees lost their jobs in the crisis years, Figure 2-3. 1924 became the 'annus horribilis' in BBC's company history for another reason: Charles Brown, 61 and Walter Boveri, 59 died during that same year. Brown had become seriously ill over time, while Boveri lost his life in a car accident in the Netherlands. The death of these highly appreciated company founders finished the build-up phase.

Figure 2-9 *First 'continental', serially produced steam turbine, BBC Baden 1902* © *ABB*

In the wake of WW I, Brown Boveri suffered heavy losses in its newly created network of successfully operating subsidiaries in Austria, France, Germany, Italy, Norway and the Balkans due to the devaluation of the French Franc and the German Mark. On top of this, the Swiss manufacturing costs in the domestic market grew considerably at the same time, while the domestic sales remained unchanged, causing further company losses. Consequently, BBC devalued its capital by 30 percent in 1924, and the promising license agreement with Vickers ran out and was not renewed in 1927. Serious economic problems prevailed until the early 1930s, even the turbine production in Baden was put into question and discussed in favour of a potential shift to Le Bourget or Mannheim. Finally, the supervisory board decided to proceed with the development centre in Baden and also to maintain a substantial development budget, otherwise *'we would lose our lifeblood'*.[15]

In view of popular business doctrines, the BBC development is a classical case study because it was one of only a few multinational corporations with subsidiaries that were larger than the parent company. So it is not too surprising that at times the Swiss central business unit ran into difficulties to maintain managerial control over some of its larger subsidiaries. For the next few decades especially, the relationship between BBC's Swiss headquarters in Baden and the expanding, increasingly mighty and independently operating German daughter company in Mannheim generated problems.

In 1934/1935 BBC had finally crossed the economic lowlands. BBC Switzerland saw a duplication of order intake out of the blue, though at marginal prices. In 1936 BBC share prices sky-rocketed, not the least thanks to the artificial German economic upturn based on straightforward war preparations; intermediately accumulated losses were completely recovered. During these years, BBC Mannheim surpassed the parent house considerably, Figure 2-10, and on the basis of a joint-stock capital of 12 Mio Reichsmark achieved actual revenues in the order of 112 Mio Reichsmark; consequently, the stock capital had to be duplicated to 24 Mio RM in 1938.[16] In reference to the starting values in 1933, the revenues had grown by 400 percent seven years later. Though BBC Mannheim was not directly involved in weapon production, its manufacturing of electric power generation equipment had of course considerable relevance in this respect. In the power regime BBC held the 3rd rank in Germany in 1939 with 17 percent market share, behind Siemens-Schuckert and AEG, both with 40 percent.[17] Soon the order intake exceeded the actual production capacities, so that e.g. customers for power plant equipment in Mannheim had to wait up to two years at the beginning of WW II. This gap between demand and actual production capability grew considerably during the war years, first throttled by Baden's opposition against further huge investments, after 1943 by significant destruction of the Mannheim manufacturing plants by allied bombing raids.

Since 1937, BBC Mannheim had become a dedicated 'marine plant', first by delivering steam turbine propulsion units up to 200,000 shp for several battleships, then as a prime source for German submarine electric drives. The other strong support out of Mannheim for the German Air Force since 1941 came from a dedicated department for aircraft turbo propulsion, in short TLUK; details on BBC's aero engine activities in Germany and Switzerland before and after WW II are planned to be described in a follow-on book project. A third fascinating area of 'dual-use technology' was wind tunnels for scientific research as well as for immediate military purposes. After the first installation of that kind, Ackeret's unique supersonic wind tunnel at ETH Zurich, 1932-1934, BBC was active in this field throughout Europe. While Baden specialised in these high-speed facilities for aerodynamic testing, Mannheim became practically single source for all kinds of high-altitude test facilities for piston and jet engines. Further details on these widely unknown areas of BBC's technology and product history have been collected in Section 4.2.4 as part of the early compressor development story.

---

[15]   See Catrina, BBC, p. 54: quotation from VR (supervisory board) protocol, dated Oct. 26, 1934.
[16]   See Catrina, BBC, p.55
[17]   See Ruch, Geschäfte und Zwangsarbeit, p. 81

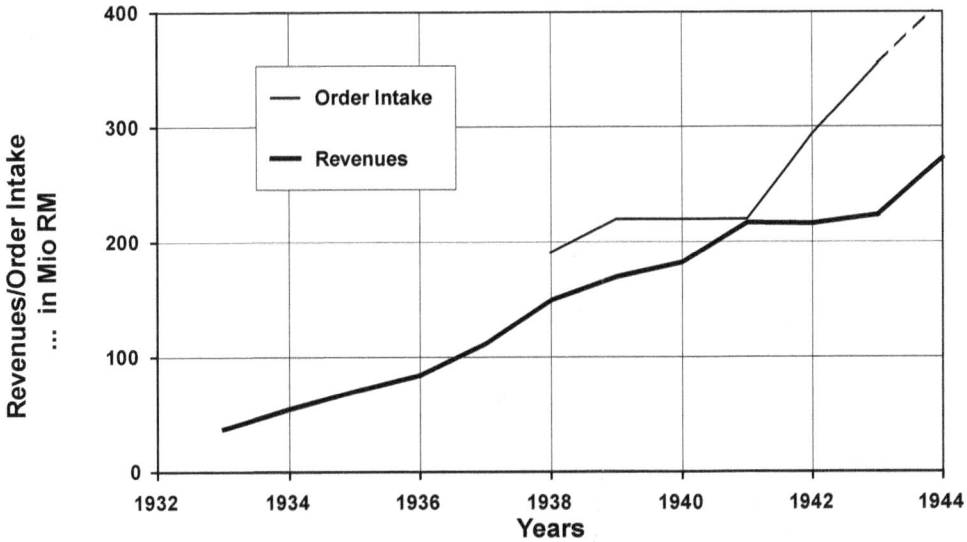

Figure 2-10   *BBC Mannheim, development of revenues and order intake 1933–1944*[18]

As an immediate consequence of the outbreak of WW II, in Switzerland 50 percent of the BBC personnel was mobilised to Swiss Army service, though over time the demand of the army decreased and some of BBC's employees were able return to the workbenches. The identification of the employees (and their families) with their Country and Company was extraordinary, Figure 2-11.

Currency exchange and strategic material problems hampered Swiss business activities during wartime, nevertheless BBC developed successfully and steadily as illustrated in Figure 2-3. BBC's workforce in Switzerland grew rapidly to 6,000 – with 1,500 'white collar' jobs in the offices.[19]

Beginning in 1939, the whole Swiss machining industry experienced a lasting war boom, both Germany and Italy as well as the Allies were treated as major customers. Since there are no suitable BBC data for this period, we refer here alternatively to Switzerland as a whole, assuming that these figures are also representative for the largest Swiss engineering company at that time. Figure 2-12 illustrates that the war parties dealt with the Swiss very differently. Dominating imports from Germany certainly also comprised war-related production orders, but to the same extent coal and iron/steel for Swiss needs and other exports[20]. German-Swiss trade agreements from July 1941 caused the British government to extend the sea blockade to Switzerland (with the exemption of food and feeding stuffs) while the USA continued to deliver to Switzerland, even after the outbreak of the German-American hostilities, thus also strengthening Switzerland's independent position towards Germany. The interests in the bilateral Swiss-German relationship were very balanced. The

---

[18] See Ruch, Geschäfte und Zwangsarbeit, p. 83 and 91
[19] See Catrina, BBC, p. 63
[20] See Eichholtz, Kriegswirtschaft, pp. 481–484

Figure 2-11   *'Proud To Be Swiss', BBC 'Halle 30'* [30], *the public day at BBC's 50th anniversary in Oct.1941*
saw 30,000 visitors, three times the population of Baden at that time   © *ABB*

Swiss Achilles' heel with respect to coal and steel imports was more than compensated by
the dependence of South German industries on Swiss electricity deliveries and the need for
unhindered Italian-German railway connections through Switzerland. The right-hand side
of Figure 2-12 correspondingly quite clearly illustrates the trend to increased export figures for war-related goods from Switzerland to Germany up to 1942. It can be assumed that
BBC covered a considerable share of the shown category 'Ironware'.[21] Though the Allies
already started to produce a permanently updated 'company blacklist' of those 'trading
with the enemy' early in the war, Brown Boveri was not listed. Obviously – and there will
be some proof in this direction – BBC was too important for both sides to be denounced
too severely. Strained capacities finally even led to the unique decision on the side of BBC
to limit the order intake from the Soviet Union in 1940 to 20 Mio Swiss Francs, not to become too dependent just from one customer.[22]

---

[21]   The data from the German Statistics Office as reproduced by < Eichholtz, Kriegswirtschaft, p. 483 > had not
been publicised during war-time. The focus on war-related goods is underlined by the fact that in the same
observation period e.g. the Swiss export of cheese to Germany shrank from 4.1 (1938) to 0.3 Mio RM in 1943.

[22]   The sources are somewhat vague in this context. The 20 Mio SFr limit is a quote from < Catrina, BBC, p. 62 >.
Internet sources make believe that the Baden deliveries towards the SU until the outbreak of the German-Russian
war in June 1941 comprised amongst others at least one of the planned three 200,000 shp steam turbine drives for

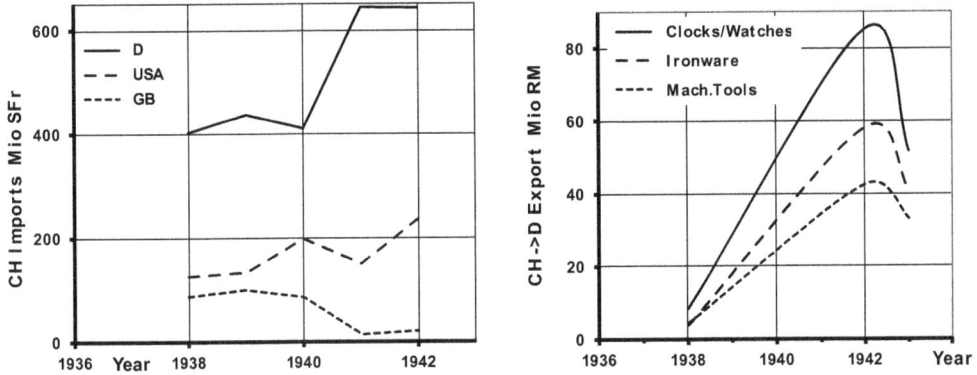

Figure 2-12   *Swiss imports from D, USA, GB (l) and exports to Germany (r)* [20]

The BBC Baden board started its post-war planning in time. The stock shares in most European subsidiaries had been wisely already written off to one SFr in March 1944; nevertheless, it was possible to invoice considerable license fees that had remained unchanged still then. End of 1945, a cash treasure of 47 Mio SFr and enormous material stocks put Baden in an excellent position for the upcoming, huge international reconstruction programmes. This capital stock was mostly used for the building of new fabrication plants between 1947–49, and for the improvement of production processes in general, Figure 2-4. By the end of the 1950s, a kind of series production had been established for key steam turbine parts, leading to a reported cost reduction of 25 percent. However, shortly thereafter, a certain ageing of the upper and middle management and of their style of handling businesses signalled negative consequences. The whole company accounting still only existed in rudimentary form.

One illustrative report in this direction came from Piero Hummel, who began in 1949, aged 26, and finished in 1988 as last CEO of the BBC group.[23] He started as mechanical designer in the steam turbine department, deliberately as he claimed and against the popular stream towards the more fashionable Gas Turbine department. When the renowned technical director Claude Seippel retired in 1965, Hummel became head of the Thermal Department. One of his first initiatives was 'vertical accounting' to assess the profitability of all major products in his eight sub-sections. This unveiled surprising results: BBC's total profit was smaller than that of the Thermal Department alone! Similarly, the turbochargers were identified as 'cash cows', while the electric departments – generating 70 percent of BBC's total revenues – contributed losses throughout. Thermal products and especially the highly profitable turbocharger business indirectly subsidised large sections of the group. Typically for the business organisation in the early 1950s, sales and production departments were not linked and consequently, without setting up a formal budget, decisive business figures were mutually unknown.[24] Strangely however, these revelations had no organisational consequences, yet.

---

the 75,000 ton battleships of the 'Sovetskiy Soyuz' class. It appears that also Mannheim produced heavily for Russian orders up until the very last moment. < Ruch, Geschäfte und Zwangsarbeit, p. 85 > quotes from BBC Mannheim supervisory board meetings that five altitude test facilities and 13 steam turbo sets had been ordered from Moscow; until June 1941 the (obviously unfinished) SU order backlog had grown there up to 23.3 Mio RM.

23   See Catrina, BBC, p. 82

24   See Wildi, Organisation und Innovation bei BBC, p. 19

Figure 2-13   *Proud of work and company, hydro generator delivery, Baden 1952*   © *ABB*

The post-war upturn of BBC is reflected in the growing personnel numbers, Figure 2-3, that to a large extent were attracted from foreign countries. This also became visible in the surrounding, logistically well-positioned canton Aargau (AG), Figure 2-14. Except for the short recession in 1975 and 1976, the canton's population grew steadily due to immigration, largely attracted by BBC's workforce demands. Percentage-wise, the share of foreigners rose from three percent in 1941 to 18 percent in 1970 and surpassed the 100,000 mark (19 percent) in 1996. This then corresponded to 28 percent Italians (compared to 57 percent in 1974), 32 percent from the Former Yugoslavia and roughly 10 percent from Turkey. Most recently, the Alstom acquisition of the power generation business and the signing of bilateral agreements between Switzerland and the European Union has further boosted the share of the workforce from EU countries, especially from France and Germany.

By the mid-1970s, more than 50 percent of the gross income produced in Baden was generated by BBC in one way or another. Although this dominating influence has diminished markedly, ABB and Alstom are still the most important employers and economically of key importance in the Aargau region.

From a global perspective, there was a huge demand for high-quality goods such as power stations, switchgear and transmission lines. The bulk of business was in steam and gas turbine-driven turbogenerators. The outputs of power units leaving the BBC shops took on undreamt of proportions. In 1962 BBC won out over tough US competition to deliver a 550 MW steam turbine, which was claimed to be a 'world record' in the annals of the industry. However, merely ten years later Brown Boveri advanced into the gigawatt range, producing units with outputs of 1,300 MW.

Figure 2-14 *Social consequences: Tower buildings to accommodate the immigrants of the 1970s, Limmat Valley, shopping centre Spreitenbach[25], AG, CH*

In 1970, BBC began a comprehensive reorganisation; the company's subsidiaries were split into five groups: Germany, France, Switzerland, the medium-sized production network with seven plants in Europe and Latin America and Brown Boveri International. Each of these groups in general had five product divisions: power generation, power distribution, railways, electronics and industrial equipment.

Throughout the 1970s, BBC struggled for access into the US market. E.g. the company discussed a joint venture with North American Rockwell to combine forces on Rockwell's 'sodium-cooled breeder' technology, but the companies could finally not agree on the financial terms.[26]

New lines of business, namely electronics and nuclear energy, were added to the traditional product portfolio in these years. BBC grew to become the largest electrical engineering company in Switzerland, and four out of the five Swiss nuclear power stations operate using steam turbine generators from BBC. But nuclear power failed to live up to the original expectations. Several of the advanced demonstration projects including the HTR high-temperature reactor in Germany had to be given up in the early 1980s, when it was realised that the technical readiness was not achievable according to original planning. Both Siemens and BBC had

---

[25] Source Wikimedia Commons http://de.wikipedia.org/wiki/Spreitenbach

[26] See Catrina, BBC, p.110 ff., interesting, though today nearly unbelievable, the speculation on the planned delivery of 5×1,000 MW nuclear power stations to the US market – per year!

to write off research and development expenditures in the order of several billions of Deutsche Mark (DM).[27]

BBC entered the global recession following the 1973 oil crisis with full order books. The large-scale power plants on order ensured that company capacities would be fully utilised for years to come. At the same time, sales shifted to countries less affected by the cyclical downturn. In the mid-1970s, growing power demand in the Middle East distracted BBC from its push into North America. Riyadh 8 in Saudi Arabia, one of the largest crude oil-fired gas turbine power plants in the world, is a prime example of that trend. In 1982, BBC landed a relieving turnkey order to build this 800 MW project to supply energy for the country's capital. 'Turnkey' in this context means, that the main contractor takes care of all of the client's issues surrounding the project. For Riyadh 8, Figure 2-15, BBC handled planning and development, provided financing, supported local manufacturing, trained client personnel and set up effective service networks. Even the building of a water tower and a mosque for the power station had been contracted to BBC.

Figure 2-15   *Riyadh 8, 800 MW – Gas turbine power station in Saudi Arabia   © ABB*

Changes in the global economic structures struck Brown Boveri with full force in the years around 1980. First problems had started in 1973, when the international system of fixed currency exchange rates had to be given up. The Swiss Franc subsequently gaining

---

[27]   Sometimes the Chernobyl disaster in 1986 is mentioned as major cause for the failure of the various nuclear projects. This may hold true for the public opinion, but e.g. the helium-cooled 300 MW 'Thorium High-Temperature Reactor' project with BBC participation at Hamm-Uentrop, Germany showed unforeseen technical problems with the pebble bed reactor ('Kugelhaufen') concept. After a lengthy construction period from 1970–1983 with costs overshooting excessively, the operation that had to be interrupted several times was finally terminated in 1989. The technical inheritance comprises 390 t of radioactive materials that can be dismantled earliest in 2027 when the critical radiation values will be diminished. [Wiki 'Kernkraftwerk THTR 300']

strength in combination with the 'oil shock'[28] led to the most severe recession of all industrialised countries in Switzerland. In due course the attempt was made to compensate the reduced export revenues caused by the 'cheap dollar' of that time with extraordinary company acquisition activities in North America, Figure 2-2. After several years, the BBC corporate group of North America comprised no less than six operational business units with 20 companies and 50 sales offices spread across the country. The initial buy was the takeover of the gas turbine division of Turbodyne in St.Cloud, Minnesota in 1977, which later on was renamed to Brown Boveri Turbomachinery Inc. However, the presumed cheap acquisition showed no lasting benefit. Quality problems caused more damage and the planned technology transfer (e.g. to stabilise the US production quality) from Switzerland to Minnesota never materialised properly, Section 5.1.3. As an immediate consequence of the oil crises, the order volume for gas turbine power generation in particular shrank considerably, the sales flattened out and the company's earnings declined. When the GT market cycle recovered in the US in the early 1980s, BBC had already given up and had decided to close the Turbodyne plant. Consequently, the logical idea of investing in an existing BBC licensee did not pay off. In many business areas orders fell off sharply due to the recessive trends in buyer countries. The all-important power station segment was hit particularly hard. Excess capacities in electricity production around the world created tough competition for the few available orders and narrow profit margins.

Besides these economic difficulties, the BBC management was additionally strained by the de-facto 'compulsory purchase'[29] of BBC France (C.E.M.) in 1982. BBC owned 79 percent of this company, which traditionally produced two thirds of its output for the French power generation market. Indirectly this development had also been induced by a reorientation of the government-controlled French nuclear energy programme, replacing General Electric reactors with those from Westinghouse. For secondary equipment this meant that all further national activities should be focussed solely on Alsthom-Atlantique and Framatom, without exception for the established No.3 in France: C.E.M. In Section 4 the successful, long-term partnership between C.E.M. and BBC, Figure 2-1, will be honoured; it includes many excellent engineering contributions especially from Georges Darrieus, whose role for the emerging technical leadership of BBC in the 1930s cannot be rated highly enough.

In 1983 BBC Mannheim, which was responsible for nearly 50 percent of all power sales, recovered somewhat. Nevertheless, short labour weeks and layoffs were the consequences of this economic downward plunge from 1983 onwards. In spite of an increase in orders, the inherent cost structure kept earnings down. In desperation, BBC set itself ambitious goals to become more profitable and efficient. The group trimmed capacities and instituted certain reorganisation measures that indicated some positive effects. However, competitive price decreases, made more severe by unfavourable shifts in currency exchange rates, compensated the achieved gains by and large. Finally in 1986, the Swiss parent company acquired a significant block of shares in the German subsidiary, bringing its total stake up to 75 percent again.

In 1987, the BBC Group comprised 159 subsidiary companies on all five continents. BBC still had an innovative image, but with less and less cost-effectiveness, the products were

---

[28] Amongst others, the oil price rose by a factor of four as a consequence of the first energy crisis.

[29] See Catrina, BBC, p. 173

correspondingly more expensive than those of the competition. On top of all this, the inherent rivalry between the Baden headquarters and the legally independent daughter company in Mannheim cast increasingly darker shadows over the relationship in general. Billions of DM investment in nuclear technology since the early 1970s had to be written off. Finally, on 31 December 1987 nearly 100 years of the technically impressive industrial history of Brown, Boveri & Cie. ended.

2.1.2   ABB Asea Brown Boveri Power Generation Ltd., 1988–2000

It was a great surprise when news agencies announced the merger of BBC Brown Boveri & Cie. and ASEA Allmänna Svenska Elektriska Aktiebolaget on 10 August 1987; on 1 January 1988 the two partner companies merged their many group companies to the newly founded ABB Asea Brown Boveri Ltd., the shares of which they held on a 50/50 basis. Two medium-sized international groups with strong national roots had been transformed into a single European technology group as the assumed answer to future world economic challenges. In sheer numbers BBC was larger than Asea – 25 percent in sales and 33 percent in personnel, but the Swedish trumped BBC in stock market value. In due course, BBC had to increase the share capital by 800 Mio SFr and to accept that several Asea affiliates remained outside the venture.

As a visible sign of the transformation the new ABB logo was unveiled in July 1988; in Baden it was prominently placed on the new engineering building Konnex, which was opened in 1995, a masterpiece[30] of architectural design and function by Theo Hotz, Zurich, Figure 2-16.

The initial phase following the merger was anything but easy for the new company. It even reported a loss in 1989. Only after ABB had concentrated once again on business areas with core competences and removed duplicate structures, was it able to return to sustainable profitability. This restructuring caused a considerable number of layoffs, but fortunately, this phase was short. Still in 1989, the volume of new orders had grown to such an extent that workers began to be rehired, with an inherent profound change. The personnel stock was tuned to highly qualified applicants in all areas, while less demanding shop activities were reduced. This implied a short-term change from labour-intensive, heavy machinery

---

[30] Appreciation for advanced architecture and for adequate industrial design has a long tradition in Baden, see Affolter, Architekturfuehrer, and correspondingly in the power companies; best known example is the 'AEG-Turbinenhalle' of Peter Behrens (1868–1940) at Berlin-Moabit, 1909, an icon of modern industrial architecture which today belongs to the Siemens energy sector. Since its foundation in 1891, BBC had asked for iron structures for their production halls. One of these rare, remaining historic buildings is the 50 m long 'Alte Schmiede' (Old Forge) from the year 1906 at ABB Area West. BBC's largest 'Halle 30' was erected in 1927 by MAN on the site of the present Konnex building; it was 150 m long and 20 m high –, see Figures 2-16 and 4-2. For a long time this was the largest hall building in Switzerland, so that it was also used for the canton Aargau's direct-democratic assemblies, see also Figure 2-11. This hall became the measure for a total of 8 buildings designed and built by Roland Rohn according to his general plan from 1944, e.g. the 'Hochspannungslabor' (High Voltage Laboratory), in 1942, today the TRAFO corner building, the 'Zentrallabor' (Central Laboratory) in 1956, along Haselstrasse, and todays ABB Turbo Systems main building in 1947 to 1952, along Brugger Strasse. A further architectural landmark is the 'BBC Gemeinschaftshaus' built in 1952/1953, on Martinsberg, which saw the BBC management and workers united, not only for lunch, but also in the evenings, say, to attend a piano concert by Arthur Rubinstein.

production to a smart and flexible engineering enterprise with dominating research and development characteristics. In this respect, an astonishing number of 3,000 newly defined jobs were created on short notice, again to a considerable extent from in-house (Mannheim, Nuremberg) and external engineering sources in Germany.

Long overdue, one bid farewell to the old BBC practice of producing entire systems; the in-house production depth was in most cases cut in half – from original values above 80 percent. The vertical range of manufacture was reduced, while the proportion of purchased components was increased. An immediate consequence was that highly qualified purchasing departments had to be established. E.g. complex gas turbine blading travelled at times up to 8'000 km amongst highly specialised production centres, distributed all over the world. These new procurement processes demanded the newly established SCM supply chain management for both cost and time reasons. Rolling out a reliable quality and process control regime over the whole supplier network was of utmost importance. Continuous Cost Cutting, beating lead times and Concurrent Engineering[31] became the names of the game for the coming decade. As a result of sometimes painful adaptation processes, the produced power generation equipment was significantly more cost-effective and regained competitive positions. Overall systems even gained in value for the customers, since the purchase of parts and sub-systems from specialised suppliers could generate added value. The prudent use of 'could' in this context indicates that this process of production outsourcing was not automatically and by all means a success. Especially, high-temperature gas turbine parts represent an amount of production complexity, individual suppliers could not cope with alone. Therefore, the network capabilities had to be built up jointly – a cumbersome effort with drawbacks. This was an experience that sometimes was even new to the upper management, who had gained personal, practical experience in the field at 'another times'. Consequently, to 'manage the management' in some instances became an additional task for those fighting problems under pressure in all-new territories. On a broader scale this trend for specialisation was also implemented in the internal production factory network, where the new Swiss production site in Birr qualified as a 'rotor manufacturing plant' (gas and steam turbines, generators), whereas Mannheim took 'stator manufacturing' and assembly tasks.

In hindsight, the period of the power generation division with de-facto bankruptcy of BBC in 1987, then the successful recovery under the newly founded ABB regime followed by the Alstom takeover in 2000 appears like a dynamic outbreak of long-term stored management energies in comparison to the earlier, apparently somewhat complacent business developments in foregoing decades. A direct comparison of that period as expressed in the personnel development numbers of Figures 2-2 and 2-3 illustrates on the one hand the stunning performance in the early 'era Barnevik' from a global perspective, and on the other hand that during these transactions the Swiss share of the workforce in these companies could not maintain the volume achieved in the 1980s and is today – expressed in sheer numbers – back to a level of the mid-1950s.

---

[31] Concurrent Engineering stands for the parallel, collaborative execution of design, mechanical engineering and manufacturing development tasks for a certain design part instead of the traditional, sequential execution. This saves time and trades design demands against aspects of manufacturing feasibility and cost. In combination with an international supplier network, this demands a very flexible management of highly qualified and at the same time broadly trained engineering resources.

Figure 2-16   *ABB engineering building 'Konnex', 1995, designed by Theo Hotz, Zurich*

The turning point was in or shortly after 1995. The foregoing ABB years had been characterised by an unprecedented rush of activities. The main products of the ABB group were
clearly power plants, mainly gas turbines, steam turbines and the combined cycles of both,
locomotives, turbochargers and electric switchgear. All these areas were immediately driv-
en to a comprehensive renovation and modernisation of the product portfolio and of the
organisation – very often even in an internally competitive mode. Armin Meyer, President
and CEO of the now independently managed ABB Power Generation Ltd. from 1992, states:
*"Looking back, I remember three things from the fusion period: first aggressiveness, the dy-
namics of action and the unbelievable, sweeping spirit of heading to new horizons."*[32] The field
of gas turbines was critically reviewed to decide whether to either close it down or push it
to the technological forefront. Of course, for enthusiastic engineers the latter was the
chance of a lifetime, so it was done. *'Leap frog the competition'* was the rallying cry; over
time a completely new family of gas turbines for the 50 and 60 Hz markets emerged, later

---

[32]   See Catrina, ABB, p. 49. Armin Meyer, *1949, entered BBC as development engineer in 1976, Head of electrical
      drives R&D in 1980 and of the international business unit 'Electrical Generators' in 1984, General Manager of
      ABB Drives Ltd. in 1988, President and CEO of ABB Power Generation Ltd. from 1992, Executive VP of ABB
      Ltd. and member of the ABB Board between 1995–2000, Chairman of the Board and CEO of CIBA Ltd., 2001–
      2007, member of the Board of Zurich Insurances Company, since April 2010 non-executive Director of Amcor
      Ltd., PhD in Electrical Engineering from the Swiss Federal Institute of Technology (ETHZ), served 12 years as
      Professor for Electrical Engineering and Drives at ETHZ.

to be known as GT 24 and GT 26. Details of this development history will be outlined in Section 6. Prototypes of the new gas turbine were ready for operation in 1994, in 1998 approx. 300 Mio SFr had been spent on development improvements. For aero gas turbines, where the development costs are more transparent in multi-national consortia, it was and is established knowledge that each new, full development programme requires financial resources in the order of 2 BUSD. The ABB management became nervous, when the 300 Mio SFr development mark was recognised in the late 1990s. Questionable management and engineering decisions further hampered the problems with the new gas turbines. Some engineering problems came to light quite naturally over time and their extent could be considered rather normal, given the exposed uniqueness of the approach. The management counter-measures were not too rational in any case. But the resulting financial loads of several billions of Euros were in the long run shifted onto the shoulders of the new owners of the power generation division, the French Alstom Ltd. Gas turbine development was only one of several ABB problem zones at that time; the management dynamics of them all obviously got out of control and/or were beyond the capacity of those in charge at the turn of the millennium.

As clearly visible in Figure 2-2, company acquisitions on a grand scale continued to reign Barnevik's term. In the early 1990s, ABB purchased CE Combustion Engineering, a leading US firm in the development of conventional fossil fuel power and nuclear power supply systems, to conquer the North American market. Still not saturated, ABB purchased Elsag Bailey, a process automation group in 1997, which included established companies like Bailey Controls, Hartmann & Braun, and Fischer & Porter. This was the largest company transaction in ABB's history to date. While the acquisition wheel was still spinning, CE was confronted with huge asbestos liabilities from former workers. Unexpected GT development costs, accumulated acquisition debts and unforeseen liability claims, this negative trilogy nearly brought on the collapse of ABB. Barnevik left the ABB supervisory board in November 2001 with a 'golden handshake' of 148 Mio SFr, which at that time was considered a record sum and heavily criticised accordingly. In the meantime, after the excesses of global fortune makers in the past years, apparently a rather modest final compensation for the 'manager of the year'[2] several times over.

End of 1996, Göran Lindahl followed his Swedish fellow countryman Barnevik as ABB's CEO; his management style is still remembered for divesting large parts of the hardly consolidated company. Rail and transportation engineering went to a joint venture with Daimler Benz beginning 1996 – Adtranz ABB Daimler-Benz Transportation. In 1998 ABB retreated from this joint venture and sold the 50 percent share to Daimler-Benz. Similarly, ABB Power Generation Ltd. was transferred into a 50/50 joint venture with Alstom, becoming ABB Alstom Power Ltd. effective 30 June 1999. Following the well-known pattern, ABB retreated here after one year as well and sold all of power generation to Alstom.

In retrospect, these millennium years and their fragmentary leftovers in memory appear to be somewhat 'unreal'. Junior ABB strategists trying to explain what was going on declared the whole power generation business as part of 'old economy', while the presumed bright future of the 'New ABB' should comprise *anything with the Internet*. This short-sighted enterprise generation soon acclimatised to the familiar heavy machinery environment of Alstom, together with railways and shipbuilding Alstom's third product leg.

2.1.3   ALSTOM Power Ltd., since 2000

Alstom is a large French multinational industrial conglomerate with interests in the power generation/transmission, transport and renewable energies markets;[33] the headquarters are located at Levallois-Perret, a northwestern suburb of Paris, F, Figure 2-17.

Present Chairman and Chief Executive (Président-directeur général) is Patrick Kron (*1953), who managed Alstom's successful turn-around since 2003. Alstom's global business activities spread across 100 countries at present. The total number of employees is close to 92,000 (2012), recent annual global revenues (April 2011–March 2012) were € 19,934 billion with a profit margin of 7,1 percent.[34] At € 52 billion on 31 December 2012, the backlog represented 30 months of sales.

In Switzerland Alstom occupies a workforce of 6,000 in the business areas Power, Transport and Grid with annual revenues of 4,2 billion SFr (2011/2012), thus belonging to the biggest Swiss industrial enterprises. The corresponding German numbers are 9,000 employees at 24 sites with annual revenues of € 2,6 billion.

The global headquarter of Alstom's Power sector is at Baden (AG) – with the corresponding production site at Birr (AG). The Grid activities are located at Oberentfelden (AG), while the Swiss Transport branch is at Neuhausen am Rheinfall (SH), with a regional office at Lausanne (VD). Philippe Cochet (*1961) is President of Alstom's Thermal Power Sector and Executive Vice-President of Alstom since 2011, following Philippe Joubert (*1955), who kept this position since 2009. The new Alstom Thermal Power Sector has sales of over € 9 billion and 38,000 employees. It covers Gas, Steam and Nuclear power generation as well as the Service and Automation & Control activities.

Power activities comprise the design, manufacturing, services and supply of products and systems for power generation and industrial markets. The group covers all energy sources – gas, coal, nuclear, hydro and wind. Alstom supplies and maintains all components of a power plant and provides complete turnkey solutions. The company has a leading role in environmentally friendly power solutions based on advanced low emission combustion and special CO_2 reduction technologies. The Power sector in Switzerland is global lead centre for gas and steam turbine research and development – with special focus on combined cycle power plants and in addition, carries out the system planning for hydro power plants and is responsible for turbine and generator component manufacturing. The Service branch deals with global power plant maintenance (combined cycle, gas and steam turbine plants), with plant operation optimization, equipment refurbishment and maintenance/upgrade activities in Swiss nuclear power stations.

---

[33] Alstom, the builder of the Queen Mary 2 cruising liner (gross tonnage 150,000 t, length 345 m, passengers 2'600) sold its unprofitable shipbuilding business to Åker Yards of Norway in early January 2006. Alstom's CEO Patrick Kron had it made clear since taking over in 2003 that he planned to quit shipbuilding to focus the company on its power and rail divisions. In 2010 Alstom re-acquired the electric power transmission division of Areva S.A. (previously sold in 2004), creating Transmission as Alstom's third main business area, called Alstom Grid. Finally in 2011, Alstom reshaped its operational activities into four sectors: Thermal Power, Renewable Power, Grid and Transport.

[34] The successful management turn-around becomes evident in comparison to the 2002 numbers: then annual revenues of € 21,35 billion were accompanied by losses of € 1,35 billion.

In parts the company history reaches even further back than that of Brown Boveri & Cie., Figure 2-1. The name of the company was derived from the French region Alsace and the name of Elihu Thomson (1853–1937), who founded several electrical companies in the USA, Great Britain and France. Alstom with the original spelling 'Alsthom' in 1928 evolved from the merger of SACM Société Alsacienne de Constructions Mécanique, originally founded in 1879, and Thomson-Houston with its first and still existing factory in Belfort. The company developed into the leading railway and power generation equipment manufacturer in France, especially for nuclear power plants in the 1970s. In 1976 CGE Compagnie Générale d'Électricité took over the majority of shares and C.E.M. was integrated (see Section 2.1.1). In 1984 Alsthom acquired the Chantiers de l'Atlantique, a famous shipyard at St.Nazaire, figuring first as Alsthom Atlantique and later on solely under the name Alsthom. 1989 saw the formation of GEC Alsthom from the merger of the power and transport activities of CGE and the UK GEC. France's market was no longer sufficient, so the merger was to enable Alsthom to export into Europe. In 1998 GEC and CGE (Alcatel Alsthom since 1991) separated from GEC-Alsthom, sold off their stakes in the capital, and the company continued as Alsthom, before the name was simplified to Alstom in 1998.

Reacting to progressive concentration processes in the power generation business, Alstom and ABB decided to merge the corresponding units of both companies in 1999: first on a 50/50 basis to form ABB Alstom Power Ltd.; one year later Alstom acquired the remaining ABB shares. These and other acquisitions increased Alstom's debt level so that the company fell into financial difficulties when technical and contractual problems with the former ABB gas turbines, some stalled orders in cruising shipbuilding and a breakdown in the power generation market in the wake of the collapsing 'US bubble' superimposed. As a short-term remedy the industrial gas turbine business up to 50 MW power had to be sold to Siemens. Additional bank loans were secured by a generous debt guarantee of the French government. In 2006 the French industrial group Bouygues, diversified in construction, property and telecom/media, took the 21 percent share from the French government, which it sequentially increased to 30,07 percent until October 2007.

Figure 2-17   *Alstom headquarters at Levallois-Perret, a suburb of Paris, France – on the Seine river bank, ~3 km north-west of Arc de Triomphe*

Alstom also managed the technical turn-around for the ABB gas turbine heritage. In 2006, the GT 24 and GT 26 families combined achieved more than 1.5 Mio OH (Operating Hours). Major sales to countries such as Italy, Germany, Spain, Great Britain and Thailand demonstrate that technical teething troubles of this competitive, technologically advanced product have finally been overcome.

2.2   Gas Turbine Types

A 'gas turbine', also referred to as a 'combustion turbine', is a rotary machine that extracts energy from a flow of combustion gas. At the heart of the machine is a combustion chamber for heat addition (or, alternatively, a kind of heat exchanger). Upstream is the compressor, coupled downstream to a turbine component. A gas turbine is a heat engine which converts thermal energy into mechanical output. Energy is then extracted in the form of shaft power (like in power generation by means of a driven AC generator), compressed air or thrust to power aircraft, and any combination thereof.

The principle advantages of the gas turbine are:

1. the high power density; it is capable of producing large amounts of useful power based on its relatively small size and weight,
2. the GT's long mechanical life and relatively low maintenance costs,
3. the relatively short start-up-time,
4. the wide fuel versatility,
5. atmospheric air as working fluid, without further coolant liquids etc.,
6. the relatively low costs for cheap and quick plant erection.

2.2.1   Stationary Power Generation Gas Turbines

The gas turbine in its simplest form has a single shaft configuration with one or more compressors, one or more combustors where – once the GT has been started – the energy-carrying fuel is mixed, ignited and burnt with part of the compressed air (while the remainder of the compressor air is used for cooling purposes in the combustor and turbine area as well as sealing) and a turbine group (again of one or several partial components) that drives the compressor group and provides the resulting net power to the output shaft.[35]

The gas turbine has found increasing service in the power industry over the past 40 years, both in utilities and merchant plants. Its fuel versatility is remarkable, also in view of future hydrocarbon fuel limitations; today there are gas turbines that run on natural gas, diesel fuel, naphtha, methane, low-BTU and biomass gases. The years since 1990 – partially influenced by global political changes with less emphasis on military-led developments – has seen a large growth in gas turbine technology. Higher compressor pressure ratios and turbine entry temperatures have increased the GT thermal efficiency beyond 40 percent. This increase is a consequence of the application of improved

---

[35]   For this and the following basic description, see Boyce, Gas Turbine Engineering Handbook and, see Langston, Introduction to Gas Turbine.

computational fluid dynamics (CFD) for turbo-component design, equally supported by improved materials and manufacturing technologies, new temperature-resistant and wear-resistant coatings and more efficient cooling schemes. The economics of power generation depend on the fuel cost, running efficiencies, maintenance cost and first cost – generally in that order.

The gas turbine is classified broadly in five groups:

- Heavy-duty or frame type gas turbines in the upper power range until 400 MW,
- aero-derivative gas turbines, adapted to electrical generation industry by removing the bypass fan and adding a power turbine at exhaust, with a typical power range up to 60 MW,
- industrial type gas turbines, ranging up to 25 MW, mostly applied in petrochemical plants and for pipeline compressor drive trains,
- small gas turbines up to 2.5 MW find e.g. applications as APUs auxiliary power units,
- micro turbines, typically below 350 kW, with an impressive market upsurge due to the trend to distributed generation in the past 10 years.

In principle, approximately two thirds of the turbine power is used to drive the compressor, the remainder is used as output shaft power to turn an attached electrical generator (or e.g. a ship's propeller). As an example of a typical heavy-duty power generation gas turbine, Figure 2-18 illustrates the 'thermal block' of Alstom's most advanced gas turbine family. With the introduction of the GT24 (60 Hz) and GT26 (50 Hz) gas turbines between 1995–1997, a technology level was introduced to the power market that meets the requirements for extraordinarily low emissions, high total efficiency and unique operational flexibility. The GT cycle parameters, as indicated in Figure 2-21, illustrate the superior design principles of this GT family with (for the 50 Hz, 296 MW GT 26), an overall total pressure ratio of PR = 33.3, max. turbine inlet temperatures TIT > 1,700 K and a gross electrical efficiency of 39.6 percent.

Alstom's new gas turbines are characterised by a unique design feature, the sequential combustion, that distinguishes them from conventional machines. Downstream of the air intake, the 22-stage subsonic axial compressor pushes the design mass flow of 650 kg/s to a pressure level of nearly 35 times the inlet ambient pressure. This record level for power generation gas turbines is achieved on a single shaft with four rows of variable guide vanes. At full load, approximately half of the total amount of fuel is burnt in the first EV (EnVironmental) combustor. A first expansion occurs in the single stage HPT high-pressure turbine. The remaining fuel is introduced and burnt in the SEV (second EV) or 'reheat' combustor, followed by a second expansion in a four-stage LPT low-pressure turbine. At over 600 °C, the exhaust gas temperatures are ideal for combined gas/steam turbine cycle applications. The sequential combustion concept has a long tradition that dates back to the 1948 Beznau, CH engine from BBC, Section 5.1.5.2, with an interim revival in the unique air storage gas turbine in Huntorf, D, Section 5.5.3. During ABB's GT design re-launch in the 1990s it appeared to be an interesting vehicle to limit the challenges of high material temperature; in the meantime the operational versatility of this concept has become a clear marketing advantage in view of combined-cycle part-load performance and low emissions.

Besides the described 'open cycle' process of the constant-pressure gas turbine, there are also 'closed cycle' and 'semi-closed cycle' arrangements. The working fluid in a closed cycle GT facility can be air or other gas that is continuously recycled (back to the compressor entry) by cooling the exhaust air through a compressor pre-cooler. Because of the confined, fixed amount of gas, the closed cycle is not an internal combustion engine. Here, the normal combustor is replaced by a second heat exchanger. The heat is supplied by an external source such as a nuclear reactor, the fluidised bed of a coal combustion process or any other external e.g. crude oil combustor. The closed cycle gas turbine that is discussed in detail in Section 5.2.1, was invented by J. Ackeret and C. Keller in Switzerland; a first prototype of this gas turbine according to their AK process was built and presented by Escher Wyss Zurich in 1939. Inherent advantages can be listed as follows:

1. The turbo-components are relatively small, due to the low specific volume of the pre-cooled compressor air; a cycle pressure rise is used for a simple and cheap power level adaptation.
2. Very beneficial for the early developments was the fact, that there is no corrosion and blade contamination due to the external combustion; therefore, it is also possible to burn sulphuric crude oil and low-BTU fuels.
3. Plant efficiencies above 50 percent can be achieved e.g. by using an inert gas like Helium as the working medium. This allows the use of high-stress materials like Molybdenum alloys at elevated temperatures.

A related semi-closed gas turbine concept that tried to merge the advantages of open and closed cycles was proposed and demonstrated by Sulzer Winterthur, CH in the 1950s. A demonstration plant was built at Weinfelden, CH with an output of 20 MW, Section 5.2.2.

So far, abundance of cheap, clean fuels such as natural gas, the development problems of gas-cooled nuclear reactors as well as the general difficulty of realising a reliable and financially attractive fossil-fired gas heater with exhaust temperatures of 1,050–1,100 K have hampered the introduction of these configurations; this situation may change in the future.

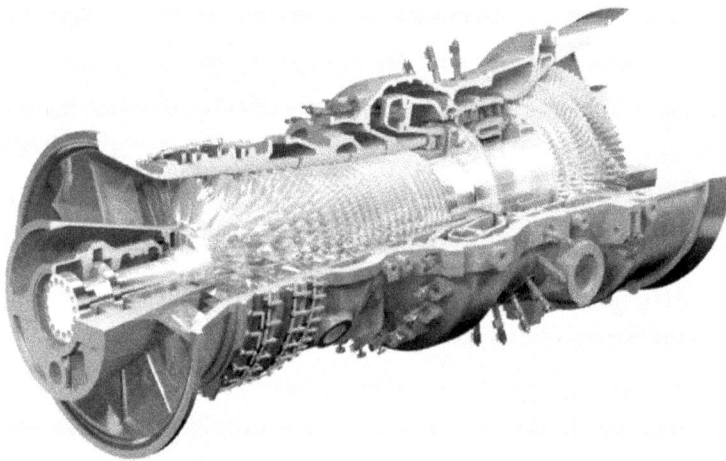

Figure 2-18   *Thermal block of the Alstom gas turbine GT24/GT26*

2.2.2   Gas/Steam Combined-Cycle Power Plants

A combined-cycle gas turbine power plant, here mostly identified by the abbreviation CCPP (combined-cycle power plant), is essentially an electrical power generation unit in which a gas turbine and a steam turbine are used in combination to achieve greater efficiency than would be possible independently, see Section 2.3.3. As illustrated in Figure 2-19, the classical Alstom arrangement is to place the generator between the GT 'cold end'/inlet and the steam turbine, so both can drive either generator side. The GT exhaust is then used to produce steam in a heat exchanger, called HRSG (heat recovery steam generator), to supply the steam turbine to generate additional electricity.

The historic tracking of the combined-cycle power plant leads without detour directly back to N.L. Sadi Carnot (1796–1832) and his ground breaking 1825 essay 'Reflections on the Motive Power of Heat', where he stated:

'... Air, then, would seem more suitable than steam to realise the motive power of falls of caloric from high temperatures. Perhaps in low temperatures steam may be more convenient. We might even conceive the possibility of making the same heat act successively upon air and vapour of water. It would only be necessary that the air have an elevated temperature after its use, and instead of throwing it out into the atmosphere immediately, to make it envelop a steam boiler, as if it were issued directly from a furnace.'[36]

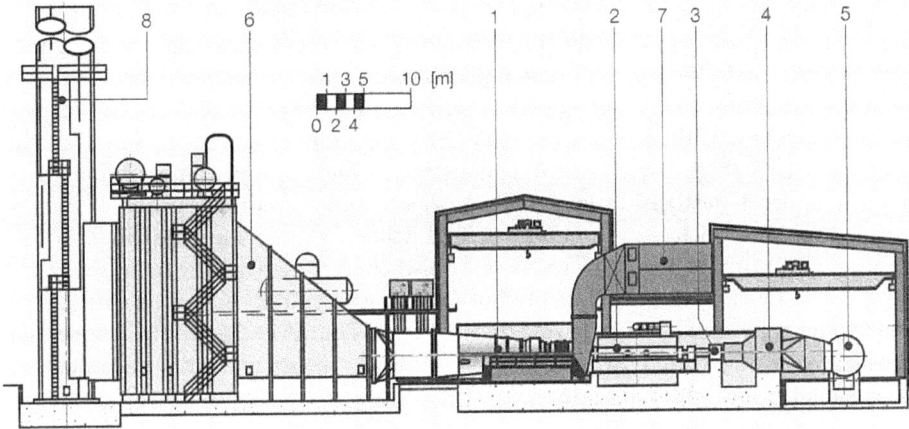

Figure 2-19   *Combined-cycle power plant KA26-1 with single shaft GT and a triple-pressure steam cycle: 1 – gas turbine, 2 – generator, 3 – clutch, 4 – steam turbine, 5 – condenser, 6 – HRSG, 7 – air intake, 8 – exhaust stack*

---

[36]   See Carnot, Réflexions p. 61, where the quotation reads in the original: '*L'air semblerait donc plus propre que la vapeur à réaliser la puissance motrice des chutes du calorique dans les degrés élevés: peut-être, dans les degrés inférieurs, la vapeur d'eau est-elle plus convenable. On concevrait même la possibilité de faire agir la même chaleur successivement sur l'air et sur la vapeur d'eau. Il suffirait de laisser à l'air, après son emploi, une température élevée, et, au lieu de le rejeter immédiatement dans l'atmosphère, de lui faire envelopper une chaudière à vapeur, comme s'il sortait immédiatement d'un foyer.*'

As a result of the described technology-driven performance jump, the combined-cycle gas turbine to date is fast replacing the steam turbine as the base load provider of electrical power throughout the world. This is even true in Europe and the United States where the large steam turbines were the only type of fossil base load power for a long time.

Where appropriate, the realisation of CHP (combined heat and power plants), also known as cogeneration plants, would raise the thermal efficiency beyond 80 percent. CHP district heating e.g. uses the CCPP reject heat for large housing areas. Other forms of cogeneration comprise process steam export for industry, desalination plants, etc. Between 1954 and 2010 Alstom and the foregoing companies have successfully delivered more than 200 combined-cycle power plants for an installed power output of more than 90 GW, i.e. in reference to the present product portfolio the advanced versions KA26 and KA24 for the 50 Hz and 60 Hz markets respectively, as well as the KA13E2 (50 Hz) and the KA11N2 LBTU (50/60 Hz), especially for low calorific fuels.

2.2.3   Aero Propulsion Gas Turbines

In an aircraft gas turbine, the output of the turbine(s) is used to turn the compressor(s) that may also have an associated turbofan or turboprop(eller). The hot air flow leaving the turbine is accelerated into the atmosphere through an exhaust nozzle to provide thrust (propulsion power). Jet GT engines are differentiated as low and high BPR (bypass ratio) configurations. The BPR designates the ratio of air mass flow drawn in by the front fan but bypassing the core engine to that of the air burnt in the core engine. The low-bypass turbofan is more compact, but the high-bypass turbofan can produce much greater thrust, is more fuel efficient, and much quieter. Present commercial turbofans of the leading engine manufacturers General Electric, Pratt & Whitney and Rolls-Royce produce static TO (take-off) thrusts of up to 500 kN per engine. BPRs range from 6-11, the technology of HBPR (high bypass ratio) demonstrator 'ducted prop' configurations with BPR < 18 is under investigation in parallel to open rotor 'propfan' concepts.[37] These propfans with BPR < 30 have the potential for further considerable reductions in fuel consumption, but with critical, still unknown noise characteristics. Thrust is generated by both the cold bypass air as well as the hot core engine output. A turbojet has no bypass stream and generates all of its thrust from air that is burnt in the gas turbine engine. Turbojets have smaller frontal areas and generate peak thrusts at high speeds, making them most suitable for fighter aircraft.

In the context of this book it may be worthwhile selecting one of the first turbojets for illustration. Figure 2-20 shows an early version of the German BMW 003A turbojet, which together with the Jumo 004B was one of the two engines in series production during WW II. BBC (Mannheim) provided alternative compressors for this engine which would have increased the TO thrust to 9 (C version) and 11.5 (D) kN, i.e. by nearly 50 percent. The quality of military engine design is traditionally measured by the T/W (Thrust-to-Weight) ratio which was about 1.42 [kgf/kg] to start with. Progress in engine design and technology over the following decades is clearly marked by the T/W ~ 10 achieved in the meantime.

---

[37]   See Geidel, Gearless CRISP and – see Eckardt, Future Engine Design Trade Offs

Figure 2-20 *BMW 109-003A German turbojet 1943: 7.8 kN TO thrust, length 3.53 m, diameter .69 m, planned –003 C/D versions with improved BBC 7- and 10-stage compressors* [38]

Over the first 50 years of gas turbine history, the aero GT engines (turbojets) were the leaders in most of the GT technology areas. The design criteria for these engines are high reliability, high performance with many starts and flexible operation throughout the flight envelope. Engine service lives of about 3,500 hours between major overhauls were considered good. Increases in engine thrust/weight ratios are achieved e.g. by higher turbo-component stage loading, by the development of high-aspect ratio blades[39] in the compressor as well as by optimising the pressure ratio and TIT (turbine inlet temperatures) for maximum work output per unit flow and the introduction of lightweight materials (composites, TiAl titanium-aluminide cast blading) in general.

2.3 Technical Basics

2.3.1 Gas Turbine Thermodynamics

Every thermodynamic system exists in a particular thermodynamic state. When a system is taken through a series of different states and finally returned to its initial state, a thermodynamic cycle is said to have occurred. In the process of going through this cycle, the system may perform work on its surroundings, thereby acting as a heat engine. A system undergoing a Carnot cycle is called a Carnot heat engine, although such a 'perfect' engine is only theoretical and cannot be built in practice.

The Carnot cycle was proposed by the aforementioned Nicolas Léonard Sadi Carnot[40] in 1824, see Section 2.2.2, and expanded by Benoît Paul Émile Clapeyron in the 1830s and 40s. It is the most efficient existing cycle capable of converting a given amount of thermal energy into work. The Carnot cycle when acting as a heat engine consists of the following steps:

− isentropic work input (compression),
− isothermal heat addition (at hot temperature T_H),

---

[38] Picture source: Gersdorff, Flugmotoren und Strahltriebwerke, p. C XI
[39] Meaning a high ratio of blade height to blade chord length
[40] See also Section 3, Footnote 14.

- isentropic work output (expansion),
- isothermal heat rejection (at cold temperature T_C)

This leads to the simplest form of the definition of efficiency:

$$\eta = W / q_H = 1 - T_C / T_H$$

i.e. the ratio of useful work (W) done by the system to the heat input q_H can be expressed by means of the temperature ratio between 'Hot' and 'Cold' conditions or in other words for fixed ambient conditions, the maximum heat engine efficiency is a direct function of the realistically achievable T_H level in the combustor. The ideal Carnot cycle serves mainly as basis for the comparison to assess real engine cycles, like the Joule cycle for gas turbines and the Rankine cycle, correspondingly, for steam turbines. In any case, Carnot's theorem applies: *No engine operating between two heat reservoirs can be more efficient than a Carnot engine operating between those same reservoirs.*

The GT simple-cycle Joule process, in the literature sometimes also named after the Boston engineer George Brayton (1839–1892) and shown in graphic form in Figure 2-21a as a *h – total enthalpy/T – total temperature vs. s – entropy* diagram, is a representation of the properties of a fixed amount of air as it passes through a gas turbine in operation. The compression work W_C (with losses) is followed by a constant pressure heat addition process; then the hot gas is expanded in a turbine (again with losses), generating the turbine work W_T. Finally, the cycle is closed by a constant pressure heat rejection process. In this respect the Joule/Brayton gas turbine cycle is very similar to the steam turbines' Rankine process, with the main difference being that the latter refers to a two-phase (liquid and gas) substance with phase changes occurring during the two constant pressure processes.[41] Since the turbine drives the compressor, $W_T - W_C$ represents the net power output. For an optimum design in view of a high power output **and** efficiency, the increase in turbine inlet temperature has to be accompanied by a higher compressor pressure ratio. To give an example for TIT = 1,750 K, optimum power requires a compressor PR ~ 20, while the best efficiency values can be expected at PR ~ 40. The final fixation somewhere in between is part of the design evaluation and optimisation process.[42]

The gas turbine belongs to the category of IC (internal combustion) engines employing a continuous combustion process. This differs from the intermittent combustion occurring in diesel and automotive IC engines, but also from that in an 'explosion gas turbine' according to the German engineer Holzwarth, Section 3.3. This 'constant (combustion) volume' engine concept was actively investigated by BBC before the path to the present 'constant pressure' concept was started in the early 1930s; especially valuable experiences in high intensity heat transfer could be applied to the following development of the present days' standard GT configuration.

---

[41] The naming of the corresponding cycles after Brayton and Rankine is somewhat disputed in the Anglo-Saxon literature – see e.g. Potter, The Gas Turbine Cycle and Haywood, Analysis of Engg. Cycles – with the basic arguments, that the elementary GT concept was already disclosed in the 1791 patent application of John Barber, while both the ideal and practical gas turbine cycles were revealed in the 1851 paper by James P. Joule (1818–1889), if one is willing to substitute turbomachinery for the reciprocating compressor and piston engine. The same applies for the ideal Joule cycle in correspondence to the ideal Rankine cycle – after William J.M. Rankine (1820–1872). Both Brayton and Rankine – it is said – never suggested their cycles explicitly.

[42] See Simon, Entwicklungen für grosse Gasturbinen

Alstom's sequential combustion has been illustrated with the characteristic double peaked process in Figure 2-21b as the so-called 'GT reheat cycle'. Reheating occurs in the turbine zone and is a way to increase on a relative basis the turbine work ($W_{HPT} + W_{LPT}$) more than the compressor work ($W_{LPC} + W_{HPC}$), thus limiting the thermal load of the turbines. As already shown in the foregoing Section 2.2.1, a row of SEV combustors was inserted between the HPT high-pressure and LPT low-pressure turbine for reheat, where the extra heat q_{SEV} is added in addition to q_{EV}. As opposed to the simple cycle, the expansion reheat cycle runs at three pressure levels. This process staging increases the GT thermal efficiency in principle by one to three percent. A comparison of both cycles shows the reheat process less sensitive towards deviating component efficiencies, with the mentioned advantage. In addition, the reheat cycle has the general benefit of some 15 percent higher specific work. Summarising, the processes with intermediate reheat represent a few essential advantages in comparison to the simple Joule process[43]:

1. The reduced sensitivity towards component efficiencies is in general reflected in a slightly higher GT thermal efficiency (assuming the same component performance).
2. The double expansion raises the specific work considerably above that of the Joule process, which increases the relative power output for same mass flows.
3. For equal thermal efficiency, the GT exit temperature of the reheat configuration is principally higher than that of the simple cycle. This has significant benefits, especially for combined-cycle (part-load) operation.
4. The splitting of the fuel input into two combustors generates also a few control options that cannot only be exploited favourably in combined-cycle part load, but also in view of a limitation of the turbine entry temperature. The advantages in view of material demands have been already discussed; however, in view of increasing environmental demands, the following aspect has become more and more important.
5. By injecting fuel into two combustion systems in series, it is possible to increase the output and cycle efficiency without significantly increasing the emissions at full and part load as the firing temperature in the first combustor can be kept relatively low and the second combustor does not contribute substantially to the critical engine NOx emissions. In this respect, the recently adapted sequential combustion operation principle in favour of lower NOx values has shifted the temperature peaks of
6. Figure 2-21b by unloading the NOx-sensitive first row of EV burners towards the SEV burners, where the incoming hot gas has a considerably lower O_2 content and thus, less oxygen is available for NOx formation.

---

[43] See Lechner, Stationäre Gasturbinen, p. 44. In addition to the listed thermodynamic and control-related advantages, the sequential combustion of course also has a few disadvantages:
In general, there is a need for higher compression ratios to exploit the advantage of the better reheat process efficiencies, thus increasing the number of compressor and turbine stages.
The thermally highly loaded SEV combustor is another costly GT component.
The additional mechanical design effort for the sequential combustion concept is accompanied by increased complexity e.g. for fuel supply and GT controls.

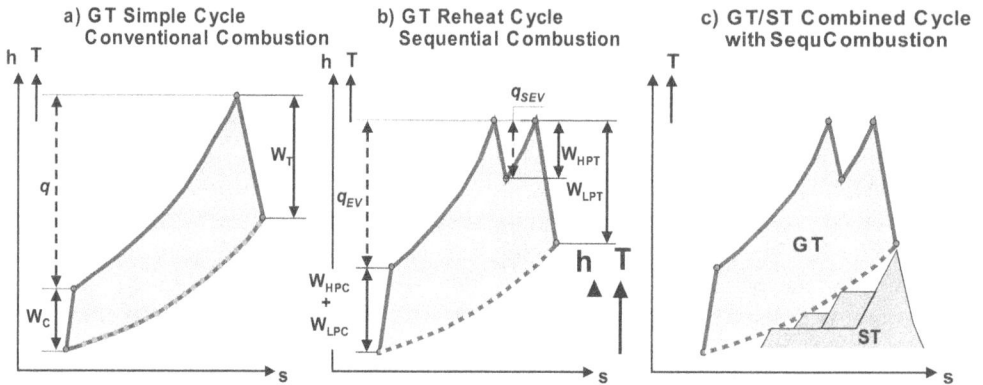

a) GT Simple Cycle
Conventional Combustion

b) GT Reheat Cycle
Sequential Combustion

c) GT/ST Combined Cycle
with SequCombustion

Figure 2-21   *GT simple, reheat and GT/ST combined cycles, h/T-s charts*

2.3.2   Gas Turbine Component Performance

The industrial gas turbine has always emphasised long life with – to give an example – MTBO (mean time between overhaul) values of 24,000 EOH[44] or longer; this rather conservative approach in the past has sometimes led to certain compromises against high performance and in favour of rugged operability. This has all changed in the past 15–20 years, either by direct competitive pressure from aero-derivative gas turbines or by a transfer of corresponding aero know-how to the heavy-duty GT configurations. Put in corresponding numbers, one could estimate the increase in the average engine pressure ratio from approx. 25 to more than 40, while the trend of turbine entry temperatures increased from a modest 1,250 K to impressive 1,750 to 1,800 K. Consequently, the former performance gap towards aero engines has been nearly closed, while the superiority in parts' life has increased even further.

The reason why the first gas turbine realisation efforts, like those of Stolze and Armengaud-Lemâle in the early 1900s, Section 3.2, did not prove to become a success can be seen in Figure 2-22 that shows GT thermal efficiency as a function of GT specific work – both for an early design standard of 1920/30 and for present state-of-the-art. With a turbine inlet temperature TIT[45] of about 830 K upfront of turbine vane 1 achieved by injecting water into the combustion chamber, the turbine was just able to supply the mere power for compressing the air. The cause of this disappointing result must be sought in the great volume of compressed air required to reduce the combustion temperature of 2,200 to 2,270 K to the value admissible for the gas turbine blading.

As already shown, one of the major disadvantages of the gas turbine in the past was its lower efficiency and hence its correspondingly higher fuel consumption when compared to

---

[44] EOH equivalent operating hours refer to the number of operation hours, supplemented by the equivalent life consuming effect of cyclic loads, differentiated by number of events and thermal gradient over time.

[45] There are different TIT definitions, besides the principle difference between absolute temperatures in 'degree Kelvin' (K) and those in 'degree Celsius' (°C), T (K) = T (°C) + 273.1 (deg C). **Thg** refers to the 'hot gas temperature' at turbine vane 1 inlet, while **Tmix** represents a theoretical 'mixed-out state' of hot gas mixed with all added cooling and leakage air; the latter is approx. 100–200 K below Thg.

other IC engines and to steam turbine power plants, see also Figure 2-23. However, during the last 70 years, continuous engineering development work has pushed the thermal efficiency from 18 percent for the 1939 Neuchâtel gas turbine to present levels of about 40 percent for SC (simple-cycle) operation and 60+ percent for CCPP (combined-cycle power plants). Even more fuel-efficient gas turbines are in the preparatory stage, with predicted SC efficiencies of 42 percent and especially large CCPPs considerably beyond the 'magic' 60 percent mark; these values are significantly higher than those of any other prime mover, such as steam turbine power plants.

2.3.3   Gas/Steam Combined-Cycle Power Plant

The h/T, s diagram of the KA26, the 'Kombi-Anlage' (combined-cycle plant version) of Alstom's GT26 gas turbine has been sketched in Figure 2-21c; the net electrical output for the (1 on 1) single shaft version is > 500 MW with > 60 percent electrical efficiency and for the schematically shown triple pressure reheat steam cycle.

The key advantage of the single-shaft arrangement as already outlined in Figure 2-19 is its operating simplicity which raises reliability – as much as 1 percent above multi-shaft blocks. Operational flexibility comes from the fact that the steam turbine can be disconnected, using a self-synchronising clutch during start-up or for simple cycle GT operation.

The combined cycle efficiency η_{CC} can be derived fairly simply from the equation :

$$\eta_{CC} = \eta_{GT} + \eta_{ST} - \eta_{GT}\,\eta_{ST}$$

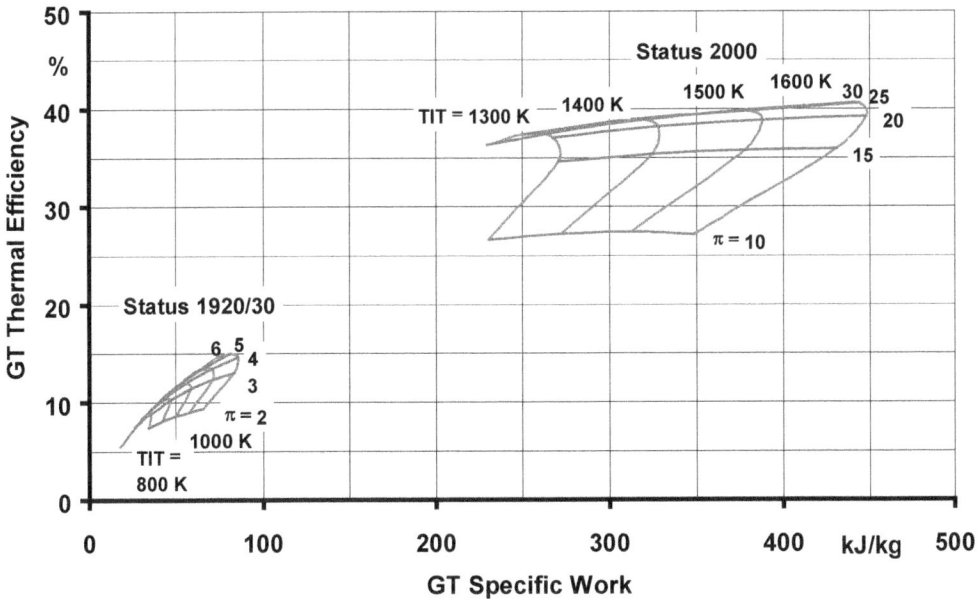

Figure 2-22   *Gas turbine thermodynamics [approx. performance model, TIT Turbine Inlet Temperature (Thg), π total pressure ratio [46]]*

---

[46] π – or mostly used PR – expresses the total pressure ratio between compressor discharge and compressor entry.

in other words, the sum of individual GT and ST efficiencies, minus their product. This remarkable equation reveals why CCPPs have become so successful. With reasonable upper values for current high-performance gas and steam turbines η_{GT} = 0.4 and η_{ST} = 0.3, the combined efficiency would amount to η_{CC} = 0.58, which is greater than the efficiency of either of the component engines taken separately.

At the end of this fundamental survey on the gas turbine principle and as an outlook to the following sections on early turbomachinery and gas turbine forerunners, Figure 2-23 illustrates the development of the efficiency of thermal engines over the past 300 years. Along these lines lie historically important achievements by great inventors and designers such as, to name just a few: Savary's water raising engine, 1700; Newcomen's atmospheric engine, 1712; Watt's condensing engine, 1775; the caloric engine by Cayley, Stirling, etc., 1818; the first internal combustion engines of Lenoir, 1860 and Otto, 1867 and 1888. In 1876, the combustion engine overtook the steam engine and thereafter the efficiency increased rapidly. The steam turbine by Parsons and de Laval dates from 1885; the diesel engine was introduced at MAN in 1895 to 1897 and the Buechi/Sulzer four-stroke engine turbocharging in 1913. The BBC gas turbine at the Neuchâtel utility plant in 1939 is an event of outmost importance in this context. The advent of the first BBC GT/ST combined-cycle power plant at Korneuburg in 1961 boosted the GT, CCPP performance string considerably, crossing the IC (internal combustion)/diesel with Turbo-Charging line, until the present-day best efficiency values that have come to lie at 60 percent and beyond in the meantime.

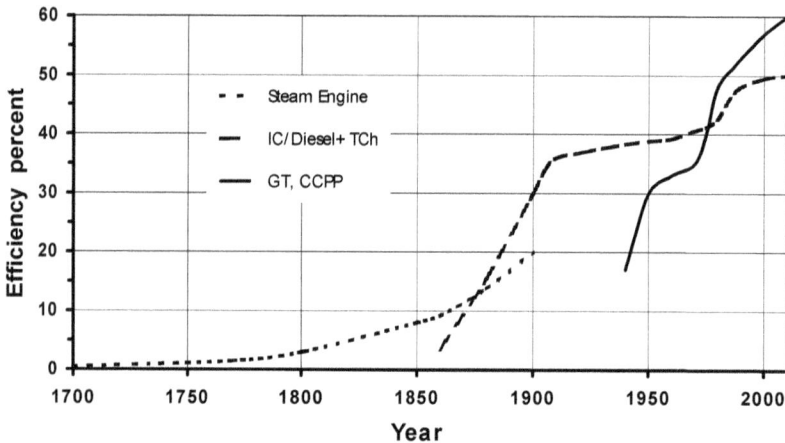

Figure 2-23   *Efficiency development of thermal engine families*

3   Gas Turbine Forerunners

Before we follow the path to the first power generation gas turbine between approx. 1927 and 1945, it is worthwhile reviewing in short the history of early turbomachinery components and in its specific form of turbochargers. Then we deal with a number of early attempts to verify the gas turbine principle before we focus on the Holzwarth gas turbine as the last significant forerunner of the power generation gas turbine as it is known today.

3.1   Turbomachines and Turbochargers

3.1.1   Early Turbo- and Fluidmachines

Of course, this short summary cannot replace specialised literature[1] on this topic. Traditionally, the history of early turbomachinery starts with the Greek inventor and geometrician Heron of Alexandria in Roman Egypt and his 'aeolipile', a simple steam-powered pure reaction machine some 2,000 years ago. A closed spherical vessel mounted on a bearing axis is fed with steam from a closed boiler below. The steam discharges tangentially at the vessel's periphery, thus put in rotation by the reaction of the hot steam jets.

Besides the Romans who present as early as 70 B.C. the knowledge of pure impulse, paddle-type water wheels for grain grinding mills, the following centuries were apparently 'dark'. The next landmark is in turbine component technology. Like in many other field of technology, it is a finding in Leonardo da Vinci's exuberant sketchbooks that in fact documents the first known hot gas (axial) turbine wheel, Figure 3-1.

The four blades overlap in the rear part to form the hot gas passage. The reaction turbine blading has a backward leaning shape with high resemblance to present advanced propfan/propeller configurations, which achieve this shape after lengthy numerical optimisation only. This fine, artistic free-hand sketch not only illustrates a turbine as a kind of forerunner for a future gas turbine component, but also remarkably early represents a landmark documentation of the turbocharger principle that officially became patented only some 450 years later. Putting another log onto the burning fire is equivalent to a forward push of the fuel throttle lever, produces higher burning temperatures and correspondingly increases the convective draught in the chimney which automatically – by continuity –

---

[1]   From a vast bulk of literature on turbomachinery and gas turbine history only three examples are presented here: see Ruehlmann, Allgemeine Maschinenlehre ... is regarded as classic founder of the descriptive history of technology in Germany; a highly valued rarity these days, see Feldhaus, Ruhmesblätter der Technik– and see Constant, The origins of the turbojet revolution.

asks for more cold air to be sucked into the fireplace. In the English literature this device following Leonardo's description is also known as 'smoke jack'.

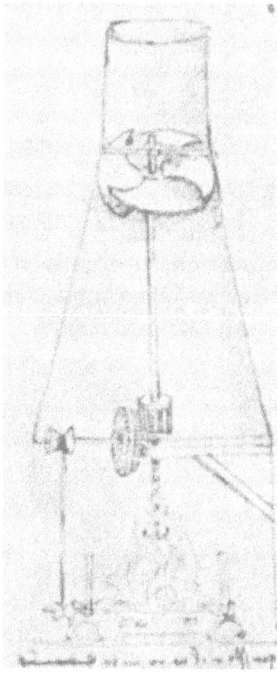

Figure 3-1   *Leonardo da Vinci: Hot gas turbine drive of a rotating spit in a chimney, Codex Atlanticus* [2], *fol.5 verso-a, approx. 1500*

Rotating spits driven by hot gas turbines apparently became popular all over Europe after Leonardo. There exist numerous descriptive reports, amongst others a foreboding sign across several centuries, a report by Montaigne[3], who saw one of these devices in 1580 in practical use in Baden, Switzerland, later the site of Brown Boveri and the follow-on companies specialising in power generation turbomachinery.

Interestingly, for the same purpose of a rotating spit drive, but following a different line of thinking, the first steam-powered roasting jack was dealt with in a work on Islamic engineering. It was described by the Ottoman engineer Taqi al-Din in 1551 in the form of a ru-

---

[2]   The *Codex Atlanticus* is the largest collection of Leonardo's papers ever assembled. It was originally put together by the sculptor Pompeo Leoni and takes its name from its large size (61 x 44 cm) as it is comparable to an atlas. The drawings represent the time-span from 1478 to 1518 and reflect every aspect of Leonardo's interest. The manuscript is now arranged in 12 volumes, comprising over 1,100 miscellaneous drawings and fragments by Leonardo. Selected parts of the *Codex Atlanticus* are exhibited at the Biblioteca Ambrosiana in Milan. The presented illustration of an axial hot gas turbine is from the book – see Piantanida, Leonardo da Vinci, p. 503.

[3]   Michel Eyquem de Montaigne (1533–1592) is one of the most influential writers of the French Renaissance. In 1578 he started suffering from painful kidney stones. From 1580 to 1581, Montaigne travelled in France, Germany, Austria, Switzerland, and Italy, partly in search of a cure. He kept a detailed journal also recording the episode with Leonardo's hot air turbine in Baden.

dimentary impulse steam turbine attached to a spit. He or more probably an independent invention may have inspired the Italian engineer Giovanni Branca (1571–1640), who proposed actually useful and working impulse steam turbines in 1629. In his device for a stamping mill, a jet nozzle directed steam onto a horizontally mounted turbine wheel. The rotation was then converted to a stamping action by means of bevel gearing that operated the mill.[4] These stand-alone inventions of steam wheels were obviously not very useful without the still undeveloped art of mechanical construction. Though there is no proven record of such a facility actually ever built, according to an anecdote, Branca built this device but, upon the explosion of the boiler, was locked up under the pretence that he must be crazy.[5]

On similar, somewhat uncertain grounds stands the claim that as early as 1543, the Spanish naval officer Blasco de Garay (1500–1552), captain in the Spanish navy during the reign of the Holy Roman Emperor Charles V, used a primitive steam machine to move a ship in the port of Barcelona.[6]

SECHSTES BUCH 175

Figure 3-2 *Centrifugal blower for exhaust ventilation of an ore-mining gallery by Georg Agricola, De Re Metallica, 6th book, 1556* [7] *©Springer*

---

[4] See Wilson, Turbomachinery, p. 29

[5] See Meher-Homji, The historical evolution of turbomachinery, p. 283

[6] See Wikipedia, 'Steam Turbine History' and 'Blasco de Garay'

[7] The merit for finding the hidden 'compressor dwarf' in the microcosm of 'De Re Metallica' goes to Robert C. Dean jr., former MIT professor, entrepreneur/founder and initiator of numerous engineering companies at the forefront of scientific R&D since the 1960s (Creare, Hypertherm, Concepts ETI, Fluent, Synergy Innovations,

3.1.1.1   Historic Compressors

The first description of a turbocompressor dates back to 1556. It was somewhat hidden in the woodcut prints attached to the 6th book on mining tools and machines of 'De Re Metallica' by Georg Agricola.[8] Figure 3-2 illustrates a hand-driven, radial impeller with four blades in a wooden casing. The function may have been somewhat mixed up by the wood-carving artist: Ideally the suction air from entry C at the casing top should bypass the wheel to be fed sideways to the wheel centre at a small radius, while the pressurised compressor exhaust duct is correctly depicted at the outer casing rim and a higher radius at D/E. Alternatively and 2nd best, the horizontal inner bar (X, letter added) might have had the function of a pressure baffle, with the suction inlet C and exhaust duct D/E shifted axially in opposite directions (as shown for the vertically mounted E duct on the complete installation drawing). The apparent 'paddle arms' H carry heavy masses at the end as a kind of flywheel to stabilise the rotating speed. The radial impeller (rotor) transfers a rotating movement to the entry air, and the centrifugal acceleration along the curved, outwardly directed flow path creates a radial pressure field. The fluid pressure rise therefore already occurs solely due to the radially outward directed passage of a fluid particle in this pressure field. This is the basic difference between a centrifugal and an axial compressor, where the pressure rise in axial flow direction is caused by flow deceleration/'diffusion' and circumferential turning. This flow mechanism determines the performance of energy transformation from rotating energy to static pressure that also played a decisive role in the realisation of the gas turbine principle later on.

Compressor design principles obviously did not change fundamentally in the following 350 years, as demonstrated by Figure 3-3, which shows a centrifugal turboblower impeller from one of the earliest BBC turbo-publications[9] in 1907.

---

etc.). He was the first who suggested and carried out research on the jet-wake flow phenomena in centrifugal compressor impellers, subject of the author's doctoral thesis – see Eckardt, Investigation of the jet-wake flow – and who used this illustration as entry vignette for an ASME paper of 1973, see Dean, Centrifugal compressor, that is still worth reading today. For the present (modified) illustration, see Agricola, De Re Metallica, VDI/Springer

[8]   Agricola, Georg (Bauer/Pawer, Georg) (1494 Glauchau – 1555 Chemnitz, D), German metallurgist who wrote De Re Metallica Libri XII (12 Books from Mining and Metallurgy) under the Latinised version of his name. He was a school teacher and physician, who settled in Joachimstal (today Jachymov, Czech Rep.), then the most important ore-mining field in Europe. The books are famous for the attached illustrations, actually an assemblage of information on all techniques and practices current at that time. As the textbook for mining and mineralogy it would have been brought to England by German miners when they were employed by the Mines Royal in the Keswick area in the late 16th century. This was a kind of early technology transfer which would become more common in turbomachinery – in both directions – in the 19th and 20th century. In addition, the book is remarkable for its reception history in the English speaking world. The translation from Latin into English was carried out by Herbert C. Hoover (1874–1964), the 31st US President between 1929 and 1933. His father, a blacksmith and farm implement store owner was of German (Pfautz, Wehmeyer) and German-Swiss (Huber, Burkhart) descent. He graduated from Stanford with a degree in geology and then went to Australia and China for a London-based mining company. There he was trapped for more than a month in the Boxer Rebellion. After founding what later became the Rio Tinto Group, he was financially independent, became a private mining consultant in 1908 and published the translation of the mining classic De Re Metallica together with his wife in 1912, see Agricola, De Re Metallica, Hoover.

[9]   See Rummel, Turbogebläse

Figure 3-3   *Brown-Boveri-Rateau dual-flow turboblower impeller for Siemens-Martin steelworks at Rote Erde, Aachen, D 1907 1.3 m diam. steel casting, pressure ratio 1.2, mass flow 14 kg/s, speed 2'600 rpm, 750 hp steam turbine drive*

Since the difficulties of a flow breakdown against a rising pressure could hardly be understood before Prandtl's 'boundary layer theory'[10] had been published in 1904, early gas turbine inventors preferred to follow the route to the simpler principle of centrifugal compression. However, it was realised only in hindsight that the performance of centrifugal turbocompression, if solely based on the naturally given pressure rise without or with only poor additional flow diffusion, is insufficient to produce the net power output of a gas turbine. Therefore, the realisation of the well-known gas turbine principle, Section 3.2, had to wait until a more efficient and at the same time aerodynamically much more demanding axial flow compressor could be realised – by Brown Boveri in Switzerland.

---

[10] Ludwig Prandtl (1875–1953), German pioneer of applied science of aeronautical engineering. His studies identified the boundary layer, thin-airfoils, and lifting-line theories. The dimensionless Prandtl number Pr was named after him as the ratio of kinematic viscosity to thermal diffusity, which describes the relative thickness of velocity and thermal boundary layers e.g. in heat transfer similarity analysis.

3.1.1.2   Water Wheels

The Swiss mathematician Leonhard Euler (1707–1783), then at the Berlin Academy of Sciences working on a contract by Frederick II of Prussia to improve the Sanssouci Castle waterworks[11], analysed Heron's turbine, the water equivalent of it, Segner's water wheel[12], and carried out corresponding experiments around 1750. He published his application of Newton's Law to turbomachinery, now universally known as Euler's Turbine Equation in 1754 and thereby immediately permitted a more scientific approach to design than the previous trial-and-error methods. None other than the renowned Swiss high-speed aerodynamicist Jakob Ackeret[13] took the opportunity to outline Euler's work on water turbines and pumps at length and in rare clarity. He praises Euler as was to be expected, not only for the analytical calculus, but also as inventor of the inlet guide vanes as a direct result of Euler's request for 'shock-free' entry flow to the turbine rotor[14] and for introducing the 3D curvature of the turbine (which by intuition had already characterised Leonardo's earlier turbine design, Figure 3-1). This achievement nearly automatically followed from the han-

---

[11] Euler is considered to be the preeminent mathematician of the 18th century, and one of the greatest mathematicians to have ever lived. He made important discoveries in fields as diverse as infinitesimal calculus and graph theory. He also introduced much of the modern mathematical terminology and notation, particularly for mathematical analysis, such as the notion of a mathematical function. He is also renowned for his work in mechanics, fluid dynamics, optics, and astronomy. Euler spent most of his adult life in St. Petersburg, Russia, and in Berlin, Prussia. http://en.wikipedia.org/wiki/Leonhard_Euler Despite Euler's immense contribution to the Academy's prestige, he was eventually forced to leave Berlin in 1766. The King also expressed disappointment with Euler's practical engineering abilities in a letter to Voltaire dated 25 January 1778: '*I wanted to have a water jet in my garden: Euler calculated the force of the wheels necessary to raise the water to a reservoir, from where it should fall back through channels, finally spurting out in Sanssouci. My mill was carried out geometrically and could not raise a mouthful of water closer than fifty paces to the reservoir. Vanity of vanities! Vanity of geometry!*' http://en.wikipedia.org/wiki/Ruinenberg

Again, it was the pump, the water-equivalent to the compressor, which caused problems. The basic flaw was that the diffusing/pressure-recovering vaned exhaust diffuser downstream of the centrifugal pump wheel, corresponding to Euler's turbine inlet guide vanes upstream of the turbine wheel, had been neglected.

[12] 'Segner wheel' designates a type of water turbine invented by the German scientist Johann A. Segner (1704–1777), using the same principle as Heron's Aeolipile and is nowadays found in all kinds of rotating sprinklers. The classic device is placed in a suitable hole in the ground or the slope of a hill. Water is delivered to the top of a vertical cylinder, at the bottom of which there is a rotor with specially bent pipes with nozzles. Due to the hydrostatic pressure, the water is ejected from the nozzles causing the rotation of the rotor. The useful torque is transferred to a powered device through a belt and pulley system.

[13] See Ackeret, Euler

[14] Deviating from Ackeret, David G. Wilson in Turbomachinery, p. 30 assigns the honour for the 'shock-free' fluid machinery rotor inlet flow to General Lazard N. M. Carnot (1753–1823). Though this is not very likely since Euler's corresponding work in Berlin dates around 1750, it is worthwhile presenting a few facts of the interesting life of 'the Carnots' in this context. Between 1773–1784 Lazard Carnot was a member of the engineering corps in the French army, on 15 January 1793, as member of the National Convent, he voted in favour of the execution of King Louis XVI and in 1795 became one of the founders of the École polytechnique, proposed the 'levée en masse' and was forced to flee to Switzerland between 1797–1799. He became War Minister for Napoléon Bonaparte, but retired after a few months to return to his scientific studies. During the period of Restoration, he was accused of murdering the King, but managed to escape to Prussia with the help of Alexander von Humboldt and died in 1823 in Magdeburg. In 1889, his mortal remains were brought to Paris and entombed in the Pantheon. His son Nicolas L. Sadi Carnot (1796–1832) was a French physicist and military engineer who, in his 1824 *Réflexions sur la puissance motrice du feu et sur les machines propres à développer cette puissance*, 65 p., gave the first successful theoretical account of heat engines, now known as the Carnot Cycle, Section 2.3.1, thereby laying the foundations of the second law of thermodynamics. He is often described as the 'father of thermodynamics', being responsible for such concepts as Carnot efficiency, the Carnot heat engine, and others. His father named him for the Persian poet Sadi of Shiraz (1184–1283/1291?), and he was always known by this third given name as Sadi Carnot (in parts from Wikipedia 'Lazard N. M. Carnot' and 'Nicolas L. Sadi Carnot').

dling of flow problems by means of velocity vector triangles at the blade row entry and exit – typical for the application of Euler's Turbine Equation. Here, circumferential speed u, relative velocity in the rotating system w and absolute velocity c are tied together by the simple term: u + w = c.

Ackeret's lifelong personal enthusiasm for the engineering beauty of turbomachinery, which later will be shown in his personal involvement in the compressor design for the high-speed wind tunnels at his ETH Zurich institute in the 1930s, Section 4.2.4.1, also becomes visible in the pride with which he presents a very rare piece of hardware of that time, Figure 3-4. The salient feature of this concept came to fruition some 200 years later: the horizontal arrangement of the turbine wheel with 3D-shaped blading.

In the 1820s there was enormous interest in making traditional waterwheels more efficient, since the new machines of the 'Industrial Revolution' required more power. The French engineer Benoît Fourneyron (1802–1867), Besançon, in 1827 built a radial-outflow 'turbine'[15], consisting of two sets of blades in a horizontal plane, curved in opposite directions to get as much power as possible from the water's motion through this stator/rotor arrangement. Contrary to its forerunners, Fourneyron's turbine used efficient blade angles and ran fully loaded, rather than in the 'partial admission' mode in the form of a single incoming jet. In the development of his turbine, Fourneyron used the brake produced in 1822 by the Baron Riche de Prony (1755–1839). The Prony brake permitted much greater accuracy in measuring efficiencies, and with the greater reporting accuracy, the efficiencies of engines also began to increase rapidly. By 1834 1-ft-diameter Fourneyron turbines ran at 2,300 rpm, producing 43 kW at an efficiency of 80-87 percent. One of these 'showcase' turbines at St. Blasien in the Black Forest, only some 45 km across the Rhine River to the north of BBC's later site in Baden, became a place of pilgrimage for engineers from all over Europe. It had a record high effective water head of 108 m. From 1834 until 1883, this high-pressure turbine powered the machinery of a cotton spinning mill; it is today on exhibition in the German Museum at Munich. Within a few years, hundreds of factories used Fourneyron-style turbines. Finally, in 1895 Fourneyron turbines were installed on the US side of Niagara Falls[16] to generate electric power. In total he produced some 100 turbines, all of the radial-outflow type.[17]

---

[15] The word 'turbine' was coined in 1822 by another Frenchman Claude Burdin (1788–1873) from the Latin *turbare* for spinning.

[16] In 1893, Westinghouse Electric was hired to design a system to generate alternating current on the Niagara Falls, and three years after that, the world's first large AC power system was created, activated on 26 August 1895. The Adams Power Plant Transformer House today still remains a technical landmark of the original system. By 1896, with financing from moguls like J.P. Morgan, John Jacob Astor IV, and the Vanderbilts, they had constructed giant underground conduits on the US side of the Niagara fall (with an approx. head of 25 m) leading to turbines generating upwards of 100,000 horsepower (75 MW), and were sending power as far as Buffalo, 32 km away. Some of the original designs for the power transmission plants were created by the Swiss firm Faesch & Piccard, a forerunner company of Charmilles and Escher Wyss, which also constructed the original 5'000 hp waterwheels (from Wikipedia 'Niagara Falls').

[17] See Wilson, Turbomachinery, p. 31

Figure 3-4   *Reaction turbine wheel with 3D-curved wooden blading, Southern France approx.1620 (AG Rieter & Co., Winterthur, CH)* [13]

The first mention of a vaned diffuser was in 1875 a patent by Osborne Reynolds (1842–1912)[18] for a so-called 'turbine pump', which was for the combination of a centrifugal impeller with a vaned diffuser. That same year Reynolds also built a multistage axial-flow steam turbine running at 12,000 rpm. In 1885 he described the convergent-divergent nozzle, an item of great relevance for future steam turbines.

3.1.1.3   Steam Turbines

In 1883 the Swedish engineer Gustaf de Laval (1845–1913) designed the first practical power delivery steam turbine. This was a single stage impulse turbine with a speed of 30,000 rpm, which was reduced to 3,000 rpm by gearing. The invention was a rather accidental by-product. In 1878 he made a new cream separator, requiring a speed of 12,000 rpm. After testing a kind of Heron reaction turbine unsuccessfully, he turned to an impulse wheel and to a convergent-divergent, supersonic nozzle and ran the turbine up to 30,000 rpm and a peripheral speed of 360 m/s.[19]

The first steam turbine that had a major impact on the engineering world was built in 1884 by the Englishman Sir Charles A. Parsons (1854–1931), whose first model of a multistage reaction turbine, running at 18,000 rpm, was connected to a dynamo that generated 7.5 kW (10 hp) of electricity. Once recognised, Parsons' turbine was soon extensively used in many fields, but it was especially suited for large power applications. The Parsons' turbine turned out to be easy to scale up. De Laval on the other hand was limited to small powers because his turbine had to use gears for speed reduction. The invention of Parsons' steam turbine

---

[18]   Osborne Reynolds (1842–1912) of Irish/English descent was a prominent innovator in the understanding of fluid dynamics. Separately, his studies of heat transfer between solids and fluids brought improvements in boiler and condenser design. The Reynolds number Re is a dimensionless similarity number that gives a measure of the ratio of inertial forces to viscous forces and consequently, quantifies the relative importance of these two types of forces for the given flow conditions and for the comparative assessment of model vs. real configurations. The concept was apparently already introduced by another Irishman, Sir George Gabriel Stokes (1819–1903), in 1851, but became known under its present name after Reynolds popularised its use since 1883.

[19]   See Wilson, Turbomachinery, p. 34

made cheap and plentiful electricity possible and at the same time revolutionised marine transport and naval warfare. His patent was licensed to Allis Chalmers and Westinghouse in 1895, followed by Brown Boveri in 1899, Figures 2-8 and 2-9. Parsons had the satisfaction of seeing his invention adopted for all major world power stations, and the size of generators had increased from his first 7.5 kW set up to units of 50,000 kW capacity; within Parsons' lifetime the generating capacity of a unit was scaled up by about 10,000 times.[20]

The success of emerging turbomachinery industries in Switzerland and Germany in the late 19th and early 20th century was owed not to a small extent to a few outstanding teachers and their new, previously unheard of combination of mathematical and mechanical principles with the aspects of practical mechanical engineering. One of these outstanding personalities was Ferdinand J. Redtenbacher (1809–1863), who after his studies at the Polytechnikum in Vienna for six years became a professor at the 'Hoehere Industrieschule' at Zurich, before in 1841, he finally became professor in mechanics and mechanical engineering at the Polytechnikum Karlsruhe, Germany. As its director between 1857 and 1862, he transformed it into a school of international standing. Redtenbacher is considered as the founder of scientific mechanical engineering, supplying previously empirical teaching with a mathematical and scientific background. Among his students were outstanding engineers such as Karl Benz, Franz Reuleaux and Emil Škoda. His books[21] received widespread recognition in Germany and Switzerland, some of them in translated form also in France.

The other engineer with a wide, nearly global influence and impact was Aurel B. Stodola (1859–1942)[22]. He was of Slovak descent, after finishing his diploma studies with distinction at the Swiss Polytechnical Institute (ETH) at Zurich in 1881, he returned as a professor for mechanical engineering in 1892 and quickly established his global reputation as the leading expert of his time for fluid- and turbomachinery. At the Polytechnikum he established the most advanced machine laboratory in Europe. He worked closely with local industries and inspired companies like Escher Wyss, Sulzer and Brown Boveri, but he gained highest influence and reputation with his book 'Die Dampfturbinen' (The Steam Turbines) in 1903 that was published in six editions (including those in English and French), each time gaining considerably in substance and volume, finally as 'Dampf- und Gasturbinen'[23] by Julius Springer Berlin in 1924, Figure 3-5. In addition to the thermodynamic issues of turbomachinery design, the book discussed integrally aspects of fluid flow, vibrations, stresses, analysis of plates, shells, rotating discs and stress concentrations e.g. at holes and fillets.

---

[20] See Wikipedia 'Steam Turbine'

[21] Redtenbacher's key publications in German were 'Theorie und Bau der Turbinen und Ventilatoren' 1844, 'Theorie und Bau der Wasserraeder' 1846, 'Resultate fuer den Maschinenbau' 1848 and 'Die kalorische Maschine' 1852, all published at Mannheim, Germany.

[22] See Lang, Stodola

[23] Stodola's book was first published in 1903. It had 220 pages and was the result of an expanded lecture at VDI Dusseldorf, Germany in 1902. Since the author consistently tried to cope with the growing know-how in his field, the 6th and last German edition published in 1924 comprised 1'150 pages, 1'200 neatly drawn steel engravings and 13 tables and weighed 3.1 kg. The 2nd edition in 1905 was also translated into English and French. The 6th German edition was translated to become the 2nd English version, appearing in two volumes simultaneously in London and New York in 1927. Three years after Stodola's death there appeared a reprint of the US edition in 1945; moreover, there is an offset facsimile print of the last German edition distributed in China. (see Lang, Stodola, p. 40)

DAMPF- UND GAS-
TURBINEN

MIT EINEM ANHANG ÜBER
DIE AUSSICHTEN DER WÄRMEKRAFT-
MASCHINEN

VON

A. STODOLA
DR. PHIL., Dr.-Ing., PROFESSOR AN DER EIDGENÖSSISCHEN TECHN. HOCHSCHULE
IN ZÜRICH

FÜNFTE
UMGEARBEITETE UND ERWEITERTE AUFLAGE

MIT 1104 TEXTABBILDUNGEN
UND 12 TAFELN

Die Dampfturbinen

und die

Aussichten der Wärmekraftmaschinen.

Versuche und Studien

von

Dr. A. Stodola,
Professor am Eidgenössischen Polytechnikum in Zürich.

Mit 119 Textfiguren und 7 Tafeln.

N° 681.1

Berlin.
Verlag von Julius Springer.
1903.

BERLIN
VERLAG VON JULIUS SPRINGER
1922

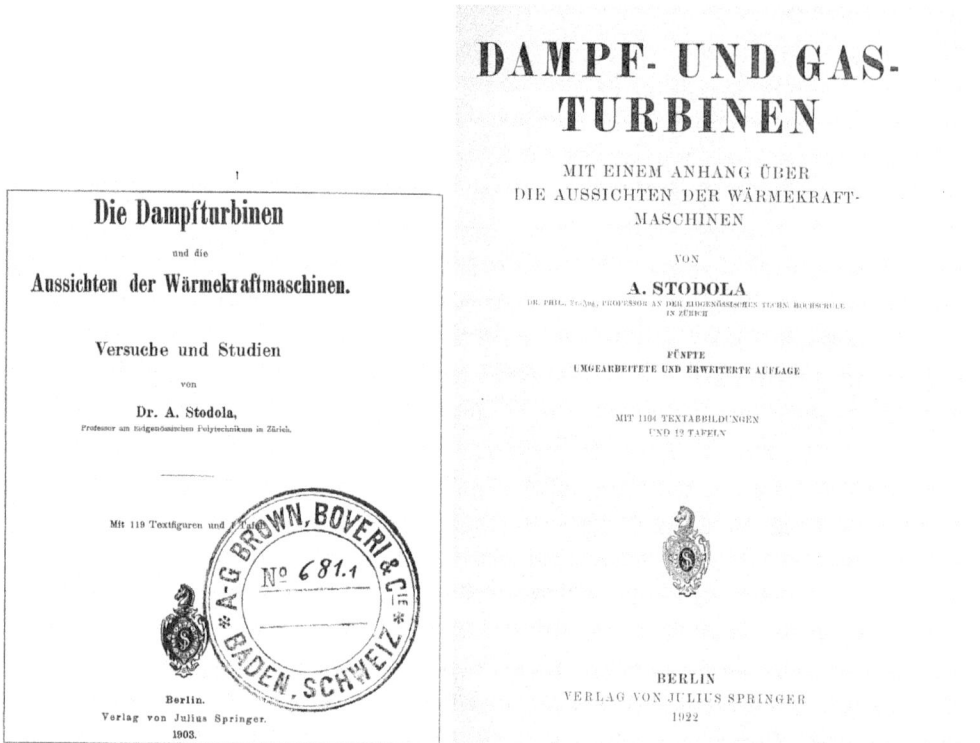

Figure 3-5   *Aurel Stodola, front pages of 1ˢᵗ (1903) edition 'Die Dampfturbinen' and of the 5ᵗʰ (1922) edition 'Dampf- und Gasturbinen*[24] *– the 'bible' of generations of students/engineers in thermo-mechanical engineering*

3.1.1.4   Turbocompressors

In 1884 Charles Parsons patented an axial-flow compressor and returned to corresponding experiments in 1897 and made an 81(!)-stage machine in 1899 which is claimed to have attained 70 percent efficiency. By 1907, his company C.A. Parsons & Co. had made or on order 41 axial compressors, but they were plagued by poor aerodynamics, so that production came to a halt in 1908. According to present understanding, Parsons' (circumferential) spacing-to-chord ratio was far too great for the rotor blading and consequently, all blade rows operated presumably 'stalled' i.e. with flow separating the blade surfaces so that the corresponding losses quickly led to an excessive overheating of this compressor. Parsons' axial compressor design had a blading with a degree of reaction equal to 1 with all flow diffusion (pressure rise) taking place in the rotor blading, while the stator guide vanes had the sole task of redirecting the flow back to axial.

The other major pioneer working on (radial) compressors at that time was the French mining and mechanical engineer Auguste Rateau (1863–1930), who contributed significantly to the further development of quite a number of turbomachines such as centrifugal pumps,

---

[24]   See Stodola, Dampf- und Gasturbinen

turboblowers, action or impulse steam turbines ('Rateau turbine' with a degree of reaction equal to 0) and turbochargers. In 1903, Rateau founded his own company 'Société pour l'exploitation des appareils Rateau' for the production and sales of these products at La Courneuve, today part of the 'banlieue' to the north of Paris, thus becoming one of the various nuclei of the present Alstom, Figure 2-1. Rateau's first turbocompressor provided a pressure ratio of 1.5 at 12,000 rpm, but the measured efficiency was disappointingly at only 56 percent. Subsequently, he designed and built radial compressors of increasing pressure ratios and mass flows, and gradually increasing efficiency.[25] In 1905 he demanded a more rigorous definition of efficiency and in 1907, as already shown in Figure 3-3, BBC delivered its first compressor according to a license from Rateau. The continuing story of compressor development is largely connected to the several decades of struggle to produce an effectively working gas turbine. Many early attempts failed simply due to poor compressor efficiency; it was again left to BBC to overcome these GT power generation hurdles in the early 1930s.

3.1.2   Turbocharger Development History

Since the steam turbine had replaced the piston-type steam engines, in hindsight, the gas turbine might have been expected to replace the piston-type internal combustion engine correspondingly. But this was not the case straight away. It was rather the merging of both concepts which in the form of the turbocharger pushed the combustion engine to extraordinary performance levels.

There were various technological lanes to combustion engine supercharging.[26] In 1885, Gottlieb Daimler (1834–1900) applied for the first supercharging patent ever with external pre-compression by means of an under-piston pump. The air was fed straight through the piston into the cylinder. Practical tests, however, proved unsatisfactory and he abandoned the idea. For the self-igniting oil engines that took the efficiency lead in the 1890s, Rudolf Diesel (1858–1913) had mentioned supercharging in a patent claim in 1896, together with the intermediate cooling of the air charge. First tests showed an output increase of nearly 30 percent, but the fuel consumption was not so impressive, and since high efficiency was his prime goal, he broke off these trials.

The Swiss engineer Alfred Buechi (1879–1959) today is considered the inventor of the turbocharger technology in general, though this does not hold true in this broad, undifferentiated sense. He certainly was the first to address the subject of increasing the performance and power output of ICE internal combustion engines[27] with the application of turbocomponents. This was expressed in his first patent in 1905, Figure 3-6. Ironically, it

---

[25] See Wilson, Turbomachinery, p. 32

[26] The terminology <supercharger> is used, whenever the drive power of the charging compressor reduces the engine output, i.e. when there is a mechanical, geared link between engine and compressor. Alternatively, the 'free-wheeling' <turbocharger> exploits the waste energy of the motor in a turbine which directly drives the compressor. Effective high altitude flight with piston engines beyond 10 km could only be realised by turbocharging – with the inherent drawback of a characteristic turbocharger acceleration lag.

[27] 'Internal' refers to combustion within a piston-pressurised cylinder contrary to external combustion for e.g. steam turbines, Stirling engines etc.

was later critically reviewed by a Buechi publication[28] in 1909, which showed a free-wheeling compressor-turbine 'turbocharger' configuration for the first time.

Consequently and strangely, though the whole area has been plastered by patent claims from a bunch of inventors, the unique and simple turbine-compressor arrangement could apparently never be patented (nor was this early Buechi publication ever referred to in later patent conflicts either by him or by other parties). On the other hand, as outlined in short at the end of this Section, the outcome of a serious patent conflict between BBC and the aero industrial complex of the 'Third Reich' might have been different, if somebody had remembered and referred to the Buechi 1909 paper. Buechi realised that conventional ICEs have low efficiencies, since two thirds of the thermal energy are lost with the exhaust flow. During a series of tests at Sulzer, Winterthur, CH between 1911–1914, Buechi's idea to use the exhaust thermal and kinetic energy by means of a turbine to drive an intake air compressor boosted the thermal efficiency of the engine to a record value of 38 percent, together with a significant rise in output power. Though these facts are impressive, the practical impact of Buechi's early invention was not meaningful; the patent was finally dropped in 1917. But the discussion was revived, when one (of approx. 50 patent families) of Buechi's later patent applications became extraordinarily successful.

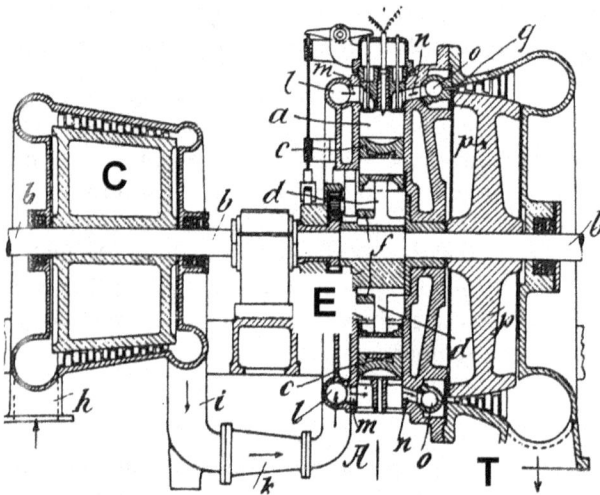

Figure 3-6   *Buechi's patented turbocharging scheme of 1905[29]: (axial!) compressor C, piston engine E and turbine T have a common shaft*

The first immediate necessity for aero engine supercharging arose during WW I, especially during 1916 and thereafter. More and more performance was demanded: greater aircraft speed, heavier loads, longer ranges and higher altitudes. The latter was important to increase flight speed with lower air density and accordingly to decrease drag and escape anti-aircraft fire. The solution of motor over-sizing or over-powering was limited; supercharg-

---

[28] See Buechi, 'Ueber Verbrennungskraftmaschinen' and 'Geschichtliches', Figure 2

[29] Swiss patent CH35259 'Kohlenwasserstoff-Kraftanlage' with a priority on '13 Nov. 1905, 7¼ Uhr p.'

ing and turbosupercharging could provide a solution by pre-compressing the less dense and colder air at altitude, while the thermal engine load and the piston engine power could be maintained unchanged. Therefore, the super-turbocharger compressor design target was to compensate for the decrease in air pressure with altitude[30]:

Altitude above S.L. [m]	Rel. air pressure $b_0/b = \pi_{Target}$
0	1.0
2,000	1.28
4,000	1.65
6,000	2.15
8,000	2.80
10,000	3.83

The main problem in aircraft turbocharging arises from this ambient pressure variation with altitude. But this challenge also has an advantage. Since the basic target is only to maintain power output with altitude, the engine is ready for turbocharging without any significant changes. The challenge was not only the required high-pressure ratio close to 4 (which during WW II could be only achieved in two stages) for altitudes up to 10,000 m and beyond, circumferential tip speeds of 500 m/s, turbine entry temperatures of up to 1,000 °C that had to be endured e.g. for the B 17 'Flying Fortress' for 2.5–3 h, but also quite complicated control systems.

The origins of aero-supercharging cannot be determined unambiguously and as a result of its importance, there are several fathers of the application for the first turbocompressor as supercharger, for the first free-wheeling turbocharger (compressor-turbine combination) and for the first meaningful series production. According to one source[31], the American 'Murray-Willat Company' in 1910 already produced a 90 hp super-charged two-stroke rotating piston engine with integrated radial compressor for applications up to remarkable 5.2 km. Professor Auguste C.E. Rateau in 1916 took out patents for turbocharging equipment that was manufactured and subjected to practical trials in 1917. The linkage of Rateau's company to the historic Alstom company network has already been outlined in the context of Figure 2-1. Rateau's decisive breakthrough to the turbosupercharger freed him from the necessity of heavy gearing, at the same time allowing for high rotational speeds for the compressor impeller with sufficient pressure out of a single-stage configuration. Rateau began tests in early 1917, and some of his turbochargers were still used by the French Air Force during WW I.

Sanford A. Moss (1872–1946) of General Electric, Lynn Ma., USA was the first to manufacture turbochargers on a regular basis. The first GE turbocharger ran in 1918 by using an impulse turbine from De Laval Company. A resulting GE – US Army Air Corps programme finally culminated in aircraft like B 17 and P-38 and even had a decisive impact on the outcome of WW II. As an ironic side note, the professorial Sanford Moss, who received the prestigious Collier Trophy in 1941 together with the US Army Air Corps, disliked airplanes and never flew in one. In 1917 the US Army also tested a turbocharger designed by Earl Hazard Sherbondy (1887–1958) that had been built by the De Laval Company for Fergus

---

[30] Quoted from Jenny, The BBC turbocharger
[31] See Schwager, Höhenflugmotor

Motors. It is said, the Sherbondy machine was essentially similar to that of Rateau. Although it showed a 'number of ingenious details', it was dropped after the construction of three models in favour of the GE unit.[32]

In Germany, Christian Lorenzen of Berlin-Neukölln had participated in early variable pitch airscrew and turbocharger development at the end of WW I. The basic setup of the Lorenzen turbocharger is explained in Figure 3-7, typically combining the centrifugal blower with a one- or two-stage axial turbine in an integral wheel with hollow turbine blades. The radial outflow of compressed air at the same time cools the turbine blading. The nominal design speed was 30,000 rpm, though he claimed a realised burst speed close to 50,000 rpm. In 1926, a turbocharger of his design was tested at DVL Deutsche Versuchsanstalt fuer Luftfahrt (German Aeronautics Test Laboratory) in Berlin-Adlershof in conjunction with a 300 hp Hispano-Suiza flight motor with attached water brake; low

---

[32]   See Constant, Turbojet revolution, p. 124. Constant refers to a 1975 interview with G.D. Wardrop, co-author of E.H. Sherbondy's 1920 'Textbook of Aero Engines', today a bibliophile rarity. Constant's laconic remark that *'the Sherbondy machine was essentially similar to that of Rateau'* triggers a maze of speculation, typical for the early military turbocharger scene. Other US sources make believe that Rateau only 'dusted off' Buechi's 1909 publication. In due course, GE received Rateau drawings via W.F. Durand, chairman of what later became NACA, who got these drawings during a trip to Europe in 1917 in the Interallied Commission. On the eve of the US entry into WW I, *'the Army had practically no material, personnel nor experience in designing, producing, or using aeronautical equipment'*. With 35 pilots assigned in April 1917, an US air component was nearly non-existent. An intensive engineering exchange was launched, more than 100 American engineers were sent to Europe to gain practical experience. In October 1917, the Army Signal Corps selected a site in Dayton, Oh. to build an aviation engineering and test centre, what became WPAFB Wright-Patterson Air Force Base. The department's mission included the evaluation of foreign scientific and technical programmes related to aircraft – the start of scientific and technical intelligence in the emerging US Air Force. Head of the Foreign Data Section and thus the first in a row of well-known commanders of the US National Air Intelligence Centre was E.H. Sherbondy. www.airforcehistory.hq.af.mil/PopTopics/histechintel.htm
Of course, later neither Sherbondy nor Moss at GE stressed this historic background, officially all their turbocharger work had been based on their own independent detailed design. Two acquaintances in this context create the link back to this company history on power generation gas turbines:
Sanford **Moss**, after finishing his studies in Cornell (where Durand had met him as professor), he started at the Thomson Houston Co. at Lynn Ma., USA (in 1892, consolidated with the Edison General Electric Co. to what then became GE General Electric). Moss later claimed he was fortunate to receive his initial exposure to one of the great pioneers of the electrical and turbine industries, Elihu Thomson (1853–1937), from whom AlsTHOM still preserved the name.
Claude **Seippel**, BBC's axial compressor pioneer, Sections 4.1.2 and 4.2, worked at E.H. Sherbondy's engineering office at Cleveland Oh., USA between 1924–1926. There, he followed Sherbondy's personal recommendation to go back to single airfoil measurements from Eiffel, Paris and NACA for a planned axial turbocharger design – with potential application in a large automotive car engine. The project remained on paper, but after his return to BBC Seippel made the first successful axial compressor happen. Sherbondy, according to Seippel *'a genius without success'*, had no wish to join GE after he had lost the US turbocharger competition. He continued as an independent inventor, and his only financial resources were provided by friends; sadly, he finally became mentally deranged.
To my knowledge, Moss and Seippel never met. However, in 1935, Moss (63), near his retirement age, toured Europe for turbocharger 'news' and wrote an intriguing 8 p. travel report (amongst others about visits at RAE and RR) that was found at the National Air and Space Museum, Washington DC and was reproduced in 2009 in the Derby branch magazine of the RR Heritage Trust (by courtesy of Dave Piggott, RR). With reference to BBC, Switzerland, Moss wrote: *'Buchi's scheme ... is now in use in many exhaust superchargers for diesel engines made by the Brown Bovery Company for motor ships and rail cars. I tried to visit the Brown Boveri Company but could not obtain permission. However, I was informed that the Brown Boveri Co. have never made exhaust superchargers for aviation engines. ...'* At about that time RR and BBC had started secret turbocharger collaboration, which amongst other things brought about first film-cooled turbine blading, Section 6.2.2.1, 25 years before similar aero engine designs became known. Following the Moss report, none of these activities leaked out at Derby and the Swiss BBC had already lifted the draw-bridge in time. Moss – Sherbondy – Seippel, the gentlemen knew each other, more or less.

back pressure at the turbine exit side was simulated by a suction blower, which was coupled to a 300 hp Rolls-Royce flight engine.[33] The compressor pressure ratio was in the order of rather modest PR=1.25. In 1928, these tests continued by installing this turbocharger into a four cyl. 10 hp Mercedes-Benz compressor car engine, replacing the standard Roots blower. The 1928 DVL report ended with a positive outlook and the announcement that in a next step the realisation of a 'large' constant pressure gas turbine was planned. In fact, Lorenzen's own publication in 1930[34] deals with a 7.5 MW recuperated constant-pressure gas turbine based on this principle. The fate of the Lorenzen turbocharger in Germany was sealed, however, by another DVL report[35] in 1937, where the major drawbacks of this concept were listed: the compressed air short cut flows directly to the turbine duct and the heating of turbocharger air in the turbine. Finally, after evaluating the DVL test results, the concept was considered not acceptable for further flight application.

One feature of Lorenzen's work was beneficial, though, as Constant[36] states correctly, *'his work on air-cooled blades proved invaluable to the Germans (jet engine developments) during WW II'*. This view was generally accepted up to the point when we traced the little Footnote reference[37] to Lorenzen's (presumed) gas turbine patent DRP 346 599 dated 3 August 1920. To our own astonishment, the DRP Deutsches Reichs Patent and the corresponding Swiss priority patent no. CH 92250, filed on 21 July 1920, belonged to AG Brown, Boveri & Cie, Baden (Switzerland)! Thanks to another reliable side note of the excellent Constant[38], we finally found a note more than 20 years after the event in a 1947 reference paper[39], entitled '**Recent developments in gas turbines**' by Adolf Meyer, head of BBC's thermal department since 1923, that generously explained the situation: BBC had the priority patent on compressed turbine-air cooling, Figure 3-7, and Lorenzen (potentially) operated with a BBC licence in the case that he would have marketed the corresponding gas turbine and turbocharger concepts successfully. The whole context will be thoroughly reviewed in Section 6.2.2, where turbine component and turbine cooling history will be described with a few more 'surprises' of that kind.

After this short excursion into turbocharging in Germany and BBC's involvement, it is worthwhile having a look at direct, motor-driven supercharging as well. Motor over-sizing first tried by Junkers and BMW had to be given up due to weight and performance problems. In 1916/1917 with a critical development in the war, the engine manufacturers turned down substantial design changes. On the other hand, the increasing effectiveness of anti-aircraft defence, especially against the relatively slow 'giant aircraft' (which had replaced the Zeppelins over the battlefield) urgently demanded to fly at greater altitudes, Figure 3-8. The German Walter G. Noack, Section 4.1.2, like Buechi an incessant advocate of engine supercharging, had to report to arms in 1915. After a few intermediate posts he came to Berlin, where he was soon awarded the technical responsibility for the machinery equipment of the 'giant aircraft' fleet. In total, there were 37 giants built by several manufactur-

---

[33] See Heller, Die Gasturbine von C. Lorenzen

[34] See Lorenzen, The Lorenzen Gas Turbine

[35] See Leist, Laderantrieb durch Abgasturbinen

[36] See Constant, Turbojet revolution, p. 148

[37] See Heller, Die Gasturbine von C. Lorenzen, p. 1869; the apparent mistake of Heller will be discussed further and clarified in Section 6.2.2.1.

[38] See Constant, Turbojet revolution, p. 147

[39] See Meyer, Recent Developments

ers, amongst these 18 of the Zeppelin-Staaken type, the rest by Siemens-Schuckert and Gotha. Eighteen R.VIs were built serialled 'R25' to 'R39' and 'R52' to 'R54'. Exceptional, the 'R30' was used exclusively as a supercharged engine test-bed.

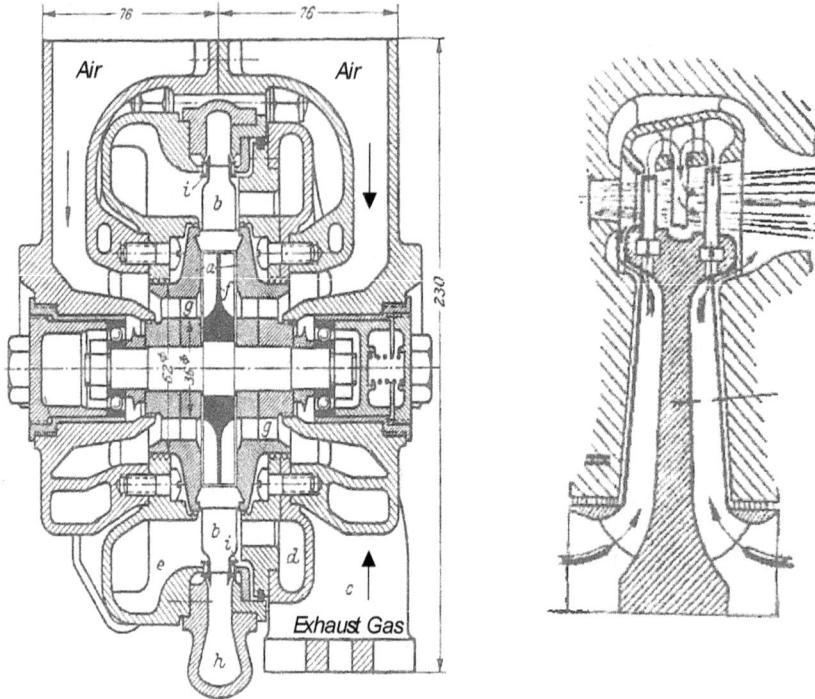

Figure 3-7   *Lorenzen turbocharger,1924 (l)* [40] *and BBC Patent CH 92250,1920 (r) a – turbine disks, b – hollow turbine blades, c – exhaust gas entry, d – gas collector turbine inlet, e – gas collector turbine exit, f – dual-flow compressor wheel, g – air duct, h – air collector/diffuser, i – seal rings*

Due to Noack's initiative, BBC Mannheim built a four-stage radial supercharger[50] in 1917 to be installed in the 'R 30' aircraft; the installation took place by driving the blower with a separate Mercedes D.II engine located immediately behind the pilot in the central fuselage[41] and by putting the pressurised air piping to the two tandem engine nacelles and to the blower motor, Figure 3-9. In parallel, Noack carried out the first altitude tests to determine the decreasing motor power output and the possibilities of compensation by supercharging in a low pressure test chamber at Zeppelin Airships at Friedrichshafen, Lake Constance,[42] an early reference to BBC's

---

[40]   Figure source: Heller, Die Gasturbine von C. Lorenzen, Fig. 10

[41]   As a further novelty, the same aircraft was fitted with four of the first mechanically adjustable, variable-pitch propellers from Helix Propeller Ges. according to a design by Prof. Hans Reissner, TH Berlin-Charlottenburg. Besides the BBC turbocharger, the new propellers were instrumental for achieving the new record height of 5'900 m on 2 April 1918. A similar engine-powered supercharging system with additional large charge air coolers was developed by Daimler-Benz in World War II for the German Henschel Hs 130E and Dornier Do 217 P high-altitude fighter aircrafts, then known as 'Höhenzentrale' (HZ-Anlage); Do 217 P achieved 14,000 m during test flights in June 1941.

[42]   See Noack, Lebenslauf, p. 15

later unique experience in building high altitude test facilities. The first flight of the experimental giant aircraft R 30/16 with a supercharged motor plant took place on 2 April 1918. The supercharger produced a pressure ratio PR = 1.75, lifting the operational ceiling from 3,800 to 5,900 m. Enthusiastically, a BBC report commented the achievement: '*While before the supercharger was installed the full aircraft required two and a half hours to achieve the ceiling altitude of 3,900 m, it reached the 6,000 m level with continuous blower support within only one and a half hours !*'[43] The speed difference between drive motor (1,450 rpm) and BBC blower (6,000 rpm) was bridged by a carefully manufactured ZF gearbox. Actually, BBC Mannheim developed two compressors for the R central blower unit and two more to be coupled directly to 260 hp motors, e.g. for a twin-engine Gotha bomber configuration.[44] In total, according to Noack, some 75 supercharger sets were manufactured; in his own comprehensive paper, he describes not only the BBC superchargers but also three more blower configurations from Schwade at Erfurt, AEG in Hennigsdorf and Siemens-Schuckert in Berlin-Siemensstadt.

Figure 3-8   *German Zeppelin-Staaken R. VI (R Riesen=giant) aircraft without/with BBC supercharger in central fuselage, 1917/1918*[45] *Wing span 42 m, prop. diam. 4 m, service ceiling 3,800/5,800 m, max. flight speed 135/160 km/h, 4 × Mercedes D.IVa 6cyl. water-cooled inline, 4 × 260 hp TO power + 1 × 120 hp Mercedes D.II SC drive motor, crew 7, flight endurance 7–8 h*

---

[43] See BBM, BBC Flugzeug-Geblaese...1919, p. 122

[44] See BBM, BBC Flugzeug-Geblaese...1919, p. 125

[45] Source/courtesy: http://www.thomasgenth.de/html/zepp_-staaken_rvi.html (top) and http://hoffstrizz.typepad .com/hoffstrizz/2010/12/biggest-planes.html (bottom)

Figure 3-9   *BBC four-stage radial compressor for 1,200 hp R aircraft engine equipment, 6,000 rpm, 0.47 m impeller diam., PR = 1.75 and installation impression (r), viewed from the rear towards the flight deck.*[46]

On 24 May 1919, one month before the signing of the Treaty of Versailles on 28 June 1919 would stop (at least officially) any of such activities completely, the plane R 30/16 took off on a test flight piloted by Hauptmann Georg Krupp (commander of Rfa 501 – **R**iesenflugzeug-**a**bteilung – and later chief of the 'Deutsche Luftfahrtsammlung' at Berlin-Moabit, a collection of salient German aircraft developments), Leutnant Offermann, Dipl.-Ing. Walter G. Noack and five technicians. During this flight, one engine and presumably also part of the wing canvas caught fire at an altitude of 3,300 m. A steep descent kept the flames small, but the fire was still burning until Noack managed to extinguish the flames with a standard fire extinguisher. He did this according to accounts[47] by crawling cold-bloodedly to the burning nacelle over the exposed lower wing and by removing the smouldering canvas flapping around. Then the R 30/16 landed safely.

After this in-depth excursion into military aircraft turbo- and supercharging during and after WW I, we continue the following short report on further developments only to the extent necessary for the understanding of power generation GT history. After WW I, interest in commercial turbocharging first arose in the maritime community. In 1923, the Vulkan shipyard at Stettin, D (today Szszecin, Pl) ordered two large passenger ships, each to be powered by two turbocharged ten-cylinder, four-stroke MAN engines. The turbochargers

---

[46]   See Noack, Flugzeuggeblaese, Figures 27 (r) and 28 (l)

[47]   A comprehensive survey of historic WW I and WW II aircraft information can be found on the website of Thomas Genth, http://www.thomasgenth.de/index.html. It provides the details of Noack's dramatic flight, re-quoted from the book – see Hundertmark, Phoenix – as well as the miraculous survival of just the one R 30/16 engine nacelle: In due course and after 'Versailles', the aircraft had to be scrapped and only one of the engine nacelles went to the Deutsche Luftfahrtsammlung at Berlin-Moabit. With the onset of heavy allied air raids on Berlin in 1943, the unique air collection was partially destroyed in the night of 22/23 November 1943. The rest of approx. 30 aircraft was brought to Czarnikau/Czarnkow, 35 km south of Schneidemühl/Pila in Pomerania, east of the river Oder, an area that became Polish after WW II. Consequently, this and other pieces of early aircraft and engine history are now on exhibition at the Polish Aviation Museum, Cracow since 1963, still in Polish custody.

were designed and built under Buechi's supervision. Launched in 1926, these two ships were the first in maritime history to have turbocharged engines.

For BBC things changed in favour of turbocharging in general in 1923 with the publication of a report on successful low-pressure supercharging trials with a four-stroke diesel engine at MAN Augsburg, D.[48] Brown Boveri, under the lead of its new director Adolf Meyer, now made the decision to apply its extensive know-how in building turbines and compressors to the development of turbochargers. This may have been the prime interest, though there must have been ulterior motives, the gas turbine. For a long time, the high gas temperatures of the gas turbine appeared to be an insurmountable hurdle. However, with a compound solution like Figure 3-6, the diesel motor could be used as a combustion chamber and with sufficient work done, the diesel exhaust temperatures would be low enough (say 600 °C) to be acceptable for the subsequent turbine blading. Under the assumption of technological progress e.g. in material temperatures, this development strategy would allow the diesel share to be reduced and the work of the turbo-group to be increased steadily. This development would have been close to Buechi's original compound considerations of 1905, but the reality developed differently.[49]

In 1923, Swiss Locomotive and Machine Works (SLM) were testing a two-stroke experimental engine that needed bringing up to a higher power level with better fuel consumption. Brown Boveri recommended using an exhaust gas turbocharger that would feed into the scavenging blowers, and SLM subsequently placed an order for such a machine. In June 1924, turbocharger VT402, the world's first heavy-duty exhaust gas turbocharger, left the Baden works of Brown Boveri, Figure 3-10.

In 1925, Buechi led Sulzer's patent department and was allowed to take out a patent in his own name on the 'Buechi-Duplex turbocharging system' for diesel engines, which was later to make him wealthy and world-famous. This system, coming into force as Swiss patent no. 122 664 at noon on 30 Nov. 1925 employed collective piping for certain cylinder groups, tuned-length exhausts and inlet and exhaust valve timing modifications to assure complete cylinder scavenging and smooth turbocompressor operation. But while the Buechi system was technically successful, it required complex, heavy exhaust piping and was efficient only over a relatively narrow range of engine operation conditions, so that as a direct consequence it was never applied in aircrafts.

Nevertheless, this system approach of advantageous patent exclusivity, tailor-made design solutions and accordingly provided hardware was the breakthrough that everyone had been waiting for. A joint venture, named the 'Buechi Syndicate' and comprising Alfred Buechi's engineering office, SLM Schweizer Lokomotiv- und Maschinenfabrik, both at Winterthur, CH and Brown Boveri Cie., Baden, CH, was set up the following year. Buechi was put in charge of engineering and customer relations, Brown Boveri was to build the turbochargers (except for the USA) and SLM would provide the diesel engines for tests and

---

[48]   See Riehm, Leistungserhoehung; he described tests at MAN in preparation of the Vulkan ship orders. The turbochargers were manufactured in the Vulkan Works in Hanover, with BBC Mannheim supplying the compressor. BBC Mannheim had gained corresponding manufacturing experiences for Noack's aero engine superchargers in 1918/9. The ship installation happened in September 1926 so that the delivery for the SLM order of BBC's VT 402 in Baden was presumably a few weeks earlier.

[49]   See BBM, Leistungssteigerung von Dieselmotoren...1937, p. 176

trial runs. The 'Syndicate' was in place until 1941, when BBC finally decided to carry out future turbocharger activities solely in its own name.

Figure 3-10   *VT 402, the first BBC turbocharger [50] for a large diesel engine*   ©*ABB*

An improved, larger turbocharger designated VT 592 was supplied to SLM in 1927 for a second experimental engine with Buechi's Duplex system. The impressive results were the breakthrough for low-pressure turbocharging:

- 50 percent more power under full load,
- less volume and weight in comparison to an un-charged configuration of the same power output,
- very high overload capability,
- excellent fuel consumption values,
- no increase in thermal stress, less wear and maintenance requirements,
- less noise, due to the exhaust turbine noise absorption,
- overall, a considerably cheaper and more efficient engine.

Licensing agreements were now being reached between the Syndicate and leading engine manufacturers. First test runs on diesel-electric locomotives took place. Turbochargers were also putting themselves forward for more economic operation of stationary diesel power plants.

---

[50] The letters VT for the first BBC turbochargers stands for VT Verdichter-Turbine (compressor turbine). On the initiative of W.G. Noack, the first VT 402 turbocharger for SLM was obviously derived from electrically driven scavenging blowers for two-stroke diesel engines that BBC had been supplying to Sulzer and other customers since 1915. In principle, the electric drive motor was replaced by the turbine component. The two-stage compressor provided PR = 1.35, the impellers had 0.4 m diam. This explains the short delivery time, the SLM order dates from September1923 and testing (with steam) and final delivery were already carried out in June 1924. Due to this sequence in timing, it is also likely that Noack – in his function as the responsible engineer for the R (Riesen = giant aircraft) division of the German imperial flying troops in Berlin-Charlottenburg in the final stages of WW I – used certain fundamental design parameters of this earlier BBC electric blower for the four-stage aero supercharging compressor with PR = 1.75 and 0.47 m diameter wheels in 1918/9, manufactured at BBC Mannheim. Figure 3-10 from Jenny, The BBC turbocharger, p. 46.

This obvious success story was overshadowed by a decade-long personal dispute amongst some of the key players. Part of the confusion resulted from the originating patent itself, Figure 3-6. Actually, the patent illustration showed a 'compound motor' configuration, where intake air compressor C, combustion engine E and exhaust gas turbine T have a common shaft. In later disputes, Buechi was reproached for describing only the highly supercharged compound motor and not the low-pressure turbocharging that started its industry success some 20 years later. Though this statement was apparently confirmed by the patent drawing, the decisive two patent claims left the kind of compressor/turbine arrangement completely open. Consequently, both the direct drive compound configuration as well as the conventional turbocharger with compressor and turbine wheel on one shaft, gas-coupled, but mechanically separated from the piston engine would have been covered. There were no immediate consequences in 1905 and the following years, but the dispute re-appeared BBC-internally with the rapid success of the Buechi Syndicate in the early 1930s. It was mainly Walter G. Noack, who repeatedly pointed out that in his opinion BBC's unshakable adherence to the agreed Syndicate principles was not justified since Buechi's success was basically the result of BBC's practical support and implicitly of Noack's personal involvement.[51]

True, Alfred Buechi had been a friend in (Noack's new boss) Adolf Meyer from their student days as Stodola's scholars in the early 1900s, which may have influenced some generosity from BBC's side towards the independent, experienced self-marketer. The license fees for using the Buechi patents for turbocharger production by BBC were in the range of 2.5 to 8.2 percent and thus quite considerable. On the other hand, the success of turbocharging had really been initiated to a large extent by Buechi's initiative.

So far, the argument follows established understanding of specialists in the field, as outlined in detail in Jenny's book on the BBC turbocharger.[52] Another aspect might have influenced Meyer's thinking in view of the decade-long patent struggle on the subject and Buechi's well-known experience in this field. In the 1930-1932 time frame, BBC laid the foundation for later extraordinarily successful new products, the Velox boiler for rapid steam generation and the axial compressor as part of the Velox turbocharging group and as an indispensable component of the coming power generation gas turbine. As will be discussed in detail in the future BBC aero engine history, a BBC patent application in Germa-

---

[51] See Noack, Lebenslauf; this document from 1942 – written three years before his unexpected death at 64 – is a technically inspiring document with a wealth of detailed information on historic background and interdependencies for the most decisive thermal engine developments in the period of roughly 1890–1940, BUT one would hesitate to use it as a characterisation of his personality. The selfish egomaniac style is hard to digest even today and must have been a shocking provocation for his contemporary, long-time colleagues. It appears that the outbreak of World War II also strained the established relationship with the company and his Swiss environment for the German Noack. A more balanced, comprehensive assessment of his technical achievements and his personality have been attempted in Section 4.1.2. As an indicator of his feelings in respect to the discussed time frame (and BBC's continuing support of the flashy and commercially successful Buechi) a quotation from Sir James Swinburne [1858–1958, a prominent British engineer and entrepreneur in the electrical and plastics industry (Bakelite), who coined many new electrical words] shall be reproduced which Noack put (in English) in his memoirs – see Noack, Dr. h.c. Walter Noack... – at the beginning of the chapter entitled <My work at Brown Boveri after my re-entry in Baden 1920> *'Looking at modern industry as a whole, one realises that it is based on invention and depends on a great class of technical men, who are largely behind the scenes, and in proportion to their knowledge and ability do not get much limelight or swollen bank accounts. They seem happy all the same, because **they like their work and have the spirit of the artist.**'* (**bold passage** in the preserved copy presumably from the author himself).

[52] See Jenny, The BBC turbocharger, pp. 38–56

ny in 1932 (inventor Cl. Seippel) in view of the Velox turbocompressor design and regulation caused extraordinary industrial commotion and may have even influenced part of the
German rearmament activities for aero engines. The industrial opposition was finally in
vain, when the German Patent Office in Berlin decided in favour of BBC in February 1936,
against massive industrial-political pressure (including Swiss Escher Wyss in lead opposition). In hindsight, this decision was somewhat astonishing, since the key patent claims
under consideration a) the self-regulating power characteristics of a turbocharger and b)
the airfoil-deduced design principles for the compressor blading were no longer new at
that time. But there is the general impression for all these early patent conflicts that thorough and comprehensive investigations – e.g. with respect to pre-released information –
were rare and especially in Germany nearly impossible, given the short, formal patent ob-
jection period of only three months. Correspondingly, the presented Escher Wyss opposition material was weak and irrelevant. But, e.g. a hint on the direction of Buechi's first turbocharger patent, Figure 3-6, with an all-axial, multi-stage turbomachinery arrangement,
supplemented by some additional materials on intermediately, broadly used free-wheeling
turbocharger arrangements[53] might have changed the case considerably. In this respect,
BBC may have had an underlying reason for the purported, generous treatment of Buechi,
other than personal friendship and premature acceptance of questionable patents.

Except for the continued turbocharger developments at GE, similar activities in Europe
started slowly in the 1930s. A target specification of the newly emerging German Luftwaffe
in May 1927, one year after the Versailles aero limitations had been finally removed, demanded *'pre-compression equipment and variable-pitch propellers for high altitude flights
above 10,000m'.*[54] This in part may explain the initial German reaction to the BBC axial tur-
bocharger/Velox patent between 1932–1936. Moreover, two contacts in reference to aero
engine turbocharging were noted within BBC for the year 1934: Discussions took place between the French daughter company C.E.M. and Hispano-Suiza on turbocharging their 12
cylinder flight engine and in addition, first contacts with Rolls-Royce happened to prepare
a BBC Baden turbocharger for what later became the Merlin engine, widely used for Spitfire and Hurricane aircraft applications. The turbine of this latter BBC product is of extraordinary interest since it represents the first application[55] of turbine air-cooling in
turbomachinery. Before that turbine air-cooling was applied in a simplified form in Ger-

---

[53]   The history of the free-wheeling turbocharger is complicated, to say the **least**, and also reflects the reckless-
       ness in dealing with patenting issues. Buechi's 'scavenging patent' of 1915 explicitly showed the free-
       wheeling configuration (and Adolf Meyer confirmed this as priority setting later on), though there was an
       apparently overlooked Buechi publication from 1909 with a separate, non-compound turbocharger. Since
       WW I with aero-applied, free-wheeling turbochargers e.g. by Rateau in France, as well as by Moss (GE) and
       Sherbondy in the USA, the turbo-charger (with centrifugal compressor wheel) and its self-regulating opera-
       tion principle was a matter of fact. See Jenny, The BBC turbocharger, pp. 39–44.

[54]   See Gersdorff, Flugmotoren, p. 42

[55]   This statement is true even in view of Lorenzen's earlier turbocharger that in this respect used 100 percent of
       compressor air for turbine cooling. Moreover, as outlined in the context of Figure 3-7, the turbine cooling as-
       pects of Lorenzen's turbocharger relied on earlier BBC patents. The BBC turbocharger used 10 to 20 percent
       of the compressed air for the specific turbine cooling task (with the upper value near surge margin) and was
       therefore close to later front row cooling values. German jet engine cooling, both at Junkers and BMW, is of-
       ten claimed to have built on Lorenzen, though his patent situation was clear to DVL and must have been clear
       to RLM as well, but after mid-1940 the original BBC patent CH 92250, dated 21 July 1920, had expired any-
       how. The RR-BBC turbocharger cooperation broke up in 1939/40, when RR insisted on a fully-licensed pro-
       duction in the UK, while BBC wanted to keep these jobs in Switzerland, a strategy that materialised immedi-
       ately after the war.

man war-time jet-engine developments of Junkers Jumo 004B and BMW 003. BBC's development of turbine film-cooling or more specifically 'blade showerhead cooling' some 30 years before less sophisticated versions became popular in aero jet engine developments of the 1960s, is still an untold chapter, which will be added to BBC's already multi-faceted technological history in Section 6.2.2.

The further story of BBC turbochargers can be described by one word: success, though a lot of development work lay ahead of BBC's engineering, as described in detail in Ernst Jenny's book.[56] On 1 April 1949, turbocharging became an independent BBC department within Thermal Machines. It was placed under the direction of Claude Seippel, and in the same year the keen-witted decision for a separate turbocharger manufacturing plant 'turbo-boosted' the further sales development. Figure 3-11 illustrates the strong rise in annual deliveries with a first peak of 4,000 delivered units in the early 1960s, then reflecting economic changes with double peaks of 6,000 annual units in the mid-1970s and -1990s. Only recently, the breakthrough of the record high of more than 10,000 units per year happened, with this culmination basically resulting from a considerable order backlog of several foregoing record years. It demonstrates the unbroken attractiveness of this great, though comparatively simple turbomachinery concept.

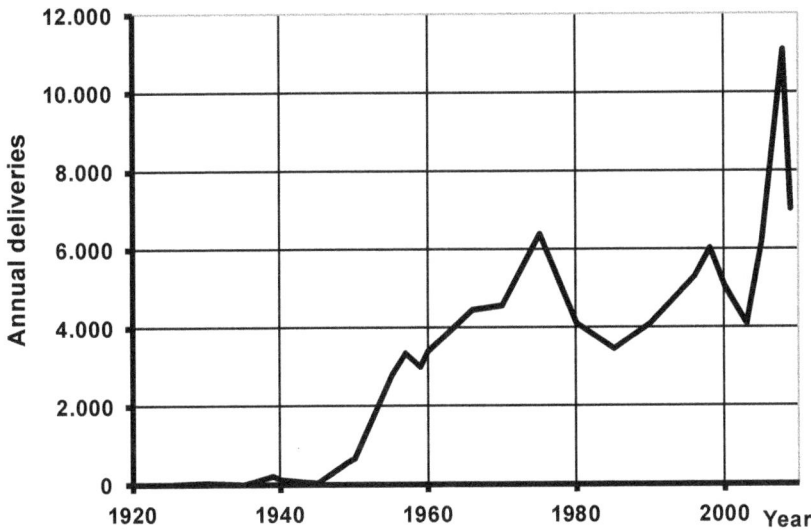

Figure 3-11   *Breakthrough in turbocharging – the development of annual BBC and ABB Turbo Systems deliveries* [57]

---

[56]   See Jenny, The BBC turbocharger, p. 89 ff. The turbocharger 'success' in Baden after 1945 has to be put into perspective due to the general Allied decision to prohibit gas turbine manufacturing in Germany that excluded Mannheim completely from this booming market. There are not many details known about turbocharger development and manufacturing in Mannheim before 1945. Jenny on p. 81 refers to a total production of 1'700 turbochargers between 1935–1945 from BBC Mannheim, *'many times the production in Baden'*. The four-stroke MAN M6V 40/46 with two BBC (called Buechi) turbochargers VTA 450 for the late German submarine XXI class was considered by American post-war evaluation *'as the best diesel engine in the world at that time'*.

[57]   See Jenny, The BBC turbocharger, Fig. 5.1 and more recent data, courtesy to ABB Turbo Systems

3.2   Early Attempts with the Gas Turbine Principle

The gas turbine looks back on a long and frustrating history. In comparison, the steam piston engine or turbine was easy to design and manufacture. Little effort is required to force water into a boiler, to heat it and when steam is formed at high pressure and led to either an expansion piston or turbine, it will produce more power than required by the feed pump.[58] The same applies for internal combustion/piston engines. Although much more work is required to compress air in a piston than is needed to get water into the boiler, their maximum process temperatures are much higher than those for gas turbines. Therefore, for the piston engine again, the expansion work is much greater than the compression work, even if considerable losses have to be taken into account. Since gas turbines could not use such high process temperatures, the demand for a positive work output immediately requested low losses, especially on the compression side. Consequently, for decades many inventors produced machines that either never ran self-driven at all or the output was so small, that the overall effort was not justified in view of other competitive concepts. Therefore, following a thought from Claude Seippel[59], later undisputedly referred to as the 'father of the axial compressor', one can roughly differentiate three development waves, before the successfully power-producing gas turbine was realised:

–   First, the early concept demonstrations until approximately 1900, an era of great inventors and of men acting by intuition more than by reasoning and computation. This phase already saw almost perfect diesel engines, but no gas turbine of any resonance.

–   Second, self-propelled engine trials with some, though relatively small positive work output up until the early 1930s. More precise theoretical thinking supported intuition. The period saw the growth of steam turbines to 100 MW and the first operational gas turbines with small scale power output. The reason for this failure in general was still a lack of aerodynamic understanding, especially between the differences of accelerating (turbine) and decelerating (compressor) relative flows and limited material properties for high-temperature applications; intensive research activities in both areas were necessary to overcome these hurdles.

–   Third, beginning in 1930, BBC's early one- and two-stage axial blower developments followed by Velox boiler development activities, which latest until 1936 had handled most gas turbine problems successfully, demonstrating positive output power as a rather surprising by-product of the axial turbosets in the Houdry refinery process, Section 4.3.2, – until in 1939, the first 4 MW utility gas turbine was installed at Neuchâtel in Switzerland. The 3rd era, in Seippel's words '...*is a period of systematic experimentation to support theoretical thinking which in turn guides intuitive action. It is the era of wind tunnels, of metallurgical laboratories, of research on combustion, of physicists and chemists assisting the engineer. ... The gas turbine is a very pretentious machine that, unlike the combustion engine and the steam turbine, had to await the advent of the 3rd era before it could be made a success.*'

John Barber (1734–1801) of Nuneaton, England is the undisputed inventor of the gas turbine. Originally financially independent by heritage, he continued patenting (even after he

---

[58]   Since the power requirement of turbocompressors is proportional to the high volume flow per unit of time, and the process takes place in the gaseous phase, compressors are thermodynamically 'handicapped' in comparison to e.g. feedwater pumps, which work on a small liquid volume at a fraction of the compressed air volume.

[59]   See Seippel, Gas Turbines, p. 122

had gone bankrupt) with the best result in 1791. He received the British patent #1833 for '*A Method for Rising Inflammable Air for the Purposes of Producing Motion and Facilitating Metallurgical Operations*'. In it, his 4th and most important patent, he describes an engine that could be applied to the '*grinding of corn, forging, ...turning of mills for spinning, directing the fluid stream into smelting furnaces, ... for propelling ships*'. In principle, there may have been two motives behind this invention at that time:

 — to replace the complex steam plant with steam boiler, steam engine and condensing plant with simpler equipment and
 — to generate the engine rotation that was required for power transmission anyhow, immediately by means of a turbine (and as such getting rid of the crank and connecting rod mechanism of the steam engine).

The engine of the patent description, Figure 3-12, comprised two retorts for heating coal or oil to produce an inflammable gas, one to operate while the other was cleansed and recharged. The resultant gas, together with the right amount of air, passed through two beam-operated piston compressors to a water-cooled combustion chamber and then further to a water-cooled nozzle in an impulse turbine which drove the compressors and provided the output. Thus, the logic of the patent description clearly follows the thermodynamic sequence that later became known as Joule Cycle (or Brayton Cycle in the USA). Though it seems unlikely that John Barber was able to get his engine to work – 100 years before a continuously operating combustion chamber was achieved – the details of the description nevertheless suggest that practical experience must have been gained.[60]

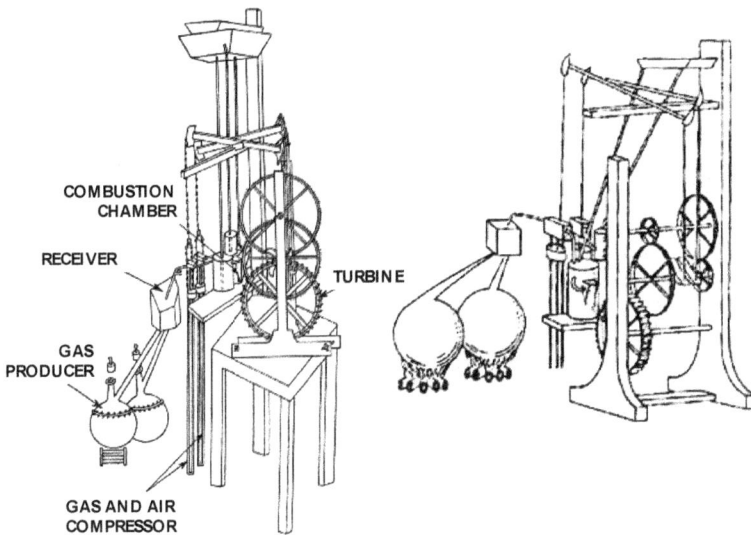

Figure 3-12   *First patent for a gas turbine by John Barber 1791*[61]

---

[60] See Day and McNeill, Biographical Dictionary, p. 844. Alstom engineers and presumably others too regularly salute to John Barber these days when they drive through Nuneaton, half-way from Birmingham Airport to Leicester, an area of highly-qualified manufacturing centres for gas turbine blading e.g. at AETC Ltd.
[61] Figure source: http://en.wikipedia.org/wiki/John_Barber_%28engineer%29

Franz Stolze (1836–1910) of Berlin, Germany was obviously the first, who came up with a convincing gas turbine concept, which at first sight has great similarity with present gas turbine configurations, Figure 3-13. He relied completely on a one-shaft, two bearing rotor and an all-axial flow turbomachinery set for both compressor and turbine (K – for 'Kompression' and E for 'Expansion' in Figure 3-13). His design shows – varying depending on his patents and publications – many more stages in the turbine than in the compressor, as he incorrectly probably thought necessary due the higher energy head acting on the turbine. Stolze can even be considered the inventor of the annular combustion chamber. His standard concept of a 'fire turbine' showed a vertical combustion chamber with compressor air passing upward through a perforated grate and a kind of fluidised bed of burning anthracite. But in his British patent no. 7398 with application in 1898, he states '*If gaseous or liquid fuel be made use of – the combustion may take place directly within the cylindrical mantle or casing, connecting the mantles or casings of the turbine systems, thereby obviating the use of a separate furnace.*' [62] Stolze's life was influenced considerably by the gas turbine and overshadowed by an unhappy patenting process in this context. During his philosophical studies in Jena, D, he came in contact with Redtenbacher's books on fluid machinery and from this resulted his first patent application of the 'fire turbine' in 1873. Convinced that he had proven a severe fault in Redtenbacher's work, he stressed the multistage concept of a turbomachine as an effective method of energy transformation. Obviously too much so, because his new gas turbine concept in general dropped out of sight. Since multi-stage axial machinery was already known, Stolze's patent application was denied by the German Patent Office and for 24 years he turned to other subjects. [63]

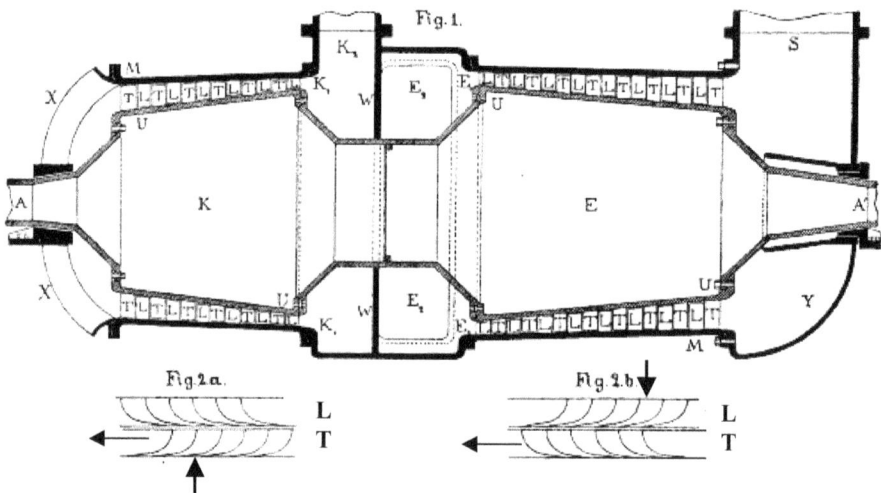

Figure 3-13   *Franz Stolze: Hot air machine, patent No. CH18721, 1899*

---

[62] See Friedrich, Dokumente, p. 81

[63] During all of his life Stolze was highly motivated to propagate his father's invention of a special stenography technique. Besides turbomachinery, his other technical interest led him to high-resolution photography and optics. In 1874, immediately after his first 'fire turbine' patent application had been rejected, he accompanied a research expedition to Isfahan, Persia to document the transit of planet Venus by means of a 'photoheliograph'. Thereafter, he supported archaeological excavations at Persepolis and in the mosque of Shiraz by taking photogrammetric measurements. He composed poems, novels and drama in Persian style.

Eleven years after Stolze's first and rejected patent application, Charles Parsons was more successful with his British patent no. 6735, comprising the same combination of multi-stage axial compressor and turbine on one shaft with a multi-fuel combustion chamber. After Stolze's second patent application for the 'fire turbine' (with an added heat exchanger) had also finally been accepted in 1897, he turned wholeheartedly to the practical demonstration of his gas turbine ideas. In 1904, he developed and built a test and pilot plant with an expected power output of 200 hp in Berlin-Weissensee, Figure 3-14. Material problems, higher pressure losses and particularly the unsuitable compressor blading profiles hampered his success; continuous self-maintained engine operation was not possible, let alone any power generation.

The decisive flaw of Stolze's concept can be seen at the bottom of Figure 3-13, where compressor and turbine, stator and rotor blading are completely 'mirrored'. The kind of sheet metal blading has no signs of aerodynamic profiling with thickness variation.[64] Moreover, as is known today, the strong blade shape curvatures would work for turbine flow, but would cause boundary layer separation and accordingly high losses for the 'diffusing', separation-prone, compressor flow. This lack of sensitivity for the aerodynamic peculiarities of diffusing flow with a pressure rise is typical for the time and, besides other prominent exponents, also caused Parsons to fail in his tireless struggle for sufficiently high compressor efficiency. Even in 1924 to 1926, the early beginnings of compressor research at BBC could have been easily overshadowed by this misleading conclusion.

Figure 3-14   *Franz Stolze, aside of his all-axial gas turbine prototype plant at Berlin-Weissensee, 1904 (compressor inlet shown in front)* [65]

Two company representatives – for BBC Jean von Freudenreich (1888–1959), head of the Baden test laboratory and – for C.E.M. Georges Darrieus (1888–1979) visited the World

---

[64] This is somewhat surprising, since Otto Lilienthal (1848–1896) carried out his glider experiments since 1894 at Berlin-Lichterfelde, some 20 km south of Berlin-Weissensee, based on profiled (stork) wing studies – using already polar diagrams, Figure 4-19, for describing the aerodynamic performance.

[65] Picture source: Gas Turbines, The Electrical Magazine (Power) 2, London 1904, p. 574 f.

Power Conference and Exposition[66] at London, Wembley in June 1924. Darrieus later remembered in a somewhat amused tone: '*In front of an open Parsons turbine with typical 45° blading and a degree of reaction of 0.5 Freudenreich developed the idea that this symmetrical blading might be suited to run this engine in reverse as a compressor by changing the direction of rotation. We ignored the fact that Parsons had already tried to realise an axial compressor himself in 1901, giving up this idea due to unsatisfactory results.*' Luckily, some aerodynamic wing theory had already been accumulated for practical usage so that the spontaneous exercise found a successful practical end.[67]

Seippel's afore-mentioned 2nd phase of GT development was initiated by the Norwegian inventor J.W. Ægidius Elling (1864–1949).[68] He started working on GT designs in 1882, filed his first patent in 1884, and realised the first constant-pressure gas turbine to produce a net output of 11 hp (in the form of compressed air) in 1903. It had an advanced six-stage centrifugal compressor with angle-adaptable diffuser vanes and water injection between the stages. A heat exchanger produced steam to be mixed with the combustion gases before the nozzles of a driving centripetal turbine. In 1904, Elling built the first regenerative gas turbine in the modern sense, in which the turbine outlet gases heated the compressor delivery air. This turbine raised the turbine inlet temperatures to 500 °C (in comparison to 400 °C of his former design) and the power output to 44 hp, this time directly to an electric generator. Elling understood that if better materials for higher temperatures could be found, the gas turbine would be an ideal power source for airplanes. He realised that economic reasons demanded gas turbine units of 1,000 hp or more, and far-sightedly he imagined 700–1,000 kW units as feasible.

Shortly after Elling, another stand-alone, net power generating GT power plant was built according to a patent of Charles Lemâle by the French René Armengaud (1844–1909) and Charles Lemâle at Paris in 1905 to 1906. In 1903, they had founded the Société Anonyme

---

66 Shortly after World War I, Scotsman Daniel Nicol Dunlop (1868–1935), a visionary working in the British electricity industry, decided to bring together leading energy experts for a World Power Conference to discuss current and emerging energy issues. The First World Power Conference was held in London's Wembley exposition ground in 1924 and attracted 1'700 delegates from 40 countries. The meeting was so successful that those attending decided to establish a permanent organisation to continue the dialogue begun at the conference. From this nucleus resulted today's World Energy Conferences, run by the non-governmental and non-commercial WEC World Energy Council.

67 See Darrieus, Hommage au Dr. Claude Seippel, 'Devant une turbine ouverte de Parsons laissant voir sa constitution caractéristique avec aubes inclinées à environ 45° et a degré de réaction 1/2 le regretté chef de la plateforme d'essais à Baden émit l'idée que cet ailettage symétrique devrait se montrer réversible et se prêter, en inversant le sens de rotation, à un fonctionnement en compresseur. Nous ignorons alors que, déjà en 1901, Parsons eût lui-même tenté de réaliser de tels compresseurs axiaux, d'ailleurs promptement abandonnés faute de résultats suffisamment satisfaisants.' Later Seippel commented dryly on the innocence in this context – see Seippel, Entstehungsgeschichte des Axialkompressors – that Parsons' disappointing compressor trials had been carried out with a degree of reaction of 1.0 blading, 'what these gentlemen did not know either'. By the way, in 1926 Freudenreich successfully realised his original idea of a reverse-turbine compressor. But before starting the tests, a small reaction turbine with some ten stages received a new set of blading with flatter, only moderately curved airfoil-like profile shapes; unexpected and presumably overestimated – the first tests showed an efficiency of 80 percent. See Darrieus, Hommage au Dr. Claude Seippel

68 See Johnson, Elling. Elling's achievements were overlooked unaccountably in most historical reviews of turbomachinery in the past, until by the relentless initiative of Jan Mowill, himself known for his excellent high-pressure centrifugal compressor designs, the turbomachinery community took notice of this hidden champion of early gas turbine history. Elling's gas turbine prototypes from 1903 and 1912 are exhibited at the Norsk Teknisk Museum in Oslo.

des Turbomoteurs. While Stolze was in search of the all-axial engine from the very beginning, the French rather looked for a proven radial compressor design from Brown Boveri.
This machine that was built in Baden and dispatched to Paris in 1906 was one of the largest
turbocompressors in the world at that time, Figure 3-15. This first experimental gas turbine
consisted of this 25-stage radial compressor (System Rateau) from BBC and a water-cooled
two-stage Curtis wheel turbine of 0.94 m diam. (again based on a Rateau design) achieved
self-sustained operation by adding large quantities of steam generated in combustor cooling and feeding it back to the turbine in a kind of early 'steam injection process'. Since
there were combustion temperatures up to 1,800 °C claimed for the carborundum-lined
combustion chamber, a considerable amount of water was necessary to cool the gas to
the 450–470 °C, permissible for the intermediate turbine stator row. The cooling tube
above the compressor can be clearly seen in Figure 3-15. The thermodynamics of this
milestone project were discussed in the context of Figure 2-22. The actual GT efficiency
should have been between two to three percent only, or 6–10 kW of equivalent power
produced.

When René Armengaud died in 1909, further experiments with the large gas turbine
were abandoned. However, the company used the experience gained to develop a 'paraffin turbine' naval torpedo that, rather than using only compressed air to drive the propeller turbine, injected and ignited paraffin into the combustion chamber to raise the
temperature and hence the pressure of the compressed air stream. This design was immediately licensed to Robert Whitehead (1823–1905) in England, the inventor of the
modern (pressurised air) torpedo together with the Austrian Johann Luppis.

Interestingly, Brown Boveri participated in the Armengaud-Lemâle gas turbine project
not only by delivering the impressive compressor, but also used the contact to learn
more about the gas turbine in general. On the 24 February 1910, Walter G. Noack, Section 4.1.2, whom we already met in Section 3.1.2 on turbochargers and later one of the
key engineers in realizing the BBC Velox boiler and subsequently the industrial gas turbine, visited the St. Denis plant to attend both a test run of the torpedo turbine and to
discuss the prospects of gas turbines with the remaining of the Armengaud brothers,
Marcel. In his travel report[69], Noack was highly impressed by the compact design of the
torpedo engine that produced 200 hp in a volume of 0.35 m diam. and 0.5 m length,
speeding the propeller shaft to 1,300 rpm in 1.5 seconds. The overall test time, however,
was also very short. Since the black exhaust smoke was led directly into the test cabin,
the visibility was apparently close to 'zero' after only 30 seconds running time. With respect to the gas turbine experience, Armengaud predominantly recommended 'to improve the turbocompressor'. He considered the circumferential speed of the Rateau compressors as too low for gas turbine applications: 'A company like Brown Boveri should try
to create something new in this field'. Noack and his masters back home in Switzerland
understood the message, but for a successful realisation of a constant-pressure gas turbine, as Noack noted later in his memoirs[70], 'there was a lack of heat-resistant materials at

---

[69] This BBC-internal travel report is one of the earliest preserved in our collection of gas turbine papers, see
BBC, Geschichte des Schweizer. Turbomaschinenbaus. Moreover, the author owes detailed information on the
'Cabinet Armengaud Frères' and its members to Jean-Michel Duc, Sept. 2010.

[70] See Noack, Lebenslauf

that time, and also the turbocomponents had not reached an efficiency level, as required for turbines to drive their own compressors (and more)'.

Figure 3-15   *Armengaud-Lemâle gas turbine, Paris St. Denis 1906 [71] with BBC 25-stage centrifugal compressor (front), PR = 4.5, η_{ad} = 65–70 percent, 3,600 m^3/h, 4,250 rpm, turbine (far left)*

3.3   The Holzwarth Gas Turbine

Interestingly, another early engine to achieve partial success also saw the participation of Brown Boveri from 1909 until 1938, this time with both branches from Mannheim and Baden. The concept was proposed by the German Hans Th. Holzwarth (1877–1953)[72] as early

---

[71]   Picture source: Pfenninger, The evolution of the Brown Boveri gas turbine, Fig. 1

[72]   After finishing his mechanical engineering studies at TH Stuttgart in 1901, Holzwarth worked for MAN for a short time before going to Hooven, Owens, Rentschler & Co. (H.O.R.) at Hamilton, Oh., USA for the period 1903 to 1908, where he developed a steam turbine, known under the name Hamilton-Holzwarth-Turbine. There, he also began his comprehensive body of inventions for the 'explosion gas turbine': His first patent dates from 1903, the last one on the same subject was published after his death in 1957. His collection comprised of nearly 200 granted applications, a life-long occupation. His aim was to develop and market the turbine for industrial use, although, in later years, at least one patent DE575'054 was filed in February 1930 proposing the use of the turbine to drive railway locomotives, and another one showed up in Germany only after WW II, DE923'337 (with priority from 2 March 1945) for a 'low-drag', counter-rotating turboprop version of the explosion gas turbine. Holzwarth's activities for the first demonstration gas turbine in 1911 were financially supported by Erhard Junghans jr. (1849–1923), a local patron and Holzwarth's cousin, who together with his brother had grown the inherited watch manufacturing plant founded in 1861 at Schramberg, some 80 km southwest of Stuttgart, into the biggest watch (and fuse) manufacturer in the world. Consequently, E. Junghans was named as 2nd inventor for the patents issued between 1906 and 1913. For all of his lifetime, Holzwarth remained in close contact with rich investors for his idea. In 1927 the Holzwarth-Gasturbinen GmbH was founded at Muelheim/Ruhr – together with August Schilling (1854–1934). The last patents are shared with August H. Schilling (1915–1998) from Woodside Ca., USA, a miraculous story in its own right. 'The spice boys', August Schilling & Georg Volkmann from Bremen, D had made a fortune with a spice firm founded in 1881 in San Francisco, distributing their top selling 'Schilling's Best' baking powder all over the US – without any additives, early forerunners of the 'green' movement. August H. Schilling, grandson of the founder, spent the spice fortune on land, convinced that San Francisco would someday become a major international port, rivalling New York. Though this expectation did not materialise, it laid the ground for the Schilling Estate Company, obviously with sufficient financial reserves for Holzwarth's explosion gas turbine project until the mid-1950s.

as 1906 as an explosion or constant-volume cycle, in which the high temperatures and pressures of gas or fuel combustion that were obtained in an explosive, repetitive firing system could compensate for the compression efficiency deficits of that time. In the alternative, finally succeeding constant-pressure or continuous-combustion gas turbine concept, the compressor has to deliver a large amount of air to mix with the combustion gases in order to reduce the gas temperature to an acceptable turbine entry level. At the same time, this compressor had to be efficient enough not to drain too much of the available turbine power. Holzwarth, obviously aware of the difficulties encountered by the Armengaud-Lemâle gas turbine, Section 3.2, sought to avoid these difficulties by the 'isochoric', i.e. the constant-volume combustion concept. The (radial) compressor for such a unit had to supply a relatively much smaller amount of air and charging pressure, and its efficiency could be low without having too negative consequences.

Between 1908 and 1913 Holzwarth built a principle demonstration plant at Koerting Brothers AG, Hanover, D and immediately thereafter an improved version for operational testing at BBC Mannheim, Figure 3-16. However, Holzwarth's 1,000 hp design produced only 200 hp. The principle had been successfully demonstrated, but after a short time of operation the setup deteriorated thermally, even with the wide use of water-cooling for all thermally exposed parts.[73] So, measured against its size and efficiency it was less favourable than available reciprocating engines of the time. Holzwarth continued to persevere with his engine concept during his time as chief engineer for gas turbines at M.F. Thyssen, Muelheim/Ruhr, D between 1912 to 1927. Thereafter, he founded his own company Holzwarth Gasturbinen GmbH, also in Muelheim.

As said before, the Holzwarth constant-volume GT concept is fundamentally different from present day constant-pressure gas turbines. The combustion takes place in a closed chamber hermetically closed off by valves, where the gas mixture is ignited electrically, an operational principle identical with that of the Otto motor. The four to five times pressure rise compared to ambient happens mainly due to the explosion, so that no demanding turbocompressor is required. The expanding two-stage Curtis turbine was of steam turbine origin; since the temperature of the gases at the inlet to the turbine was about 700 °C, various steels were tested in a 'rainbow set' for the blades of the first moving row, Figure 3-17, which was fitted with water cooling in a later, improved version, Section 6.2.2.1 and Figures 6-26 and 6-27. Typically, the whole setup is mounted vertically and supplemented by comprehensive gas generator equipment.

---

[73] Gifted with a certain PR talent, Holzwarth published immediately results in a book in 1911 – see Holzwarth, Gas Turbine – in which he stated unmistakably 'A practically useful gas turbine requires the operation with periodic-explosive combustion like in gas piston engines'. One of these early Holzwarth gas turbines is on exhibition at the Deutsches Museum at Munich.

Walter G. Noack, who – as mentioned before – was on site in Paris to inspect the Armengaud-Lemâle gas turbine, as one of his first jobs in Baden also participated in BBC's share of the Holzwarth design adaptation work, while the demonstration facility, as visible in Figure 3-16, was erected at BBC Mannheim.

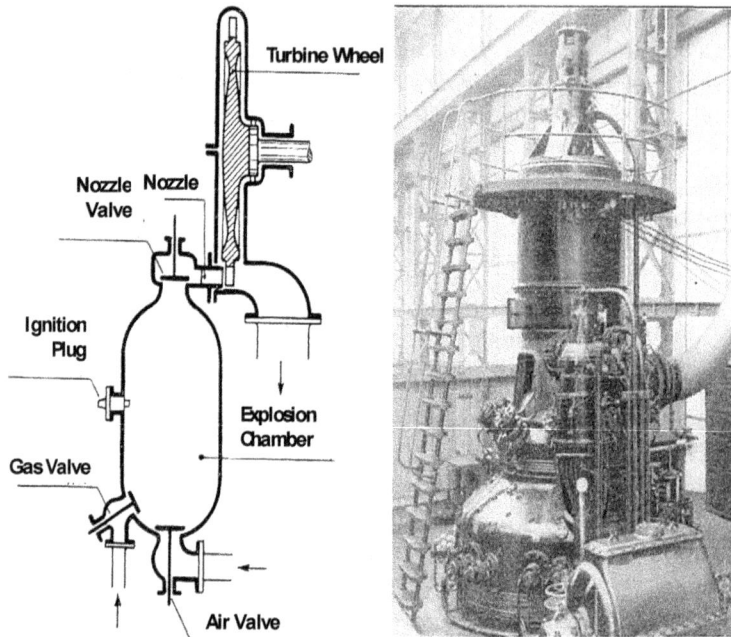

Figure 3-16   *Holzwarth gas turbine 1909–1911, principle [74] and installation with generator on top, 1,000 hp, 2nd build of BBC Mannheim [75], 2.95 m diameter at bottom, 6.36 m height*

Influenced by the fact that Holzwarth managed to get the renowned Professors Stodola and François L. Schuele (1860–1925, since 1901 Director of EMPA, Eidg. Materialprüfungs- und Forschungsanstalt, Zurich) involved in the complicated thermodynamic analysis of the engine, BBC Baden was also drawn into these development activities. In 1927, Stodola had been called upon to test a 500 kW oil-fired Holzwarth GT in the Thyssen shops at Muelheim/Ruhr. For the water-jacketed turbine it was found that an unexpectedly large amount of heat was transmitted to the water; only eight percent of the fuel energy was transformed into mechanical turbine energy, while another ten percent was available in the potentially power-producing steam generated in the jackets. From these results, Stodola predicted that an optimally designed Holzwarth GT should have the potential for efficiencies up to 30 percent.

Though BBC did not agree with these expectations, it decided to participate in the further development. In 1928 a licence agreement was signed between Holzwarth and Brown Boveri that led to the construction of an oil-fired turbine with an output of 2,000 kW. In order to undertake research into the fundamental principles of the machine, an experimental model was built in Baden, CH, that enabled e.g. tests to be performed on various types of steel in order to establish their heat-resistant properties at elevated temperatures and possible applications in real machines, Figure 3-17; at the same time combustion phenomena were also investigated in detail. Since the intention was to carry out experiments with gas, Holzwarth endeavoured to find an opportunity of installing his machine in a steelwork.

---

[74]   Figure source: Holzwarth, Die Holzwarth-Gasturbine, Figure 1

[75]   Picture source: Holzwarth, Die Gasturbine, Figure 121

Figure 3-17   *BBC Baden, 1928: Experimental model turbine wheel for the Holzwarth GT development with steel blades of different material properties   ©ABB*

So, over the years the Holzwarth GT concept improved slowly. Better efficiencies for the turbo-components and an improvement in the heat-resistant steels provided further development thrust. An order was obtained from the Prussian State Railways for a turbine to drive a 350 kW generator. This unit was built by Thyssen and put into service in 1923. It ran successfully for a number of years and some time it was claimed to be the only gas turbine in regular commercial use.[76] In hindsight, all these collected experiences were highly beneficial for BBC's learning and improved understanding in view of upcoming own activities in the gas turbine area.

Between 1908 and 1930 eight Holzwarth GT prototypes were built. After the last 2 MW version from BBC Baden working to a three-phase generator had successfully accomplished 500 operational hours continuous running at the very large steelworks of August-Thyssen-Huette at Hamborn-Bruckhausen, north of Duisburg, D in 1933, the 9th, a 5 MW single-shaft unit with a turbine inlet temperature of 930 °C was ordered by the steelworks and made by Brown Boveri in Mannheim again; it was installed to run on blast-furnace gas at the Thyssen steelworks at Hamborn, D shortly before the beginning of the war.

These two last and biggest Holzwarth gas turbines realised a new two-stroke, 2×2 chamber concept that was jointly developed by Holzwarth and BBC Baden to equalise the power output of the system. The drawback was the additional complexity. A cross section of the

---

[76]   See Kay, German jet engines, p. 193

total unit is shown in Figure 3-18; on both sides of the generator (centre), there is one turbine casing with **two** combustion chambers (of which only one is shown on both generator sides).

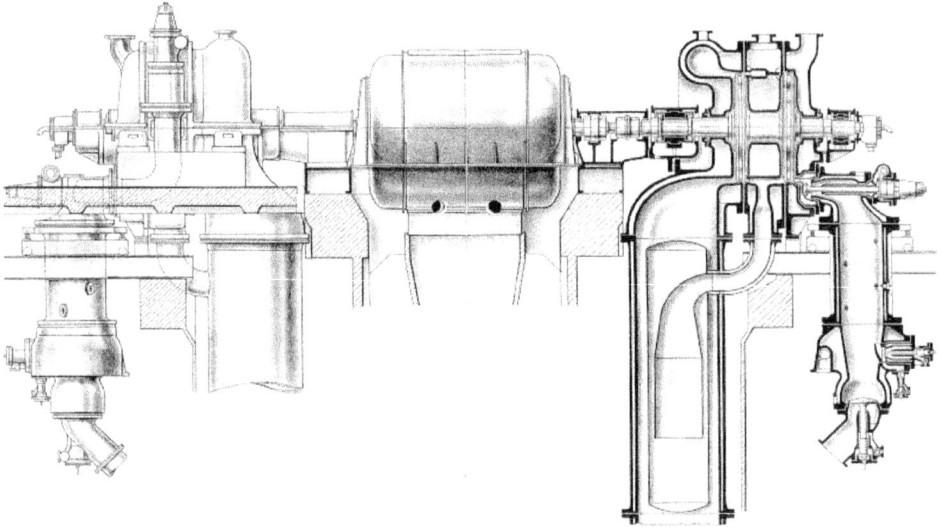

Figure 3-18   *Holzwarth-BBC two-stroke, 2×2 chamber gas turbine, 5 MW generator at centre, Thyssen Hamborn 1938* [77]

The figure shows the (outside) view of the gas turbine on the left hand side, while the corresponding cross-section is given on the right. As said in the foregoing, just one of two combustion chambers is shown on the extreme left- and right-hand sides. This so-called type 4706 gas turbine drove the 5 MW generator and had its steam supplied by a BBC exhaust heat boiler. The fuel for combustion was blast-furnace gas compressed to about 6 bar; the hydraulically operated valves worked at 60–100 Hz.

This unit was still regarded as experimental and not considered part of the regular steelworks equipment. Consequently, it was only run from time to time during day shifts, with the last run made in 1943, when it suffered heavy bomb damage. After repairs and just before further trials were planned, a bombing raid on 22 January 1945 once again damaged the unit. As late as February 1944 designs were drawn up for a large 4×5 MW unit (type 4710)[78], but the end of the war in May 1945 apparently also saw an end to the then already outdated Holzwarth constant-volume gas turbine, except for the US patenting activities after the war which have been already described in a footnote[72].

---

[77]   Figure source: Holzwarth, Die Holzwarth-Gasturbine, Figure 6
[78]   See Holzwarth, Die Holzwarth-Gasturbine, Figures 10-12

4 The Path to the First Power Generation Gas Turbine, 1927–1945

4.1 Development Survey at BBC

As previously described, the Holzwarth developments at BBC fell short of expectations during machine testing. Post-test analysis showed that an unexpectedly high amount of heat was lost through the turbine's water jacketing. This observation inspired the idea of turning this disadvantage into a benefit by building a supercharged boiler. This would become one of BBC's most successful products, the famous Velox boiler. In line with the GT principle, air was compressed and burned, but afterwards the combustion gases transferred heat at a high velocity to compact, superheating surfaces and were then expanded in the compressor-driving turbine prior to after-cooling in an 'economiser', Section 4.3.1.

Since these activities took place during BBC's development of the Holzwarth gas turbine, the first Velox boiler was conceived using the explosion principle. However, it was realised relatively quickly that problems such as scavenging, fuel injection in a large combustion chamber, ignition and stable burning were difficult to handle. The rattling of the large valve mechanisms was particularly unpleasant, not to mention the explosions which frequently banged out of the chimney.

These difficulties with explosion gas turbines were assessed again, presumably in late 1931[1] following progress in axial turbomachinery. The key engineers were Walter Noack, struggling with the Holzwarth concept, and Claude Seippel, newly responsible for axial turbomachines. Together with Seippel's deputy Kurt Niehus, the men gathered around Noack's drawing board with the thermal department head Adolf Meyer, and channelled their frustration and enthusiasm into a decisive move to switch from the explosion cycle to the constant-pressure gas turbine concept. This decision immediately implied the development of a high-efficiency turboset, with astoundingly simple beginnings consisting of a multi-stage axial compressor and a turbine with repeated 'symmetrical' blade rows, which

---

[1] See Noack, Lebenslauf, where he refers to an internal memo from July 1930 in favour of 'constant pressure' combustion in order to phase out the explosion turbine, for use in both the Holzwarth GT and the planned Velox boiler. Various patent applications from 1928 also confirm this position, e.g. US1948537. In particular 'On a Steam Generator' by W.G. Noack (16 December 1929) opens with the statement '...Among the objects of the invention is the provision of a novel steam generator of the continuous combustion type, in which the combustion mixture is burned under high pressure, and the pressure is utilised to discharge the combustion gases at high velocity along heat exchanger surfaces holding a steam generating fluid, thereby securing higher efficiency, economy in construction, space and operation, and better control.'

became the first industrial prototype of today's power generating gas turbine in the subsequent 4–7 years[2], Section 4.3.

4.1.1   Development Department and Business Environment

It is generally assumed that gas turbine development at BBC started around 1927, partly through the continuation of Holzwarth GT developments, but mainly through the initial testing for the later decisive axial compressors. The first signs of a burgeoning new form of power generation were embedded in a strong, technically prolific upturn in steam turbine development. This economically successful period had already been observed in the first peak of Swiss workforce development (Figure 2-3), when the number of employees rose rapidly from about 4,000 to 7,000.

Figure 4-1 illustrates the impressive rise of ordered steam turbine power, increasing by nearly four times its original value between 1923 and 1929. This power increase was supported by a general trend towards larger plant sizes, as indicated by an increase of unit orders by 100 percent (only). In 1929, the steam turbine order book contained almost 200 units per year, with a total power output of 1,150 MW. At the time, this raised BBC's all-time order volume for steam turbines to a new record high of 12,700 MW in total, again a sharp increase in just 10 years, and at the same time a clear confirmation of the broad market acceptance of BBC's superior product quality and reliability[3].

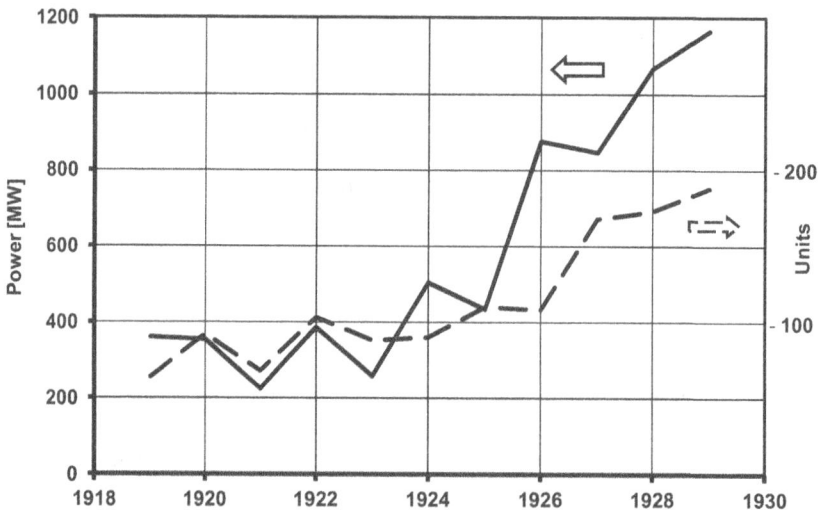

Figure 4-1   *BBC steam turbines between 1919–1929: Annually ordered power and number of units*

---

[2]   The time frame of '4–7 years' refers to two development milestones which will be addressed in detail in the following sections. In 1936, the gas turbine powered 'Houdry Refinery Process', Section 4.3.2, saw continuous net power production as a side effect of BBC turbomachinery equipment for the first time, while the first stand-alone power generation gas turbine was brought into operation by BBC for Neuchâtel, CH in 1939, Section 4.3.3.

[3]   See BBM, Die Konstruktionen in ...1929, 1930

This strong commercial success[4] quickly led to radical renovation and modernisation of the Baden plant site, as outlined in Figure 4-2. A new powerhouse (part of which still exists today) was built in 1926, immediately followed by the new 'Halle 30' for the assembly of large machinery (see also Figure 2-11).

Rapid technological advances in the generation of steam turbine power allowed for considerably improved exploitation of energy input, which in turn reduced coal consumption per kWh by nearly 50 percent overall. These savings were mainly a result of the following factors: better turbine and boiler efficiency, which typically increased from 75 to 85 percent; expansion of the usable energy head at both upper and lower ends, increased steam pressure and temperature as well as vacuum at the process tail. In addition, the steam process was improved by pre-heating feed water and reheating intermediate steam after its partial expansion in the turbine section.

Figure 4-2   *Boomtown Baden 1927 [5]: 1 – View of BBC production site along Brugger Strasse, 2 – newly erected power house, 3 – assembly hall 'Halle 30' (150m long, 20m high, today replaced by the Konnex Building), 4 – aerated wood stock for casting moulds   ©ABB*

---

[4]   In contrast to the successful Swiss economy, Germany suffered a period of hyperinflation from mid-1922 to November 1923. For BBC Mannheim this inflation phase ended with a new gold mark opening balance on 1 January 1924. The balance sum of the last paper mark-based balance sheet at Mannheim previously stood at 19,271,852,591,547,996,383 Marks. In 1923, when economic upturn occurred in the USA, the Swiss economy also started to pick up. Better incomes meant an increase in savings, the amount of which rose steadily. This situation changed rapidly following the sudden crash of the New York stock exchange in October 1929. When the Swiss population discovered how deeply the Swiss banks were involved in the international credit crisis, nationwide panic broke out. Most banks were confronted with huge decreases in savings deposits. Out of the eight biggest banks at that time, only Credit Suisse and Swiss Bank Corporation survived the crisis, one other only did so because of the financial help of the Federation, and three banks were forced to close, see Pohl, Handbook on Banks.

[5]   Figure source: BBM, Rueckblick... im Jahre 1927, 1928, Figure 102

Figure 4-3 illustrates how average steam pressure doubled between 1923–1929 from an original value of approx. 15 bar abs. to more than 30 bar, while the average steam temperature rose from 330 °C to nearly 390 °C during the same period. In 1923 BBC started the race to achieve the highest steam pressures possible, through detailed theoretical investigations that convincingly proved the economic benefits of this technological strategy. Over a short period of time live steam pressures rose from 50 bar (450 °C) to 55 bar, then to 80, 100 and finally a previously unheard of level of 200 bar. BBC was at the forefront of technology, demonstrating that reliable and economic steam turbines could be built and operated for any of these parameters.[6]

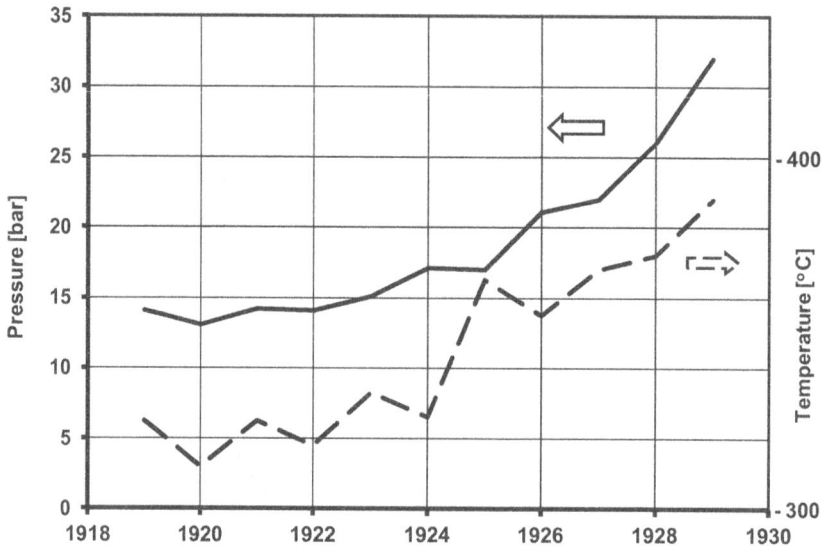

Figure 4-3   *BBC steam turbines between 1919 and 1929: Average steam pressure and temperature development*

The immediately visible net benefit of this situation was that other turbomachinery developments, such as the newly emerging axial turbocompressors and turbines, could rely on the best scientific standards of that time in aero-thermodynamics, mechanical integrity, materials research etc., and as will be shown in Section 4.2, many features of early axial compressor design followed established BBC steam turbine design principles. In the second half of the 1920s BBC's showcase for steam turbines was the Hell Gate power plant in New York, USA, a unit which briefly held the global record for installed power: 160 MW. The project for United Electric Light and Power Co., NY[7] began in 1920 and grew over the years due to the addition of several turbine blocks:

---

[6]   See BBM, Die Konstruktionen der Turbinenfabrik in 1929, 1930

[7]   By 1878, urban illumination was first established on the Avenue de l'Opera in Paris using electric arc lamps. However, arc lamps gave off a harsh light. Many inventors tried to create a more pleasant and lasting light, but none met with success until Thomas A. Edison turned his attention to the problem. Backed by financiers including J.P. Morgan and the Vanderbilt family, Edison established the Edison Electric Light Company in order to own and license his patents in the field of electric DC (direct current) light. After more than a year of

- in 1921   35 MW and 40 MW
- in 1923   50 MW
- in 1926   160 MW (record sized unit).

Figure 4-4   *BBC installed power record in 1928: steam turbine low pressure casing, 340 t, of 160 MW Hell Gate power plant, Bronx, NY* [8]   *©ABB*

experiments, Edison and his young assistant Francis Upton finally developed a carbon filament that would burn in a vacuum glass bulb for 40 hours. They demonstrated the light bulb to their backers early in December 1879 and exhibited the invention to the public by the end of the month. In 1887, Henry H. Westinghouse (1853–1933), a younger brother of George Westinghouse (1846–1914), founded with associates the company later known as the United Electric Light and Power Company to generate and distribute AC (alternating current) power throughout New York. Competition was strong, and mergers and acquisitions became common. By 1901 Consolidated Gas had acquired most of New York's electric utilities and consolidated these into a single subsidiary called 'The New York Edison Company', which also took over the Hell Gate power plant project at the Bronx (New York's Upper East Side) during the 1920s. By 1932, New York Edison's parent company, Consolidated Gas, was the largest company in the world providing electrical services.

Hell Gate power plant became prominent in another, criminal context which kept New York City in suspense for 16 years during the 1940s and 1950s. In 1931, George P. Metesky (1903–1994), later known as the Mad Bomber, was working there as a generator wiper when a boiler backfire produced a blast of hot gases. The accident left him disabled and after collecting 26 weeks of sick pay, he lost his job. Angry and resentful about the events which led to his injuries, Metesky planted at least 33 home-made, gunpowder-filled pipe bombs of which 22 exploded, injuring 15 people. This was after initially confronting Consolidated Edison and becoming increasingly frustrated at the lack of public support, particularly when visiting movie theaters. For a complete coverage of the fascinating case, see http://en.wikipedia.org/wiki/George_Metesky.

[8] See BBM, Die Konstruktionen... in1926, 1927

The plant was originally designed to operate at 18.6 bar steam pressure and 322 °C. These rather moderate parameters were established so that the plant had the opportunity to increase its operational flexibility, thus determining the huge dimensions seen in the sketch of Figure 4-4. The Hell Gate project also initiated the founding of BBC North America in July 1926.

There are not many details known about the early organisation of the departments back in Baden in the 1920s and 1930s. Like other similar companies, BBC operated a horizontally split, columnar departmental form. Various departments for development, production, sales and accounting worked side-by-side yet acted independently. Separate directorates were responsible for the production of electric and thermal machinery. The feeling of autonomy among various sections sometimes led to fierce confrontation, even between departments responsible for design and production of the same product line. In 1955 the department responsible for steam turbine development consisted of approximately 80 people, when the young Piero Hummel (32)[9] replaced his predecessor who had built up 40 years of valued service for BBC. The decision to appoint such a young candidate, made by the Technical Director Claude Seippel, was somewhat revolutionary for a system established on merit and seniority.

The size of the gas turbine department at that time (1955) was just 25 to 40 people, which leads to the conclusion that the dedicated core development team during the crucial time in gas turbine invention during the 1930s may have been just a handful of employees. Due to the close relationship between early gas and steam turbine design, working support was certainly flexible across department boundaries. One of the few examples we know about today concerned the Hell Gate installation, where due to its size (deviating from proven BBC practice) the final measurements for the turbines could not take place in Baden, but had to be carefully tested on the spot. Thorough prediction, measurement and control of thermal elongation and deformation of casings and shafts were of the utmost importance. Years later, this attention to detail also brought rewards, when Claude Seippel[10] attributed the 'getting it right first time' success of the Neuenburg gas turbine to the essential contributions of the young engineer Willy Burger, who had performed all the complex 3D thermal deformation calculations required for the reaction turbine precisely by hand in the centimetre range.

---

[9]
Piero Hummel started his professional life at BBC in steam turbine design in 1949 as a 26 year old ETH engineer. Ironically, when he resigned in 1986 after an unprecedented in-house career as chairman of the BBC Corporation, he had accomplished almost another 40 years' service, just as his predecessor as head of the steam turbine design department had before him. In his first few years he focused on engineering, characterised by many successful developments such as the introduction of the intermediate reheat concept which resulted in large US orders (230 MW, City of Los Angeles, 1959; 550 MW, Tennessee Valley Authority, 1959). On 18 June 1965 he took over the Technical Directorate for Thermal Machinery (TD-Th) as the successor to Claude Seippel. After company reorganisation in 1970 he became the first head of BBC Switzerland and in 1978 he finally reached the summit, as the chairman of the corporate board. However, in the following years clear signs of stagnating and increasingly inflexible company policy were ignored and the management team, together with VR president Franz Luterbacher, failed to launch any decisive corrective action. Hummel finally resigned after a confrontation with the new VR president Fritz Leutwiler, who initiated the Asea/BBC merger to ABB.
[10]
See Seippel, Wie vollzog sich der Uebergang...

The foundation of a specific 'Gas Turbine Department' in the BBC Baden Thermal Design Office was documented in Mitteilung A, Nr. 800, issued by Adolf Meyer and Paul Faber on 1 October 1938:

> 'The Velox boiler design group, which has existed for several years as part of Dept. D (steam), has expanded through additional business as well as additional design tasks for gas turbines and electro boilers, meaning that separation from Dept.D and a move towards an independent design department would be useful.'...'The working areas of Dept. GT today are as follows:
>
> – VELOX boilers,
> – Electro boilers for high and low voltage,
> – Gas turbine groups for power generation and other purposes.
> – For the tasks mentioned above, Dept.GT also uses blower designs from Dept.V (Compressor) as required.
>
> Current organisation of Dept. GT:
>
> | | | | |
> |---|---|---|---|
> | Head: | Herr Seippel, | | |
> | Deputy: | Herr Niehus | | |
> | Calculation: | for VELOX- and electro boilers: | Herr Bütikofer, | |
> | | for gas turbines and blowers: | Herr Pfenninger | |
> | Design: | Overall responsibility | Herr Nobs, Dept.D | |
> | | Group leader for design group GT: | Herr Burger | |
> | | Design of regulating parts: | Herr Baur, Dept.D | |
> | | Blading design: | Herr Nyfeller, Dept.D | |
>
> The D and GT design groups support each other when required.[11]

It is evident, although expected, that the GT group did not start from scratch, but grew out of the hosting steam turbine design department. This is especially noticeable in the case of Dept. D, responsible for the remaining overall design and blading design. The influence of the latter is apparent in Sections 4.2 and 6.2 in special design features such as the typical BBC blade attachment configuration, or in the use of degree of reaction R = 0.5 compressor blading, a straightforward transition from reaction turbine blading.

Another rare document dating from the early BBC days survived in the form of an internal memorandum from upper management, signed by the VR Delegates of Brown, Boveri & Cie., Baden: at the time, Max Schiesser (1937–1953) and Leo Bodmer (1934–1946). The memo, titled 'Mitteilung A, Nr. 812', was issued at Baden on 31 August 1939, one day before the outbreak of WW II. In German, the memo is entitled 'Richtlinien zur Behebung von Mängeln und Reibungen im Betriebe und zur Umsatzsteigerung, namentlich auf dem Gebiet der Normalprodukte' (see Schiesser, Richtlinien) which can be briefly summarised as 'Guidelines for overcoming faults or difficulties in operation and increasing revenue, particularly from standard products'. These 12 key statements provide insight into the constitution of the company at that time, but they also appear to represent a form of eternal management wisdom:

---

[11] See Seippel, Award Papers, a collection of materials covering various assignments, awards and birthday recognitions for Claude Seippel

1. Serve the customer in every respect!
2. Collaboration: not all watertight barrier walls between individual departments have been torn down yet, overall company welfare comes first.
3. The 'sanctity of deadlines': time is money.
4. Generosity towards our people at the forefront of sales: pettiness while solving claims is not beneficial.
5. Standard products are the basis of our economic stability, not 'specialities'. Customer focus and short-time deliveries are key.
6. Write less internally.
7. No dispute regarding internal cost accounting.
8. Objectivity in correspondence, displaying fair rationality.
9. Immediate response to customer queries.
10. Respect the prestige of our representatives.
11. Lack of operational directives must be overcome quickly.
12. Consider the possibility of your own mistakes before others (however, only hesitant non-performers make no mistakes).

The emerging role of GT development is illustrated indirectly by the number of test reports issued on the subject, shown in Figure 4-5 (compared to all turbomachinery test reports issued at 5-year intervals between 1915 and 1944). This unique representation of engineering activity is only available because the BBC TFVL/ZX test facilities have kept an uninterrupted series of test reports from 1900 up until the present day. With a certain time lag (due to time for testing, measurement evaluation and reporting) this chart also mirrors the company's economic progress, culminating in a total of 270 reports in the period from 1925 to 1929. It is noticeable that GT reporting initially occurred during a period of general economic stagnation, meaning that the first peak in the reporting of gas turbine testing, from 1935 to 1939, impressively represented one third of the total output. On the other hand, it is clearly visible that this first success period in the development of BBC gas turbines came to an end with the economic downturn at the end of WW II, after a short boom of just 20 years.

4.1.2   Prominent Engineers

In order to better understand BBC's gas turbine developments, the four key engineers will be portrayed. Another way of examining engineering activities of that time from a current perspective is to look at patents. In view of the previously mentioned small number of people involved in gas turbine development during the 1930s, it may not be surprising to realise that over 90 percent[12] of all BBC inventions in that area originated from just 4 top engineers: Adolf Meyer, Walter G. Noack, Claude Seippel and Hans Pfenninger. The IP Intellectual Property outcome of this group and of BBC in the field of gas turbine and turbomachinery issues from 1921 to 1955 comprises (referring to priority dates) approx. 150 granted patents in total.

---

[12] There are few examples of 'anonymous BBC CH patents' which do not explicitly name an inventor. Among those are the patent DE 655 698 (patent grant from 15 April 1932) which was most likely invented by Seippel and caused the German-Swiss supercharger patent conflict from 1932 to 1942. This was addressed in short already in Section 3.1 and will be described explicitly in a future account on BBC's aero engine developments.

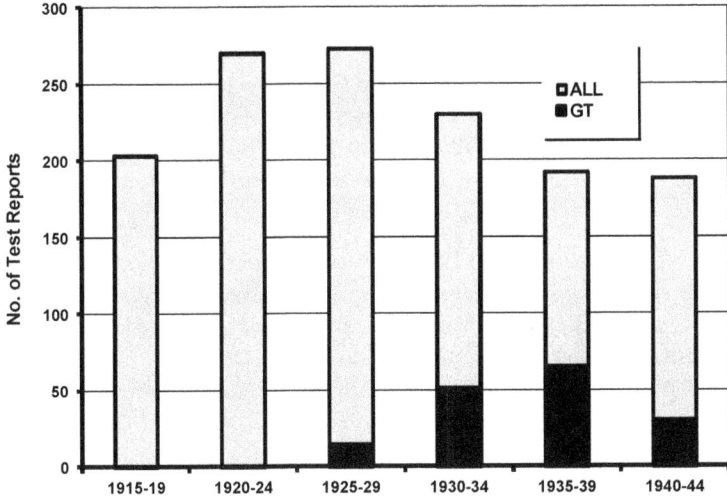

Figure 4-5   *BBC turbomachinery testing, shown through test reporting activity between 1915 and 1944*

Figure 4-6 illustrates the distribution of the inventions of the four key engineers over seven five-year periods. Again, the distribution appears to follow economic performance results; however, the importance of new gas turbine activities should not be overestimated. The 46 patents from the most productive period (1936–1940) represent just 4 percent of all BBC patent grants in that same period.

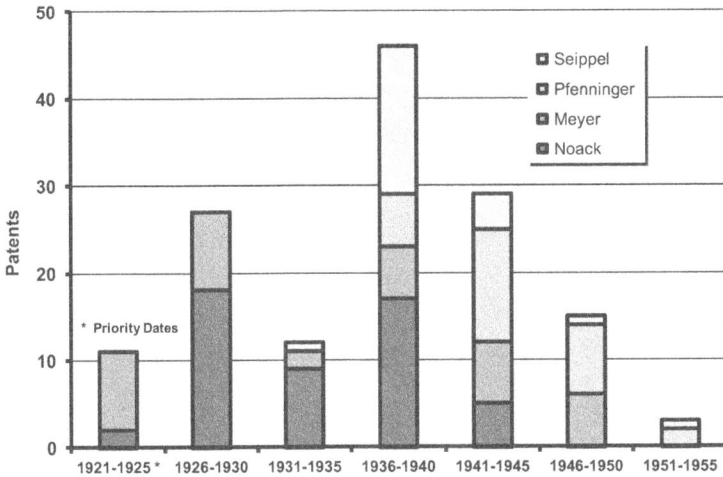

Figure 4-6   *Early BBC GT patent development between 1920 and 1955 and main contributors*

An in-depth analysis of the various personal IP contributions shows an extensive and productive period of long-term, almost equal contribution from Walter G. Noack and Adolf Meyer between 1921 and 1945/1950. For the latter this demonstrates his professional commitment as well as his successful management position.

Noack's single most prominent invention is the Velox steam-generating principle (US1948540 'Steam generator', priority date 1 September 1930, Section 4.3.1). Meyer was the first to obtain a patent for 'showerhead' film cooling to protect gas turbine blading (DE710289 'Blade with protecting boundary layer', priority date 9 February 1938, Section 6.2.2). Claude Seippel and Hans Pfenninger mostly contributed to BBC's patent portfolio in the shorter period between 1936 and 1950, though there is no clear indication that these junior engineers filed patents anonymously before.

– DE655698, 'Compressor set for pressurised firing of steam generator', priority date 15 April 1932, comes into this category. It does not explicitly name an inventor, but is probably part of Seippel's invention of the Velox/gas turbine turbo set, Sections 4.2 and 6.2. Other Seippel patents are the subsequently well-known 'COMPREX' pressure exchanger (a) and a concept for a turboprop aero gas turbine, unique at the time in Switzerland (b):
   a   DE724998, 'Pressure exchanger', priority date 7 December 1940 and
   b   US2326072, 'Gas turbine engine', priority date 28 June 1939.

– Pfenninger's most important patent is probably a combustor wall cooling concept which was later used widely in aero GT combustors:

   CH221698, 'Combustor wall cooling', priority date 12 July 1941.

A more personalised look at the four men's patents (again based on priority dates) is shown in Figure 4-7. It is apparent that Walter G. Noack alone contributed approximately one third of the patented ideas of that period, largely concerning issues of combustion and GT/Velox. Adolf Meyer follows in close second, also being the only one of the four who filed patents in all 5 categories investigated, producing the highest number of patents on other/general subjects. Similar to Noack, Hans Pfenninger shows a certain preference for GT/Velox, while as expected, Claude Seippel is equally represented in both the compressor and GT/Velox disciplines. Later in life he discovered an interest in nuclear technology and published innovative ideas in this field as well.

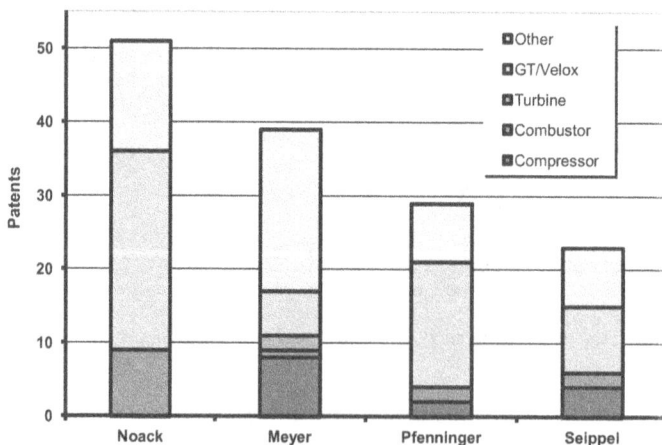

Figure 4-7   *GT patent contributions from key BBC engineers between 1920 and 1955, by engineering discipline*

4.1.2.1   Adolf Meyer (1880–1965)

Under Adolf Meyer's leadership in the 1930s and 1940s, decisive engineering developments for high pressure steam turbines and the new power generation gas turbine took place at BBC Baden. Meyer was born on 27 October 1880 in Buelach, ZH, 16 km north of Zurich. He became a mechanic at the newly founded electrical division of MFO (see Section 2.1.1), before he studied electrical engineering (qualifying as a Diplom-Ingenieur) at the ETH Zurich from 1899 to 1903. For a short time he worked at the ETH as an assistant for Prof. Albert Eugen Meyer-Schweizer and made contact with Prof. Aurel Stodola. He then toured Europe from 1904 to 1907 with short stays at the marine arsenal in Lisbon, a pneumatics shop in London and the Nuremberg Feuerlöschgeräte- und Maschinenfabrik AG. On 2 April 1907 he began working at BBC in Baden, initially working with centrifugal pumps and blowers. In 1909/10, he left BBC for Ludwig von Roll'schen Eisenwerke A.G., where as chief engineer he led the blast furnace installation at Choindez, JU (Canton Jura), CH: a Swiss technical landmark and at that time the only such facility remaining in Switzerland. After returning to BBC on 1 September 1910, he led the condenser department, was chief engineer for steam turbines and from 1 April 1923 until 31 March 1946 was director of the Thermal Department (successor to Eric Brown), responsible for all turbomachinery, steam and gas turbines as well as turbocharger activity. He was especially interested in the locomotive gas turbine, with a first world-wide prototype starting operation with a maiden journey between Basel and Winterthur on 5 September 1941.[13] During the post-war years, he first worked as a consultant to BBC until becoming a member of BBC's supervisory board (Verwaltungsrat) between 1949 and 1960. He died on 10 November 1965 in Kuesnacht, ZH, CH.

Figure 4-8   *Dr. h.c. Adolf Meyer (1880–1965, Pictures: left ~ 1905 ©ETHZ, right ~ 1945 ©ABB)*

---

[13] Claude Seippel describes the first trip of the gas turbine locomotive on 1 September 1941 in his diary. More interesting details are available at: http://www.voepelm.de/t_b/sbb5/SBB Am 4–6 1101/SBB Am 4–6 1101.htm by Thomas Brian. The GT locomotive received the SBB designation 'Am 4/6 1101', but the AM is of course no relation to Adolf Meyer, who also had a locomotive driving licence. According to the Swiss locomotive designation system, A stands for v > 80 km/h and m for diesel/combustion motor drive, see Schneeberger, SBB. The series was apparently opened by the Am 4/4 in 1939. Besides such Swiss aspects, there are indications that heavy tank gas turbine development in Germany between 1943 and 1945 was only launched after the successful realisation of the Swiss locomotive project, see Section 4.3.5.

Adolf Meyer's exceptional role for Brown Boveri is not only illustrated by his leadership in the field of steam turbines as well as decisive years for the emerging gas turbine business, but also by his multi-facetted personality, which combined the skills of a convincing sales-man, yet sensitive team-builder, with his broad personal experience as an outstanding engineer. He was honoured twice as Doctor honoris causa, in 1935 from the Stevens Institute of Technology in Hoboken, NJ, USA and in 1941 from his 'alma mater', the ETH Zurich. He maintained lifelong contact with the IMechE, the Institution of Mechanical Engineers in London, best documented by the first comprehensive presentation of the BBC gas turbine development on 24 February 1939[14] in London, including the Neuenburg plant, the gas tur-bine-electric locomotive, GT-powered (destroyer) ship propulsion and an excursus on the benefits of a 'combined gas turbine and steam plant'[15]. He received the IMechE George Ste-phenson Medal for 'best achievement of the year' in 1939 and 1943. He was a member of the ASME (American Society of Mechanical Engineers) from 1946 onwards and an honor-ary member of the American Academy of Arts and Science and the TGZ (Technische Ge-sellschaft Zurich) from 1950 onwards.

In 1924 Adolf Meyer married Georgina von Rotz (1889–1972). The couple had two sons, Adolf-Ernst (1925–1995, Prof. med., Adolf-Ernst-Meyer-Institute for Psychotherapy, Ham-burg) and Peter Otto (*1928, biologist, Kuesnacht, ZH)[16].

4.1.2.2   Walter G. Noack (1881–1945)

As previously illustrated in Figure 4-7, Walter Noack was a versatile and innovative engi-neer whose commercially successful ideas and sophisticated designs characterised the BBC product portfolio for decades. Walter Gustav Noack was born on 5 May 1881 in Nurem-berg. His father owned several water mills in the Schwarzach valley, 15 km south of Nu-remberg, as well as one of the first steam engines to drive hammer works for bronze paint manufacturing. Noack's interest in engineering was clearly nurtured from an early age. He was educated at the Königliches Realgymnasium Nürnberg (now known as the Willstätter Gymnasium[17]), a grammar school specialised in the classical humanities, which had a great influence on his clear way of thinking and speaking. After two years of experience at MAN Nuremberg, where he came into contact with the large MAN gas (piston) engine, he started his studies still at the 'Polytechnikum' in Zurich in 1902, what became in due course ETH Eidgenoessische Technische Hochschule. After finishing his engineering studies in 1906, graduating with a diploma examined by Professor Stodola, he then went to England, where he received his first job as an 'expert' for MAN gas engines at Lilleshall Co., a MAN licen-

---

[14]   See BBR, The Combustion Gas Turbine, 1939; an amusing aside: Adolph (with –ph) Meyer is also the name of a German agent character in Sir Arthur Conan Doyle's 'The Adventure of the Bruce-Partington Plans'. This may have become an issue during AM's frequent contact with Britain.

[15]   The wording implying a 'combined cycle power plant' is present, but thorough reading reveals that the GT-attached exhaust gas boiler was actually planned by Meyer for steam production only. Carnot's early idea of a CCPP, as expressed in Section 2.2.2 and the corresponding Footnote 36 in that section, was apparently still not within reach – and had to wait until BBC created the first large 75 MW CCPP at Korneuburg, A in 1961.

[16]   See Schwarzenbach, Meyer

[17]   Named after Richard Willstätter (1872–1942), a former scholar who received the 1915 Nobel Prize for (bio-) chemistry for his studies of chlorophyll, carried out at Munich University and ETH Zurich. For the engine de-signer Noack, an advantage of the biology teaching at that school was the ability to develop confident free hand-sketching with both right and left hands, a remnant of early plant determination exercises.

see. Before returning to Switzerland in 1909 to begin 36 years of uninterrupted and successful work for BBC in Baden, he was offered a position at British Westinghouse Electrical and Manufacturing Co.[18] in Manchester, to build up the gas engine department there. This connection was revived under dramatic circumstances 30 years later, immediately before WW II, in the emerging field of aero gas turbine propulsion. Finally, Noack decided to leave England and join BBC, following news from his mountain-climber friend Paul Faber[19] that somebody was needed to seriously explore the possibilities for a gas turbine. Similar to the success of the steam turbine, which had pushed the steam piston engine aside, there was the expectation that the gas turbine could overtake gas engines and open up a gap in the market for the industry newcomer BBC.

Noack realised quickly that the Holzwarth development activities which had been started were not the right approach. While Holzwarth promoted his projects publically using efficiency values of 30–32 percent, in his BBC internal proof Noack argued that 8–11 percent would be more realistic. Although BBC Baden continued the Holzwarth GT project for as long the rich Black Forest watchmaker Junghans provided funding, they refused a Holzwarth licence and looked to Noack for other opportunities.

One of these business opportunities appeared to be the Humphrey pump, a British invention[20] consisting of a combustion engine with swinging water columns, known as a 'wet

---

[18] Westinghouse in Britain was led at that time by the US-delegated German Director Philip A. Lange (1856, Rostock –1937, London). Lange came to the USA in 1882 via Siemens Woolwich UK and joined Westinghouse Electric & Manufacturing Company, East Pittsburgh Pa. in 1886 as an electrician, later becoming superintendent of the detail department. In 1906 he was sent to British Westinghouse which became Metrovick Metropolitan-Vickers. In 1909 he travelled to Zurich and hired Karl Baumann, another of Stodola's famous scholars, who was Metrovick's chief mechanical engineer since 1912 and technical director from 1927 till his retirement in1949; he played a decisive role in the development of the Metrovick F.2, the first British turbojet engine based on an axial compressor design. It was certainly Baumann who brought the F.2 designer Hayne Constant to BBC Switzerland in 1938 by contacting his friend Adolf Meyer. Lange, though seen as the honoured 'father' of Metropolitan-Vickers, had to leave the company in 1917 when war tension between Britain and Germany could not be left out of company dealings any longer.

[19] This pattern of personal job recommendation occurs several times in the early days; similarly, Jean von Freudenreich (assistant to Adolf Meyer at the time and aware that Prof. Stodola was looking for a computing assistant) hinted to his friend Claude Seippel that the famous Stodola had a vacancy, which changed Seippel's life and brought huge advancements in axial turbomachinery. At the time, von Freudenreich only knew Seippel from ARZ Aviron Romand Zurich www.aviron-romand.ch, the Swiss French community's rowing club. ARZ named one of their boats 'Le vieux Jean' as a memorial to JvF and his long term rowing enthusiasm. Von Freudenreich and Seippel had lunch together for many years at Hôtel Du Parc, at the old location (up until 1977) on the corner of Hasel-/Bahnhof- and Parkstrasse in Baden.

[20] Intelligence information on Herbert A. Humphrey and his new invention reached Baden via the 'Engineering' publisher in London, Mr. Martin. For many years, 'Engineering' was also first choice for BBC's upper management when publishing important news. Prof. Stodola, who was asked to perform and publish the Neuenburg test measurements in 1940, followed his preference to use the VDI Zeitung, Dusseldorf (aside from the 'Schweizerische Bauzeitung' for publication within Switzerland). The upper management discussed this issue at a board meeting, as it was somewhat politically sensitive at the time, and recommended that Prof. Stodola published simultaneously in 'Engineering'. The information on the Humphrey pump caused Charles Brown and his companions to travel to London immediately in February 1910, to meet Humphrey and negotiate a licence agreement. Disappointingly, they realised Siemens had signed the agreement along with a global electricity production licence the night before, leaving only the possibility of a pump application license for BBC. In hindsight, this seemingly negative event had fairly positive consequences for BBC. Siemens pursued work on 'wet gas turbines' for much longer than BBC, which turned out to be the wrong approach. After their experiences with Humphrey, Siemens conducted in-depth investigations of a gas turbine developed by Georg Stauber (1875–1952), a professor at the TU Berlin, another 'wet' concept with explosion-propelled water columns in a ring arrangement, see Stodola, Steam and Gas Turbine, pp. 1226–1228 and patent DE344550.

gas turbine' where the pumped water was charged using a water turbine to drive a generator. Noack believed that to optimise such clumsy apparatus a 2-stroke concept was needed.
Lengthy tests showed encouraging results, but the increased speed of internal combustion
engines finally pushed the Humphrey concept aside. Noack was in the middle of installing
a Humphrey pump at Argenta, in the Valle di Comacchio of the River Po, when WW I
broke out on 1 August 1914. He returned to Nuremberg to follow the call to arms, only to
realize he had to wait, as his preferred division, the newly established flying corps from
Schleissheim near Munich, was already filled with younger cadets. He returned to Argenta
before finally leaving Italy in March 1915 and delegating his work there to Adolf Meyer[21].
After an uneventful period at the French front, he came to the technical department of the
Luftwaffe in Berlin-Charlottenburg in December 1916, where he acquired overall responsibility for the engines of the fleet of 'Riesen-Flugzeuge' (giant aircraft) in a very short time.
His work there, as well as his contributions to aero engine and later land-based turbocharging in Switzerland, has been outlined in Section 3.1.2.

After WW I the excessive rise in coal prices also encouraged improvements in the efficiency of steam turbine plants. This finally cleared the way for steam high pressure and superheating, with Walter Noack an essential contributor yet again. As described already in Section 3.3, the Holzwarth gas turbine reappeared on the BBC premises for the second time in
1928. The tests in Baden showed much better results than 20 years previously, but remained unsatisfactory overall. Noack's clever thinking transformed the negative observations from the Holzwarth GT concept (excessively high heat transfer rates from the hot,
pressurised burner gases to the water-cooled turbine walls) into a beneficial design feature.
He invented the Velox boiler for rapid steam generation which exploited the high heat
transfer rates of pressurised, high velocity gas within thin, developing boundary layers at
the exchanger tube inlet. The highly efficient axial compressor was one of the key developments which finally led to the successful realisation of the power generating gas turbine
at BBC Baden in 1939. Under the coordinating guidance of Adolf Meyer, Claude Seippel
and later Hans Pfenninger timed these developments perfectly. For Brown-Boveri this occurred after 30 years of unwavering work towards engineering excellence in this field.

---

[21] Additionally delayed by the influence of WW II, the first appearance of a Siemens power generation gas turbine according to present design principles consequently had to wait until the 1970s, see Section 5.2.3.
At this point in his 1942 memoirs Noack highlights Meyer's apparent junior role. One year apart in age, the
careers of both followed a parallel course for decades, but by 1924 the elder Adolf Meyer had become Noack's
'boss'. Though both must have had a close cooperative relationship, and the somewhat hot-tempered
Franconian Noack even had the privilege of Meyer's quiet and considerate guidance on various occasions,
there was a lifelong hidden rivalry between the two. At Brown-Boveri it was impossible for the German
Noack to triumph over the Swiss Meyer, who was at that time an officer ranked 'Kaderschmiede' in the Swiss
Army corps, an indispensable precondition for any upper management career at BBC. This rivalry had its ad-
vantages and disadvantages. It was beneficial for BBC in the early 1930s, when Meyer decided to encourage
Noack away from turbocharging as he was on the brink of confrontation with Buechi. This showed Noack's
assets as an engineer and signalled the beginning of his very successful period developing VELOX and gas
turbines. At its worst, their rivalry affected the whole team, when the implications of WW II finally led to
Noack's isolation, culminating in his sometimes selfish and biased memoirs of 1942 which must have affected
Adolf Meyer greatly, and finally Noack's premature death of a heart attack on 10 December 1945 in Baden.
Noack's last publication dates from April 1945, when he announced the first two orders for the new generation of BBC's reheated gas turbines, which became the Filaret and Chimbote power plants, discussed in detail
in Section 4.3.4. See BBM, Zwei weitere Gasturbinen..., 1945

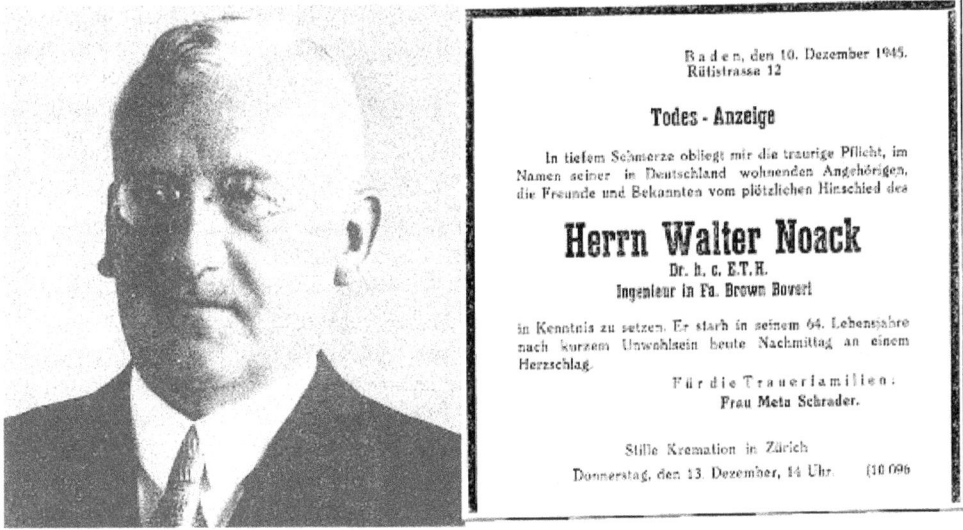

Figure 4-9   *Dr. h.c. Walter G. Noack (1881–1945, Picture ~1936 ©ABB, on the right W.G. Noack's obituary from the Badener Tagblatt)*

On 20 November 1936 Walter G. Noack was awarded the title of Doctor honoris causa from ETH Zurich, undoubtedly supported by Adolf Meyer and the senior BBC management. His activities in the remaining years until his premature death on 10 December 1945 can be seen in his late publications. They focus on Velox applications for wind generation and heating ironworks, as well as practical problems in GT coal burning under pressure. A return to his early ideas for a coal-fired 5,000 hp gas turbine can be seen in his later activities during the 1940s[22]. The memo of this earliest proposal no longer exists, but the essence can be indirectly reconstructed by means of the response-evaluation by Oswald Richter (BBC Mannheim) to Eric Brown, in which he refers to typical Noack design features: burning under pressure (9 atm) in an extremely small and compact combustion volume. Noack himself returned to these design features in his travel report of a visit to the Armengaud-Lemâle site at Paris-St.Denis on 24 February 1910: '.... *The carborundum cladding of the (Armengaud-Lemâle GT) combustion chamber was fully confirmed, ... but we had to find a suitable design for the burning of coal, lignite, etc. Even now, the GT efficiency that can be expected is no less than that of a steam turbine – for large units of 5,000-10,000 kW. But, without a doubt, the installation and maintenance costs of gas turbines are less than those for steam plants with their large boiler batteries*'[23]. He voices this considerate judgement for the last time in a BBM Editorial[24] in early 1943:

---

[22] Due to fuel shortages during war time, the technically challenging topic of GT coal burning is omnipresent in both civil research at BBC Switzerland as well as military applications in Germany (e.g. the planned tank gas turbine) with further support from BBC Mannheim. However, there is no evidence that Noack's research activities had any immediate military benefits for the Third Reich.

[23] See BBC, Geschichte des Schweizer. Turbomaschinenbaus, 1982

[24] BBM, Rueckblick... im Jahre 1942, 1943

'In a company like BBC, research and development activities are either based on per-
sonal initiatives with their own targets, or are the result of operational experiences. In
any case, they are a special effort which must be encouraged and maintained in addi-
tion to the standard production process. This entails certain difficulties which, of course,
are not always of the same origins or importance.

These difficulties are at their lowest in times of economic depression. R&D can increase
their margins in terms of reduced production with the best personnel available. Unfor-
tunately the general financial performance then restricts R&D, so that financial re-
serves from more prosperous times have to be used, resources have to be stretched and
their use is limited to short-term problems. Severe misjudgements are possible, with un-
pleasant consequences when the economic upturn comes. Times of sudden prosperity
can also have their drawbacks with a negative influence on R&D: personnel resources
can become strained and claimed 'market needs' can cause confusion. Wartime, like the
present, creates extra work due to material shortages and a lack of continuity in per-
sonnel. The result is that it is nearly impossible to gain any practical experience, as cus-
tomers prefer to satisfy their immediate needs. In short, directing R&D activities be-
comes extraordinarily demanding and requires extremely conscientious decisions. We
shoulder this extra risk, but we are aware of the possibility that some of our develop-
ments may be too premature and may need significant adaptive corrections following
feedback from broad operational experiences.'

Partially, a visionary assessment of BBC's future after 1945, as addressed in Section 5.

4.1.2.3   Claude Seippel (1900–1986)

Claude P. Seippel had a broad impact on a large variety of BBC engineering fields, but by
far his greatest influence was in turbomachinery, as a brilliant scientist, engineer and inventor. Moreover, he had a significant influence on the Swiss education scene and public
life in general. He was born on 14 June 1900 in Zurich. His father Paul Seippel (1858–1926),
a renowned writer and journalist from Geneva, was a professor of French language and literature at the Swiss Federal Institute of Technology in Zurich from 1898 onwards, meaning
Claude Seippel certainly grew up in a literary and philosophical environment. He attended
the Kantonsschule Zurich, studied at ETHZ and received his diploma in theoretical electrical engineering from Prof. Kuhlmann after passing the practical stage at the Atelier des
Charmilles S.A. in Geneva. On 1 July 1922 he started work at the BBC transformer lab, but
was transferred to the steam engine Dept. D on 2 January 1923, to support Prof. Stodola at
the ETHZ in solving disk vibration problems[25] which had been detected after severe damage to a steam plant in Copenhagen. After accomplishing this task, he did not return to
BBC Baden immediately, but left the company on 31 August 1923 for the United States, in
what would become a stay of almost 5 years:

---

[25]   In hindsight, Seippel jokingly referred to his role in this period as 'Stodola's Rechenknecht' (number crunch-
er), but the mathematical training helped him during his subsequent time in the USA, for example while handling the complex mechanical problems of automotive gear-trains by accurately applying a set of Lagrange's
equations.

- Electric Bond & Share Co. (EBASCO)[26], New York, NY,
- E.H. Sherbondy, Consulting Eng., Cleveland, OH[27],
- The Timken Roller Bearing Co., Canton, OH[28].

On 2 April 1928, he joined BBC for the second time, following the example of both Meyer and Noack. This time he began working as a liaison engineer between Dept. D and the test

---

[26] Cl. Seippel crossed the Atlantic on board the SS Manchuria, sailing from Hamburg to New York on 22 September 1923. The ship was identical to the USS Manchuria, 27,000 t displacement, a US Navy and Army troop transporter from WW I and II. In his first US job he drew up electric schemes etc. at EBASCO, 65 Broadway, N.Y., for what he called the 'Motor Columbus' of General Electric, a power plant development society. He earned 150 USD a month, just enough to live on. In July 1924, after finishing her violin diploma studies in Switzerland, his future wife Evelyne Seippel-Zangger (1901–1983) followed him to New York. This meant more pressure on Seippel to increase his salary.

[27] For Sherbondy's turbocharger activities within the emerging US Air Force, see Section 3.1.2, Footnote 32 of that Section. On 1 December 1924 Seippel started working at Sherbondy's 5 man office at no. 313 of the Davis & Farley Building, marking what 'was one of the most fruitful working periods of my life'. The working conditions must have been rather strange; Seippel describes his windowless office space and his colleague Glen Smith, a former professional wrestler, who was the team's chief draughtsman. Claude Seippel describes Smith in his US diary as a man who 'tirait ses dessins avec finesse et précision'. Meanwhile, he was married on Saturday 27 December 1924, one of the only days he could spare from work. In 1926 his wife gave birth to their son Olivier (1926–2012); often mentioned later as a source of joy for the young couple, especially during the subsequent depressing stay at Canton, Ohio.
His main project at the Sherbondy office was a newly designed, hydro-dynamic speed converter following the Foettinger concept, to be tested in combination with an existing 300 hp marine engine from Sterling Engine Co., Buffalo. The test failed, presumably due to the then unknown cavitation effects of the speed converter, and simultaneously caused the sudden demise of Sherbondy's small engineering firm. However, before the collapse, Sherbondy issued the idea of an axial-flow compressor design for charging large automotive engines, which had been unavailable since the (failed) trials by Parsons at the beginning of the century. Sherbondy directed young Seippel to prepare a basic compressor design theory based on the principles of aircraft wing lift and drag by organising wind tunnel test results from Gustave Eiffel (1832–1923) for him (see Eiffel La résistance, and for the corresponding 'NACA design tables', see Munk, Elements). In this context, a few important details from Eiffel's engineering life must be mentioned. The Eiffel family 'Boenickhausen dit Eiffel' migrated to France from the Eifel mountains near Cologne at the beginning of the 18th century. Gustave Eiffel changed his name shortly before the opening of the Eiffel Tower in 1889. In addition to his difficulties during the completion of the Panama Canal, the successful entrepreneur suffered a dramatic blow to his reputation as a renowned engineer when the 'Birsbruecke', a railway bridge at Muenchenstein near Basel, Switzerland collapsed on 15 June 1891 under the weight of a passing train, killing 73 passengers and making it the biggest railway accident in Europe at the time. The bridge had been designed and built in the 1870s by Eiffel's company for the Swiss Jura-Simplon Railway. An expert commission, which later developed into the EMPA (p. 74), identified poor-quality materials as the cause of the accident and cleared Eiffel's design from false accusations of blame. Nevertheless, the incident led to Eiffel's retreat as an entrepreneur and initiated his later career as a scientist, contributing to benefits for the axial-flow compressor development at BBC alongside Seippel.
Seippel's additional reference to the 'NACA design tables' is not easy to verify, since seemingly all standard NACA airfoil reports on this subject were published after 1930 and only summarised in the famous NACA Report No. 460 in 1935. The only exception is the Max Munk NACA TR from 1924, as referred to above. Max Michael Munk (1890–1986) was a Prandtl scholar, brought to the United States by NACA in 1920 with the permission of President W. Wilson. He is widely known for introducing the variable-density wind tunnel. He also wrote more than 40 outstanding scientific articles for NACA, mainly dispersing the thin-airfoil theory and thus the legacy of Prandtl and AVA Goettingen in the United States. After his resignation from NACA in 1927, Munk worked amongst others for Westinghouse, Brown Boveri and Alexander Airplane Company – see Anderson, A History of Aerodynamics, p. 291.

[28] Seippel's time at Timken lasted from 28 August 1926 until 21 January 1928, throughout which he apparently experienced a distressing atmosphere both at the company and 'dans l'horrible Canton, Ohio'. His task was to design a type of pressurised air chamber or 'health tank' which the owner Henry H. (Heinzelmann – acc. to his mother's German family name) Timken wanted to develop as a cure for lung disease. The outcome of this task is unknown. The Seippel family finally returned to Switzerland via Le Havre in February 1928, on board the French SS 'De Grasse', 18,000 b.r.t. of CGT – French Line.

Dept. TF-VL (Versuchs-Lokal), and shortly after became head of the emerging Velox development group. Seippel's career at BBC can be split into roughly four main periods:

From **1928**[29] **to ~1940** he managed the development, design, and testing of the multi-stage axial compressor. This work contributed considerably to the beginnings of the gas turbine in its present configuration, its first implementation being as a charging set for the Velox boiler and then used in 1939 in the 4 MW power generating GT unit for the city of Neuchâtel, Switzerland. In this period Seippel received important patent grants for the Velox turbo-set, turbocharging, GT governing, and the pressure-wave exchanger[30]. Significant publications covered the Velox steam generator, the axial-flow compressor, and heat flow in the blade root section of a gas turbine. On 1 October 1938 Claude Seippel became the first head of the newly founded Dept. GT.

From **1941**[31] **to 1954** he continued to develop the gas turbine, but was also responsible for significant improvements in the design of steam turbine blading and combined cycle investigations. He discovered and announced an important limitation to the 2nd law of thermodynamics, the importance of which was recognized only years later when the definition of exergy[32] was introduced. The discovery has become even more significant due to the current emphasis on energy conservation and environmental protection. This period was particularly productive, especially due to the development of exhaust turbochargers, for which Seippel contributed an essential patent in 1942. In a rapidly developing market these became BBC's most successful product under his responsibility. In due course, on 1 July 1945 Claude Seippel was named Assistant to the Thermal Directorate TD-Th with 'Prokura' (commercial procuration), a sure sign in the BBC hierarchy that the 'candidate elect' had the ability to rise through the ranks. Consequently, he became Director of the Thermal Department as successor to Adolf Meyer on 1 April 1946. Between 1947 and 1969 Cl. Seippel was a member of the board of the Swiss Federal Institute of Technology in Zurich (ETHZ), and from 1965 to 1966 was also president of this prestigious institution.[33]

---

[29] The couple's first daughter, Jaqueline, was born in 1931.

[30] Known as COMPREX, due to the COMPRession-EXpansion process in the wave generator channels.

[31] At the beginning of this period Seippel undertook some remarkable journeys. From 29 to 31 July 1941 he travelled to Lyon and Grenoble (then in the unoccupied Vichy Zone of France, where he met the chief engineers of C.E.M. Georges Darrieus and Paul Destival from SOCEMA, C.E.M's daughter company, to discuss and support aero engine projects unknown to the German occupants. (This trip will be addressed in detail in the forthcoming book on BBC's aero engine activities.) Secondly, from 14 to 19 September 1941 he travelled to AVA Goettingen, where a mixed Swiss university/industry delegation consisting of Prof. J. Ackeret (ETHZ), Cl. Seippel (BBC), C. Keller (Escher Wyss, participation unconfirmed though) and the Ackeret co-workers P. de Haller and G. Dätwyler met Professors Ludwig Prandtl, Albert Betz, Adolf Busemann and Otto Walchner to discuss the possibility of delivering two of the Ackeret-designed, Guidonia-type Mach 4 high-speed wind tunnels from BBC to Germany. The interesting circumstances and some revelations about these journeys are detailed in Section 4.2.4.3. In the same year Seippel's second daughter Sylvia (1941–1967) was born.

[32] The energy content which can be transferred into useful work, dependent on the environmental conditions, or Energy = Exergy + Anergy.

[33] See Speiser, Episoden. Andreas Speiser (1922–2003), who led the building of the first Swiss computer ERMETH between 1950 and 1955, reports the difficult beginnings of computer technology at ETHZ. He praises the independent, far-sighted decisions of the ETH-Board under Prof. Hans Pallmann and explicitly the later vice-president of the board, Seippel, in founding the new Institute for Advanced Mathematics under the leadership of Prof. Eduard Stiefel in 1948, and in due course acquiring the Z4, Konrad Zuse's unique relais-computer for 70,000 SFr. This extraordinarily user-friendly device gave ETHZ a world-wide head start in the development of advanced numerical mathematics methods.

Figure 4-10   *Dr. h.c. Claude P. Seippel (1900–1986, Pictures: left ~1924 ©ETHZ, right ~1955 ©ABB)*

From **1955 to 1965**, he concentrated on large steam and gas turbine projects and especially nuclear power plants, a clear BBC research preference in both Switzerland and Germany at that time. Dating back to 1941, he was a member of the Commerce Court of the Canton Aargau for 16 years in addition to his professional duties, and from 1953 to 1961 he served on the Central Committee of SIA, the Swiss Society of Engineers and Architects. After reaching retirement age, he stepped down as active senior executive on 1 July 1965 and handed the Thermal Directorate over to Piero Hummel. However, Claude Seippel continued to work as a senior consultant to Brown Boveri for almost 20 years from **1966 to 1986**. Until 1969 he was employed on a consulting contract, but colleagues still remember technical review briefings during the 1970s which Claude Seippel held regularly with admirable discipline. During this period of his career he continued his research, leading to 5 significant patents in the years up until the early 1970s, but also wrote several papers, including one on BBC's early turbomachinery history containing very valuable information which is still relevant today. The Swiss government (Bundesrat) had already elected him to the ETH board of trustees (Schulrat) in 1947; he was then elected vice-chairman in 1957 and led this prestigious board for 2 years following the sudden death of the acting president Prof. Hans Pallmann in 1965. The Swiss Federal Institute of Technology awarded Claude P. Seippel its honorary 'Doctor honoris causa' in 1959 for his contributions to turbomachinery. In 1965, he was named chairman of the Institute's board and served in that capacity for many years. The ASME American Society of Mechanical Engineers elected him as an honorary member in 1982 and two years later he became one of the few foreign associates to the US NAE National Academy of Engineering. The last traces of Claude Seippel's tireless work can be found in the BBC technical reporting archive under the entry dated 15 June 1983[34]. Active in his profession until the very last moment, Claude P. Seippel reached the end of his successful engineering life in a symbolic finish on 1 August 1986, also the Swiss National Day.

---

[34] See Seippel, Optimierung

In hindsight, the methodic development of engineering, from Stodola and Meyer to Seippel and their successors, initiated the transition to a 'scientific age'. Improved theories, developed partly in close cooperation with prominent personalities such as Prof. Walter Traupel from ETHZ, enabled more accurate fluid mechanic calculations and flow simulations. The importance of technical development departments grew steadily in comparison to the long-unchallenged position of test laboratories.

4.1.2.4   Hans Pfenninger (1903–1989)

Hans Pfenninger had a pronounced impact on gas turbine development at BBC from the mid-1930s to the 1950s, as illustrated already in the patent review (Figure 4-7). He was born on 18 June 1903 in Imst/Tyrol, Austria to a Swiss family from Canton Zurich, attended the Realgymnasium and subsequently the Technical University of Munich where he graduated as a mechanical engineer in 1928 with a Diplom-Ingenieur degree. During his studies he also worked at the foundry Kleindienst & Cie. and at MAN AG, Augsburg, D. On 15 April 1929 he began his professional career at Brown Boveri, Baden in the steam turbine department, with his first development contributions in the area of mechanical integrity calculations and mechanical designs, later supporting Walter Noack in work on the Holzwarth gas turbine and initial Velox boiler investigations.

With the launch of the new Gas Turbine Department on 1 October 1938, Hans Pfenninger took on overall responsibility for aero-thermodynamic design of gas turbines and compressors/blowers. The Neuchâtel GT plant was the first prominent highlight of his work. He was the deputy to Kurt Niehus from 1945 until becoming head of the GT department in 1948, a post which he fulfilled with success until his official retirement on 1 September 1968. However, he remained active as an assistant to the Thermal Direction, accomplishing his 50th professional anniversary in April 1979, until his final retirement in 1981, then aged 78!

In 1958 the TU Munich acknowledged Hans Pfenninger's lifelong efforts towards improved economic performance of the power generation gas turbine by awarding him the title of 'Doctor honoris causa'. Hans Pfenninger passed away on 9 May 1989 after an exceptional life as a BBC engineer and manager.

This short, in-depth view of prominent engineers from the early BBC era cannot be concluded without mentioning one special aspect: these men were not only highly qualified technical experts but also had unique personalities and characters. Adolf Meyer was described not only as an energetic, sometimes harsh manager, but also as a communicative, joyful and humorous man who knew all his foremen on the shop floor personally. Examples are illustrated in Figure 4-12, taken from the 1951 BBC 'Hauszeitung', which appeared monthly for the internal information of employees since July 1943. On the other hand regular external customer information dates back already to July 1914, when both German issues of 'Brown Boveri Mitteilungen' and of the corrsponding 'Brown Boveri Review' in English were distributed on a monthly basis as well. Though the Swiss 'BBC family' already comprised more than 10,000 members at the time, see Figure 2.3, the company maintained an intensive social life across the company hierarchy with obvious benefits for employee – company relations.

Figure 4-11 *Hans Pfenninger (1903–1989, Pictures: left: ~1965, right: with Cl. Seippel [m] and W. Thomann [r, TFVL] at Neuchâtel GT plant installation, 1939)* ©ABB

Figure 4-12 *Social engineering, Baden, CH, 1951:*
– Professional jubilee celebration with BBC orchestra (l), Hotel Linde, Baden, AG, CH ©ABB
– TFVL skiing race (r) at Alt-St. Johann near the 2,500m high Mt. Saentis, SG, CH

4.2   GT Key Component: Axial Flow Compressor

In the mid-1920s activities began at BBC Baden which culminated in the first utility gas turbine in 1939. This chapter deals primarily with the development of the axial-flow compressor, first in the historic context of the mid-1920s and subsequently by highlighting specific development achievements up until the present day. As mentioned in Section 1, a chronological order is rejected in favour of a continuous focus on this essential gas turbine component. The same approach will also apply to the combustor in Section 5 and for turbines and turbine cooling in Section 6.

In the mid-1920s, circumstances for the implementation of innovative ideas were positive. Steam turbines signalled a lasting commercial upturn with sold power doubling within 1925, Figure 4-13. On 1 April 1923 the energetic Adolf Meyer (43) took over the Thermal Directorate with a clearly visible strategy to investigate possibilities for extending business to the heat engine market. The development of the Holzwarth 'explosion gas turbine' still contained numerous challenges for the ingenious Walter Noack (then 42). In 1928 the stage was set for young Claude Seippel (28) to return with fresh enthusiasm from his US experience, which resulted in the first all-new 'axial-flow compressor' completed in record time by 1932 and in due course lasting commercial success of new products. Figure 4-13 shows the accumulation of all kinds of axial-flow compressor deliveries between 1932 and 1952, totalling 280 units. The stand-alone compressors, for example for waste gas compression, wind tunnel blowers and refrigerator plants, represent 2/3 of the compressor production volume, followed by all-axial turbo-sets for Velox steam boiler applications (approx. 80 units in that period) and 24 gas turbine power generation units in the first 13 years of production.

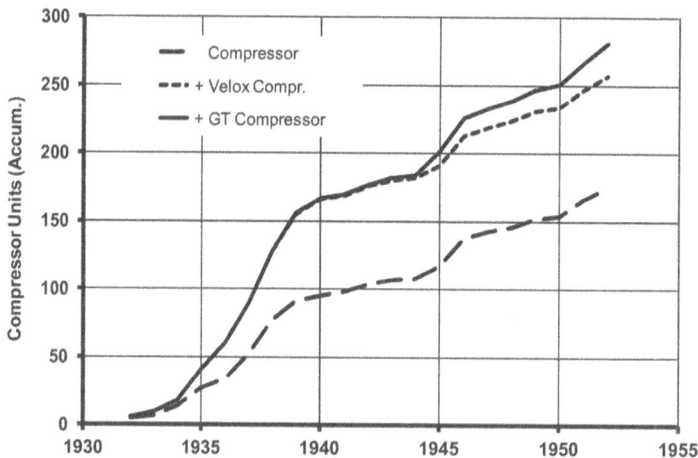

Figure 4-13   *BBC multi-stage axial-flow compressor deliveries, 1932–1952* [35]

---

[35]   See BBR, Progress...in 1952, 1953, p. 81; BBM, Rueckblick...auf die Konstruktionen im Jahre 1939, 1940, p. 6 and – see Pfenninger, Evolution; Endres, 40 Jahre.

Though the commercial success for the industrial realisation of axial-flow compressors was enjoyed by BBC, the initiation of the first technical development of this new turbomachine cannot be so easily differentiated and evaluated. Several development paths towards axial turbomachinery can be identified, with reference to the emerging aero wing theory.[36] The following section details the story of BBC industrial compressors, while early activities in England, Germany and the USA in the aero engine field will be left to a future evaluation.

4.2.1   Early Ideas and Trials

The 'magic moment' between Georges Darrieus and Jean von Freudenreich in June 1924 which can be considered as the beginning of axial-flow turbomachinery at BBC has already been described in Section 3.2 and in Section 3, Footnote 67. In the years 1925 to 1931, the C.E.M. Cie. Electro-Mécanique of Paris (the French subsidiary of Brown Boveri, Section 2.1.1) built a series of wind turbines under the supervision of their chief engineer Georges Darrieus, ranging from 8 to 20 m in wheel diameter, Figure 4-14. According to Seippel's later statements[37], Darrieus recommended the application of the airfoil theory to turbomachinery at BBC for the first time, having had the design of multi-stage, axial-flow compressors in mind from the beginning. A single sheet of paper with a profile/velocity triangle hand sketch by Darrieus dates from 26 April 1924, and his first surviving axial compressor concept from July 1926.[38] However, both Seippel and Darrieus gave credit to the German engineer W. Bauersfeld for the radical change in cascade design perspective, away from established stream tube and flow channel theory towards the lift/drag principle deduced from single airfoil measurements. Bauersfeld published his key paper on this as early as 13 May 1922 in the 'Zeitschrift of VDI' based on comprehensive airfoil measurement data from AVA Goettingen published by L. Prandtl in 1921.[39]

Work on axial compressors at BBC started quite unspectacularly; the increasingly large quantities of air necessary for the cooling of electrical generators of high output determined the need for this new type of high-flow blower. A series of single- and two-stage axial fans were developed to be directly driven by the generators, Figure 4-15. A simple double-conus wind tunnel was created in which the generator fans worked under ideal conditions, so that the measured lift/drag curves of the ventilator blading could be compared with the single airfoil measurements from the AVA Goettingen deliveries, Figure 4-18.

---

[36] Apart from BBC in Switzerland, independent axial-flow compressor development also began at AVA Goettingen (A. Betz, J. Ackeret, W. Encke) in Germany and at RAE Farnborough, UK (A.A. Griffith). Returning from Goettingen, first to Escher Wyss, Zurich in 1927, J. Ackeret and later C. Keller started a second Swiss attempt of axial-flow compressor design at ETH Zurich in 1930. In The Origins p. 110, Constant, who is otherwise quite reliable, is wrong in ignoring BBC's early lead role in deducing practical axial compressor design principles using Prandtl's lift/drag theory.

[37] See BBR, The development of the Brown Boveri axial compressor, 1940 and – see Darrieus Hommage à Claude Seippel

[38] See Seippel, Die Entstehungsgeschichte

[39] See BBR, The development of the Brown Boveri axial compressor, 1940; Darrieus, Hommage à Claude Seippel; Bauersfeld, Die Grundlagen and Prandtl, Ergebnisse. Walther Bauersfeld (1879–1959) was a German engineer, lifelong employed as a leading manager of Carl Zeiss Jena/Oberkochem, D. He is most renowned for a series of planetariums erected between WW I and II at the German Museum Munich (1923, initiated by Oskar von Miller), Wuppertal-Barmen, Leipzig, Jena, Dresden, Berlin, Dusseldorf, Rome, Paris, Chicago, Los Angeles and New York. www.en.wikipedia.org/wiki/Walther_Bauersfeld

As already described in Section 4.1.1, the decision to introduce the multi-stage axial compressor into the BBC product portfolio occurred in 1932. The period between 1926 and 1932 was consequently used to develop a fundamental understanding of the unknown axial compressor flow, especially the treacherous phenomenon of compressor flow stability (rotating stall and surge). Flow instabilities had caused the failure of Parsons' early attempts to build axial flow compressors and the early BBC experiments on single-stage axial blowers were also seriously hampered, as seen in Seippel's account some 40 years later[40]: *'The blower performed as expected against the back pressure it had been designed for. If, however, that back pressure was raised, there was a sudden increase in noise and it was found that only the outer portion of the blades delivered air while inside the air streamed back. There was not much time to observe the details of this backflow because, after less than an hour, the blades failed. What happens in this case? The rather flat blades behave like airfoils or airplane wings in the air stream. Throttling the air increases the angle of attack until it reaches stalling point. The lift collapses and air begins to flow back. This need not happen on the whole length of the blades or on all blades simultaneously. If in one area the air flows back, there may be enough air in total to produce a stable angle of attack in the remaining area. Instruments at the time could not reveal whether the inner portion of the blades had reverse flow over the total of the periphery or only on a sector of the periphery, whether the sectors rotated, showing today's well-known phenomenon 'rotating stall'.*

Figure 4-14   *Georges Darrieus (1888–1979, picture ~1940)*[41] *and wind turbine by C.E.M. ~1930, design G. Darrieus*[42]   ©ABB

---

[40] See Seippel, The evolution

[41] Picture source: BBM, Hommage à Monsieur Georges Darrieus, 1968, p. 405

[42] Georges Darrieus (24.09.1888–15.07.1979) was the son of the vice-admiral of the French Navy Gabriel-Pierre Darrieus (1859–1931), who became famous as the commander of the 'Gymnote', one of the first all-electric submarines, launched in 1888, successfully operated in approx. 2,000 sorties and significant for further ac-

Figure 4-15   *Axial-flow fans for generator cooling, ~1929 (l)* [43] *and 1952 (r)* [44]*, 22,000 kVA, 500 rpm*   ©ABB

Therefore, the noise sensorium was (and still is) of extreme importance for compressor test personnel. 'Flow instabilities' and 'backflow' are strongly linked to the fluid mechanics understanding of boundary layer flow separation, an emerging scientific concept from Ludwig Prandtl[45] and his research team at that time.

---

ceptance of this weapon. Georges Darrieus, after finishing his engineering studies at the École Centrale de Paris, worked continuously for C.E.M. from 1912 to 1958 as a versatile engineer and scientist in areas such as ballistics, fluid- and thermodynamics, turbomachinery and electrical engineering, only interrupted by WW I. For the latter, he formulated the 'Théorème de Darrieus' in 1937, for the optimum setting of transfer power and frequency in electric grids.

Today, he is mostly known for his pioneering work in wind energy. The above-mentioned 20 m diameter rotor with horizontal axis produced 35 kW of electric power at a wind velocity of 29 km/h, but was destroyed in a storm. Seippel describes these wind turbines *'as designed according to modern aerodynamic principles, thus supporting the impulse towards further developments along these lines'*. Deviating from this credible assessment, however, Darrieus' only patent for this type of wind turbine, US 1,820,529 with French priority dating back to 1927, shows only symmetric 'fish' profiles (without camber).

Of the several wind rotors which Darrieus designed, the most important was a rotor made of slender, curved, airfoil-section blades attached at the top and bottom of a rotating vertical tube. Compared to the more common horizontal axis turbine, this specific 'Darrieus turbine' has several attractive features. The blade operates at very high tension, so a relatively light, inexpensive blade is sufficient, which on top operates independently of the wind direction. By locating the power train, generator, and controls near ground level, they are easier to construct and maintain. However, a disadvantage of this type of 'Darrieus turbine' is that it is not self-starting, so an induction motor connected to the local power network is required. Picture source: Seippel, The development of the Brown Boveri axial compressor, Figure 8

[43] Picture source: BBR, The development of the Brown Boveri axial compressor, 1940, Figure 9

[44] Picture source: BBR, Progress and work in 1952, 1953, Figure 31

[45] L. Prandtl (1875–1953) became a professor of mechanics at the TH Hannover in 1901. In 1904 he delivered a groundbreaking paper, *Fluid Flow in Very Little Friction*, in which he described the 'boundary layer' and its importance for drag and streamlining. It also described 'flow separation' as a result of the boundary layer, clearly explaining the concept of 'stall' for the first time.

The effect of the paper was so great that Prandtl became director of the Institute for Technical Physics at the University of Göttingen later that year. Over the next few decades he developed the institute into a

4.2.2   First Axial Compressor Development Tests after 1926

Even before Seippel's return from the United States in April 1928, Jean von Freudenreich initiated axial compressor activities at BBC by building a 4-stage test rig, Figure 4-16. The first blading had 4 identical stages with untwisted, symmetrical blades and vanes on a cylindrical hub of 0.43 m diameter and conical outer contour variation from 0.54 to 0.53 m diameter. At the entry, an inlet guide vane prepared the flow conditions for the following symmetrical stages. An exit guide vane was planned for axial outflow re-direction, but was removed after initial testing. The first preserved document in the history of BBC's axial compressor development, dated 7 October 1926, is the Test Programme for that rig with an introduction by Jean von Freudenreich:

> 'The purpose is the application of new aerodynamic theories to compressors to improve efficiency and simplify construction'.

The next document, dated 9 February 1929, is by Claude Seippel, who describes the purpose of follow-up testing:

> 'The axial compressor of test configuration V 602 has airfoil-shaped blades. While airfoils are normally exposed to the wind stream in a single or twin configuration, the compressor represents an endless row of airfoils which interact with each other. Therefore, optimum airfoils for aircraft are not necessarily best suited for turbines and compressors too. Moreover, there is reason to presume that compressor applications require a somewhat stronger camber (bending) which shall be investigated.'

The results of this test campaign, thoroughly analysed in Test Reports[46] by W. Girsberger, were disappointing overall. The first design had taken the known sensitivity of the diffusing flow into account and selected an airfoil from the Goettingen classification system, G 265. But concerns arose quickly about the deviating degree of turbulence (Tu) of the Goettingen airfoil measurements (Tu low) and the BBC compressor tests (Tu high). Would the airfoil fulfil its energy transfer purpose in a highly turbulent compressor flow? In fact, the tests indicated no further pressure rise after an (extrapolated) seventh stage. The meridional flow distribution was very irregular due to the untwisted blading and the entry and exhaust conditions of the rig were 'chaotic', so new blading was launched in 1931[47] with convincing success (see also Points 1 → 2 in Figures 4-31 and 4-32, with in hindsight daringly high stage loading). At 5,000 rpm the compressor delivered 3.65 m^3/s at a total pressure ratio PR=1.24 and an adiabatic compressor efficiency η_{ad} ~0.83. Inter-stage pressure measurements confirmed the balanced stage loading.

---

powerhouse of aerodynamics, world-leading until the end of World War II. In 1925 the university used his research to create the Kaiser Wilhelm Institute for Flow Research (now the Max Planck Institute for Fluid Research) and the AVA Aerodynamische Versuchs-Anstalt in Goettingen. Amongst Prandtl's famous co-scientists were Jakob Ackeret (1898–1981), professor and head of the Institute for Aerodynamics at ETH Zurich from 1931, who stayed in Goettingen between 1921 and 1927 and developed a specific compressor design school at AVA together with Albert Betz (1885–1968).

[46]   For these key documents, see BBC, Axialkompressor, 1931

[47]   See BBC, Vierstufiges Axialgebläse, 1932

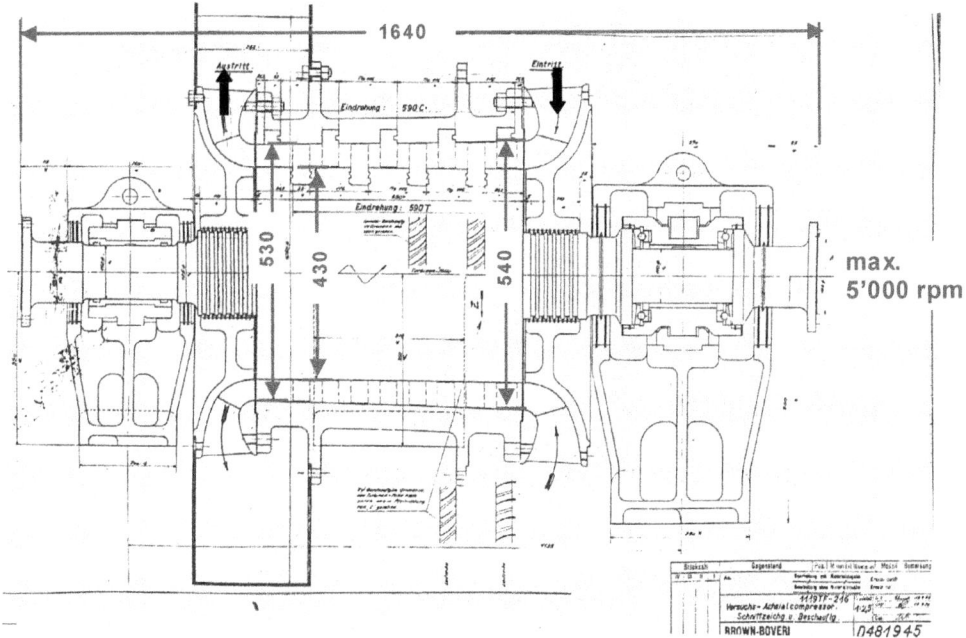

Figure 4-16   *BBC 4-stage axial compressor test rig, original drawing D481945, 1926* [48]

The use of the single airfoil data for a so-called 'compressor cascade', the repetitive arrangement of blades on the cylindrical surface of a blade row (Figure 4-27), meant a complicated mathematical calculation/drawing process to transform the isolated Goettingen profiles into a compressor cascade configuration. The selection of optimum circumferential blade spacing was highly important, in other words the right blade number to accomplish the energy transfer/ flow turning task with a minimum amount of loss. Figure 4-17 shows a corresponding BBC exercise from 1931. Seippel also contributed to public discussion [49] on this subject in 1933, shortly before the principles of axial compressor design based on airfoil theory became widely known through Curt Keller's 1934 doctoral thesis. [50]

---

[48] See BBC, 4-stage axial compressor test rig, 1926–1932

[49] See Seippel, Bemerkungen

[50] See Keller, Axialgebläse. Curt Keller, born in Berlin in 1904, studied Mechanical Engineering at ETH Zurich (1923–1927) with internships at BBC Baden and Escher Wyss. Afterwards, he became a private assistant to Professor Stodola for 6 months (see thesis CV), before he joined Escher Wyss where he led the Thermal Test Dept. from 1931. After his return from Goettingen, J. Ackeret was chief scientist and head of the Escher Wyss Hydraulics and Fluid Machinery Lab (1927–1931), before he became an ETH professor and founded the renowned Institute for Aerodynamics (IfA).
In the last 2 years at Goettingen Ackeret had participated in the development of independent AVA compressor technology in close cooperation with Albert Betz. Keller, working closely with Ackeret, specialised in axial compressor aerodynamics when compressors for two wind tunnels were required for the new institute's equipment. The two blowers for the large, open wind tunnel were of an identical Keller design, while the compressor for the closed supersonic wind tunnel, Section 4.2.4.1, came entirely from BBC.

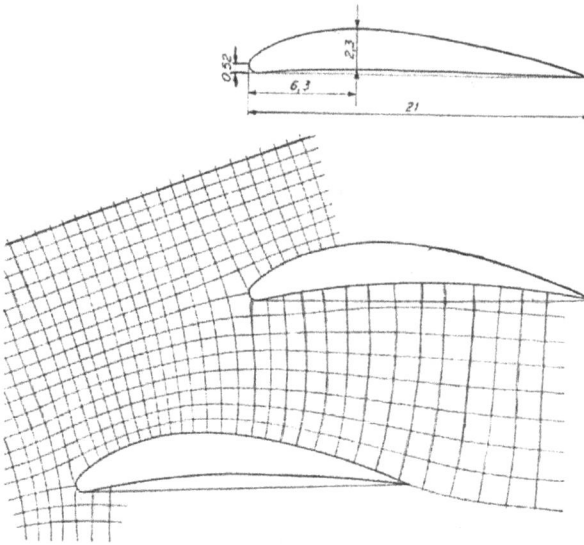

Figure 4-17   *Conformal mapping of Goettingen airfoil # 265 and transformation to BBC compressor cascade, 1931* [51]

The primary design target for the second rig set-up was to achieve regular and repetitive flow conditions with axial outflow per stage all along the blade height. This automatically meant a degree of reaction R~1, including an adapted blade twist from ϑ_{Hub} = 30° to ϑ_{Tip} = 21° to keep the radial axial velocity distribution constant. This also solved the inlet and exit flow problems. The new blading produced a pressure ratio PR=1.24 at 5,000 rpm and boasted a convincing 83 percent efficiency. Intermediate pressure measurements confirmed balanced stage loading. The test report clearly stated that the potential addition of further stages would not have any detrimental effects: the long-sought signal for the launch of a successful production series. [47]

From the very beginning, BBC compressor development had a multi-stage, axially repetitive concept in mind, but the 1926 'jump start' towards a simple 4-stage test configuration failed dramatically. Therefore, a step back to a single-stage configuration helped to confirm fundamental fluid mechanic assumptions on the basis of a specific 'double conus' compressor test rig, Figure 4-18. As a positive side effect this exercise helped to optimise the performance of single-stage generator blowers (already addressed in Figure 4-15). The special rig created clean and undistorted flow conditions for the isolated rotor. An electric drive motor of 500/600 kW respectively was located upstream and allowed speeds of up to 3,600 rpm; approx. 3.2 m downstream from the sketched exhaust diffuser was a tube-integrated through-flow measurement nozzle until the ducting was closed off after a second straight tube section of 5.7 m by the exhaust throttle.

---

[51]   See Pfenninger, Die Gasturbinenabteilung..., Bild 1

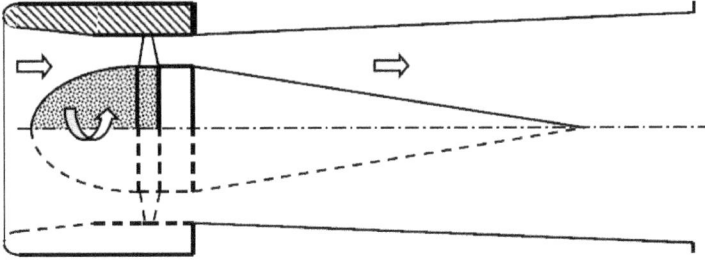

Figure 4-18   *BBC 'double cone' compressor rig, rotor tip diameter 1.19 m, rotor hub diameter 0.83 m and entry spinner/rotor hardware* [52]

As stated previously, the first axial blower wheels were based on published single airfoil (aircraft wing) measurements. It was still unknown how the aerodynamic behaviour of a rotating blade would deviate from that of an isolated airfoil. The test programme of 21 August 1930 outlines this in detail:

> *'The tests should determine the blade characteristics and the achievable efficiencies of such an axial ventilator. The Goettingen test results are only valid for the lift/drag relation of an isolated airfoil. Since we have no insight about the cascade behaviour, the tip clearance and the rotating airstream influence, the lift/drag relation of the investigated rotor must be determined. This will become the basis for the design of new rotor configurations.'*

These tests irrefutably confirmed the polar plot characteristics of the Goettingen airfoil used, as illustrated for Gö 265 in Figure 4-19 where the lift coefficient C_L has been plotted over the drag coefficient C_W. Both coefficients have no dimensions: in the first case the lift force of an airfoil is referred to the dynamic flow pressure and a reference planform area, while the drag coefficient of the investigated wing section is defined as the ratio of its drag force divided again by the dynamic flow pressure and a reference area. The Goettingen results for the profile Gö 265 show the original data for the airfoil chord/width ratio b/h = 1:6, which was then recalculated as an infinitely long wingspan ratio b/h = 1: ∞ to eliminate the induced wing tip drag by simulating a ducted rotor with zero tip clearance. Various angles of attack between α = –1.5° ... +15° are shown along the Goettingen airfoil

---

[52]   See BBC, Axialventilator, 1930

measurements (hatched area).[53] The corresponding BBC double-cone ventilator test fits perfectly into the expected range of single airfoil test data. The addressed figure also shows the corresponding curve for multi-stage axial compressor analysis.

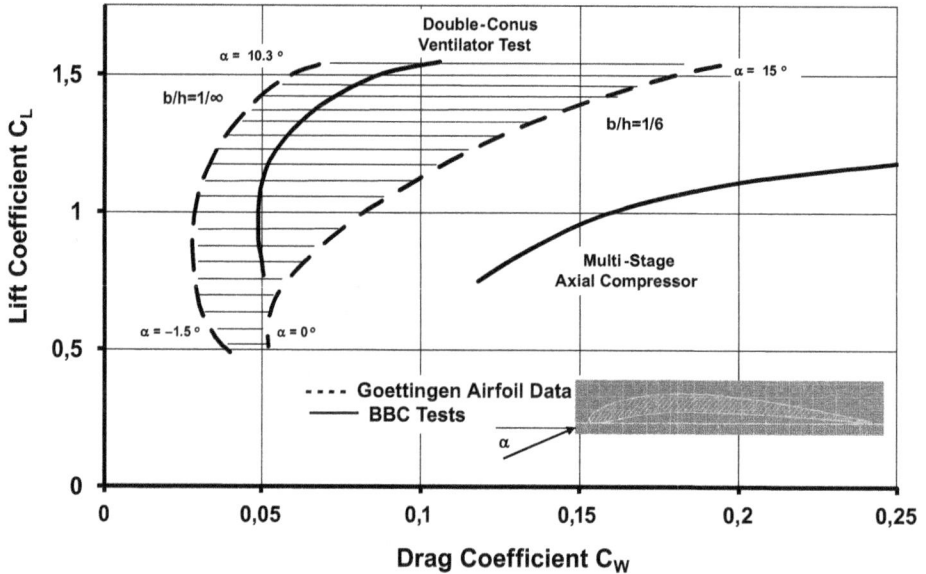

Figure 4-19   *Gö 265 lift/drag polar plots – from single airfoil via single-stage ventilator test to multi-stage axial compressor*

Though reference to the single airfoil lift/drag catalogue had already been abandoned at BBC at a relatively early point in the mid-1930s, the fundamental checks and proofs of known fluid mechanic characteristics and the use of the 'clean' double-cone test rig in this context helped to considerably improve practical compressor design expertise:

- the benefits of blade twist vs. originally cylindrical blades became manifest in significant performance increases
- variations in airfoil shape illustrated benefits of thicker airfoils in respect to off-design performance and flow stability
- a broad variation of blade numbers ('pitch/chord ratio') facilitated better understanding of loading and performance correlation (which finally led to the concept of an optimum Zweifel number, see Section 4.2.3 and Footnote 80 of this Section).

Beginning with the first commercial compressor products (Mondeville and ETHZ), all new engines in 1932 were tested in-house at Baden prior to delivery. Besides conventional 8- and 12-stage test compressors, which became the test vehicles after 1945, there was seemingly only one other test configuration implemented in a specific rig. The unique test compressor configuration with variable blades appears in a remarkable test series[54] directed by

---

[53]   This specific form of experimental airfoil test data representation dates back to Otto Lilienthal (1848–1896), the German pioneer of gliding flight.

[54]   See BBC, Versuchs-Axialgebläse, 1940–1945

O. Zweifel et al. between 1940 and 1945. The test programme comprised various combinations of twisted/untwisted blades and differently staggered vanes.

Figure 4-20   *3-stage BBC experimental axial compressor test rig with adjustable rotor blading, 1940*[55]
©ABB

The rig, Figure 4-20, was designed for speeds of 7,800 to 10,000 rpm, total pressure ratios varying between PR=1.11...1.17, cylindrical (rotor) duct dimensions D_a=0.37 m and D_i=0.274 m. The compressor characteristic showed that for a higher degree of reaction of R=0.8, the best impact of a blade angle variation was from α=10° until α= −20° at a volume flow ratio of more than 3 with a stable operation range and almost unchanged efficiencies. In this respect performance was similar or even better than with the later introduced concept of variable guide vanes.[56] The addressed report also contains a revealing result in terms of the impact of degree of reaction on the variation range of variable **vane** compressor configurations. Here the clear advantage of BBC's preferred R=0.5 concept, Section 4.2.3, is obvious:

Degree of reaction	Relative volume flow
R = 0.5	V/V_o = 2.15
0.8	1.70
1.2	1.50

The idea of a blower with variable rotor blades[57] had been brought up in 10 t Velox boiler studies, Section 4.3.1, where an additional blower with variable geometry could be run

---

[55] Picture source: BBM, Rueckblick...im Jahre 1940, 1941, p. 243. Interestingly, this picture of the special 3-stage rig with adjustable blades had already been published in the UK in G. Geoffrey Smith's 'Gas Turbines', 1st ed, 1942. The book appeared over a short period of time, with the 1st edition in December 1942, the 2nd ed. in June 1943 and the 3rd ed. in April 1944. The author clearly must have maintained a close relationship with BBC Baden, as various BBC technology highlights such as the air-cooled turbocharger turbine blading, Section 6.2.2, Seippel's turboprop patent of 1939 and details of the 1st railway gas turbine combustor, Sections 4.3.5 and 5.3 were reprinted very shortly after BBC had released the information. In due course, whether intended or inevitable, German intelligence of that time came to the conclusion that a special/friendly relationship between BBC CH and various British institutions must exist.

[56] See BBR, Axial compressor with adjustable stator blades, 1963

[57] The idea of variable geometry compressor blading may not have been so revolutionary at that time. In 1936, Walter Encke, see Section 4.2.4.3, investigated a 2-stage counter-rotating compressor with variable pitch rotor

safely with a constant speed electric motor while the necessary volume flow adaptation was achieved by blade angle variation. In this case the actual Velox turbo-group runs on its own at variable speed, while the additional, regulating blower allows for volume flow variation at constant operational speed. However, the concept found no lasting application in multi-stage compressors due to the relatively expensive blade adjustment mechanism; it therefore remained an isolated solution for single-stage wind tunnel blowers.

The first steam-generating Velox boiler, ordered in 1932[58] to burn blast furnace gas for the Mondeville ironworks near Caen, France, was already equipped with a row of inlet guide vanes adjustable before starting. There were two axial compressors of similar sizes, one for air and one for blast furnace gas, and the optimum air/gas ratio was set by the adjustable entry vanes. Even then, before 1940, the first mechanism to allow for vane angle variation during operation of turbomachinery was implemented in an axial fan for mine ventilation.[59]

In early 1958 BBC finally began investigating the implementation of variable vanes for (all) rows of multi-stage axial compressors. Two 8- and 12-stage test compressors were used to investigate the possibilities for adapting the steep and rather inflexible compressor characteristics to industrial needs, mainly furnace wind generation in large ironworks. In this respect special care was taken to make the whole variable geometry design wear-resistant, in addition to the encapsulated, dust-protected vane shift arrangement as seen in Figure 4-21. For years, adapting the required broad operation range was only possible by means of speed variation, either by large steam turbine drives for the upper power class (> 7 MW) or complex electric motor regulation. These difficulties were overcome by setting variable compressor vanes during operation, allowing for a wide volume flow variation at a high efficiency level without drawbacks in operational stability. Within 5 years of introduction BBC sold almost 60 percent of all furnace blowers (or the compressor equivalent of 165 MW total drive power) in line with the new variable stator concept. Following design principles valid at the time, all stages were equipped with variable guide vanes. An impressive example of this configuration is illustrated in Figure 4-21. In the following decade a trade-off between extra costs for the variable geometry configuration and gains in wide operation range meant a move towards the present day standard: only 1–4 front stator rows are adjustably installed. In addition to variable geometry features, the inset of this figure illustrates the characteristic annulus shape of all BBC compressors from their early beginnings right up to the 1970s. BBC opted for a

---

[58]   blading at AVA Goettingen (AVA archive no. GOAR 337 and 143). In Switzerland, Escher Wyss received a patent no. CH245,483 with 1945 priority for axial variable rotor blading.

[59]   BBC order no. 72862, dated 18 February 1932, from Société Métallurgique de Normandie

Quoted in an article from Cl. Seippel in BBR, The Development, May 1940, together with a compressor map (Fig. 12) to illustrate the effects of adjustable stator vanes during fan operation. The introduction of the concept of 'variable stators' is attributed to various sources in the aero engine industries. By far the most fascinating account of its introduction during J79 development at GE Cincinnati in 1951/2 came from Gerhard Neumann, see Neumann, Herman the German. Neumann was also named inventor for US patent no. 2,933,235 for a 'Variable stator compressor' with priority on 11 January 1955. The heavier BBC stator adjustment mechanism clearly did not interfere with these patent claims. Neumann's claim to the first usage of VIGV has become weakened after a secret German patent was recently discovered, suggesting use at least 10–15 years earlier in German turbojet projects: CGD-759 (Captured German Documents), – see Klapdor, Der Technologietransfer Deutschland – USA. This patent by Max Adolf Mueller 'Einrichtung zum Verstellen von drehbaren Leitschaufeln fuer Turbomaschinen', Hirth-Motoren Stuttgart-Zuffenhausen, priority 17 March 1942, AEL Translation 'Mechanism for Adjusting Turbo-Machinery Guide Vanes', 1945, 6 p., shows an adjustable ball bearing supported vane design, different from 'customary designs for adjusting guide vanes' (quotation of patent text). Mueller will be mentioned again several times in future dealings with BBC aero engine activities.

constant inner diameter due to the expected inherent advantages of a mechanical design with constant disk diameter. In this case the mean annulus diameter decreases in the flow direction, which results in relatively larger blade heights towards the compressor exit compared to alternative annulus shapes. This design approach somewhat reduces the relative impact of wall zone losses, but has the drawback of smaller stage works at the compressor backend.

Figure 4-21   *Blast furnace gas compressor for a gas turbine plant at Cornigliano S.p.A. Genova, I 1962.[60]*
All rows with variable stators at two ring diameters, PR = 5.4, V_0 = 110,000 m^3/h, servo motor
with actuation shaft at front left hand corner, inset – principle of encapsulated casings
©ABB

4.2.3   The Degree of Reaction and Other Fluid Mechanic Research

In the late 1920s Switzerland realised the growing importance of aeronautics. In 1929 Professor Arthur Rohn (1878–1956), then President of the 'Schweizerischen Schulrat' (ETH supervisory board), founded a Scientific Commission for Aeronautics as an institution for coordination and screening. In 1928 representatives of ETH, the KTA 'Kriegstechnische Abteilung' (Dept. of Military Engineering), military and civil aeronautics specialists had already agreed to found a Swiss Test Institution for Aeronautics with independent test facilities. Finally, the 'Schulrat' recommended establishing the said facility at ETH Zurich. In 1930, an extension of the ETH Maschinenlaboratorium was planned under Stodola's leadership. Ackeret was invited to contribute proposals for a new aerodynamics lab and he took the chance to suggest two (!) impressive wind tunnels. In an open, low-speed wind tunnel with a 2.1×3.0 m square nozzle, two 275 hp blowers generated wind speeds of up to 250 m/s. However, Ackeret's design of the second, supersonic wind tunnel was revolution-

---

[60]   Picture source: BBM, Rueckblick... in den Jahren 1961 und 1962, 1963, Figure 30

ary. It allowed **continuous** operation up to Mach 2 or 2,300 km/h in a 0.4 × 0.4 m test section. The closed circuit arrangement reduced the required driving power to less than 700 kW and allowed an independent Ma, Re setting[61] via pressure variation. The core piece of the new concept was the 13-stage BBC axial compressor for PR = 2.2 and 55 m³/s at max. 3,800 rpm, Figure 4-33. Only this brand new axial compressor allowed the elegant combination of 90° flow turning and wind tunnel propulsion. Both the customer Ackeret and the contractor BBC took a considerable risk by using an untrialled concept. It is not known if Ackeret's plan already existed in such detail in 1930, as shown in Figure 4-22, when he presented the plan to Stodola who rejected it![62] But the enthusiasm of 32 year old Ackeret was so convincing that Rohn gave the plan the go-ahead.[63]

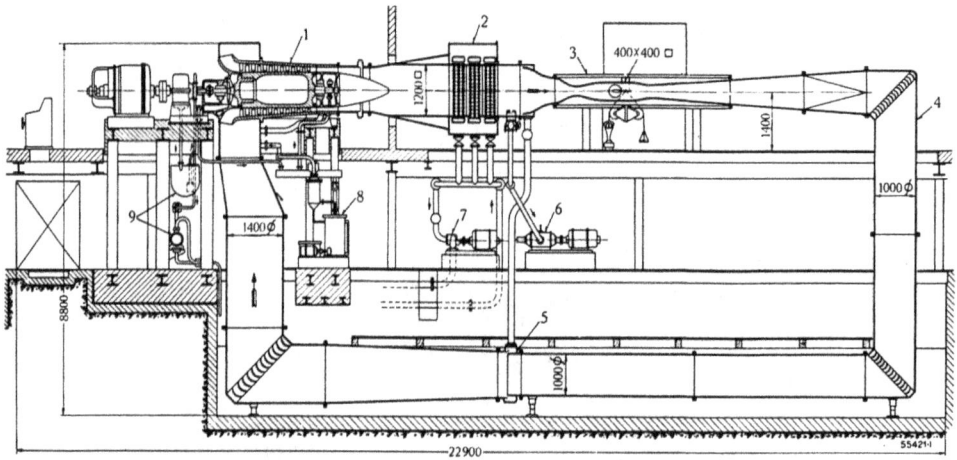

Figure 4-22   *Jakob Ackeret: plan of the supersonic wind tunnel at IfA, ETH Zurich 1931* [64], *operational start spring 1935, Ma$_{max}$ = 2, 1-axial compressor, 2-air cooler, 3-test section, 4-turning corners, 5-ejector bypass, 6-vacuum pump, 7-cooling water pump, 8-compressor oil supply, 9-gearbox oil supply*   ©ABB

For BBC, this was also not a simple situation. The early 1930s were a period of commercial recession in Switzerland, see Figure 2-3, making costly development activities unthinkable; but they bravely exploited their limited opportunities. As for the Keller compressors[65], the BBC designs for the ETHZ supersonic wind tunnel compressor most likely also began in autumn 1931. It is highly probable that the decision to change the Velox concept from unsteady explosion burning to steady, constant pressure operation occurred during the

---

[61]   Non-dimensional flow similarity parameters: Ma – Mach number (flow velocity/speed of sound), Re – Reynolds number (inertial force/viscous force).

[62]   See Eckert, The Dawn of Fluid Dynamics, p. 208 contains a hint that Ackeret's novel idea of a closed-circuit supersonic wind tunnel had already been tested (with centrifugal compressors? DE) and *'had not been successful in even the most advanced aeronautical research laboratories abroad'*, so Stodola's hesitation may have been somewhat justified.

[63]   All information in this paragraph from http://www.library.ethz.ch/exhibit/ackeret/ackeretindex.html – an excellent 'virtual scientific exhibition' on Jakob Ackeret (1898–1981) by Rudolf Mumenthaler, 2001

[64]   Figure source: BBC, Geschichte des Schweizer. Turbomaschinenbaus, 1982

[65]   See Keller, Axialgebläse, p. 5 introduction

Mondeville order, received on 18 February 1932. Towards the end of 1932 many decisive steps were made towards developing axial compressors using the 10–11-stage Mondeville compressor as a cheaply disposable test vehicle. The timing shows the frequent milestone decisions made during these months:

03 December 1931   Ackeret sends request for proposal to BBC

24 December 1931   BBC offer for 'Axialgebläse für das Institut für Aerodynamik, E.T.H.'[66]

18 February 1932   Order intake for Mondeville plant, later first Velox boiler with all-axial turboset

24 May 1932   Final test report for 4-stage test compressor with improved blading

16 June 1932   Test programme (Seippel) for Mondeville I, axial compressor with 10 stages, PR=2.6, 11,000 rpm

18 November 1932   Test report for Mondeville I compressor, insufficient volume flow, recommendation to add an 11th stage

30 November 1932   New blading drawings for 11-stage Mondeville II compressor approved by Seippel, Figure 4-23

22 December 1932   Test programme for Mondeville II, axial compressor with 11 stages and a variety of blade/vane settings, i.e degree of reaction R = 0.5...1.0...1.25

8 May 1933   Test report for Mondeville II, successful test with obvious advantages for R = 0.5

An example of a typical manufacturing drawing for the rotor blading of Mondeville II is shown in Figure 4-23. The manufacturing information for all 11 rows is given on this one sheet. In the upper left-hand corner the profile tip section is illustrated, followed by the profile root section deduced from the Goettingen profile 262 B. The text block in the middle of the upper frame limitation says:

'*Profile from L (approx. 30 percent blade height) to M (blade tip) to be increasingly ground off and from blade mid height to tip (mechanically) increasingly flattened*'. The tabulation in the lower right-hand corner contains, amongst others, the mechanically imposed blade twist from hub to tip, typically for a delta angle of 8 deg for row 1 and 6 deg for row 11.[67] There are consistently(!) 57 blades/row and the blade annulus height varies between 55 and 40 mm from entry to exit.

---

[66] See BBC, Offer, 1931–1933. In an overly positive conclusion the offer text states: '*We may point out that Brown Boveri has carried out comprehensive tests with 1-, 2- and 4-stage axial blowers, partly using dimensions similar to the project, so we can fully ensure that machine target performance will be attained*'. In contradiction to this dynamic offer phase the signing of the actual delivery contract between the 'Direktion der Eidgenössischen Bauten, Bern und der Aktiengesellschaft Brown, Boveri & Cie., Baden' was then delayed until 29 September 1933.

[67] This kind of mechanically induced, ductile deformation of the blade material was a long-term 'nightmare' for any department responsible for mechanical integrity. The problem cropped up occasionally in the 1980s, but in the meantime it has been eradicated due to thorough quality control.

The high development pressure until the end of 1932 did not prevent Seippel from making the following Mondeville II test campaign into something of extraordinary importance, with far-reaching implications for future aero GT compressor activities in Germany and England both before and during WW II. The prime test targets were described in a straightforward manner:[68]

—   *'increase the pressure rise per stage,*
—   *push the surge area back,*
—   *compare the bladings for large and small volume flows'.*

Interestingly, the most important outcome of surprisingly large efficiency variations in the dependence of degree of reaction (i.e. the rotor/stator staggering) was not mentioned in the steam turbine department. In light of the economic restrictions the invention of an all-new engine concept was neither intended nor desired. Development risks had to be minimised by every possible means, and consequently the first axial compressor was very similar to BBC's established product, the reaction steam turbine, more specifically its best working part, the intermediate-pressure section. Fundamental for any turbomachinery design, the degree of reaction is the ratio of the specific work change in the rotor to the specific work change over the whole stage, or simplified in reference to incompressible pressure changes:

Degree of reaction $R = \Delta p_{Rotor} / \Delta p_{Stage}$

The following blading was prepared for the 11-stage Mondeville II compressor tests:

—   Rotor A   with stagger angle $\vartheta=42°$ (ϑ referred to the circumferential direction), Figure 4-23 and Figure 4-24

—   Rotor B   with $\vartheta=22°$,

—   Stator A1   with $\vartheta=42°$,

—   Stator B1   with $\vartheta=90°$,

—   Stator C1   with $\vartheta=125°$.

The following rotor/stator combinations were investigated:

Variant   1)   Rotor A/ Stator A1   R = 0.5

         2)   Rotor B/ Stator A1   R < 1.0

         3)   Rotor B/ Stator B1   R = 1.0

         4)   Rotor B/ Stator C1[69]   R > 1.0

---

[68]   See BBC, Axialgeblaese fuer Veloxkessel, 1932–1933

[69]   According to Seippel, The evolution, this variant 4) had been proposed by Keller to produce 'excess compression' in the rotor blading, followed by expansion in the slightly convergent/accelerating stator channels, and was thus conceived to favour flow stability. In fact a direct comparison between variants 1) and 4) revealed considerably higher through-flow coefficients for the latter case, but no significant enlargement of the stable operation range $\Delta V/V_o$.[68] However, the information in Figure 4-25 was kept absolutely secret and only revealed 25(!) years later in Horlock, Axial Flow Compressors, Figures 4.11 and 4.12, with the explicit comment

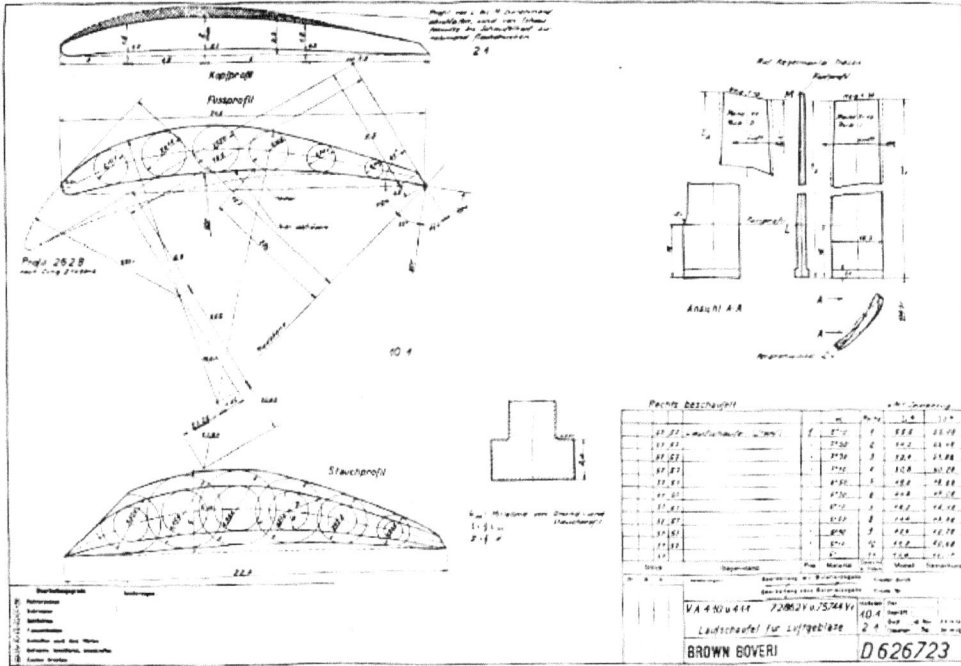

Figure 4-23   *Manufacturing drawing #D626,723 for 11-stage Mondeville test compressor blades, dated 30.11.1932 (checked Se Seippel)* [70]

Finally, Seippel reminded the BBC test department '... *These tests are vital for the design of further* (Velox) *boiler blowers and a blower for the ETH Zurich.*' The speed was kept and the extensive axial compressor test programme for Mondeville II was carried through and reported until 8 May 1933.

'*Courtesy Cl. Seippel, Brown Boveri Ltd., unpublished data, March 1957*'. In private correspondence with the author, Sir John replied on 1 Dec. 2008: '*With regard to the reaction data I quoted in my compressor book [fifty years old this year!] it was given to me personally by Claude Seippel when I visited BBC as a consultant for CEGB* UK in about 1955 or so. He learned I was writing a book on compressors and encouraged me in this venture and to use the data unpublished, I think. He was very kind to me and I had much respect for him – a gentleman engineer of the old school.*'

*CEGB Central Electricity Generating Board. Under a nationalised structure, CEGB was responsible for electricity generation in Britain between 1957 and 1990. The reasons for keeping reaction-dependent performance data such a secret may become understandable in view of German jet engine development in the 1930s and 1940s, which relied consistently on an AVA Goettingen compressor design philosophy with R=1. A detailed account of BBC aero engine technology amongst the conflicting military parties of WW II is planned. The importance of this information might be indicated by the fact that a change in reaction of the compressor blading from R=1 to 0.5 alone would have increased the combat performance (endurance, interception time) by 15–25 percent, for the example of an Me 262 fighter aircraft with an accordingly modified Jumo 004B jet engine.

[70] Figure source: BBC, 4-stage axial compressor test rig, 1926–1932

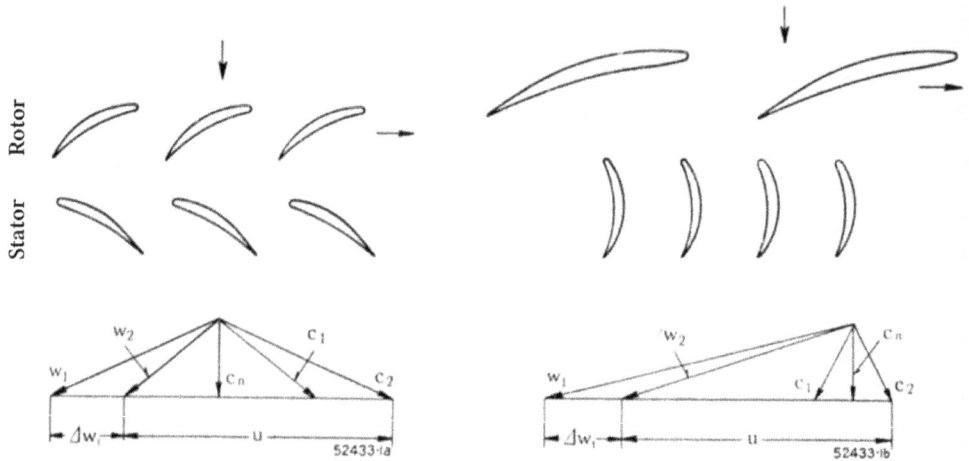

Figure 4-24   *Fundamental compressor test configurations for differing degrees of reaction – variant 1,
R = 0.5 (left) vs. variant 3, R = 1.0 (right), 1933*[71] *c – absolute velocity, w – relative velocity,
u – circumferential speed, 1 – blade row inlet, 2 – blade row exit*   ©ABB

A comparison of variants 1 (R = 0.5) and 3 (R = 1) as illustrated in Figure 4-24 is most interesting. The blade staggering of rotor blades (above) in relation to stator vanes (below) is
shown together with the velocity triangles for equal work transfer u Δc$_u$. For the R = 0.5
configuration the rise in stage pressure is split equally per definition, 50/50 between rotor
and stator, resulting in a 'symmetrical blading' with identical profile shapes in the rotating
and stationary systems. On the contrary, R = 1 uses only the rotor blading for the flow diffusion in the relative/rotating system, while the stator only turns the flow back to the original rotor entry direction without further flow deceleration/ increase in static pressure. It
appears that the principle drawback of R = 1 blading becomes evident just by analysing the
velocity triangles, given the much more balanced velocity distribution between rotor and
stator for R = 0.5 and the especially striking difference in the relative velocity vector
lengths. The obvious advantages of an R = 0.5 design are the relatively low velocity levels
even at circumferential speeds in the supersonic regime. In a corresponding R = 1 arrangement, shock losses and related separation losses can occur. Early airfoil aerodynamics
typically showed a significant high-speed sensitivity beyond Ma = 0.6–0.7.[72]

---

[71] Figure source: BBR, The development of the Brown Boveri axial compressor, 1940, Figures 7a and 7b

[72] A. Raymond Howell was obviously the first to publish a detailed assessment of compressor cascade losses
on a theoretical basis. The publication of his two widely known papers was presumably delayed until 1945,
when the aerial war against Germany was decided. Howell, Design of Axial Compressors makes the point
explicitly in Figure 95 there, where the efficiency drawbacks due to profile losses of an R = 0.9 design (ex-
emplified by the Jumo 004 engine) are compared with a clearly superior R = 0.5 design. With his theoretical approach Howell also confirms the measured efficiency differences of BBC in the order of 5–10 percentage points,
Figure 4-25. Howell's figure was first reproduced in German literature in 1960 in the book – see Kruschik, Die
Gasturbine, 2nd ed., p. 144, hinting at the drawbacks of the Jumo 004B's performance. A detailed discussion of the
R-impact, especially on the design of axial compressor mechanics, has to wait until the planned historic account
of BBC's aero engine activities.

The corresponding test results[73] have been replotted in Figure 4-25. It shows two settings for the 11-stage Mondeville II compressor bladings, a) with the classic 'symmetrical' degree of reaction R = 0.5 (solid lines) and b) for R = 1 with pressure rise in rotor only (dashed lines). The diagram combines polytropic efficiency curves (top) and pressure rise characteristics $\Delta p/(\rho\, u^2)$ with the entry through-flow coefficient c_{ax}/u. While the pressure data confirmed the trend towards a higher stage pressure ratio for R ~1, the results confirmed BBC's long-term preference for the R = 0.5–0.6 airfoil setting due to the impressive (polytropic) efficiency advantages of 8–12(!) percentage points in comparison.

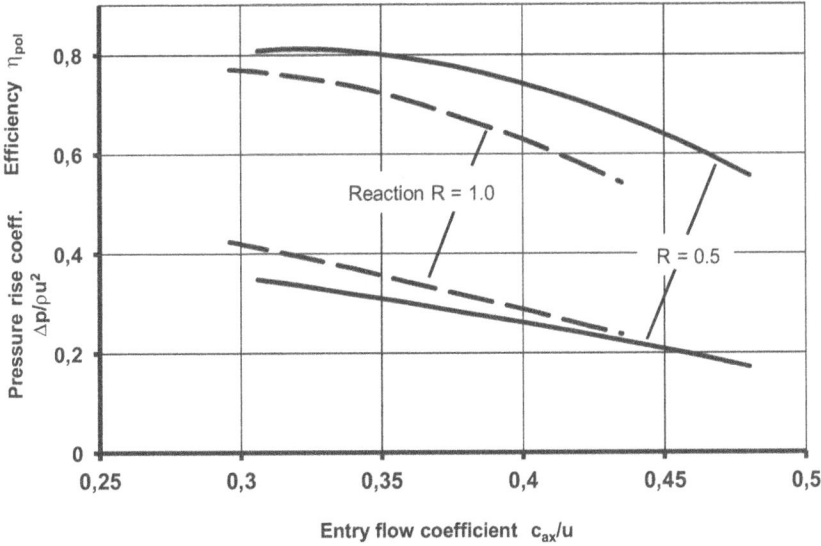

Figure 4-25   *BBC experimental characteristics of 11-stage Mondeville compressor for the differing reaction R=0.5 vs. 1.0, 1933*

Besides presumed compressibility effects with steeply increasing profile losses for the relatively high Ma_{rel} numbers in the case of R = 1, the achievement of high stage pressure ratios is restricted by the maximum admissible circumferential speed, Mach number and pressure coefficient. The choice of a suitable blade form enables the profile losses to be reduced to a minimum value determined by the required mechanical blade strength. Reaction is not necessarily constant along the radius. In a 'free vortex' velocity distribution $r c_u$ = const, the reaction increases from rotor hub to tip. The distribution of reaction is related to the vorticity, which can be expressed by a dimensionless coefficient with c_u – circumferential component of absolute velocity c, u – circumferential speed, ω – angular speed and r – radius. 'Free vortex' has q_{rot} = 0, while a 'constant reaction design' in radial direction is characterised by q_{rot} = 0.5.

$$q_{rot} = \text{rot } c_u / \text{rot } u = 1/2\omega\; d(r\, c_u)/(r\, dr)$$

[73] See BBC, Axialgeblaese fuer Veloxkessel, 1932–1933

Radial deviations of axial mass flows basically depend on the energy transfer, the radial vorticity distribution and the degree of reaction, especially for 'forced vortex' configurations. In this case the largest impact can be seen for large pressure coefficients and lower degrees of reaction. The only exception is the 100 percent reaction for forced vortex design: the AVA approach.

'Radial equilibrium' designs – with the radially acting forces on the fluid in equilibrium – were verified relatively early by BBC.[74] However, these tests showed that the method for calculating the resulting radial blade twist only had a small influence on the compressor efficiency and did not strictly need to be fulfilled. This considerably facilitated the task of the designer and allowed for simpler and cheaper blade forms. Service reliability was always first priority, while a compromise between efficiency and manufacturing costs was seemingly justified. Unchanged, compressor stability represented the largest development obstacle. It was known how to achieve stability on two stages, but there was a great deal of apprehension as to how 10 or more stages would perform. Theoretical studies had already revealed that 'stability' is not only a function of the flow conditions inside the compressor, but depends largely on the whole system volume connected to it. This was confirmed by experience, when the axial compressor was exposed to the real world after satisfactory laboratory tests, for example with an attached Velox boiler system. Considerable flow and combustion instabilities overshadowed the first test period, until it was realised that the addition of 1–2 stages worked wonders, avoided all instabilities and saved the compressor from being scrapped due to heavy, surge-induced vibrations like numerous predecessors. Already the first commercially sold axial compressor for the Mondeville plant had to be 'boosted' by an 11th stage[75]; the 10-stage test configuration is shown in Figure 4-26.

Figure 4-26   *Mondeville 10-stage test compressor rotor, PR = 2.4, n= 10,000 rpm, η_{is} = 0.83, R = 0.5, ~January/February 1933* [76]

---

[74]   See BBC, Versuchs-Axialgebläse, 1940–1945, and – see BBC, Ventilator, 1929. Particularly the latter case showed a successful application of the airfoil theory for a single stage generator blower for the Geertruidenberg, NL project as early as 1929. Later on these test experiences served as a reference for subsequent multi-stage compressor designs. The introduction of a corrective, radial blade twist for the blower wheel helped to increase efficiency from 20 to 45 percent. 'Free vortex' designs found early application in the US and UK, see Hawthorne, The Early History. He outlines in a Footnote on p. 59 of the ref. book: *'Free vortex or constant circulation blading for steam turbines was introduced to O.G. Tietjens in the late twenties, Griffith had used it on his model rotor and Constant on the axial supercharger'*.

[75]   The design data for the first commercial axial compressor worldwide, the Mondeville 11-stage configuration, was extrapolated from the BBC TFVL 963-I tests to PR = 3.3, V = 3.5 m³/s at 12,000 rpm. High efficiencies of 83 percent were achieved for 10 stages at PR = 2.4 by using temperature measurements. Torque measurement produced approx. 78 percent in comparison.

[76]   Picture source: Pfenninger, The evolution of the Brown Boveri gas turbine, Figure 11

The preference for symmetrical reaction blading meant that specific flow treatment at the compressor inlet was necessary. The velocity distribution of repetitive stages with residual rotation (rot) had to be steadily induced (and back to an axial outlet stream). This was accomplished using one or more rows of entry guide vanes with the extra possibility of reducing excessively high relative inlet Mach numbers. BBC kept a French patent FR781182 (inventor G. Darrieus) with first priority in Switzerland from 3 February 1934[77], which firstly claimed a degree of reaction R=0.5 blading in general and secondly 3–4 rows of entry vanes to steadily redress the inlet flow from the axial entry direction to comply with R=0.5 standard blading. Present designs try to combine these principles by using a row of inlet guide vanes to diminish the relative Mach numbers at entry and to steadily reduce the degree of reaction from somewhat higher values of R = 0.7 to the most efficient R = 0.5 in the following stages, then increasing the degree of reaction again for swirl free exit conditions.

Immediately after the design rules for optimum blade/vane staggering had been investigated, BBC successfully resolved and implemented another fundamental cascade principle, with the advantage of exclusive usage for a decade. Suggested by their young employee Otto Zweifel[78] (1911–2004), BBC has used an aerodynamic loading coefficient ψ for design calculations and cascade test evaluations since 1936. Today this is well-known and still useful for compressor and turbine design in an adapted form: the Zweifel number. Deduction with the target to determine an optimum blade count appears relatively easy. High blade numbers would create overly high friction losses, while small blade numbers would be threatened by even higher flow separation losses. Therefore, an optimum blade pitch should be dependent on a dimensionless loading coefficient. In a very short time Zweifel realised that the standard lift coefficient $C_L = f(w_\infty)$ was inappropriate in this respect, since its optima vary strongly with flow angles and grow excessively especially for large turnings. Alternatively, the new loading coefficient ψ referred to the stagnation pressure w_2 of the cascade **exit** flow.[79] Surprisingly, it was shown that optima for small and large turnings occurred almost always at values between ψ = 0.8...1.0. Consequently, with the help of some graphical tools the designers achieved much faster, more acceptable blade shapes and forms of flow channels by firstly determining the blade pitch t from inner to outer rotor

---

[77] There is some likelihood that the R = 0.5 patent might have additionally alienated the German side in the patent conflict with BBC between 1932 and 1942 on high-altitude 'combustion engine turbocharging', which will be reviewed in the future account on BBC aero engine activities.

[78] After a year assisting Prof. Eichelberg at the ETHZ, Otto Zweifel began his professional career at the BBC test department in 1936 where he was responsible for turbine cascade testing. Numerous blade designs (profiles, stagger angles, pitch chord ratios) were investigated and amazingly, Zweifel came up with a consistent loading data interpretation in an extraordinarily short time. He left Brown Boveri in 1946, first to lead the technical office and then to become a director of SIG Schweizerische Industrie-Gesellschaft at Neuhausen near the 'Rheinfall' in 1948, specialised in railway car manufacturing. SIG developed an international reputation, especially in the field of bogie manufacturing. In the 1990s they developed electrical tilt technology in competition with FIAT's hydraulic tilt technique. Sold to FIAT in 1995, SIG Rail returned to Alstom in 2000 together with FIAT Ferroviaria.
After his professional stay at SIG, Otto Zweifel became a full-time professor at the ETH Zurich from 1953 to 1976, specialising in 'machinery for construction works and transportation' from 1956 onwards, which also implied spectacular cable car projects important for Alpine Switzerland.

[79] According to its definition the Zweifel number is the ratio of the actual circumferential force to a 'reference force' in a two-dimensional blade cross section, which is thought to be induced by an 'ideal' blade pressure distribution. The reference distribution assumes a constant pressure equal to $p_{t,1}$ on the pressure surface and a constant pressure equal to p_2 on the suction surface.

diameter and thereafter by calculating the axial cascade width b under the usage of $\psi_{opt} = 0.8$ (see nomenclature of Figure 4-27):

$$\text{Aerodynamic loading coefficient ('Zweifel number') } \psi = 2 \sin^2 \beta_2 (\cotg \beta_2 - \cotg \beta_1) \, t/b$$

Zweifel was awarded the task of designing the turbine for the first power generation gas turbine project at Neuchâtel, CH in 1938. He kept $\psi = 0.8$ for all rows from root to tip and produced the best turbine efficiency of all BBC designs up until then.

As stated before, the economic benefits of using such a practical loading coefficient were known exclusively to BBC and therefore kept secret. It was only when in-house information travelled further that similar considerations were made public in 1943 by Maurice Bidard[80] from the French daughter-company C.E.M., meaning the BBC management *'wanted to document Baden as the origin of this coefficient'* which resulted in Zweifel's article of 1945, *'cautiously, only in Brown Boveri Mitteilungen[81]'* (quotes from the Zweifel letter as ref. in the following).

In private correspondence, Otto Zweifel wrote on 14 December 2002: *'The paper was mainly aimed at those readers who only knew a few publications by Briling, Christiani and Keller[82] at that time. Therefore, it was demonstrated for these three examples how well the ψ coefficient could be used to optimise blade cascades with constant (Briling), accelerated (Christiani) and decelerated (Keller) flows'.*

Indeed, the later naming of the loading coefficient ψ as the well-known 'Zweifel number' happened somewhat coincidentally. Professor Traupel's famous textbook apparently never took notice of the Zweifel number but used Bidard[83] instead. Even Claude Seippel tried to maintain the hidden existence of the Zweifel number in a related SBZ-article[84] in 1959, where he only showed Bidard's 'm,n-Diagram' (without further reference) but not the more practical Zweifel approach. It was only in late 1976 when Seippel[85] pushed the 'Zweifelsche Kennzahl' and its inventor into the limelight in Switzerland, at a time when the Zweifel number was already established in Anglo-Saxon design circles, for example through the widely-read book by Sir John H. Horlock on 'Axial-Flow Compressors'[86].

In the past 75 years understanding of the fluid-mechanics of optimum loading has considerably improved, especially for low pressure turbines (lpt) in aero engines. Zweifel's original approach of using test results from cascade wind tunnels and from wind tunnel measurements at low Re numbers, typical for lpt flows at high altitude flight conditions, are seen more and more critically. The turbulence modelling in wind tunnels is clearly not representative of real turbomachinery flows in general and as such, especially for low Re testing, modeling considerably overestimated the laminar boundary layer transition and separation losses. In recent years improved fluid-mechanical insight has revealed that unsteady interaction of the laminar boundary layer on the profile suction side with periodic flow distortions from upstream blade rows initiates rapid and smooth transition (without loss-

---

[80] See Bidard, Quelques considerations
[81] See BBM, Die Frage der optimalen Schaufelteilung, 1945 and – see BBR, The Spacing, 1945
[82] See Briling, Verluste; Christiani, Experimentelle Untersuchung and – see Keller, Axialgebläse.
[83] See Traupel, Thermische Turbomaschinen, 3rd ed. p. 220
[84] See Seippel, Dampfturbinen von heute (SBZ – Schweizer. Bauzeitung)
[85] See Seippel, Die Zweifelsche Kennzahl
[86] See Horlock, Axial-Flow Compressors, reference #189

increasing transitional separation bubbles), along with rather advantageous loss characteristics. This explains why current UHL 'ultra-high lift' turbine blading today achieves its
optimum rather at higher Zweifel numbers of $\psi = 1.1...1.2$[87].

As previously discussed in Figure 4-19, the behaviour of an 'airfoil cascade', Figure 4-27,
can be represented by a polar diagram, fixing the relation between the angle of incidence
and the lift and drag coefficients. Experience showed, however, that this method only had
basic benefits for widely spaced blades and small deflections, whereas the approach failed
for the closely spaced cascade of an axial compressor.

The isolated airfoil approach was not precise enough to give accurate predictions of compressor performance. The limitations of the isolated airfoil approach were first recognised
by British aerodynamicists of the Royal Aircraft Establishment (RAE), where A.R. Howell
published his two landmark papers on cascade theory in 1945.[88]

For the calculation of a turbomachine, it is first necessary to know the optimum deflection
which can be obtained with the cascade. For this reason, a method was applied to analyse
the results of the tests which is in principle analogous to Howell's famous publication of
1945[88], i.e. the exit flow angle β_2 from a given cascade was determined as a function of the
entry flow angle β_1, pitch to chord ratio t/s and stagger angle ϑ; and the angles β_1 and β_2
and the total and static pressures in front of and behind the cascade were measured in the
plane of symmetry over a length of one or two pitches. This description of the 'new' BBC
turbo-machinery evaluation concept in the post-war years will be followed by a comprehensive account of events in 1950.[89]

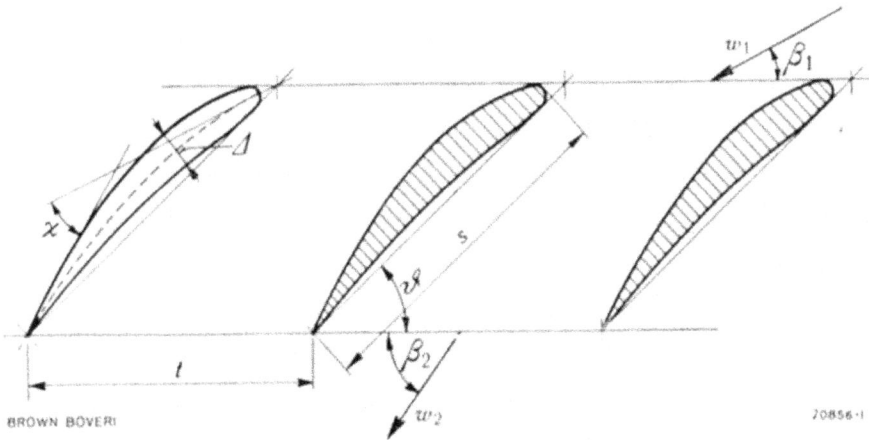

Figure 4-27   *Compressor blading cascade, parameter definition* [90]   ©*ABB*

---

[87]   See for example patent description EP 0937862, Volker Schulte, BMW Rolls-Royce, 20 Feb. 1998

[88]   See Howell, Fluid Dynamics of... and Design of Axial Compressors. A. Raymond Howell (1917–1988) was one
of the leading compressor scientists at the RAE Royal Aircraft Establishment in Farnborough, UK. He played
a major role in the evolution of successful axial compressor design methods which would be used by the first
post-war UK gas turbine engines. In 1946 he was appointed Head of Aerodynamics Dept. of the RAE successor organization NGTE, which he led for more than 20 years. See Dunham, Howell

[89]   See BBR, Aerodynamic methods, 1950

[90]   Figure source: BBR, Aerodynamic methods, 1950, Figure 3

Relatively quickly it became clear that test results from stationary blade cascades were not directly applicable to multi-stage turbomachinery bladings (3D flows, turbulence levels, etc.). Therefore, a development process was organised in stages: first the blade cascade, then, if appropriate, the single stage rotating machine and finally, the multi-stage compressor or turbine. For this concept BBC shared common resources from early on, for example a cascade wind tunnel which was shared with its French daughter company C.E.M and its 'Centre d'Essais de Mécanique des Fluides' in Paris.

The year 1946 marked a new stage in the development of BBC's aerodynamic equipment at Baden. A supersonic wind tunnel was installed at the BBC test laboratory, in principle similar to the one built for Prof. Ackeret's high-speed wind tunnel in 1934. Compressed air delivered by an axial flow compressor flowed in a closed circuit through a cooler (which at the same time worked as a flow straightener) into the supersonic test chamber and via a measuring nozzle back to the compressor.

By means of an attached reciprocating compressor, the wind tunnel could either be evacuated or pressurised, so that it was possible to achieve a Reynolds number variation by a factor of 10 for a given velocity. The whole installation was very flexible and allowed both measurements and observations of stationary high-speed turbomachinery cascades in the test chamber or the replacement of the test section with a cylindrical duct in the axial compressor tests of multi-stage blading. The 8-12-stage axial test compressor was driven by a 750 kW DC motor through a 2-stage 1500/11,200 rpm gearbox, which was mounted on swivel bearings so that the compressor drive power could be directly determined, Figure 4-28.

Figure 4-28   *BBC test compressor for supersonic wind tunnel propulsion or alternatively, multi-stage compressor testing, 1946* [91]   *©ABB*

---

[91]   Picture source: BBR, Aerodynamic methods, 1950, Figure 2

Numerous tests paved the way for a comprehensive understanding of influences and inter-dependencies, such as the efficiency of advanced, fluid-mechanically optimised axial compressors and the impact of:

- Reynolds and Mach numbers
- Stage pressure and velocity coefficients
- Blade pitch t, blade length l and radial clearance
- Profile form and twist of the blade
- Pressure losses in the inlet and outlet casings
- Sealing and gland losses.

The aerodynamics of an axial compressor and a reaction steam turbine are fundamentally different due to the different behaviour of the passage flow boundary layer in decelerating vs. accelerating flow. Despite this, the two turbomachines share much of their mechanical technology, which explains why steam turbine manufacturers were at the forefront of axial compressor development.[92] Most mechanical problems, such as designing blades to withstand vibrations under the impact/excitation of high-speed air/steam flows, or designing a blade root section to resist large centrifugal forces, were common to both steam turbines and the new axial-flow compressors.

In 1927, Stodola wrote the first sentence of chapter XI 'The Gas Turbine' in the newest issue of his book with a clear view[93]:

'The most obvious working cycle for a gas turbine is as follows: air and gas are separately compressed to a more or less high pressure, are burned at constant pressure in a chamber, and are led directly to the turbine. The type of construction of the turbine is theoretically immaterial. The burned gases are invariably expanded in the turbine to atmospheric pressure.'

A few lines further, in the Introduction:

'The hopes for the success of the gas turbine are based mainly upon the possibility of utilising this **complete** *expansion (compared to the 'explosive', constant-volume gas turbine) and thus obtaining a greater amount of available energy. ... On the one hand, the efficiency of the turbine is in itself smaller than that of the gas engine, and on the other hand, the work of compressing the charge is greater for a centrifugal compressor* **(which is the only compressor that can be considered for turbine drive)...'**

Less than 10 years later, a completely new compressor concept was established at BBC: the highly effective axial compressor. The path towards gas turbine development was clear.

---

[92] See Ponomareff, Principles, mentioning in this context (besides Brown Boveri and Escher Wyss of Switzerland), Metropolitan-Vickers in Britain and General Electric and Westinghouse in the USA.

[93] See Stodola, The Steam and Gas Turbine

4.2.4   Successful Compressor Reference Projects Beginning 1932

The commercial success of the axial compressor and related products since 1932 has already been addressed (in accumulated form) in Figure 4-13. A closer look at stand-alone compressor units for the same time frame (1932–1952) is given in Figure 4-29, but with additional references to annual quantities of delivered units and installed volume flow. At first glance it appears that the annual delivery rate allows clear differentiation between three pronounced peaks: directly following the introduction of the compressor market between 1934 and 1938; most surprisingly, the period immediately after WW II between 1945 and 1947, which apparently created a large backlog in demand, and finally the period after 1951. The data also reveals the trend towards bigger volume flows over time, which by the mid-1950s had already grown with compressor units of more than 300,000 m³/h. At the same time total pressure ratios reached values of up to PR = 5.5 and individual compressor driving power up to 14.5 MW.

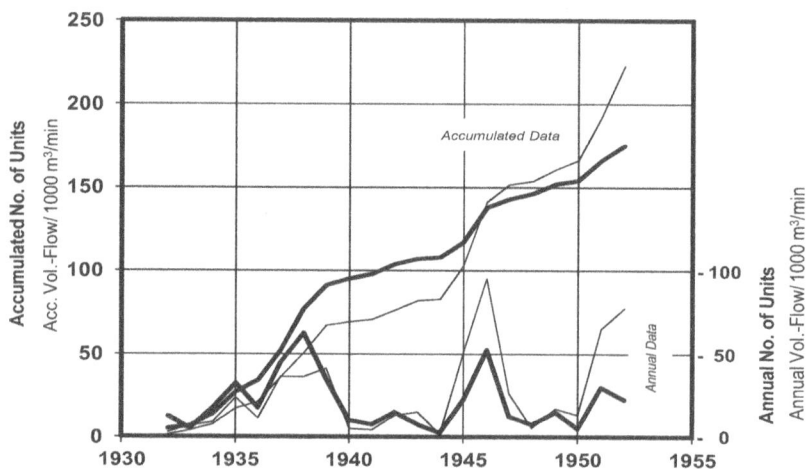

Figure 4-29   *Number and flow capacity of BBC multi-stage, stand-alone axial-flow compressors with deliveries between 1932 and 1952* [94]

Besides impressive volume flows and quantities delivered, BBC's work continued to include many individual technical highlights: an observation valid until the company ended in the 1980s and certainly not always economically justified. As an example, the production year 1943 is illustrated in Figure 4-30, showing an all-'blisk' (integrated **bl**aded **disk**) axial blower for gas compression in the chemical industry, a design concept for high mechanical integrity which emerged in military jet engines only 50 years later. The two stage compressor was operated at the limit of the upper material temperature of that time, – 530° C. For blisk manufacturing cast high-temperature steel blades were welded to the forged carrier disks.

---

[94]   See BBR, Progress and work in 1952, 1953, p. 81

Figure 4-30 *2-stage axial blisk blower for the chemical industry, 1943* [95]   *©ABB*

The development of axial-flow compressor technology at BBC until 1960 is shown in Figure 4-31 in a traditional, though somewhat too simplified form on the basis of the average Stage Pressure Ratio SPR, deduced from compressor total Pressure Ratio PR and stage number z by the relation $SPR = PR^{1/z}$. An aerodynamically more meaningful comparison of a speed-related Stage Pressure Coefficient ψ_{St} follows in Figure 4-32, based on the definition:

$$\Psi_{St} = \kappa/(\kappa-1)\ R\ T_{t1}/(u_{m1}^2\ z)\ [PR^{(\kappa-1)/\kappa} - 1]$$

for dry air, ideal gas – isentropic exponent $\kappa = 1.4$, gas constant $R = 287$ J/(kg K),

with the entry conditions: total temperature T_{t1}, – area mean circumferential speed u_{m1} and compressor total pressure ratio PR.

Eighteen industrial BBC projects have been selected for this visualisation of technical progress, divided into 3 main categories:

A.   **Experimental rigs**

1)   4-stage test rig, 1926

2)   4-stage test rig, 1932

18)   8-stage experimental compressor, 1960

---

[95]   See BBM, Rueckblick... im Jahre 1943, 1944, Figure 135

B.   **Industrial products**

3)   11-stage Mondeville compressor, 1932

4)   13-stage 'typical product compressor', 1933

5)   13-stage ETHZ supersonic wind tunnel blower, 1934

6)   20-stage compressor Sun Oil, 1936

7)   11-stage Velox compressor, 1936

8)   23-stage Neuchâtel GT compressor, 1938

9)   20-stage RAE compressor, 1940

10)   19-stage wind tunnel compressor, 1944

11)   20-stage 'typical product compressor', 1956

C.   **Aero engine projects**

12)   7-stage aero turbocharger RR, 1938

13)   15-stage SOCEMA/C.E.M. TGA-1 compressor, 1941

14)   10-stage BBC compressor for BMW –003D, 1944

15)   9-stage BBC compressor for tank GT 102, 1944

16)   7-stage BBC compressor for BMW –003C, 1944

17)   8-stage SOCEMA/C.E.M. TGAR-1008 compressor, 1945

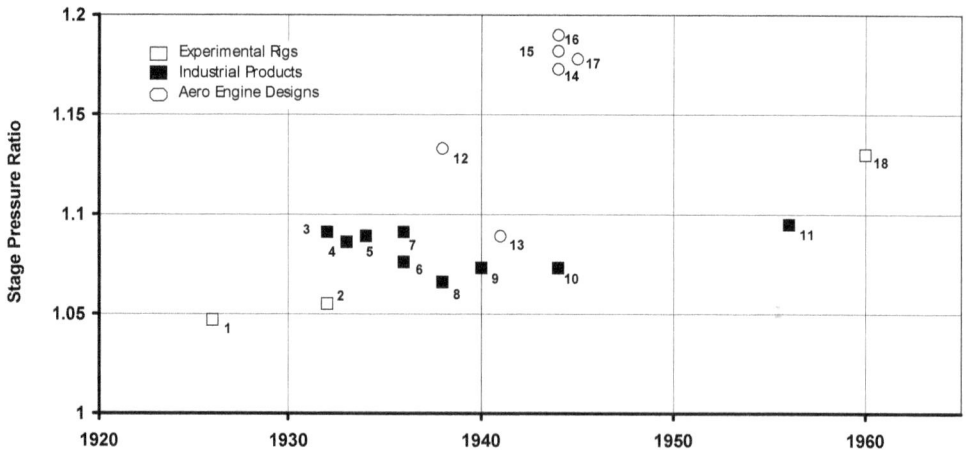

Figure 4-31   *History of BBC axial compressor development, 1926–1960*

Over a time span of 30 years the stage pressure ratio of BBC designs increased between first trials (1,2) and an 8-stage experimental rig (18) in 1960 by about 7.5 percent. Initial experiences with a 4-stage experimental rig have already been addressed in Section 4.2.2. A closer look at Figure 4-32 reveals however, that the initial aerodynamic stage loading was obviously too ambitious, even in view of later military projects.

It appears that improved design expertise must have been acquired mainly in the area of single- and two-stage axial blowers (not shown). The first commercial products (3–5) between 1932 and 1934 significantly outperformed 'ad hoc' the 4-stage test rig forerunners. Thereafter, the industrial production standard (6–11) appears widely unchanged with regards to stage pressure ratio, while the flexibility of higher design speeds was seemingly used between 1935 and 1945 even with a certain reduction in aerodynamic load. In this respect the compressor for the Neuchâtel power generation gas turbine with a fixed 50 Hz generator speed was certainly a challenging design task, which resulted in the highest number of compressor stages: 23.

An intriguing investigation is whether the various relations between BBC and aero/ military gas turbine projects until 1945 show prominent compressor characteristics in comparison to standard industrial projects. Instead of the marked cluster of visibly protruding stage pressure ratios for the aero/military projects (12 and 14–17), the corresponding ψ_{St} values are only above the industrial standard in a few cases. The two compressors for the French SOCEMA/C.E.M. jet engine projects appear to be loaded slightly higher than BBC's industrial project average. Finally, compared to the second build of the 4-stage rig (2), the planned 9-stage compressor (15) for the German tank engine GT102 has an isolated, excessively high loading parameter which indicates that the geometrical limitations would probably have caused severe performance drawbacks.

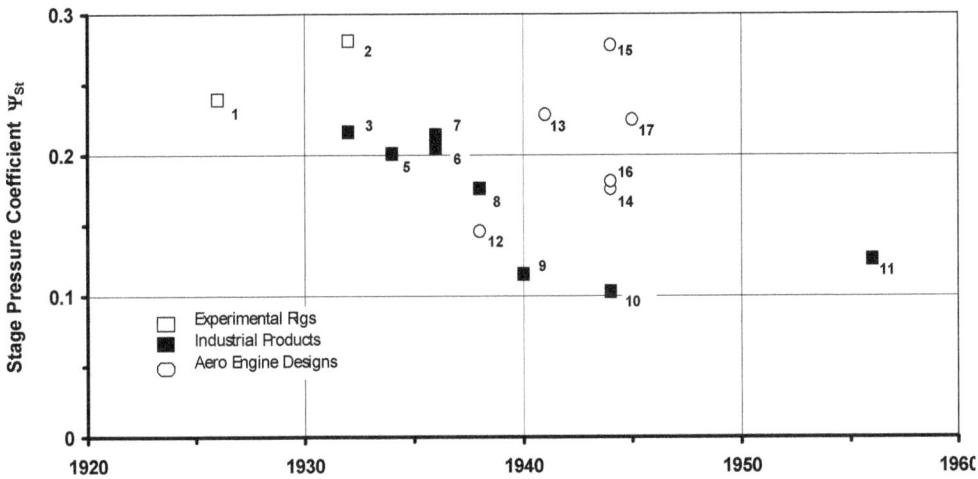

Figure 4-32   *BBC axial compressor stage pressure coefficients ψ_{St}, 1926–1956*

A prominent point in the history of BBC's axial compressor development was the propulsion system for the world's first supersonic wind tunnels, realised for Mach numbers 2–4, and with serious planning for up to Ma = 10. Though most of these were part of military research programmes, e.g. for ballistics and missile aerodynamics, there is generally more project information available than for comparable industrial projects at that time, especially via the scientific community. Interestingly most of these projects are also internationally interrelated. The following history of the BBC-built series of six supersonic wind tunnels between 1932 and 1952 and the single incomplete hypersonic wind tunnel project in between therefore reveals surprising links and unexpected details. According to a general status report[96] from 1952 there were six supersonic wind tunnels delivered at the time, although it does not provide further details. After some investigation the following list details the locations, numbers 1 to 5 will be addressed in detail in the following sections:

1. ETH Zurich, Institut für Aerodynamik, Prof. Ackeret, 1934,

2. Aeordynamic Test Facility at Guidonia near Rome, I, 1935,

3. AVA Göttingen and LFM Munich 1941, in 1944 partially delivered for LFM only, AVA delivery stopped – see Footnote 119. WVA Kochel, hypersonic wind tunnel project (unfinished),

4. LRBA C4 Vernon France 1950, Ma = 4, 1949[97],

5. AEDC Tunnel 'A' Tullahoma, Az, USA, Ma = 7, 1949, possibly with former BBC deliveries to LFM Munich and WVA Kochel,

6. VKI S1 supersonic wind tunnel, Ma = 2.2, 1950[98],

7. Politecnico di Torino, Ma = 2.2, 1950–1954, not counted in BBC listing of 1952.

4.2.4.1   The First Supersonic Wind Tunnel at ETH Zurich

As indicated above, the signing of the delivery contract between the 'Direktorat für Eidgenössische Bauten, Bern' in the name of ETH Zurich and BBC (order no. 7965) was delayed until 29 September 1933, but looking at the design execution it can be assumed that this was actually carried out earlier and thus represents the first commercial axial compressor design. Seippel praised Prof. Ackeret several times for this courageous step, without significant experience from BBC at that time.[99] The order intake document confirmed an average total pressure ratio PR = 3 at 3,800 rpm, a maximum intake volume flow of 50–55 m³/s and at the design point a typical pressure rise from 0.125 to 0.275 bar at 45 deg C, 3,400 rpm and a compressor efficiency of 75 percent. Reference point 5) shows the ETHZ compressor in Figure 4-31 and Figure 4-32. The actual measured compressor map, as depicted in Figure 4-33, together with the opened 13-stage compressor, confirms these guarantee values showing excellent agreement. This outstanding result for the first-ever design of a multi-stage axial-flow compressor impressively illustrates how well Seippel's design team prepared for such a success. Apparently not only the airfoil aerodynamics fulfilled expectations, but also the complete ducting system both up- and downstream of the compressor were carefully taken into account. The peak performance of η = 0.75 appears unimpressive at first glance, but the copy of this compressor immediately produced afterwards for the Guidonia wind tunnel revealed the impact of low Reynolds numbers for the ETHZ case. At Guidonia, the increase in driving power to 2,100 kW, Figure 4-35, allowed for a raise in inlet pressure and the Reynolds number (by a factor of 3 accordingly), the

---

[96] See BBR, Power utilization, 1952
[97] See Kingsbury, World Directory, p. 134
[98] See Kingsbury, World Directory, p. 84
[99] See Seippel, Die Entstehungsgeschichte... and BBC, Offer, 1931–1933

significant benefits of which were shown in the compressor efficiency by an addition of 6–8 percentage points.

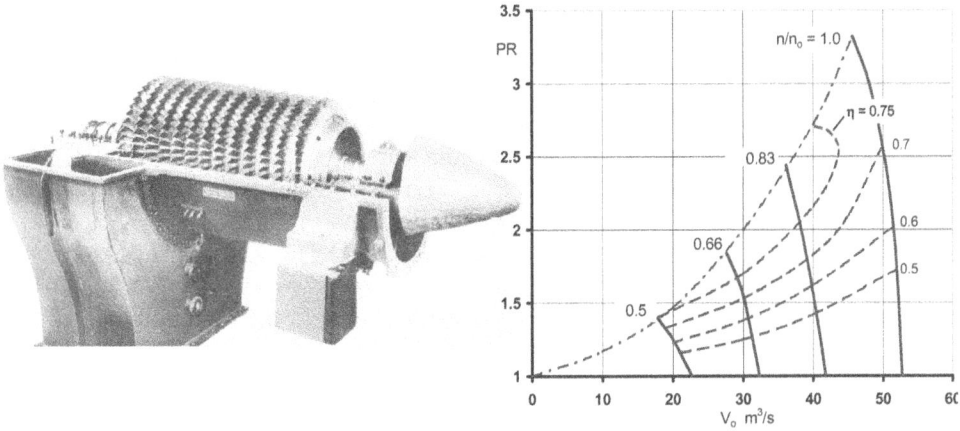

Figure 4-33 *BBC compressor for ETHZ supersonic wind tunnel facility [100] and measured compressor map, 1934/5 ©ABB (l)*

The BBC delivery content[101] for the ETHZ supersonic wind tunnel comprised as illustrated already in the general arrangement of Figure 4-22:

- a 700 kW DC motor, 1,200 rpm
- a transmission gearbox to drive the compressor with gear ratio 3.17
- the 13-stage axial-flow compressor with removable rear stages for testing at lower volume flows than designed at higher pressures or vice versa and vacuum-tight by oil sealing
- a 3-stage, individually regulated finned cooler to reduce the heated test air from approx. 120 deg C compressor exhaust level back to standard test conditions of 45 deg C
- the complete oil supply unit
- the test section with wooden nozzle contour inserts; following Ackeret's first design the 0.4 x 0.4 m square dimensions[102] of the test section characterised all other supersonic wind tunnels thereafter. The power-saving closed-circuit wind tunnel concept had already been invented during Ackeret's time at AVA Goettingen.

---

[100] Picture source: BBC, Geschichte des Schweizer. Turbomaschinenbaus, 1982

[101] See BBM, Rückblick auf die... Konstruktionen im Jahre 1935, 1936, G Aerodynamik und Flugwesen (Aerodynamics and Aviation) pp. 57–60 and – see Ackeret, Das Institut, pp. 5–33

[102] Ten years later when the von Kármán evaluation team interviewed A. Busemann, presumable on 9 May 1945 at LFA Luftfahrt-Forschungs-Anstalt Braunschweig-Völkenrode, this was still an issue in comparison to established circular wind tunnel cross sections of the Allied facilities: 'A. Busemann is of the opinion that a rectangular test chamber has the advantage that it is easier to avoid shock-wave formation. ... However, a price is to be paid in compressor power as the pressure recovery is lower in the diffuser.' – see Tsien, Tech. Intelligence Supplement p. 35

The charged cost of 100,000 SFr for the whole package was extraordinarily low, even for those days. The BBC upper management speculated about the 'propaganda'[103] effect in the international scientific community and the emerging aviation industries, but a dispute with the production plant had to be settled internally as the official bill did not even cover their own manufacturing costs. In fact, since Ackeret had acquired the Guidonia project which was to follow immediately, almost an identical copy of the ETHZ facility, this price was somewhat justified as a kind of demonstration. The guarantees had almost been accomplished for the first project. Following established steam turbine practice, the blade tips of the 3 compressor front rows displayed damping wires first in fear of inlet distortions. However, this precaution was not necessary and removed in due course as a source of extra loss. Ackeret pressed forward to ensure the new wind tunnels at his institute were ready for operation. A picture in the 1934 Annual Report (dated 12 November 1934) shows the BBC compressor mounted to the baseplate. The subsonic wind tunnel with a large concrete structure was successfully operated for the first time in the evening hours of New Year's Eve, and the inauguration of the supersonic wind tunnel installation presumably followed in the spring of 1935. A comprehensive survey of supersonic wind tunnel use at Ackeret's institute over the following years and the achieved fundamental research results was provided by N. Rott[104] in 1983. This interesting paper also details Ackeret's considerations on introducing the term 'Mach number' for the velocity ratio of flight speed to the speed of sound for the decisive similarity parameter of high-speed flows.

4.2.4.2   The Guidonia Wind Tunnel

The start of wind tunnel operation brought Ackeret and his new institute at ETH Zurich international attention. The ideal, highly prestigious discussion forum was the 5th Volta Conference[105] in Rome on 'High Velocities in Aviation', from 30 September until 6 October 1935.

The Volta conference was concluded with an excursion for all participants[106] to Guidonia, approx. 30 km north-east of the conference location at Villa Farnesina on the road to Tivoli. The first 'Campo di Aviazione di Montecelio' was founded in 1915. It became **the** Italian centre for the development of aeronautical technologies in the 1920s under the direction of

---

[103] The BBC Baden public relations organisation in the 1930/40s was actually called 'Abt. Propaganda'.

[104] See Rott, Jakob Ackeret. Ackeret suggested the Mach number definition in his ETHZ inauguration lecture on 4 May 1929, entitled 'The aerodynamic drag at very high speeds', after the Austrian physicist Ernst Mach (1838–1916). Here Ackeret followed his mentor Ludwig Prandtl who similarly introduced the naming of the 'Reynolds number'.

[105] These were international conferences in Italy organised by the Royal Academy of Science in Rome and funded by the Alessandro Volta Foundation. In the 1920s/30s they typically covered alternating topics in science and humanity. The first conference, which took place at Lake Como in 1927, became known for the public introduction of the uncertainty principle by Niels Bohr and Werner Heisenberg. The second conference was organized in 1932, officially to celebrate the 10th anniversary of the fascist 'March to Rome', and was hosted by Benito Mussolini himself. Its topic was 'Europe' and it was most notable for the participants, such as the German delegation consisting of personalities such as Hermann Goering, sociologist Alfred Weber (brother of Max Weber), Alfred Rosenberg and Werner Sombart. VOLTA III in 1933 was on 'Immunology' and the fourth conference in 1934 was on 'The Dramatic Theatre'.

[106] It is not known on which day the excursion actually took place and if all the conference participants attended the tour to the test site on air force territory in Guidonia-Montecelio. After the afore-mentioned Italian declaration of war in Abyssinia/Ethiopia on Thursday 3 October 1935 the English participants had received governmental instructions only to participate in the remaining scientific sessions, while no similar official restriction for the French side existed.

Alessandro Guidoni and after his death on site during advanced parachute testing in 1928, by General G.A. Crocco, his son L. Crocco and A. Ferri. It is assumed that the installation of the BBC supersonic wind tunnel equipment was unfinished in October 1935 at the time of the Volta excursion, completed in 1936. The highlight of the tour was a visit to the hydroplane model test facility, Figure 4-34.

Figure 4-34   *Hydroplane model test basin at Guidonia, 1935*[107] *Basin length 500m, max. speed of 'aircraft carrier' 250 km/h*

The theme of the Volta Conference 'Le Alte Velocita in Aviazione' in 1935 had certainly been influenced by Italy's international speed world record the previous year. In 1937, the whole complex of test facilities in the vicinity of the 1.8 km long air field with characteristic north-south orientation was inaugurated officially as Città dell'Aria by Benito Mussolini and named 'Guidonia' after Alessandro Guidoni. It is presumed that the supersonic wind tunnel installation was finished by TIBB[108] during 1936, Figure 4-35. The facility was to a large extent a 1 : 1 copy of Ackeret's Zurich wind tunnel, except for the drive power which was more than doubled and expanded the researchable Re number range considerably. However, the facility required an additional air cooler circuit (from 3 stages at the ETHZ to 4 stages). At Guidonia (in comparison to the ETHZ wind tunnel), the ejector bypass duct after the cooler was more pronounced, which allowed an increase in the pressure drop across the test section for small test area settings and thus increased the Ma number measuring range.

As a reference to the Guidonia facility, Figure 4-36 depicts a BBC-TIBB compressor tag which was actually delivered in 1950 to Politecnico di Torino (DC drive motor in 1954). But it can be assumed that the engine parameters of both facilities were almost the same, for example the indicated pressures on the tag from 0.125 ata to 0.375 ata were equal or close to those of the Guidonia wind tunnel. In comparison, the operational pressure rise for the first wind tunnel at ETHZ was only from 0.125 ata to 0.275 ata.

---

[107] Picture source: 'Ackeret Files' at the ETH-Bibliothek, archives and personal papers: Galleria stratosferica ultrasonora Guidonia, 12. XII.1935, ETH-Bibliothek, Archive, Hs 552b: 17.

[108] TIBB Tecnomasio Italiano Brown Boveri, founded in 1871 as TI in Milan to build electric machinery, in 1903 fully acquired by BBC and renamed TIBB, which played a dominant role over the decades in the electrification of the Italian railway network. In 1988, TIBB became ABB Tecnomasio, 1996 – Transportation-Adtranz Italy, 1999 – Daimler-Chrysler Rail Systems (Italia) S.p.A., 2001 – Bombardier Transportation Italy.

Figure 4-35   *Supersonic wind tunnel at Città dell'Aria, Guidonia, 1937[109] TIBB deliveries: 1 – Ward-Leonard variable speed drive, not visible: 2,100 kW DC motor, transmission gearbox, 2 – air collector and 3 – 13-stage axial-flow compressor, 4 – 4-stage air cooler with 5 – ejector bypass duct, 6 – 0.4 × 0.4m Mach 2 test section ©ABB*

The further fate of Guidonia (and of the BBC supersonic wind tunnel) can be reconstructed indirectly from a report about Antonio Ferri's further activities. The airfield, parts of the test facilities and the nearby city of Guidonia-Montecelio were already severely damaged during the Allied bombing campaign in and around Rome between 16 May and 13 August 1943.

Figure 4-36   *BBC-Tecnomasio compressor tag, similar to that of the Guidonia wind tunnel 1936, from facility #7 at Politecnico di Torino, 1950[110]*

---

[109] See BBM, Rueckblick... im Jahre 1944, 1945, Figure 101

[110] Picture source: Courtesy of Prof. Gaetano Iuso, Politecnico di Torino, 16 March 2010

BBC's third and fourth supersonic wind tunnel projects were probably the most spectacular and also mysteriously interlinked. They were #3 from 1941 for AVA Aerodynamische Versuchs-Anstalt Goettingen and LFM Luftfahrt-Forschungsanstalt Muenchen at Munich-Ottobrunn, and after the war up until 1952 #4 for the LRBA C4 trisonic wind tunnel at the Laboratoire de Recherches Ballistiques et Aérodynamiques in Vernon, F. Considerable difficulties had to be overcome in order to investigate these historic facts, not only due to the general secrecy surrounding such projects in Germany and France, but also by a general lack of reliable information about the late phases of the 'Reich' and the following dark post-war period.

4.2.4.3   The AVA/LFM Wind Tunnel Projects[111]

Since the first 4 year plan, and more so after the onset of war in 1939, basic aeronautical research in Germany had been hampered by urgent, short-term military planning. The RLM realised this and after the Battle of Britain[112] and intensifying Allied bombing raids became increasingly concerned that aeronautical research needed new, independent facilities for the development of aircraft which could fly faster and higher. It became clear very early on that Germany would otherwise lose the war, solely due to the sheer Allied industrial power in the aftermath of mass production. Continuing shortages in wind tunnel capacity and the brain-drain of research establishment personnel for military and industrial development activities encouraged plans for a completely new start with independent, bigger and better research facilities. The mastermind for this radical, although finally futile move was Adolf Baeumker, head of the RLM research department. The target: LFM Luftfahrt-Forschungsanstalt Muenchen, an all-new facility with a number of advanced, predominantly wind tunnel test facilities at Munich-Ottobrunn and wherever power supply requirements recommended exploitation of water power resources in the Alpine region south of Munich.

In line with Baeumker's critical assessment of the German wind tunnel situation in 1935, in comparison to the world leading supersonic facilities at the ETHZ and Guidonia, the situation in 1941 had not fundamentally improved[113]:

- The most advanced 0.4×0.4 m supersonic wind tunnel at the HVA Heeres-Versuchs-Anstalt Peenemuende allowed the short time realisation of Ma = 3.2 in intermittent operation (after the move to Kochel this increased to Ma = 4.4) – but was simply not accessible for projects other than Wernher von Braun's rocket development.
- Next was the LFA closed supersonic wind tunnel A 9b for continuous operation, which was situated after the war at the NAE National Aeronautical Establishment in Bedford, UK until 1950. It was known then as the '3 ft × 3 ft Supersonic Wind tunnel', for a Mach number range of Ma = 1.1...1.8; but A 9b only became operational in 1944, too late to even accomplish the necessary calibration measurements.
- At the AVA Goettingen the situation in 1941 was similarly precarious, the only generally accessible 11 × 13 cm high speed wind tunnel[114] for Ma = 1.2 ... 3.2 worked inter-

---

[111] If not explicitly mentioned, the description of this section relies on documentary material from the 'Ackeret Files' from the ETH-Bibliothek, Archive Zurich CH and from corresponding AVA documents from the DLR Archive, Goettingen, D.

[112] According to the British between 10 July and 31 October 1940, from a German perspective continuing until May 1941, before the attack against the Soviet Union forced new priorities.

[113] For a comprehensive survey of all relevant low- and high-speed wind tunnels, see Meier, Pfeilfluegel-entwicklung pp. 49–86

mittently by sucking air from ambient into a vacuum tank (and thus worked at relatively low Re numbers) for 10 to 20 seconds.

In the first half of 1939 AVA received permission to build a large supersonic wind tunnel similar to the A 9b at LFA Braunschweig. A 12 MW double motor was already on order, which AVA later officially passed on to LFM. At the beginning of the war and in a typical change of mind the 'supersonic wind tunnel Goettingen' project was put on hold until it was revived urgently in 1941 which triggered a visit from the Swiss delegation. Over the years Ackeret maintained close contact and correspondence with his former AVA colleagues, in particular a lifelong friendship with Albert Betz. However, regarding the following development the initiative was clearly on the side of the RLM, very likely driven by Adolf Baeumker, head of the RLM research department LC under Ernst Udet[115] in a critical phase of German air warfare.

Figure 4-37 illustrates the planned B 3 installation in the existing salt mine powerhouse. The agreed delivery time for the Swiss content was 16 months, i.e. the delivery target was August 1944. Due to a shortage in suitable workforce at AVA, BBC was also contracted for the final assembly of the test facility.

Most of the LFM research institutes with their numerous test facilities were erected on the main site on the southern outskirts of Munich, except one huge-high speed wind tunnel in the Oetztal[116], approx. 170 km south of LFM headquarters and 50 km west of Innsbruck. Construction works for the new Aerodynamic Institute at LFM Munich-Ottobrunn started in 1942. Far from complete, there were three main wind tunnels[117] under construction:

- A **3m diameter sonic wind tunnel** was completed at the end of the war and calibration just started. Power input was 10 MW. A unique feature of the tunnel was the ability to disconnect the entire large end and to roll it aside on tracks, thereby converting the tunnel into an open return type, e.g. for running jet engine tests in the throat.
- A **25 × 25 cm supersonic wind tunnel** was still under construction, planned for intermittent operation by a suction tank and a design Mach number 3.2.

---

[114] It was in this facility where Hubert Ludwieg carried out the first swept wing measurements from the high subsonic to the supersonic regime in Autumn 1939.

[115] Ernst Udet (1896–1941), http://en.wikipedia.org/wiki/Ernst_Udet was 'Generalluftzeugmeister' (Luftwaffe Director-General of Equipment) responsible for the coordination of aero product development and manufacturing from 1 February 1939, under increased pressure following the Battle of Britain defeat, committed suicide on 17 November 1941.

[116] The Oetztal wind tunnel was supposed to have an 8m diameter nozzle with operation up to high subsonic Mach numbers at atmospheric operation. This – the biggest wind tunnel in the world at the time – was designed for 76 MW driving power, provided directly by means of Pelton water turbines to two counter-rotating 15 m diameter axial-fan wheels. The hydraulic power was to be furnished by a flow of water of 18 m^3/s with a head of 530 m. The project of a nearly finished hydraulic power station with a barrier lake had been accordingly modified on short notice. The wind tunnel was intended for high-speed aerodynamic tests on complete aircraft models at significantly higher Re numbers than any other existing facilities, on full-scale component parts of aircraft and on full-scale nacelles with operational propulsion systems. According to different sources the tunnel was between 30 (Thiel) to 70 percent (Tsien) completed and planned to become operational in the second half of 1945. After the war it became situated in the French occupation zone in Austria. In addition, the company Dingler, which led the project till the end of the war, was located in Zweibrücken in the Saar area, part of the French zone. Consequently between December 1945 and June 1946 French authorities ordered Dingler to dismantle all movable Oetztal material and transport it in 13 freight trains together with company-stored materials to Modane-Avrieux in the French Alps, where the set-up was reconstructed and finished and has been operational since 1952 as the ONERA wind tunnel S1 MA. See Thiel, Oetztal and Tsien, Technical Intelligence Supplement, p. 96.

[117] See Tsien, Technical Intelligence Supplement, p. 99.

$-\mathcal{B}, 3,$

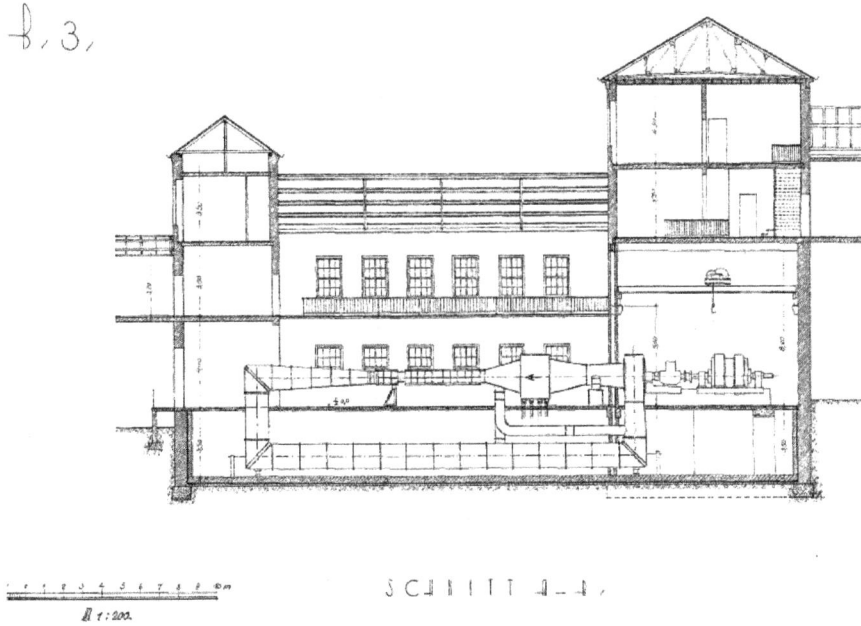

$\mathcal{B} 1 : 200.$ $\int C \mathcal{J} \mathcal{I} \mathcal{I} \mathcal{T} \mathcal{J} - \mathcal{I},$

Figure 4-37   *AVA-Reyershausen Mach 3 supersonic wind tunnel 'B 3', project drawing 1943* [118]

– Finally, a **0.4 × 0.4 m supersonic wind tunnel** for continuous testing up to Ma= 3 was planned to be installed in the finished building. The axial-flow compressor was driven by a BBC DC-motor, by means of a Ward-Leonard motor generator with a total power requirement of 6.5 MW. This latter unit was actually the one ordered by LFM from BBC in 1943 at the same time as the AVA-B3 equipment.

Three to four months before the planned delivery date the German side was hit by an unexpected, unannounced development in Switzerland. Swiss authorities had stopped the works for the AVA-B3 facility at BBC, but allowed (although slowed down) the LFM project[119]!

---

[118]   Figure source: B3 construction description 'Neubau eines Ueberschallwindkanals in Reyershausen' 17.3.43, 18 p. DLR Archive Goettingen, D, GOAR 2420

[119]   In a memo dated 4 April 1944 and classified as secret, Mr. Wiese from AVA administration noted about a meeting with Prof. Peters on 30 March 1944 at LFM Munich-Ottobrunn: '*As far as we know the preconditions for the delivery for the Goettingen facility from Switzerland (safe-guarding of valuta cash, cash transfer and allocation of the raw material contingents) have been fulfilled. Nevertheless, the facility has still not been authorised by Swiss authorities and consequently, BBC has not started the works.*' Further circumstances were not known, but the only logical interpretation of this situation was to assume that the Allied secret service and the Swiss authorities had coordinated the early termination of BBC activities for the AVA-B3 facility (by far more dangerous to the Allied side than LFM, due to the readiness of infrastructure for what were presumably 'V-3' aero model tests to be carried out by experienced personnel) and the delayed, but still contract-compliant delivery for the LFM (still without necessary infrastructure). This was the unclear situation about the AVA-B3 Mach 3 wind tunnel until very recently, as a surprising revelation occurred (if the conclusions are correct), – in direct quote, see Gorn, Harnessing the Genie, p. 25/26: '*...Drs. Tsien, Wattendorf and Dryden prepared to return to America around mid-June (1945), but not before arrangements had been made to ship to the US a great prize: a complete, uncrated Swiss-made wind tunnel, destined originally for Germany. Despite the high priority given at the time to personnel aboard cargo aircraft, Dryden insisted upon "immediate action" to transship this invaluable*

In the Brown Boveri Mitteilungen of January/February 1945[120], the following text can be found on page 56 under the header 6. Aerodynamik ...b) Ueberschall-Windkanäle with a picture of the Guidonia wind tunnel, Figure 4-35. *'We built the first supersonic wind tunnel for the Aerodynamic Institute of ETH Zurich, a further facility of the same kind has been produced by us for the Aerodynamic Test Establishment at Guidonia, Italy (ref. to picture). In the last year we delivered the axial compressor for a third facility with a driving power of 2,900 kW and in the test section a pressure ratio of approx. 3* (ref. to BBC figure no. 69865, here as Figure 4-38). *This pressure ratio allows for Mach numbers up to approx. 2.6 which corresponds to air velocities of more than 3,000 km/h.'*

Figure 4-38   *BBC axial compressor with transmission set for LFM supersonic wind tunnel, 1944,*
           $V_{0,max} = 48$ m /s, $PR_{max} \sim 3$, $Ma_{max} \sim 2.6$ ©ABB

---

equipment from its hangar at Orly Field (Paris) *to Wright Field, and late in the month a B-17 was made available for the purpose. The Swiss wind tunnel, as well as the interviews with the European scientists, the boxes of documents and laboratory equipment, and the regular technical intelligence reports assembled by the von Kármán group, all added luster to the SAG's (Scientific Advisory Group) reputation.'* See also Section 4.2.4.6.

In addition, the isolated air raid on Reyershausen (27 June 1944, 01:20 local time, 19 inhabitants killed and heavy destruction) in the vicinity of the B3 site appeared to be part of a special bombing action between November 1943 and August 1944 mainly directed at the V-3 (super gun) construction site at Mimoyecques in the northern Pas-de-Calais region of France. The importance of Allied counter-action against the V-3 also became evident in comparison to the general situation in Goettingen. After the very first air raid on Goettingen on 7 July 1944 (1 death), the situation only escalated during repetitive raids on and after 23/24 November 1944. The huge AVA test ground within Goettingen was apparently spared completely of any direct attacks. In his memoirs Th. von Kármán confirms this as a fact by claiming the Goettingen KWG/AVA facilities had intentionally not been listed as Allied bombing targets. See Karman, Wirbelstrasse, p. 335. However, this assessment must be put into perspective, as British bombing records (see Middlebrook, The Bomber Command War Diaries) actually name Goettingen railway workshops as the target of 35 Mosquitos at the time of the Reyershausen raid: '...*The raid was carried out from a medium altitude – 4,000 to 10,000 ft – but the marker aircraft experienced difficulties in locating the target and the bombing was scattered. ...*' So, a coincidental navigation error of 15 km might explain what would otherwise remain a mystery.

[120]   See BBM, Rueckblick... im Jahre 1944, 1945

[121]   Picture source: BBM, Rueckblick... im Jahre 1946, 1947, Figure 76

Reference point 10) marks the LFM compressor in Figures 4-31 and 4-32. The 19 compressor stages used reflect the technical standard at that time with a typical stage pressure ratio of approx. 1.07. An increase in the compressor entry diameter of up to 1 m while keeping the speed level at 5,000 rpm led to the lowest stage loading coefficient Ψ_{St} in the evaluation period from 1920 to 1960. This compressor shipment has not only been documented by BBM, but by two other independent sources. According to the von Kármán inspection team who were on site at Munich-Ottobrunn in early May 1945: *'The axial compressor system had just been delivered to the site and had not as yet been uncrated. The rest of the tunnel had not been assembled.'*

No further traces exist of the LFM supersonic test facility at Munich-Ottobrunn except for a principal sketch, taken during an on-site visit by the von Kármán inspection team in early May, reproduced here as Figure 4-39.

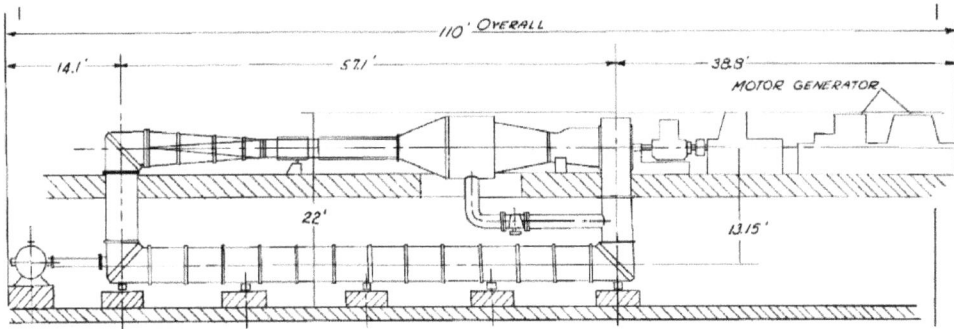

Figure 4-39   *LFM supersonic wind tunnel, principal sketch from US AAF scientific advisory group 1945* [122]

4.2.4.4   The Mach 4 Wind Tunnel at LRBA Vernon, France, 1947

Only 2 years later BBC picture no. 69865 was reprinted in BBM January – March 1947[123], this time under the header 5. Aerodynamik... b) Ueberschall-Windkanaele with the accompanying text: *'Last year one of our concessionaires[124] received the turn-key order for a supersonic wind tunnel for high Mach numbers up to Ma = 4; the entire machinery has been ordered from us'*. This BBM quotation was illustrated with a *'scheme of a supersonic wind tunnel for*

---

[122] See Tsien, Technical Intelligence, Figure 42

[123] BBM, Rückblick... im Jahre 1946, 1947, p. 41

[124] The author owes numerous details about the erection of the C4 facility to Christian Vanpouille. According to a preserved report by the consulting engineer in charge of the project, Émile Poulet, the C4 assembly was within the responsibility of S.O.C.E.M.A. (SOciété de Constructions et d'Équipements Mécaniques pour l'Aviation, Paris, a daughter company of C.E.M. and as such owned by BBC Switzerland). In November 1948 material from BBC had been intermediately stored at Mulhouse. Poulet writes: *'It is regrettable that the second rotor had not been intermediately stored like the first one in the building S3 (where C4 was to be erected); SOCEMA asks for thorough heating of rotor #2 before assembly'*. Chr. Vanpouille is convinced that DEFA (Direction des Études et Fabrication d'Armement) ordered TWO compressors for France from BBC which were to be delivered to Vernon in 1948. In December 1948, the foundations for high- and low-pressure compressors were cast. In December 1949 all other, intermediately stored BBC materials arrived directly at Vernon, except for the two (compressor) stators which were to be shipped either via Paris or via Rouen. In May 1950 *'the low pressure compressor installation was finished, the high pressure compressor installation well advanced, the same applies to the assembly of the air ducting from BBC. The manufacturing of water piping at SOCEMA Corbeham has not started yet and will delay the whole installation'*. In July 1950 all machinery was installed except for the still delayed water circuits.

MA = 4. The machinery for this wind tunnel is manufactured in our workshops at present. The *combined power of both multi-stage axial blowers together is 13,000 kW.'* The original 'scheme' is reproduced here as Figure 4-40. Given the complete survey of all European wind tunnels[125], it was easy to identify the sketched facility as the unique LRBA[126] C4 Trisonic wind tunnel at Vernon, France – 75 km north-west of Paris.

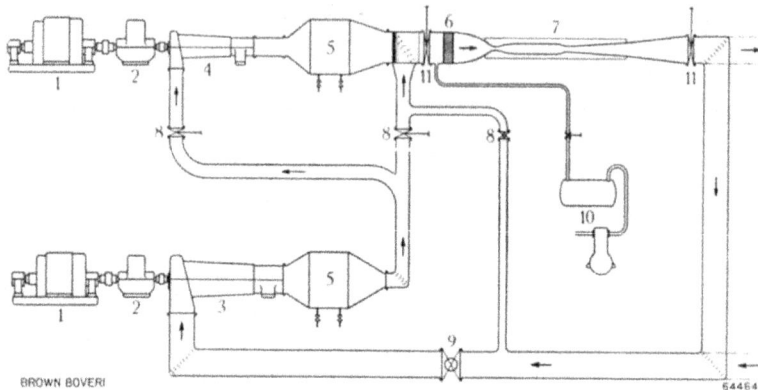

Figure 4-40   *Scheme of LRBA C4 supersonic wind tunnel*[127]*, Vernon, F; 1 – DC Motors 2 × 6,500 kW,* *2 – gearboxes, 3 – lp compressor, 4 – hp compressor, 5 – air coolers, 6 – flow straightener,* *7 – 0.4 × 0.4 m test section, 8,11 – shut-off valves, 9 – control valve, 10 – dry air tank with* *compressor   ©ABB*

Brown Boveri Review[128] announced the completion of the facility but still did not identify the actual location: *'The most interesting event ... during the period under review* (1951) *was* *the commissioning of a supersonic wind tunnel for Mach numbers up to 4, used for ballistic* *investigations* (ref. to what follows here as Figure 4-41)'. The two multi-stage axial-flow blowers with a total power of 13,000 kW were connected in series as high- and low-pressure stages for operation at high Mach numbers with an air mass flow of 35.5 kg/s at an absolute pressure of 10.5 bar; for lower Mach numbers, however, they could be run separately. The blowers were driven through gearing (visible in Figure 4-38) by induction motors, speed controlled from 75 to 100 percent by liquid resistances.

C4's realised top speed was said to be Ma = 4.4. Plans to install a third compressor-motor-unit for Ma_{max} = 6.6 appeared with the German team from Kochel (see Figure 4-45), but had to be

[125] See Kingsbury, World Directory

[126] The LRBA Laboratoire de Recherches Ballistiques et Aérodynamiques was the leading French (now European) missile development centre located at Vernon (Eure), F on a DGA Direction Générale de l'Armement site. Missile testing first took place in the French Sahara and was shifted afterwards to Kourou in French Guyana. The centre was opened in 1946/7 to host a group of 28 German V-2 scientists, with the trisonic wind tunnel C4 as their decisive working tool for the subsonic, transonic and supersonic regime up to Ma = 4. In August 1946 this group sketched out the development steps that would lead to the ARIANE rocket of the 1980s. After first exploiting available V-2 hardware the team focused on what was called Super V-2 between 1946 and 1948, an intermediate range ballistic missile which was a developed version of the original A 9, Figure 4-44. Written off as too ambitious, it was followed by the VERONIQUE (VERnon électrONIQUE) which started development in March 1949, remaining in production till 1975, and thereafter by a whole series of missile development projects until the ARIANE IV.

[127] Figure source: BBM, Rueckblick... im Jahre 1946, 1947, Figure 75

[128] See BBR, Power utilization, 1952, p. 85

given up due to power restrictions. EDF Électricité de France had already had to build a 90 kV line for 200 A (for a maximum power supply of 18 MW) for the C4 supersonic wind tunnel, which still meant that C4 could only be operated during night shifts. An intermittently operated 'C2' wind tunnel for Ma = 2 was built in addition to C4, and both were put into operation on 1 December 1952. C4 was not only used for aerodynamic missile and shell testing, but in 1964 the variable geometry air intakes of CONCORDE were also investigated there.[129]

Figure 4-41   *Coupled lp/hp axial compressor station for LRBA C4 supersonic wind tunnel, Vernon, F 1951*[130]
©ABB

4.2.4.5   The Hypersonic Wind Tunnel for Kochel/Tullahoma 1943–1960

German V-weapons V-1 and V-2 were developed at the Army Test Facility (HVA) Peenemuende on the Baltic coast. The origin of missile use for military purposes dates back to the 1920s, when existing Versailles regulations limited the development of long-range guns. The test site itself, on a secluded peninsula in the north of the island Usedom, was selected by Wernher von Braun (1912–1977) in the 1930s, who knew the area through family connections[131]. A British air raid on Peenemuende in the night of 17/18 August 1943[132] forced initial V-2 missile production to relocate to the Harz mountains and the AI Aerodynamic Institute to a newly founded, neutrally designated WVA[133] WasserbauVersuchsAnstalt G.m.b.H. (Hy-

---

[129]   See Trost, Auf dem Weg ins All

[130]   Picture source: BBR, Power utilization, 1952, p. 85

[131]   After an unsuccessful search on the island of Ruegen, von Braun's mother in fact suggested the Peenemuende area during WvB Christmas break 1935, as it was known to her relatives as a remote fishing and hunting area.

[132]   Due to expected high air defence, the 'Hydra' air raid at Peenemuende (with more than 500 bombers) was camouflaged by another, immediately foregoing 500 bomber raid 'Whitebait' on the same route from the north to Berlin. See Fetzer, Windkanalanlagen

[133]   According to the AI director Rudolph Hermann, construction works for the hypersonic test facility at Kochel had already started in 1942, see www.wolaa.org/files/Summer 2007 OHS – Windtunnels.pdf (Part 2). The designation WVA was chosen to increase secrecy and to confuse intentionally. Since 1926 there had been a

draulic Engineering Test Facility) at Lake Kochel, some 60 km south of Munich in Upper Bavaria. The transfer goods comprised 300 railway cars, including two 0.4×0.4m intermittently operated, supersonic (Mach 4) wind tunnels and one small hypersonic experimental tunnel. An 18×18 cm continuous flow tunnel was dropped at LFA Braunschweig. In April 1944 the last measurements were taken at Peenemuende and already by 15 November 1944 the first of the 0.4×0.4 m tunnels was back in operation at Kochel.[134]

Plans for a continuously operating hypersonic wind tunnel for Ma=7...10 had already emerged by the late 1930s[135] together with plans for long range missiles (5,000 km two-stage 'Amerika-Rakete', Figure 4-44), presumably with the involvement of BBC Mannheim which was known already to be specialised in high-altitude engine test facilities.

From 1939 the 'workhorse' wind tunnels at Peenemuende were two 0.4×0.4m Mach 4 vacuum wind tunnels, meaning that ambient air was sucked for approx. 20 s test time through a dehumidifier and the following test section into a vacuum tank, which could be evacuated relatively cheaply by a low power vacuum pump set – ready for the next 'shot' within 3 min. The importance of dehumidification[136] to prevent condensation shocks in the test section of high speed wind tunnels had been discussed by Wieselsberger[137], Prandtl and Busemann at the Volta Convention in 1935 and in due course led to the fact that no high-speed facility was installed without such air dryer equipment.

After the first 0.4×0.4 m intermittent supersonic wind tunnel was transferred from Peene-muende to Kochel[138], Siegfried Erdmann, one of Herrmann's assistants from TH Aachen,

---

'Forschungsinstitut für Wasserbau und Wasserkraft e.V.' at the south-west corner of nearby Walchensee (to-day Versuchsanstalt Obernach and part of TU Munich), founded by Oskar von Miller with support from Lud-wig Prandtl, Figure 4-43. The FWW had received international attention between 1930 and 1934 for carrying out, amongst others, regulation tests of Hwangho river in China.

[134] See www.wolaa.org/files/Summer 2007 OHS – Windtunnels.pdf (Part 1)

[135] Typical of German military planning, there was almost no coordination between the various armed services and scientific institutions. There was also none between the LFM supersonic wind tunnel project (RLM's ini-tiative) and the HVA plans at Kochel; presumably, the participating suppliers sometimes had a better over-view of the problem in order to coordinate their own limited resources and solve priority conflicts. In this re-spect BBC in Mannheim, Germany and Baden, Switzerland were certainly used by both parties as valuable, highly competent suppliers, especially for wind tunnel equipment.

[136] Consequently, in 1945, besides the BBC turbomachinery equipment the silica gel dehumidification unit for the hypersonic wind tunnel was the most highly regarded part of the US booty at Kochel. For high speed testing the target value for the remaining specific humidity was 0.5 g/kg, compared to an average middle-European climate value of 12 g/kg. It was rumoured that the higher specific humidity values at the new tunnel site in Tennessee, USA represented considerable adaption challenges.

[137] Carl Wieselsberger (1887–1941) finished his PhD studies at TH Munich in 1912. Afterwards he came to the Aerodynamische Versuchsanstalt (AVA, Model Test Establishment), Goettingen as an assistant to Prandtl, where he carried out decisive trip-wire tests on laminar-turbulent boundary layer transition of spheres. In 1922 he left Goettingen following the invitation of the Imperial Japanese government to build up the aeronau-tical research institute at the University of Tokyo and to complete several wind tunnel projects. In 1929 he wrote an important paper on the impact of engine installations on the change of aerodynamic forces on a wing. In 1930 he became head of the Institute of Aerodynamics at TH Aachen as successor to Th. von Kármán. Under his supervision the first supersonic wind tunnel with a variable Mach number was built by R. Hermann, later technical head of the Kochel test facility. In 1940 he received additional responsibility for the new LFM establishment with facilities at Munich-Ottobrunn and the then largest wind tunnel project in the world at Oetztal in Austria. His early death in 1941 meant Adolf Baeumker succeeded him at the top of LFM administration. For more details, see Hirschel et al., Aeronautical Research in Germany, p. 179

[138] Kochel, today a community of 4,000 inhabitants, is located in Upper Bavaria, 60 km south of Munich. The deci-sion to use Kochel was based on the **hydraulic** power due to the 200 m altitude difference between upper Lake

modified the other tunnel on site by inserting a new wooden nozzle arrangement which allowed hypersonic Schlieren pictures to be taken for the first time ever, Figure 4-42.[139] According to the general definition the supersonic Mach number range comprises Ma=1.2 ... 5.[140] There are large differences in aerodynamic design of aircraft designed to fly at supersonic speeds because of radical differences in the behaviour of flows above Mach 1. Sharp edges, thin aerofoil-sections, and movable tailplane/canards are typical. In order to maintain low-speed handling, modern combat aircraft must compromise: 'true' supersonic designs of the past include the Lockheed F-104 Starfighter and the BAC/Aérospatiale Concorde. Hypersonic speeds are characterised by Ma = 5 ... 10 flows or in metric terms 6,150 ... 12,300 km/h, respectively 1,710 ... 3,415 m/s. Aircraft designs are highly integrated (due to domination of interference effects, non-linear behaviour is dominant which means superposition of results for separate components is invalid) with small wings and a cooled nickel-titanium skin. An early example is the NAA X-15 experimental aircraft, see Figure 4-47.

Figure 4-42   *First Schlieren picture of hypersonic flow Ma = 8.83, Siegfried Erdmann, HVA Peenemuende, 17 April 1944*[141]

Walchensee and lower Kochelsee, both 1.6 km apart. This was exploited for energy production by the Walchensee hydraulic power station planned by Oskar von Miller (1855–1934), put in operation in 1924 after 6 years construction with 124 MW installed power. At present the annual energy production is 300 GWh. The Walchensee is fed from the River Isar and for the WVA plant the Rissbach from Austria was planned to be additionally directed through a 7 km tunnel to the Walchensee reservoir, which was finally completed in 1950.

[139] See Wegener, Peenemuende wind tunnels, describing this solely as a scientific achievement, but it can be assumed that these tests had practical long-range applications in mind, so long as the hypersonic tunnels at Kochel were not ready for operation.

[140] This split between 'supersonic' and 'hypersonic' velocities follows the US terminology after WW II. The corresponding words in German during WW II were 'Ueberschall' and 'Superschall', so the Kochel Mach 7...10 facility was officially called 'Superschall-Windkanal der Wasserbau-Versuchs-Anstalt Kochelsee G.m.b.H.'

[141] Picture courtesy of Beate, Thekla & Verena Erdmann, August 2012; special thanks to ir. Willem J. Bannink, Dept. of Aerospace Eng., TU Delft, NL. The date is from the Wegener files at Deutsches Museum, Munich. According to a Peenemuende organisation chart of 1940, then still Dipl.-Ing. Siegfried F. Erdmann (1916–2002) led within the Aerodynamic Institute (Dr. R. Hermann) the 'Versuchsgruppe II' (Test Group II) – with co-workers Dr. Peter P. Wegener (his successor at WVA Kochel) and Dr. Pascual Jordan (1902–1980), a theoretical physicist who made significant, pioneering contributions to early quantum field theory. http://de.wikipedia.org/wiki/Siegfried_F._Erdmann   http://en.wikipedia.org/wiki/Pascual_Jordan

The demand for continuously operating super- and hypersonic wind tunnels arose for various reasons. One known reason was to learn about the thermal boundary layer conditions in the nose zone of a Mach 4 ... 5 rocket such as the German anti-aircraft 'Waterfall' A-4b, Figure 4-44. The information was desperately required for the development of a sufficiently sensitive infra-red detector based on measurement conditions in thermal equilibrium. Test requirements for long-range rocket projects with Mach 10 re-entry conditions came out on top (which became nuclear 'ICBMs' Intercontinental Ballistic Missiles after the war). In general, the mentioned non-linearity effects left continuous testing beyond Mach 5 as the only reliable option for hypersonic rocket/vehicle development.

Figure 4-43 illustrates the Walchensee power station on the Kochelsee, 197 m below the Walchensee reservoir. It is connected by 6 tubes of 450 m in length[142] which direct 4 tubes of water to 4×18 MW Francis turbines for a 3 phase current, and that of the remaining 2 tubes to 4×13 MW Pelton turbines to generate single phase railway traction power. The WVA sites are shown in Figure 4-43; recent research has revealed with some likelihood that there were actually **two** WVA sites, the one at Kochel-Herrenkreuth WVA I, where the ex-Peenemuende 0.4×0.4 m Mach 4 tunnel being re-erected and intermittently operated here[143] and the other one, WVA II, at Kochel-Altjoch where the new 57 MW hydro-powered Mach 7...10 hypersonic wind tunnel was planned for continuous operation on a much bigger area.[144]

As mentioned before, the Aerodynamics Institute (formerly at Peenemuende) was moved to Kochel in the latter part of 1943 and began operating again in 1944. After Kochel was reached by US troops on 1 May 1945, the WVA establishment was visited by Drs. Hugh L. Dryden, Frank L. Wattendorf and Hsue-Shen Tsien of the so-called 'von Kármán Mission, AAF Scientific Advisory Group to General Arnold' on 15 June.[145] Most of the following information, especially about the WVA 1×1 m Mach 7...10 hypersonic wind tunnel project with its extraordinary BBC compressor equipment, is from this travel report[146] and related WVA literature, which was found and recovered by S. Klapdor at various US archives in 2001.[147]

---

[142] The author thanks Prof. Uwe Gampe, TU Dresden for rare drawings of the Walchensee power station piping system. The tube's inner diameter decreases in the first 60 m from a maximum value of 2.27 m at entry near the 'Wasserschloss' down to 1.85 m, while the tube wall thickness correspondingly increases from 15 to 30 mm, max. gradient in the mid altitude section is 40°50'21".

[143] In 1945 this tunnel was transferred to NOL Naval Ordnance Laboratories, White Oak, MD USA and used there until approximately 1992.

[144] To allow for a max. of 6h continuous operation, the water supply to the Walchensee was increased amongst other by re-directing the Rissbach from the Austrian border in an impressive tunnel/dive culvert civil engineering construction which was finally completed between 1947 and 1950. The WVA I site covered less than 5 ha, expropriation under way for WVA II in 1944 addressed 16 ha (40 acres).

[145] General Henry H. 'Hap' Arnold (1886–1950), who received by act of US Congress the permanent five-star rank as general of the Air Force 3 years after his official retirement in 1946, the first and unique such commission ever granted.

[146] See Tsien, Technical Intelligence Supplement; Klapdor, Technologietransfer Deutschland – USA and – see Zwicky, War Research

[147] See Klapdor, Der Technologietransfer Deutschland – USA. In December 2011, S. Klapdor kindly forwarded a large amount of BBC-related documents on the Kochel facility to the author.

Figure 4-43   *Walchensee power station with WVA wind tunnel sites at Kochel; FWW Obernach also marked   © 2012, (l) E.ON Wasserkraft GmbH, (r) Google and GeoBasis-DE/BKG*

At the US inspection in May/June 1945 the hypersonic wind tunnel (order) was approx. 60–80 percent finished[148]. This tunnel had previously unheard-of dimensions of 1 × 1 m[149] and required 57 MW hydraulic drive power from the nearby Walchensee reservoir. Near-term continuous operation at a Mach number of 7.0 was planned with the possibility of extension to Mach 10. The US team immediately recognised that the purpose of the tunnel was the development of long-range missiles, such as the transatlantic version of the V-2, officially designated in German planning which had existed since the late 1930s in various 2-stage combinations. Figure 4-44 clearly marks the significant increase in size in comparison to the A-4/V-2 already in production and use then.

After the war US specialists estimated the actual cost of the Kochel facility rather close to 50 million RM. From a BBC point of view, the following description of the turbomachinery section of the hypersonic wind tunnel equipment is of special interest. Here is the original wording from the 'travel report'[150]:

---

[148] The relevant US report (see Tsien, Technical Intelligence Supplement) mentions 25 million RM project costs, of which 15 MRM was invested. The project started in 1942 with expected completion by autumn 1945, but construction works were actually stopped prematurely on 29 September 1944. Though the general location at the southern outskirts of Kochel has been identified, relics such as an approx. 90 × 50 m complex consisting of a machine hall, measuring hall, dehumidifier and laboratory buildings close to the Kochelsee have yet not been confirmed unambiguously.

[149] According to the actual plans, the hypersonic tunnel had four test sections in parallel, switchable to the turbomachinery field. The tunnel could be run as a direct flow type, in which case the entry pressure was roughly atmospheric, or could be converted to a closed circuit, return flow type with inlet pressure variation of between1 and 4 bar.

[150] See Tsien, Technical Intelligence Supplement, pp. 99 and 104.

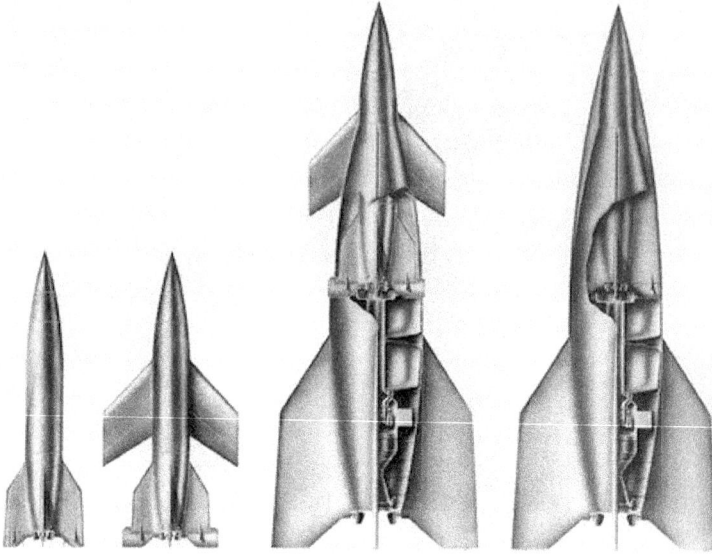

Figure 4-44   *Size comparison of A-4 (V-2)/winged A-4b ('Waterfall') rockets and projected 2-stage, long-range rockets (A-10+A-4b)/ (A-10+A-9)* [151]

'The power was to be furnished by seven compressor aggregates, with each of the first four aggregates driven by a 12,000-kw hydraulic turbine and each of the last three aggregates driven by an electric motor, the electric power for which was to be furnished by a 3,000-kw hydraulic turbine. The compressor system has an inlet suction quantity of 1,200,000 cu m/hr at a vacuum of 1.4 mm of mercury.' and *'The turbines were to be furnished by Voith Heidenheim and the vacuum pumps [sic]* [152] *by Brown Boveri of Mannheim'.*

The following Figure 4-45 is a principle excerpt of the detailed floor plan of the Kochel hypersonic 'Wind tunnel A' with 1×1 m test section [153].

---

[151] Picture source: http://www.project1947.com/gfb/a-9.htm Work on 2-stage A-10/A-9 combinations was officially prohibited after 1943, when A-4 production took priority, but certain A-4b tests were considered as preparatory steps. The wing structures especially still required thorough wind tunnel testing for atmospheric re-entry behaviour. Known approximate data for A-10/A-9 rockets assume a gross mass of 85–95 t of which 70 percent were fuel, payload 1 t, height 41 m, diameter 4.12 m, range 5,500 km, altitude flight speed < Mach 10; corresponding underground production sites were apparently under construction till 1945 at Ebensee/Traunsee, Austria. After the war US and Soviet improvements in rocket structures and engine efficiencies made it possible to design pure ICBM vehicles with cut-off velocities that could not be intercepted (twice as high as those of the A-10/A-9) and 10,000 km ranges.

[152] Should read 'compressors' instead of 'vacuum pumps'. The misnomer could be explained by the fact that the compressor stages IV-VI (Figure 4-45) were also to be used as boost pumps for an upgrade of the intermittently operating Mach 4, 0.4×0.4 m blow-down tunnel, later reconstructed at the NOL Navy Ordnance Lab at White Oak, MD, USA. The basic vacuum pumps, actually rotating piston compressors, came from Demag, allowing for approx. 15–20 sec test time at Mach 4. With additional boost pumps/compressors from BBC and an additional storage cylinder the test time was increased to approx. 4 min.

[153] An accompanying facility description (dated 25 June 1945, unsigned, by Dr. Rudolph Hermann, WVA director and (Dr.) Dipl.-Ing. Gerhard Eber) claims that the installed turbomachinery and dryer capacity would allow for a 1.4×1.4 m test section at Mach 10 in a further stage of facility expansion. Both Hermann and Eber moved

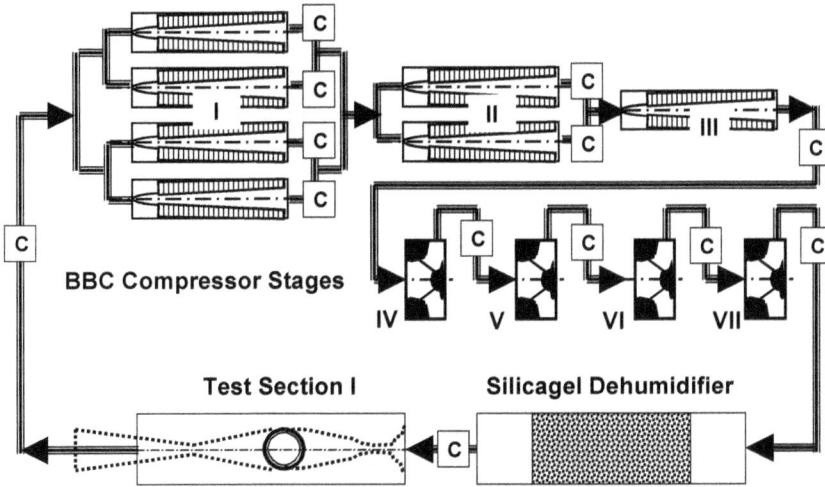

Figure 4-45   *Hypersonic wind tunnel project at WVA Kochel, D: BBC compressor field, nozzle PR=875*,*
57 MW drive power, 11-stage axial and 4-stage radial compressors, C – air coolers/heaters,
** Ref. stations: nozzle entry p < 4 bar/ test section p > 0.0046 bar*

The BBC compressor field comprised – Stage I 4 × 11-stage axial compressors Type VA 1511, PR = 2.25, 3,000 rpm, – Stage II 2× Type VA 1511 PR = 2.25, 3,000 rpm – Stage III Type VA 1511 PR = 2.25, 3,000 rpm – Stage IV 4-stage centrifugal compressor Type V 1604 PR = 3.25, 3,000 rpm, – Stage V 4-stage centrifugal compressor Type V 1104p PR = 3.17, 4,340 rpm, – Stage VI 4-stage centrifugal compressor Type V 704p[154] PR = 3.08, 6,390 rpm, – Stage VII 4-stage centrifugal compressor Type V 404 PR=2.42, 9,400 rpm. The Stage I compressor volume flow was 85 m^3/s – with a maximum end pressure of 1.4 bar abs. at 25 °C intake temperature. All stages had BBC gearboxes, Stage I 2× Type NS 180/2 Ri, Stage II and III/IV 1× Type NS 180/2 Ri, gear ratio 300/3000 (1:10) connecting each Voith 12 MW twin impulse turbine with 4 nozzles, 7.3–7.8 m^3/s water flow, with a pair of axial compressors.[155] Stage V was driven by a BBC 3.8 MW three-phase current slip-ring induction motor, 985 rpm, – Stage VI correspondingly by a 1.3 MW motor, 1,475 rpm and Stage VII by a 0.54 MW motor, 1,470 rpm.

In addition to the silica-gel drying system, with an inlet suction quantity of 780,000 kg of air per hour, an average inlet humidity of 12 g/kg and a discharge humidity of 0.5 g/kg, there were air coolers installed at the exit of each compressor and an air heater installed upstream of the nozzle to prevent partial air liquefaction. The Schlieren apparatus, mount-

---

to the USA after 1945. Hermann (1904–1991) was an USAF consultant at Wright Field until 1951, thereafter Professor for Aeronautical Engineering at the University of Minnesota and after 1962 became the first Director of the Research Institute at the University of Alabama in Huntsville (UAH), a position he held until retirement in 1970. Eber is said to have belonged to the scientific staff of NOL White Oak, MD, USA since 1946/7 as one of a total of 12 German rocket scientists.

[154] As stated before, Stages IV-VI were also used as boost pumps for the Peenemuende/Kochel Mach 4 blowdown tunnel. Status 15 June 1945, one compressor Type V 1104 and one Type V 1604 were reported in stock at Kochel railway station (together with BBC gearboxes and electric motors), in addition one Type V 1104 was on its way to Kochel. Presumably one V 704, reloaded at Goettingen in March 1945, was lost in transit. See Klapdor, Technologietransfer Deutschland-USA, p. 50.

[155] Figure 4-38 gives an impression of one such branch of a (twin) compressor-gear-train.

ed on a rail system to be moved along the tunnel axis, utilised two mirrors, each 1.2 m in diameter with a focal length of 9.25 m. Throats were fixed and interchangeable and the diffuser was adjustable. Four 6 m long test sections and three air filters were planned. Test sections and associated parts such as the diffuser, three component balances and Schlieren equipment, except for the optical parts from Zeiss, had been supplied by the wind tunnel-specialised company Dingler, Zweibrücken, D. Part of the described test section equipment (Schlieren and rail system, adjustable diffuser etc.) can be identified in the picture of a tunnel model which had been prepared for the decision and planning period after 1939, presumably still at Peenemuende, Figure 4-46.

Figure 4-46   *WVA Kochel hypersonic wind tunnel, model of test section ~1942* [156]

Fritz Zwicky of Switzerland, who belonged both to a CIOS[157] as well as to the von Kármán inspection team, was at Kochel the longest. He ordered the German specialists to write documentation reports on individual subjects in addition to the comprehensive documentation from Peenemuende. This had been brought in the meantime to the V-2 underground production site at Mittelbau Dora near Nordhausen, where it was successfully recovered and transferred to Kochel in spring 1945.[158] The fate of the facilities was undecided for some time. 'Blowing up' the whole complex appeared to be an option until Zwicky and von Kármán succeeded with their suggestion to remove all the high-speed wind tunnel equipment and documentation to become as the nucleus of an all-new USAF test center on the site of POW Camp Forrest (which became AEDC Arnold Engineering Development Center) in Tullahoma, Tn., USA.[159] The accompanying German personnel had to wait until these decisions had been made before they received US contracts; in the meantime a group of 28–30 Kochel specialists led by Dr. Herbert Graf followed French offers and moved on, first for

---

[156]   Picture source: UAH, WVA Archive No. 66/182, Attachment 3

[157]   CIOS – Combined Intelligence Objectives Sub-Committee; other common abbreviations in the context of Allied post-war reporting are BIOS – British Intelligence Objectives Sub-Committee and FIAT – Field Information Agency Technical (United States Group Control for Germany)

[158]   See Wegener, Peenemuende wind tunnels

[159]   The 0.4×0.4 m intermittently operated Mach 4 Peenemuende/Kochel supersonic tunnel along with other equipment was brought directly to NOL Naval Ordnance Laboratories, White Oak, Md., USA.

a few months to Emmendingen near Freiburg im Breisgau, before continuing to the emerging French missile development centre at Vernon in 1947. The US-directed group moved as part of 'Operation Paperclip' from Kochel via Landshut to the shipping port Bremerhaven. On 11 January 1946 the Kochel specialists and the removed equipment[160] left Germany.

Figure 4-47   *1/18 scale model of NAA X-15 experimental aircraft in AEDC's 1×1 m hypersonic wind-
'Tunnel A', ~1958, Source: AEDC*

In a 1971 book (see Lasby, Project Paperclip) the value of the transferred expertise was estimated with 1950 as a reference year:

– Supersonic wind tunnel research = savings of five years.

The hypersonic 'Tunnel A' in AEDC's von Kármán Gasdynamics Facility has been regularly used for all US high speed tests, from the space shuttle in the 1950s up until today. As an example, Figure 4-47 shows a model of the Mach 7 experimental X-15[161] in the 1×1 m test section which underwent airworthiness and stability testing in the late 1950s. There are hints

---

[160] At AEDC Tullahoma, Tn., USA the complete 'Herbitus' engine altitude test facility from the BMW site at Milbertshofen, north of Munich centre, was re-erected and the 1×1 m hypersonic tunnel completed (unknown if with original hardware) on the basis of existing, though adapted German planning materials under the designation 'Tunnel A' (the old Peenemuende project name). One of the decisive differences was unlimited use of e-motor technology, so hydraulic turbine drives were no longer required for the compressors. However, R. Hermann later highlighted the relative independence of the water drive concept, while e-installations at Tullahoma quickly required operation scheduling for the various test facilities. See www.wolaa.org/files/Summer 2007 OHS – Windtunnels.pdf (Part 2). Together with the famous 'Herbitus' plant, several altitude test facilities were delivered by BBC Mannheim during WW II, the story of which will be covered in the future book on BBC's aero engine activities.

[161] The X-15 was based on a concept study by Walter Dornberger (1895–1980), former military head of the Peenemuende facilities, written for the NACA for a hypersonic research aircraft. North American Aviation was contracted for the airframe in November 1955, and Reaction Motors were contracted to build the engines in 1956; in total three aircraft were built. The main rocket engine for later versions was a XLR-99 rocket engine delivering 57,000 lb$_f$ (250 kN) at sea level, and 70,000 lb$_f$ (310 kN) at peak altitude. The X-15 was designed to be carried aloft under the wing of a B-52. The fastest flight was #188 on 3 October 1967 at an altitude of 58 km with 7,273 km/h (Mach 6.7), piloted by Pete Knight.

that BBC Baden and/or BBC's licensee Allis-Chalmers participated in the erection of the huge AEDC test site at Tullahoma in the 1950s and possibly even the 'Tunnel A' compressor drives. Figure 4-48 was reproduced in the 1954 Brown Boveri Review with the accompanying text '*Maximum output 510,000 m³/h with a pressure ratio of 3.2:1. The air flows is axially, the discharge connection is located horizontally on the right-hand side*', but without any further clarification of the delivery target. The pressure ratio comes close to the estimation of the lp front row compressors, Figure 4-45, and the compressor size correlates nicely with the following interview sequence, presumably between von Kármán and Seippel.

Figure 4-48   '*Casting of upper part of an axial-compressor casing for a wind tunnel, max. output 510,000 m³/h, PR~3' BBR 1954*[162]   ©*ABB*

4.2.4.6   The von Kármán Team at BBC Baden CH, June 1945

At the end[163] of an in-depth inspection tour of 'mysterious', high-technology German research establishments and corresponding test facilities from 1 May till 19 June 1945 the von Kármán

---

[162] Picture source: BBR, The axial-flow compressor, 1954, Figure 10

[163] An assumption, though based on natural priorities and some credible information on the travel route of the von Kármán team, sometimes also designated as LUSTY (LUftwaffe Secret TechnologY) team. According to these claims, H.S. Tsien found himself interviewing Wernher von Braun and other members of the V-2 Rocket Team on 5 May in Kochel. Aviation Week named Tsien its 'Person of the year 2007' and made reference to that interrogation: '*No one then knew that the father of the future US space program was being quizzed by the father of the future Chinese space program.*'
W. von Braun, together with his English-speaking brother Magnus, General Dornberger et al., left Peenemuende in time to move to what would become the US occupation zone. They contacted US troops on 2 May near Oberjoch/Hindelang, (others claim it was 35 km further east at Reutte on the German/Austrian border) from where they moved to Oberammergau, located only 35 km west of Kochel. The von Kármán team presumably moved on from there to AVA Goettingen and LFA Braunschweig-Völkenrode where the revolutionary swept wing concept and related wind tunnel measurements were detected. Despite attempts to destroy the most secret documents the team is said to have recovered 3 million pages (15 t) of technical papers, laying the ground for US post-war jet aircraft development. As outlined before, the swept wing concept was first made public by Alfred Busemann in 1935 at the Volta Convention and was advanced by others such as Alexander

team came to Brown Boveri in Baden, Switzerland. The summary chapter of the Technical Intelligence Supplement of the von Kármán team has been entitled by the author Frank L. Wattendorf: 'Reports on Selected Topics of German and Swiss **Aeronautical** (sic) Developments' and the specific BBC section on pp. 114–117 is called 'Conference at Brown Boveri, Baden Switzerland'. Though the focus was announced as aeronautical issues, these were only indirectly addressed in the context of BBC's involvement in supersonic wind tunnels.

In hindsight it can be speculated, with some justification, that the von Kármán team was not able to reveal all the complex interdependencies and connections amongst the at times chaotic German organisations at such short notice. The Swiss BBC management neither had a special reason to add clarity in this respect nor can be assumed to have had transparency about the other side of the border themselves.

Taking the example of 'wind tunnels', the traveling team certainly received information about general BBC origins several times, but it remains doubtful if the different compressor sourcing at LFM Munich-Ottobrunn from BBC Baden and at WVA Kochel from BBC Mannheim was fully understood. This certainly applies to the HVA/WVA hypersonic wind tunnel project at Kochel, whose existence was perfectly hidden, and there are no indications so far that the RLM (Luftwaffe, GAF), responsible for the huge LFM projects in Southern Germany and Austria, were fully aware of what was organised in the same field in their immediate vicinity by German Army organisations (Wehrmacht, HVA). It goes without saying that any joint use of these unique facilities amongst the various branches of the German armed forces was completely out of the question.

Taking the example of 'compressors for jet engines', briefings at BMW and Jumo proudly revealed extraordinarily high performance values for certain BBC products (for example the BMW–003 C/D versions). But even if one knew the origins of these to be BBC Mannheim and Hermann Reuter's team, it was well known that Reuter and his team were excellent mechanical designers (but not developers of compressor aerodynamic design theories). Speculation most likely continued, certainly regarding Claude Seippel's Baden team who of course knew that they had not been involved, but was it worth an official denial?[164]

Another somewhat psychological aspect should not be underestimated. The BBC team probably had no specific knowledge in advance about the actual purpose of the visit.[165] Switzerland had of course been neutral during the war, but contributions from the Swiss BBC to the German side were obvious for the US team. On the other hand the independence of BBC Mannheim from the Baden headquarters and the growing rivalry between both sites that grew during the course of the war was certainly unknown to foreign observers. According to the minutes the 'Conference at Brown Boveri Baden Switzerland' was attended by Drs. <names>, as written in the minutes [and corrections]:

Lippisch (1894–1976) during the war. On 28 May Lippisch briefed the LUSTY team in Paris on his work on the Me-163 tail-less rocket plane. Consequently, the date of the von Kármán team's visit to Switzerland can be narrowed down to the first days of June 1945 (back at Paris on 8 June 1945). See Gorn, Harnessing the Genie, p. 25.

[164] As mentioned before, details of BBC's aero product are withheld here intentionally and will be used for the future book project.

[165] In a parallel action of the USNR US Naval Reserve and RCAF Royal Canadian Air Force at Mannheim in June 1945 of which both parties at Baden were presumably not aware of, the local BBC upper management (comprising 'President and Managing Director, H.L. Hammerbacher; Herr Deichmann, Director of Engineering; Dr. Caspari, Director of Sales; Dr. Saenger [Senger], Chief Engineer Turbine Design' etc.) were forced to be available, after some hesitation, for information and cooperation.

Scientific Advisory Group	**Brown Boveri Representatives**
Th. von Kármán[166]	A. Meyer
H.L. Dryden[167]	C. Seippl [Seippel][168]
H.S. Tsien[169]	G. Dätwyler
F.L. Wattendorf[170]	G. Darrieux [Darrieus]
Col. F.L. Glantzberg	van Rŷswŷck [W. van Rijswijk]
Lt. Col. G.T. McHugh	

---

[166] Theodore von Kármán (1881–1963) was a Hungarian-American aerospace engineer and physicist who was primarily active in the fields of aeronautics and astronautics. He is responsible for many key advances in aerodynamics, notably his work on supersonic and hypersonic airflow characterisation. After graduating in 1902 he joined Ludwig Prandtl at the University of Göttingen and received his doctorate in 1908. He then taught at Goettingen for four years. In 1912, he took a position as director of the Aeronautical Institute at TH Aachen. He left Aachen in 1930 and accepted the directorship of the Guggenheim Aeronautical Laboratory at the California Institute of Technology (GALCIT). During the first half of 1943 he received reports from British intelligence sources describing German rockets capable of reaching more than 100 miles (160 km). In 1946 he became the first chairman of the Scientific Advisory Group which studied, as in the context above, aeronautical technologies for the United States Army Air Forces. He also helped found AGARD, the NATO aerodynamics research oversight group (1951).

[167] Hugh L. Dryden (1898–1965) was an aeronautical scientist and civil servant. He became Director of Aeronautical Research for the National Advisory Committee for Aeronautics (NACA) in 1946, where he supervised the development of the X-15 rocket plane, see Footnote 161 of this Section. He served as NASA Deputy Administrator from 1958 until his death.

[168] Strangely, Seippel's diary has no entry about this meeting in the considered time frame. – On 18 May 1945 he wrote '*Dübendorf visite d'un Düsenjäger*' which refers to an officially organised Swiss Me 262 inspection. In the early hours of 25 April 1945 Hans-Guido Mutke escaped with a brand-new Me 262 A-1a from Fuerstenfeldbruck airfield, 20 km west of Munich. However the fuel tanks were nearly empty, so an emergency landing on the Swiss military airfield outside Zurich was his only option. This event also attracted the interest of Ackeret, who in those days dealt with flight performance calculations such as the Me 262 vs. the British Meteor fighter. In 1957, the well preserved Me 262 was handed over to the Deutsches Museum Munich as a Swiss contribution to the newly installed aero exhibition. – On 22 June 1945 Seippel recorded '*Nommé successeur d'Adolf Meyer, diŕecteur, quitté dept. GT*', but the end of the war and technical issues received no mention; it appears that private sorrows were clearly a priority at that time.

[169] Hsue-Shen Tsien or Qian Xuesen (1911–2009) made important contributions to the missile and space programmes of both the United States and the People's Republic of China. In 1935 Tsien left China on a Boxer Rebellion Indemnity Scholarship to study mechanical engineering at the Massachusetts Institute of Technology and earned a masters degree from MIT a year later. He went to CalTech to pursue his studies under Theodore von Kármán, who was the doctoral advisor for his thesis on slender body theory at high speeds in 1939. He remained part of the Caltech faculty until his departure for China in 1955, becoming the Robert H. Goddard Professor of Jet Propulsion in 1949 and establishing a reputation as one of the leading rocket scientists in the United States. After their contact at Kochel, see Footnote 163 of this Section, von Braun prepared a seminal report for Tsien called 'Survey of Development of Liquid Rockets in Germany and Their Future Prospects', which provided the guidelines for future space vehicle development in the United States. Returning from Germany, Tsien edited the leading findings of the tour in the 800-page 'Jet Propulsion' report, which became the classified technical manual for post-war aircraft and rocket design and research (for whoever had access to it). By 1949 Tsien had applied knowledge he learned to the design of practical intercontinental rocket transport. Soon after Tsien applied for US citizenship in 1949, allegations were made that he was a communist and in due course he became the subject of 5 years of secret diplomatic negotiations between the US and China, held in a state of near house arrest. Finally in 1955, his departure was traded for 12 US downed fliers from the Korean War. Immediately upon his arrival in China he went to work as the head of the Chinese missile development programme.

[170] Frank L. Wattendorf came to BBC as a representative of the Engineering Division of Air Material Command, Wright Field, Dayton Oh. USA. Later on he became the director of AGARD Advisory Group for Aerospace Research & Development, a NATO organisation, as successor to Th. von Kármán who died in 1963. As von Kármán's assistant Wattendorf considerably influenced the foundation of the aero research establishment of VKI Von Kármán Institute at Rhode-St.Genèse in Belgium in October 1956. He made innumerable contributions to the advancement of ground testing facilities and capabilities throughout the world.

Besides the impressive guest list it is interesting to see who was selected to represent BBC at this meeting. Adolf Meyer and Claude Seippel were clearly the undisputed leading technical managers in the field of gas turbines and turbomachinery.

Gottfried Dätwyler was a participant of the Swiss delegation to AVA Goettingen in September 1941 and at that time still a member of Ackeret's IfA staff at ETH Zurich. Since Dätwyler was running his own engineering office after 1946, it is not clear if he was an BBC employee in the meantime or if he had been selected due to his personal acquaintance with von Kármán (and Tsien?) during his stay at GALCIT from approx. 1933 to 1939.

Georges Darrieus was not expected in this group. Although certainly a scientist with an international reputation, he was also a credible witness of BBC's wartime support for the 'Résistance', see Footnote 31 of this Section.

The Dutchman Willem van Rijswijk rounded off the BBC 'welcoming team', known as humorous and communicative and also familiar with supersonic wind tunnel design since his corresponding involvement with the first ETHZ project. In fact the three and a half pages of conference notes show no signs of difficult questions, on the contrary the conversation appears somewhat reduced to true but unspectacular facts. Another explanation of the mutual reservations might have been that industrial leading companies are and were not used to openly revealing internal technical matters to visiting delegations.

This fact was aggravated for BBC not least by the impact of military aeronautical technology, as the global market leader for gas turbines and related turbomachinery at that time which would be pushed into second place by General Electric in a few years. Both teams were probably exhausted by time pressures and the emotional events of the weeks before. However, this all remains speculation.

Nothing is known about the whereabouts of Fritz Zwicky at the time of the described visit to Brown Boveri. He would have made an interesting addition to the line-up, but there were probably good reasons for his absence.

In the following passage the major topics of the conference minutes are repeated – with the original text in *italics:*

1. Wind tunnel compressors
   a *'The axial compressor built by BBC for the Zürich T.H. supersonic wind tunnel about 10 years ago has been more or less used as a standard.'*

This is true and can be confirmed by a comparative look at Figures 4-31 and 4-32 with data point 5) for the ETHZ wind tunnel compressor as a reference.

   b *'The largest axial compressor they are now building has a mass flow of 30,000 kg/h at a pressure ratio of 4:1. The compressor is about 4 ft in diameter, has a maximum rotational speed of 3,000 rpm, and absorbs about 15,000 kw.'*
   c *'Concerning supersonic wind tunnels, they believe that a test section of about 2 × 2 ft is the largest size practical for one compressor. For larger tunnels, they believe it would be better to have several compressors since one large compressor would be more costly and difficult to build.'*

Figure 4-49   *US inspection team representatives – Fritz Zwicky (l), Theodore von Kármán (m) and Hsue-Shen Tsien (r)* [171]

This is true in principle, although with no indication that an ax/rad compressor solution like the one BBC Mannheim designed for the Kochel hypersonic tunnel, Figure 4-45, was seriously considered. But the foregoing Figure 4-48 may prove that the BBC's cost concerns of 1945 were overcome nearly 10 years later and the four-axial-compressors-in-parallel arrangement of the Kochel lp stage planning, Figure 4-45, was presumably replaced at AEDC by one large compressor unit.

2. Axial compressor design principles

   a   *'BBC stated that the stage pressure ratio of their standard units was 1.1, and has not been increased over the past 10 years because there had not been any particular reason for needing higher stage pressures for commercial, stationary gas-turbine plants.'*

This is true: compare the statement with Figure 4-31.

   b   *'They realised, however, that aircraft application did require higher stage performance, and for that reason were now becoming interested in developing basic improvements.'*

Again no hint that anything was known or intended to be directed towards BBC Mannheim achievements in the military sector, Figure 4-31 and Figure 4-32, data points 14) to 17).

   c   *'They have tried blades with more camber, but get an earlier decrease of efficiency with increased speed. However, they have not made enough investigations to see whether this premature drop could be eliminated by proper aerodynamic design'.*

This is a remarkable statement, since it actually refers to the very first test series under Seippel's guidance, finished on 23 July 1931, Footnote 46 of this Section. This was fourteen

---

[171] Picture sources – on F. Zwicky http://scienceblogs.com/startswithabang/2009/06/15/the-last-100-years-the-1930s-a/ ... and his detection of 'black matter', – on Th. von Kármán http://www.daviddarling.info/encyclopedia/V/von_Karman.html and on H.-S. Tsien http://www.visitchn.com/2009/11/tsien-hsue-shen-dies-article-at-california-institute-of-technology-official-website.html

years after the tests had been run. This apparently indicates strong reservations about re-
leasing anything of any substantial value, for Americans at least. It is also in striking contradiction to the supposed generous support of English aero engine specialists from Royal
Aircraft Establishment in the years immediately before WW II by BBC's upper management; contacts presumably arranged through the friendship between former Stodola scholars Karl Baumann and Metrovick on the English side and Adolf Meyer on behalf of BBC
Baden. A similarly harsh reaction towards an offered US contact was already recounted in
the year 1935. A missed opportunity for the two prominent turbo-machinists Sanford Moss,
GE and Claude Seippel as Moss's offer of a visit to Baden during his European tour was
turned down, see Section 3, Footnote 32.

> d   *'BBC's present standard design method calls for a 50/50 reaction maintained over the*
> *whole blade length. This method results in unbalanced centrifugal forces, which,*
> *however, BBC does not consider serious since they attain an 82 % stage efficiency and*
> *86 % over-all adiabatic efficiency.*
>
> *Dr. von Kármán believes that balancing of the radial forces might increase the stage*
> *efficiency to 92 %.*
>
> *BBC admitted that the problem would become increasingly important with higher*
> *stage pressures. Dätwyler is making preliminary investigations on radial flow in tur-*
> *bine blades.'*

Nothing was mentioned about the available measurements of considerable performance
advantages of a symmetrical 50 % reaction compressor blading in comparison to an alternative 100 % reaction design. True, the question was not brought up, ... and so the information had to be kept secret for another 13 years before the Englishman John Horlock obtained the privilege to publish the corresponding BBC data in his 1958 textbook on 'Axial-
Flow Compressors'.

3. Turbochargers and blade cooling
> a   *'Small turbosuperchargers were shown having a radial compressor of about 5–6 in.*
> *diameter and a turbine of about 4-5 in. diameter. The rotational speed is about 45,000*
> *rpm. This unit is designed to operate on truck engines of about 100–150 hp.'*[172]
> b   *'Water cooling of turbine blades is not favored since the circulation of water is easily*
> *interrupted due to collection of sediment near blade tips by the action of centrifugal*
> *force. They prefer to run at reasonable temperature where cooling is not required.'*

Alas, water cooling or no water cooling, the reader may wonder again, why the sensational
air/film-cooled radial turbocharger of 1935 (Section 3, Footnote 32) or the marvellous 7-
stage watch-maker-type axial turbocharger, data point 12) in Figures 4-31 and 4-32, both
made for Rolls-Royce, have not been mentioned at all.

Our judgment some 65 years later may be unbalanced, but the impression remains that this
historic meeting at Baden in June 1945 was not the 'finest hour' for both sides.

---

[172]   Presumably VTx 95, with 95 mm rotor diameter and VTx 110 (110 mm diam.) for turbocharging wood generator gas engine plants with 70–120 hp output, the first road vehicle turbochargers investigated at BBC since
1941 to tackle the increasing shortage of liquid fuels. See Jenny, The BBC turbocharger, p. 75 ff.

4.2.5   Outlook – Further Compressor Development

The past 50 years of BBC-ABB-Alstom[173] axial compressor development are highlighted in Figure 4-50, as a continuation of what was already shown as average stage pressure ratio over time in Figure 4-31. The limitation of this parameter for representing aerodynamic loading had been indicated there already, but it was kept for the benefit of historical continuity. Considerable improvements are clearly visible, thoroughly prepared by a number of aerodynamically advanced experimental test rigs. A data point is shown for BBC's 8-stage compressor rig in 1960, followed by a series of transonic compressor rig builds in 1980 with relative tip Mach numbers > 1.3, at that time an important landmark in industrial development. The milestone tests had an immediate impact on the unique GT10 and GT8 compressors.[174] This tradition of advanced transonic design is preserved by the new 12-stage GT8C2 compressor with PR = 17.6. At the same time, the PR = 30 compressor for GT24/GT26 in 1994 represented another unique and unprecedented 'first' for single-shaft configuration, though the stage pressure ratio was kept at a relatively moderate level as the basis for excellent performance values. The high pressure ratio overall was selected as optimum for sequential combustion, Section 5.3.5. Finally this graph shows the most recent upgrades of the 22-stage GT24/GT 26 compressor with PR > 34, raising the average stage pressure ratio in line with the long-term trend to 1.175. This compressor has a polytropic efficiency η_{pol} > 0.91 and an excellent part-load behaviour, achieved in the recent version with four rows of variable inlet guide vanes. The higher loaded GT24/GT26 lp compressor (with 16/17 stages) is offered also for the GT11N2.

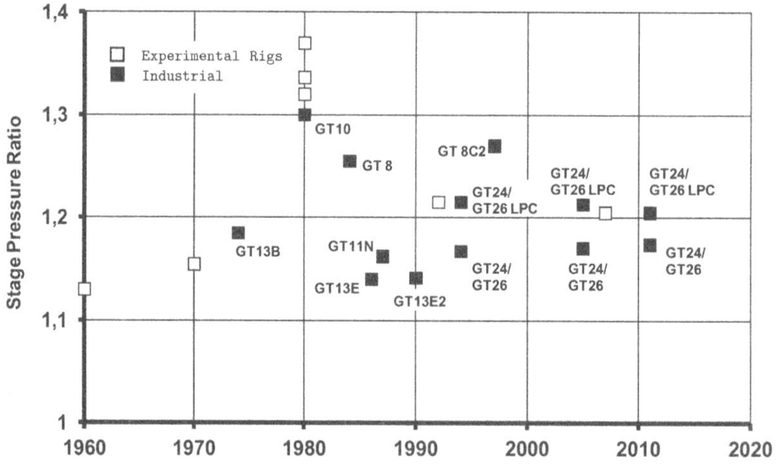

Figure 4-50   *BBC axial compressor development history 1960–2011*

---

[173]   In line with the introductory remarks of Section 2.1.1, the BBC era ends in 1988, followed by the ABB phase till 2000 and since then continuously under Alstom leadership. Important for axial compressor development in general together with positive side-effects for the standardisation of design tools was the collaboration phase between BBC and Sulzer as BST Brown Boveri Sulzer Turbomachinery between 1968 and 1974, see Section 5.1.3.

[174]   See Farkas, The Development and – see Thoren, Gas Turbine Development in Sweden

In the 1950s the traditional BBC compressor design philosophy, originally based on Goettingen-type airfoils, was replaced with the NACA 4-digit airfoil series. This statement clearly indicates – to put it bluntly – that Baden did not continue where Mannheim and Hermann Reuter's successful aero engine design team had finished in 1945: with the design methods from Eckert/Schnell's textbook.[175] Though the book was of course well-known, the Baden designers preferred to follow new, widely available French and US methods and design examples. This NACA 4-digit concept was still consistently used in the 1960s for all industrial compressors as well as for GT compressors of the types 10, 11L (60 Hz) and 12/8 (50 Hz). The performance computations were based both on published NACA cascade measurements as well as tests in 8- and 12-stage model compressors, Figure 4-28, in BBC's turbomachinery laboratory.[176] From these tests characteristics for the influence of variable rotor and stator staggering originated for the repetitive **blading concept 'VV'**, as well as for Mach number influences and the effect of varying aerodynamic loads due to differently converging ducts. It was only with GT13S, with 4 journal bearing supports for a single shaft arrangement, that a new blading concept was introduced.[177]

The beginning of the 1960s emphasised the trend towards increased gas turbine power by means of higher air mass flows, still exclusively accomplished in the subsonic regime by '**N blading**'. In principle, these were repetitive stages with slightly modified thickness distribution on a circular camber line, compared to NACA-65 (increased TE trailing edge thickness). After preparatory testing in the 8-stage experimental rig 'SILVIO'[178] at Baden in the late 1960s, Figure 4-28, the 'N blading' was used for the first time in the GT13S and subsequently in the GT9B, GT11B and GT13B.[179] Following widespread industrial principles of the time the same blade designs were simply scaled to identical Mach numbers. The height of the bladed annulus decreases from front to rear in line with the compressibility effects. Correspondingly, the chord lengths decrease according to the channel height to keep the blade aspect ratios (blade height/chord length) equal.[180] A series of comparative blading tests was carried out during the BST era in the 6-stage model compressor EIGER at Winterthur. They also confirmed the applicability of 'N blading' for variable guide vane compressors.[181] The 'N blading' steadily replaced preceding designs from BBC, Sulzer and Escher Wyss, not least due to the beneficial short axial space requirements.

The 'N blading' found successful application throughout the 1970s and 1980s in the gas turbine series GT9C, -D; GT11C, -D, -E (the latter not built) and -N; GT13C, -D and -E[182] and

---

[175] See Eckert and Schnell, Axial- und Radialkompressoren; the 1st ed. appeared in 1953. Until 1945 Eckert had developed the methods by and large during his stay at FKFS Forschungsinstitut fuer Kraftfahrwesen und Fahrzeugmotoren Stuttgart (Prof. W. Kamm). Reuter was hosted at FKFS for some time and had applied the methods in simplified form to the advanced BBC aero engine designs.

[176] See BBR, Axial Compressor with Adjustable Stator Blades, 1983; BBM, Versuchsarbeiten, 1963, and see Kruschik, Die Gasturbine, Chapter V, Section B 9. BBC's turbomachinery lab, in Swiss-German 'Versuchslokal', gave name to the 'V' blading series.

[177] See Endres, Operating experience

[178] See Pfenninger, Axialverdichterbeschaufelung and – see Baumann, Messung

[179] See Endres, Three gas turbine types

[180] An approach which had already been established in previous design periods.

[181] See Suter and Spaeti, The surge limit

[182] See Keppel and Meyer, Entwicklung eines Axialverdichters

could still be found in industrial compressors at the beginning of the 1990s.[183] A comprehensive experimental data base with information on stagger variation, influence of the hub/tip ratio, aspect ratio, Mach number and pitch/chord ratio can be used for 1D off-design calculations of stage characteristics.[184] At the same time the measurements were used to calibrate semi-empirical loss models for row-by-row off-design computations.[185] Notwithstanding a larger number of stages in comparison to more modern blade designs, the economical advantages of standardisation carried 'N blading' production through until the 1990s.

At the end of the 1960s the aerodynamic limits of NACA-65 profiles became more and more visible, meaning **DCA Double-Circular-Arc** bladings and their potential for transonic compressors was investigated. Typically for transonic compressors, the diffusion from supersonic to subsonic flows in the outboard blade section happens through a compression shock pattern, which could be exploited to raise the pressure ratio but could become problematic due to corresponding shock loss. In general, given the same flow turning task and equal blade thickness, DCA airfoils show fewer total pressure losses here than the previous generation of NACA-65 profiles. The transonic compressor therefore became attractive due to its capability to swallow higher mass flows in comparison with subsonic designs of the same dimensions.[186] The first transonic test compressor was the 4-stage 'PALUE 1' with a constant total temperature over blade height.[187] This blade configuration was somewhat difficult to manufacture, so a second design with constant total pressure over blade height was completed: the 'PALUE 2A'.

Multi-stage transonic axial compressor development accelerated in the 1970s thanks to GT power reaching the 150 MW (60 Hz) and 200 MW (50 Hz) range.[188] Higher compressor front end Mach numbers were achieved through **MCA Multi-Circular Arc** blading. Development activities at BBC Baden followed the established NASA approach, where 1-stage, 5- and 8-stage transonic compressors were tested in consecutive order. Baden started with PALUE 2A with 4 + 2 repetitive stages, followed by BERNINA 2A with 4 + 5 repetitive stages.[189] The stage-wise calculation process using repetitive blading was abandoned and replaced by individual row characteristics, with deviation and loss coefficient as a function of incidence angle over radius, opening up the possibility of meridional flow calculations. Results from PALUE and BERNINA again helped to cover the whole operation range. These experimental rigs were built and tested during the BST joint venture period at Winterthur. Scaled versions of this design were later used in the gas turbines GT10[190] as 10-stage and in GT8[191] as 12-stage compressor configurations. In 1984, the GT8 had the first transonic compressor in the crude oil, heavy-duty market for the Shell Pernis plant at Rotter-

---

[183] See Aicher and Schnyder, Umbau von Turbokompressoren

[184] See BBR, Computer program, 1983 and – see Goede and Casey, Stage matching

[185] See Keppel and Meyer, Entwicklung eines Axialverdichters; Novak, Streamline curvature; Johnsen and Bullock, Aerodynamic design; Casey and Hugentobler, The prediction

[186] See BBM, Die Entwicklung..., 1977

[187] See Hirsch and Denton, AGARD PEP WG 12 (Attachment II by F. Farkas)

[188] See N.N., BBC 120 and 210 MW gas turbines and – see Auer, Entwicklung der Kraftwerksgasturbine

[189] See Farkas, The development 1986

[190] See Marriott, Eine moderne umweltfreundliche Industriegasturbine and – see Traupel, Thermische Turbomaschinen II, 3rd ed., Section 14, Fig. 14.1.3, p. 123

[191] See Endres, The medium size gas turbine Type 8; see Wicki and Farkas, Entwicklung und Erprobung einer 45-MW-Gasturbine

dam, NL. The off-design performance of MCA/DCA blading was deduced from test data with determined mass flows and efficiencies, thus calibrating total pressure losses and flow turning.[192] Simple 1D-[193] and more demanding 2D-computational methods[194] were applied. A strong effort was invested in advanced MCA/DCA compressor blading for 11-stage axial compressors for the new gas turbine projects GT15 (150 MW, 60 Hz) and GT17 (210 MW, 50 Hz) which were cancelled during the 1980s for commercial reasons, see Section 5.1.5.4.

In the mid-1980s BBC's compressor design developed further to **CDA Controlled-Diffusion-Airfoils**. A revised version of 'PALUE 3A' emerged from the MCA design of 'PALUE 2A' (GT8) which made use of these inversely computer-designed CDA blade shapes[195]. However they could not be tested at Winterthur any more. A survey of the aforementioned basic blade forms is given in Figure 4-51.

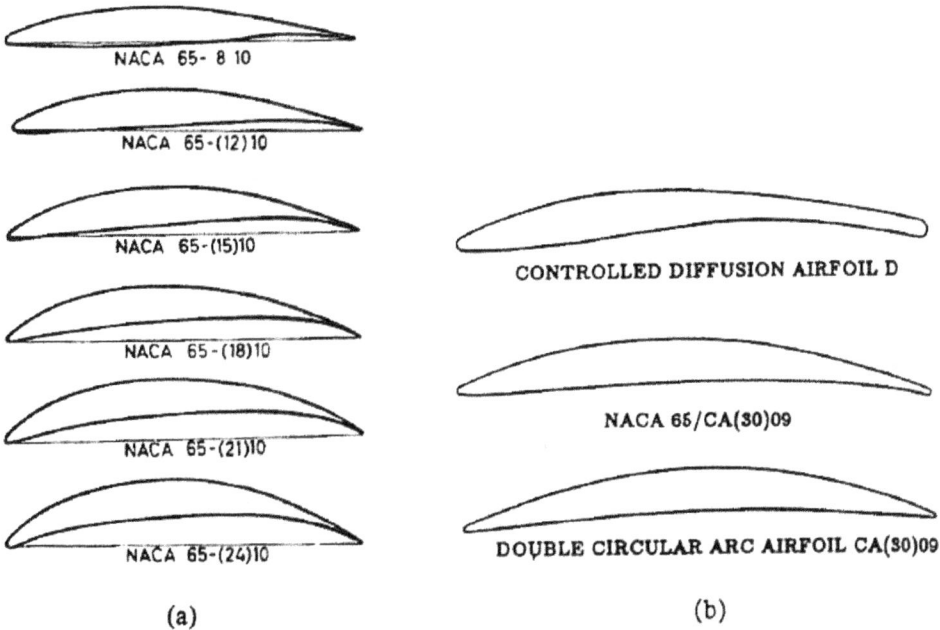

NACA 65- 8 10

NACA 65-(12)10

NACA 65-(15)10

CONTROLLED DIFFUSION AIRFOIL D

NACA 65-(18)10

NACA 65/CA(30)09

NACA 65-(21)10

NACA 65-(24)10

DOUBLE CIRCULAR ARC AIRFOIL CA(30)09

(a)   (b)

Figure 4-51   *Compressor blade profiles (a) NACA 65 series, (b) trends after 1970* [196]   © *ASME*

GT15 studies reappeared at the end of the 1980s for a 60 Hz/150 MW GT application in collaboration with Rolls-Royce. Scaling of the existing GT8 compressor was envisaged initially, until later on a reduction in mass flow favoured an upscaled version of the shorter compressor from the V2500 turbofan engine with adapted front stage, introduced by Rolls-Royce, Section 5.1.5.4. Completion was hampered by considerable cost problems, meaning

---

[192] See Stoff and Ebner, Geeichte Wirkungsgrad-Nachrechnung
[193] See BBR, Computer program..., 1983
[194] See Hearsay, A revised computer program
[195] See Stoff and Waelchli, Aerodynamische Entwurfsverfahren, Figures 1 and 5
[196] See Lakshminarayana, Fluid Dynamics, Figure 2.13 (Original from Hobbs and Weingold, 1984)

this effort (also using an 11-stage ABB compressor in the end) was stopped before reaching production status.

After a coordinated effort to introduce the GT11N2, GT8C2 and the top-selling GT13E2 compressors[197], subsequent development activities for GT24/GT26 compressors in the early 1990s in collaboration with MTU Motoren- und Turbinen-Union Munich again focused on relatively low aerodynamic loadings to support the drive for peak efficiency. Pressure ratios of PR = 15...17 were generated by 2/3 of the compressor stages at the front end at an entry Mach number level with the former 'PALUE I' blading, while rear repetitive stages push the total PR > 34, a record level for industrial compressors.[198] Figure 4-52 visualises GT26 front stages with advanced CDA blading design. For cost reasons an optimum compromise was sought between individually tailored front stages and repetitive blade shapes in the middle and rear sections.

Figure 4-52   *GT26 transonic compressor inlet region – CDA rotor blading*

The GT24/GT26 compressor's design performance, operational flexibility and starting behaviour targets were confirmed by a test compressor 'Rig 250', built[199] and tested at MTU Munich in 1993/1994, well ahead of the first GT24 operation in 1995, Figure 4-53. The rig represented the first 15 stages (of 22 in total) of the GT24 in a 1/3 scale size. The project se-

---

[197] See Jacobi et al., Modernisation of GT 11N2; see Chellini, ABB introduces...; see Viereck et al., GT 13E2

[198] See Meindl et al., The development of a multi-stage compressor

[199] For cost reasons, manufacturing of the rear part of the 2,260 single airfoils was subcontracted to TRUD, Samara, Russia, home of Nikolaj Kuznetsov's famous NK engines. In 1993 one of the remaining 'heroes of Socialist labour' produced the order by copy-milling alone, with one light bulb, for eye control between the master gauge and the actual profile.

crecy is indicated on this original document by the (fake) project designation 'ASP60/15', i.e. air storage plant 60 Hz, 15-stage compressor. The rig had a design pressure ratio of PR=17.1. Special focus was given to the simulation of variable inlet flow during start-up by 3 rows of VIGVs in combination with a bleed port after stage 5, the remaining exit swirl was then handled by a tandem row of exit guide vanes. Since the inauguration of the Birr GT power plant in December 1996, Figure 6-38, full scale compressor tests have been carried out there in conjunction with engine testing.

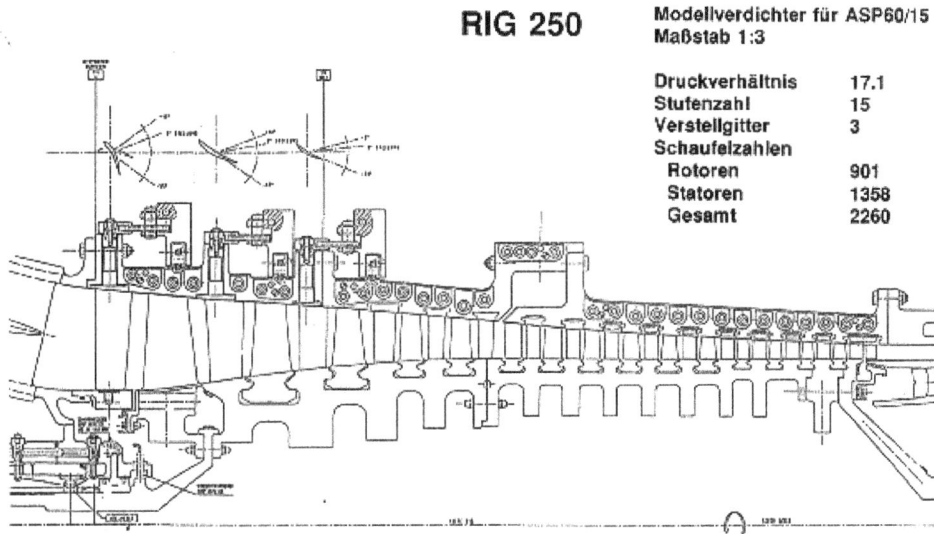

RIG 250   Modellverdichter für ASP60/15
                   Maßstab 1:3

Druckverhältnis	17.1
Stufenzahl	15
Verstellgitter	3
Schaufelzahlen	
Rotoren	901
Statoren	1358
Gesamt	2260

Figure 4-53   *Original drawing of GT24 test compressor, 1/3 scale, PR 17.1, 15 stages, 3 VIGV*   © *MTU*

In due course the 15-stage rig was modified for Alstom tests of a 5-stage version, Rig 253[200], and a 4-stage configuration, Rig 254[201], which now belongs to the DLR, Institute for Propulsion Technology, Cologne, D.

Test facilities which have been used for Alstom fundamental compressor research are – amongst other – at:

- TU Hannover, Institute for Turbomachinery and Fluiddynamics, Hannover, D 4-stage high-speed compressor rig (18,000 rpm, PR = 2.8, m = 9.4 kg/s, P = 1050 kW) for 3D front stage blade design and sidewall effect verification.
- RWTH Aachen, Institute of Jet Propulsion and Turbomachinery, Aachen, D 3-stage compressor rig (17,000 rpm, PR = 2.03, m = 13.4 kg/s, Ma_{max} = 0.9) for rotor/stator interaction and unsteady aerodynamics.
- TU Dresden, Institute for Fluidmechanics (Turbomachinery and Jet Propulsion), Dresden, D 4-stage low speed compressor (1,000 rpm, D_{max} = 1.5 m) for 3D rear stage blade design and tip clearance investigations.

---

[200] See Lecheler et al., Experimental and numerical investigation of the flow in a 5-stage transonic compressor rig

[201] See Johann, Back to back comparison of a casing treatment in a high speed multi-stage compressor

- EPF Lausanne, Group of Thermal Turbomachinery, Lausanne, CH 1.5-stage compressor rig (15,000 rpm, m = 2.25 kg/s, air & freon) for investigations of the influence of axial and radial gaps, rotor/stator interaction, annular cascade with Laval nozzle (m < 10 kg/s, PR = 3.5, Ma_1 = 1.6) for the investigation of highly loaded compressor stages.
- UniBw University of the German Armed Forces, Institute of Jet Propulsion, Munich, D Annular cascade test facility (0.3<Ma<0.7, 4-12 percent bleed flows) for bleed flow aerodynamic investigations.

4.3   The Early BBC Gas Turbines

Emotions are rare in the technical heritage of Brown Boveri. One of these rare exemptions happens in spring 1932, when the author, most likely Walter Noack, finished a short section about recent achievements of the (Holzwarth) gas turbine with a deep sigh *'Dealing with gas turbine problems – indeed a difficult problem! – has brought us (besides the Holzwarth prototype) a number of challenges and inspirations, which resulted in a new product, the Brown Boveri Velox boiler, a steam generator with turbocharged combustor.'*[202] We remember the extraordinary high heat transfer rates observed in the water-cooled jackets of the Holzwarth turbine components (Section 3.3). Here was the opportunity to change an obvious disadvantage of the unloved Holzwarth project into a striking success.

4.3.1   VELOX Boiler

4.3.1.1   Concept Settlement

The Velox steam generator is a boiler fired under pressure, the pressure being produced by a compressor driven by a gas turbine (component), actuated by the exhaust gases of the boiler. Part of the pressure produced in the compressor is used to maintain high gas velocities in the heat-transmitting parts of the boiler, thus ensuring high rates of heat transfer. The remainder of the pressure head is used to drive the gas turbine. This application as an auxiliary unit rendered the creation of a compressor set with a high efficiency essential, otherwise the exhaust gas turbine would be unable to develop the power required for driving the compressor, and the deficiency would have to be supplied by another source, which would seriously impair the efficiency of the entire steam generating unit. This problem was solved by the development of a four- or five-stage reaction turbine and a ten- to twelve-stage axial compressor, the design taking into account the results of the latest research in the field of aerodynamics at that time. Figure 4-54 shows the steam generating Velox equipment in the early stages of its development, already in the prevailing constant pressure version with all-axial turbomachinery components.

Air is compressed and burnt; the combustion gases give their heat at high velocity to evaporators and superheaters with extremely reduced surface areas, expanded thereafter in a turbine driving the compressor and then after-cooled in an economiser. Comparable to a

---

[202] See BBM, Die Konstruktionen der Turbinenfabrik in 1931, 1932, p. 41

turbocharged internal combustion engine, fuel and combustion air are delivered in compressed form so that the combustion chamber takes larger quantities than under ambient conditions. In addition to the simple density-increasing turbocharging function, the compressor generates a pressure head to the heat-transmitting combustion gas in order to generate correspondingly high gas flow velocities. In due consequence the reaction speed in the combustor and the heat transfer rates surpass standard ambient experiences considerably. Part of the Velox principle is an electric starter motor, which changes the function and becomes a generator of surplus electricity in the expected case that the turbine power surpasses the compressor work. High delivery standards in 1939, presumably for a British war ship application, had gas temperatures of between 550 and 600 °C and pressures of up to approximately 4 bar.[203]

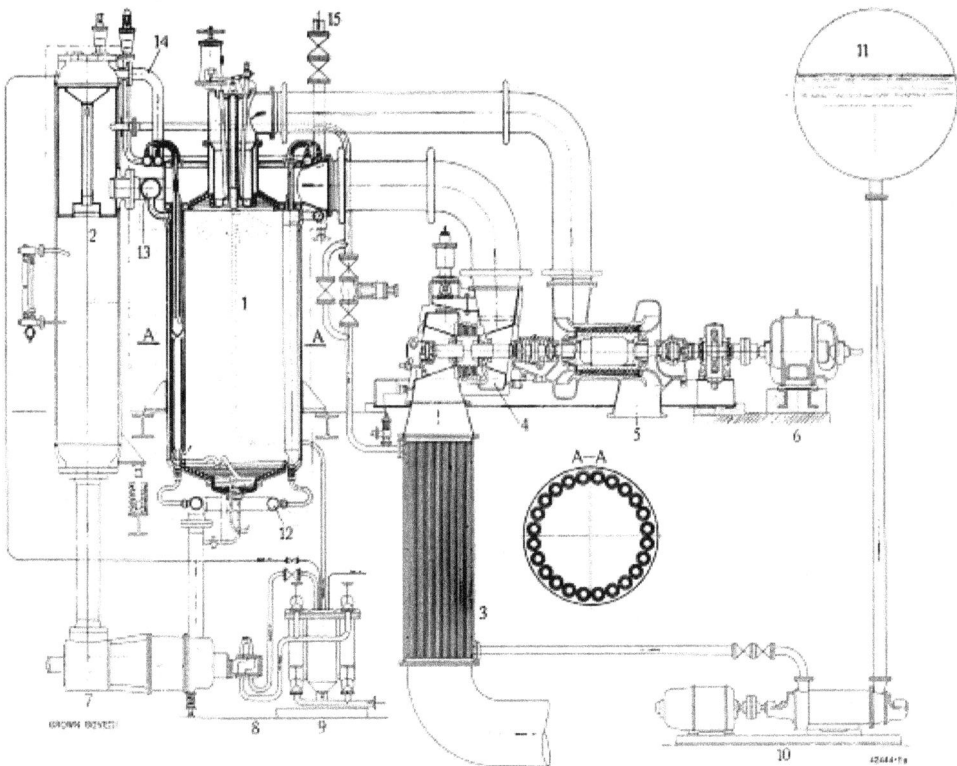

Figure 4-54   *Schematic diagram of a Velox boiler with all-axial turbo-set, 1933 [204] 1 – combustion chamber, 2 – water separator, 3 – economiser, 4 – gas turbine, 5 – axial compressor, 6 – starting motor, 7 – circulation pump, 8 – fuel pump, 9 – fuel pre-heater and filter, 10 – feed pump with motor, 12 – water supply to the boiler tubes, 13 – water/steam mixture to the separator, 14 – saturated steam pipe to superheater elements, 15 – live-steam pipe*

---

[203]   See Footnote 217 of this Section.

[204]   Figure source: Pfenninger, The evolution of the Brown Boveri gas turbine, Figure 10

The dimensions of a Velox boiler decrease with increasing charging pressure and gas velocities of both heating gas and water/steam. Theoretically this trend could continue to extremely high pressures and velocities, if there was no through-flow drag (rising with the square of the through-flow velocity) or material temperature (as a function of charging pressure) as limiting factors. A pressure head of approximately one quarter of the gauge pressure (6,000 mm WS) and through-flow channel streamlining results in sufficiently high velocities, or more precisely, gas mass flow density parameters w x ρ and related functions for through-flow area and resistance as well as heat transfer areas. Velox steam generation areas are characterised by w x ρ ~ 100 kg/(m^2 s) whereas boilers with atmospheric combustion would only have 2 kg/(m^2 s). An advantage of pressurised operation is also the combustor heat load itself which increases by a factor of ten in comparison to foregoing ambient burning. Consequently, the hot gas cross-sections are reduced to 1/50 and the boiler heating areas to 1/22 of a comparable ambient boiler, while the through-flow resistance rises by a factor of 800 on the basis of a typical charging pressure of 3 bar. The stunning increase in specific power of the Velox boiler unit is immediately evident.

The development improvements in axial compressor performance pay off regarding the success of the Velox concept. Compression work and through-flow drag are compensated by the turbine work, sometimes even overcompensated with the benefit of additionally generated electricity. The first axial compressor for a Velox boiler with furnace gas combustion was already shown in Figure 4-26. The largest ever Velox unit was manufactured by BBC Mannheim in 1941, also for steelworks in Germany, Figure 4-62. As a result of the lower heat capacity of furnace gas/air mixture, the combustor dimensions are somewhat bigger than for oil combustion. The compressor group comprises two compressors of the same size, one for combustion air and the other for furnace gas. The drive turbine is of the double flow type to compensate individually for the turbine axial thrust, while the thrusts of both compressors neutralise each other.

The efficiency of a Velox boiler with furnace gas combustion was close to 90 percent, consequently the performance for oil or gas combustion was counted as 2-3 percent higher and the small performance variation between 1/3 load and overload was appreciated. A special advantage of Velox operation was the superior operational readiness, i.e. the possibility to start and shut down steam generation at short notice. Adolf Meyer, also highly talented at promoting BBC products, summed this advantage up by stating that Velox (and GT) have a starting time in so many seconds, what a similar sized steam power plant would require in minutes ![205] Despite this convincing argument the Velox boiler (and GT) did not succeed in (UK) marine applications as hoped by the BBC management.

The Velox boiler concept was promoted by its operational simplicity, inherent reliability and lack of special maintenance demands, leading to impressive sales figures of approximately 75 units in the first 10 years of delivery, Figure 4-13. Figure 4-55 (l) shows a sort of early series manufacturing.

In hindsight, until 1933 the development of the Velox boiler with attached all-axial turbo-set, as shown in Figures 4-54 and 4.55 (r), was not as straightforward as one might have as-

---

[205] See BBR, The Combustion Gas Turbine (A. Meyer) 1939, p. 138. This paper to IMechE London on 24 February 1939 outlines the space advantages of GT installations on board destroyers in depth. Similar initiatives for Velox marine applications were certainly run in the UK in the pre-war years.

sumed. Noack remembered the early steps for combining the explosion GT concept with high flow velocities for steam generation back in 1927. In January 1928 the first patent applications followed. At the same time a small boiler was adapted for initial heat transfer experiments that was followed at the end of 1928 with the 10 t/h test boiler, fired by heavy fuel oil. A smaller test boiler was also run at Choindez with furnace gas, where the last Swiss blast furnace had been in operation since Adolf Meyer's time in 1909/10. So both originating concepts were concrete, the constant-volume explosion gas turbine after Holzwarth as well as conventional turbocharging with two-stage centrifugal compressors, Figure 4-56. That was until the all-axial concept with its superior performance prevailed – unchallenged.[206] Besides the change in compressor configuration the most significant development step was shifting the fuel injection from the bottom position of the earlier versions to the fuel nozzle on top of the vertical combustor, a move which was retained from then on for all 'silo combustor' configurations.

Figure 4-55   *Two Velox units at Baden, CH during assembly and integrated all-axial turbo-set (1936)*[207]
                   ©ABB

The Velox concept, necessary calculations and the mechanical design were carried out by Walter G. Noack in continuation of his Holzwarth activities. Without a doubt the most significant scientific contribution of Noack's early experiments was the determination of reliable heat transfer data in a workshop environment. The whole possible pressure, temperature and velocity range of steam boiler operation were investigated. Heating gases were

---

[206] See Noack, Lebenslauf p. 38. This is the only source about this interesting transition phase, sometimes questionable in its egotism. Noack claims that he had suggested the switch to constant-pressure Velox operation with moderate pressures and high through-flow velocities in a memorandum to the Management as early as July 1930. But the Directorate remained sceptical. Together with his loyal assistant Schmid he therefore decided to modify the existing test vehicle into the first constant-pressure set-up, giving it the name 'Velox'. At first it still used a 2-stage radial compressor, but shortly thereafter used Seippel's axial compressor which was a success from day one.

[207] Picture source: Pfenninger, The evolution of the Brown Boveri gas turbine, Figures 12 and 13

generated by the combustion of illuminating gas and heavy fuel oil. The highest combustor pressure was close to 4 bar, the highest gas velocity up to 400 m/s, fully in the compressible range, and the max. starting temperature was 1,300 °C. The deduced empirical formula strictly followed the Nusselt[208] heat transfer law, the constant parameters of which were adapted to his own test results. As a general observation the measured heat transfer values were always higher than the theoretical prediction, presumably due to the important influence of the length of the thin, starting boundary layer in the heating tubes. Figure 4-57 illustrates these core tube elements which turned out to be essential for the success of the Velox principle.

Because of the high heat transfer rates on the gas side, it was also necessary to introduce a circulation pump on the water/steam side. Due to the extraordinarily high steam generation rate of 1,000 kg/m^2 achieved over the initial 'starting length' (on average over the complete tube length 400–700 kg/m^2) the tubes became relatively short and the water pumping power remained low. Consequently, the pump could also carry out the steam separation task for which Noack introduced the centrifugal separator effect in a swirling steam/water flow for the first time.

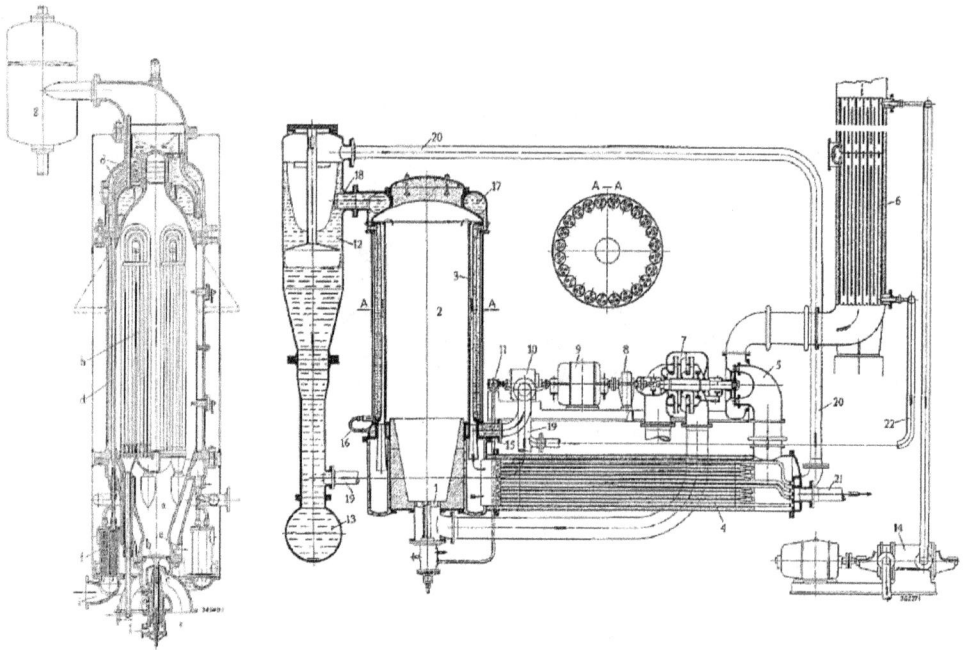

Figure 4-56   *Velox boiler development steps: Steam generation on the basis of (left) – constant-volume GT concept, 1931 [209] and (right) – constant-pressure GT concept, 1932 with 2-stage centrifugal blower [210]   ©ABB*

---

[208]   Wilhelm Nusselt (1882–1957), from Noack's hometown of Nuremberg, was a German physicist who laid the foundations for similarity theory in heat transfer.
[209]   Source: BBM, Die Konstruktionen... in 1931, 1932, p. 44

As a consequent design principle, Noack made all heating surfaces exchangeable from the very beginning to overcome potential concerns about the durability of these parts. These concerns became superfluous with the collection of longer operational experience.

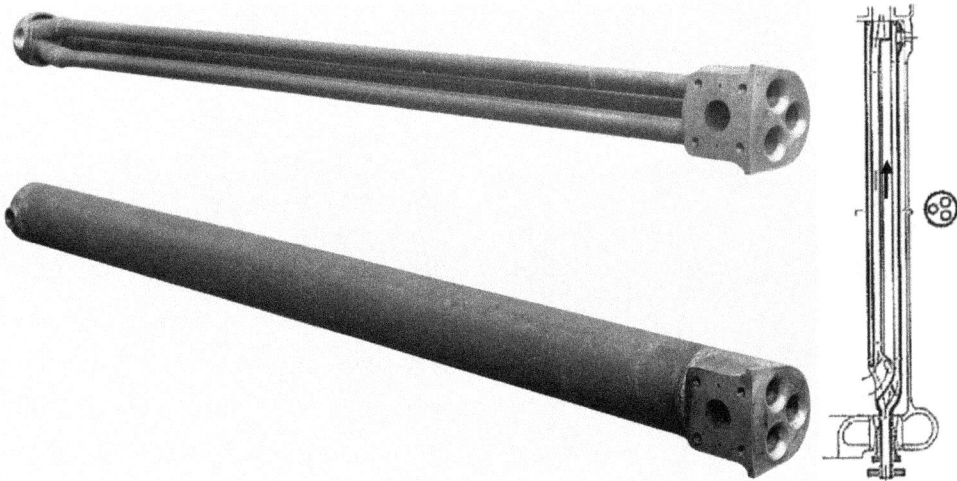

Figure 4-57   *Velox core technology* [211]: *'Heating tubes' (top) ready for insertion into 'boiler tubes' (below) with cross section arrangement; integrated boiler tubes ready for assembly into the Velox combustion chamber   ©ABB*

The first public Velox demonstration was at the VDI Scientific Conference at Berlin on 15 October 1932 and the first Velox paper was printed in VDIZ 1932, p. 1033 ff.[212] Noack also initiated most of the related BBC patents,

a   for steam generation under pressure based on the explosion gas turbine, Figure 4-56 – left, e.g. US1948940 'Steam Generator' with priority in Germany since 16 December 1929 and

b   for heat transferring tube arrangements e.g. US2046530 'Steam Generator' with priority in Germany since 15 November 1933.

After 75 Velox units with an aggregate sales volume of 25 MSFr had been successfully sold within 10 years, Figure 4-58, Noack nevertheless expressed some disappointment as his intermediate expectations had been much bigger. He felt this drawback was not due to any shortcomings of the Velox concept, but primarily by the presumption at times to launch considerable sales without sufficient operational experience, simply based on a few initial

[210] Source: BBM, Rueckblick... auf Konstruktionen im Jahre 1932, 1933, p. 39

[211] Source: BBM, Rueckblick... auf Konstruktionen im Jahre 1932, 1933, p. 40

[212] See Noack, Lebenslauf p. 47. A corresponding US information initiative was launched by Adolf Meyer as the 2nd Calvin W. Rice Memorial Lecture, delivered at the ASME semi-annual meeting, Cincinnati Oh., 17–21 June 1935. See Meyer, The Velox Steam Generator

laboratory tests.[213] This is a mistake which nearly every new generation of development teams is apparently prone to repeating.

Figure 4-58   *30 t/h Velox steam generator on the way to international shipment at Baden, CH railway station, October 1939*[214]   ©ABB

4.3.1.2   Multi-Purpose Velox Deliveries to UK, 1940

From 1933 onwards several marines launched orders or at least started experimental tests, for example the British Admirality who ordered Velox boilers with a steam capacity of 50–55 t/h, typical for what was known as a 'torpedo boat destroyer'. The reference unit for the UK was built in cooperation with Yarrow & Co., Glasgow and the English licensee Richardsons Westgarth & Co., Hartlepool. The tabulated data in Figure 4-59 clearly illustrates the advantages of an alternative Velox boiler installation on board a 40,000 t ship. Adolf Meyer[215] had already stated in his speech at ASME Cincinnati in June 1935: '*By taking full advantage of the technical means offered by the Velox boiler, war vessels with previously unseen striking power can be built. For obvious reasons no information about Velox plants in warships can be provided except the fact published in British papers that one unit has been built for the British Navy by Yarrow and Richardson Westgarth, the former building the boiler proper, the latter the gas-turbo blower and other auxiliaries.*'

Without further information the entire process is not very clear. One can speculate that the planned licence production at Richardsons Westgarth did not materialise or was considerably delayed. At the same time the need to learn more about BBC's all-axial gas turbine in Britain grew, culminating in an RAE Royal Aircraft Establishment delegation visiting Baden in the summer of 1938 with the intention of negotiating a compressor licence agreement. Though this intention could not be fulfilled, the parties must have agreed on a hardware delivery from Baden to Britain.

---

[213]   See Noack, Lebenslauf p. 48. In fact enthusiasm for the Velox concept had already cooled off within the BBC management in 1945, when a new steam-generating electro boiler was presented to the visiting US delegation led by Th. von Kármán.

[214]   Source: BBM, Rueckblick... im Jahre 1941, 1942, front page

[215]   See Meyer, The Velox Steam Generator

Steam Generation Plant	Conventional plant	BBC Velox plant
Boiler type	Water tube boiler	Velox
Number of boilers	10	8
Steam, max. continuous t/h	362	450
Steam temperature °C	385	385
Steam pressure kg/cm^2	32	32
Evaporator heating area m^2	10 × 1,250	8 × 110
Superheater area m^2	10 × 660	8 × 132
Steam generation kg/m^2	29	510
Efficiency, cont. operation %	85	92
Boiler weight w. aux. & tubes kg	2,470	630
Boiler weight/kg steam	6.8	1.4
Boiler space m^3	9,750	4,950

Figure 4-59   *Comparison of boiler plant data for a 40,000 t ship* [216]

It appears that a Velox turbo-set which had been under consideration for potential marine applications since 1935 was sent to Britain, first directly to RAE Farnborough in 1939 for comprehensive testing, Figure 4-60. The following is taken from communications with John Dunham in November 2000 and with Frank Armstrong in June 2009, both ex-RAE officials:

> *'According to notes written in 1976 by Raymond Howell, former head of R.A.E., the unit was regarded by Brown Boveri as commercially very confidential, therefore no details (except for installation) were given to R.A.E., and no information about it was given to Whittle or to Metropolitan Vickers. The contract for this unit was placed on 8 November 1938 with Richardson Westgarth and acceptance tests were run successfully at Baden from 11 to 16 August 1939. First run at R.A.E. was on 3.-6. May 1940, but due to problems with a brake dynamometer real running under load did not take place until 1. Aug. 1941. By this time, however, the R.A.E. research team's own experimental compressors were providing considerably higher pressure ratios per stage, and much higher gas temperatures were being used in the combustion and turbine components. Operation of the BBC unit was therefore limited to the gaining of some practical running experience. After about 70 hours of running, a first stage compressor rotor blade failed in Nov. 1941, and as a result the first stage was removed entirely. In late Sept. 1942 the turbine casing became too distorted to run further without considerable modification, and in 1943 the partly dismantled unit was sold back to the suppliers.' – (Richardsons Westgarth)*

The extraordinary short delivery time of less than 10 months can be either interpreted as a sign that the order was under preparation for quite a while, or that the order was a BBC standard product at that time which could be delivered almost straight from the shelf.

In the context of Adolf Meyer's sales initiative aimed at the United Kingdom, summarized in his IMechE paper in London on 24 February 1939, it is worthwhile to note that the relevant Figure 18 of that paper has the comparative ship installation of a 2 × 16,000 hp steam

---

[216] See BBM, Rueckblick... im Jahre 1933, 1934, p. 36

turbine vs. a 2 × 18,000 hp combined diesel and gasturbine propulsion unit, justifiably the first such CODAG concept drawing ever.

Figure 4-60   *1.6 MW gas turbine set at RAE Farnborough, UK, 1939*[217]   ©*ABB*

Besides general disappointment about the low number of Velox plants sold, the BBC management, with Walter G. Noack[218] as spokesman, was increasingly frustrated that the strong sales effort in military and civil marines, especially in the UK before the beginning of WW II, did not result in noteworthy orders. Noack felt the lack of coal dust burning capability hampered the marine breakthrough considerably. In this respect continuous coal burning, also for gas turbines, remained his unresolved task up to his premature death in 1945, sometimes highlighted by short phases of success (Figure 4-72).

4.3.1.3   Blast Furnace Gas Extension of the Velox Principle

During 1940, BBC Mannheim began construction of a gas turbine for one of the biggest steel works in Germany, which specialised in 'sour iron ore' found in the Salzgitter area. In principle, this low quality ore smelting was economically unattractive as one ton of 'Thomas raw iron' could only be produced at the additional cost of 1.25 tons of waste slag,

---

[217] See Pfenninger, The evolution, Figure 16. Pfenninger accompanied the shown photograph with the following text: '*Gas-turbine set rated 1,600 kW at the generator terminals after trials on the works' test-bed. Just before the war, this machine was installed at the Research Establishment of the British Air Ministry and was the first of its kind in Britain. Output of gas turbine 6,050 kW, blower rating 4,400 kW, pressure ratio 4.1:1, gas inlet tempera-ture 560 °C, speed 5,180 rpm.*' The compressor characteristic data of this '20-stage RAE compressor 1940' are part of Figures 4-31 and 4-32, marked as point 9).
There is no absolute clarity as to whether just the turbo-set was delivered, or a complete gas turbine or Velox boiler with combustor equipment, generator etc. But Seippel in his 1967 ASME paper '*The evolution...*' appears to provide the final argument in favour of the turboset-only theory: '*Later, the centrifugal compressor re-entered the field of gas turbines, particularly in Whittle's aeroengines. Under time pressure, Whittle chose to maintain the well-known radial compressor in spite of its lower efficiency and in spite of the fact that the British Air Ministry had managed* **to obtain a Brown Boveri axial compressor** *to become acquainted with its behaviour.*'
[218] See Noack, Lebenslauf p. 49

but the effort was considered worthwhile as part of the drive towards German industrial autonomy. While the Velox system contained an element of steam generation, the blast furnace gas turbine represented a further variant in the history of BBC's gas turbine development. Prior to the new concept, air was supplied to the blast furnace by means of a gas piston pump or a steam-driven turbo-blower. The blast furnace was the central element of the iron and steel industries of that time. Once charged with iron ore, coke and limestone and brought to working temperature, a continuous blast of air through the furnace was all that was needed to produce the liquid 'pig iron' which could be further transformed into various steels. A large quantity of gas was produced in the process together with a good deal of slag. The whole process could not be speeded up by higher temperatures or more air blow. The furnace then reacted either by operating irregularly or increasing its coke consumption.

BBC's prime aim to improve the process was therefore directed towards reducing plant space. It was foreseen that the boiler and condensation plant could be dispensed with if gas from the blast furnace could be used to drive a gas turbo-compressor. The air generating plant could be located directly beside the blast furnace, saving material and space and eliminating duct flow losses at the same time.

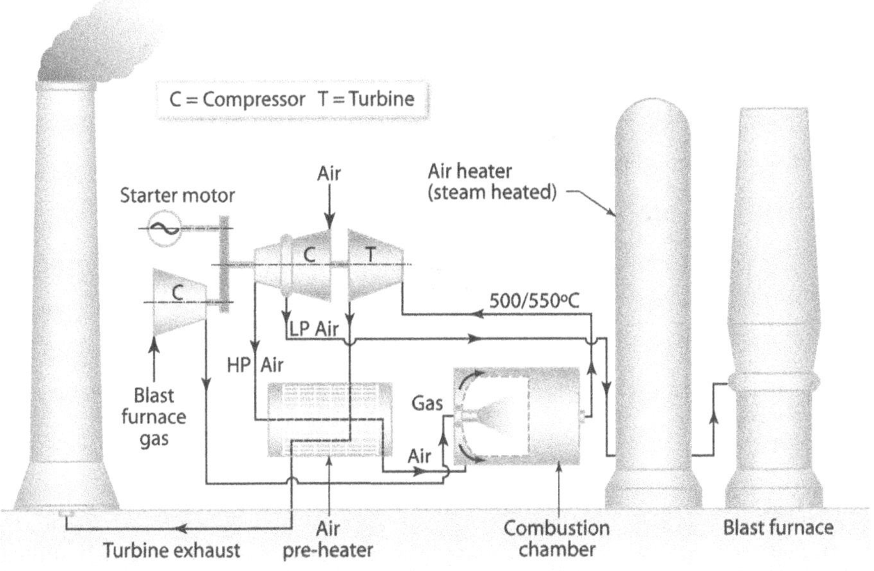

Figure 4-61   *Velox blower system for blast furnace* [219]   © *The Crowood Press*

Figure 4-61 shows the correspondingly applied GT blower scheme for blast furnaces. In principle all elements of the Velox system could be adapted, except that a special combustion chamber was required for the gas turbine. Two axial compressors, driven by one turbine, delivered air and blast furnace gas to the combustion chamber, Figure 4-62. The

---

[219] Source: Kay, German Jet Engine p. 198 f.

gas/air ratio was chosen so that a certain amount of excess air cooled the combustion gases down to a 500/550 °C level on the way to the turbine. After the turbine the hot gases passed through a pre-heater to raise the air temperature between the compressor and combustor. From the same compressor, air was tapped off at an intermediate pressure port and passed on to the blast furnace via a steam-supplied heater. Normally, the air flow to the furnace comprised 100,000 m³ of air per hour at a pressure of 1.2–1.85 bar above ambient, controlled by compressor speed. By 1945 two of these gas turbine blower units had been built by BBC Mannheim, each about 5,350 hp. They were destined for the Hermann Göring steelworks at Watenstedt near Braunschweig-Salzgitter, but actually only one was partly installed before the end of the war.[220]

Figure 4-62   *Furnace gas compressor for largest plant according to the Velox principle from BBC Mannheim, 1941 for H. Göring Steel Works at Watenstedt, Braunschweig-Salzgitter, D: Air and gas compressor 100,000 m³/h, double flow drive turbine at the rear, starter steam turbine in front*

4.3.1.4   Velox – the Final Stages

The Velox production comprised some 110 units in total. One of the last erected is shown in Figure 4-63. Early units achieved more than 150,000 operation hours, based on 5,000 h annually. From the 1960s onwards the boiler building business was licensed to the companies Duerr in Ratingen, Germany and Waagner Biró in Graz, Austria, while BBC only provided the Velox turbo-sets.

This promoted the development of the open-circuit constant-pressure gas turbine. The high thermodynamic efficiencies of both turbine components and the compressor increased GT entry temperatures to 550...600 °C in the short-term. As material expertise grew, this

---

[220] Source: BBN, Die Betriebskennlinien, Figure 14

was increased to more than 1,000° C. Accompanied by a better understanding of combustion processes at higher pressure levels, allowing large fuel quantities to be burned in relatively small combustor volumes without smoke, this ended investigations of the Holzwarth principle at BBC.[221] It also promoted the constant-pressure gas turbine as a simple, reliable and economically viable power source, primarily for the Houdry process, covered in the next section.

Figure 4-63   *Post-war Velox boiler installation at the municipal power station of the city of Basel, CH 1957*[222]*, steam generation 32,000 kg/h, from left to right: feed water heater, turboset and boiler – superheater – water separator group   ©ABB*

4.3.2   Houdry Refinery Process

4.3.2.1   BBC Equipment

The Velox operational principle of combustion at raised pressures was also used in chemical engineering for the first time in 1936, when demand for hot compressed air in connec-

---

[221] The Holzwarth-BBC contractual relationship was formally finished in 1949, see Seippel, Wie vollzog sich..., p. 5b

[222] Picture source: Pfenninger, The evolution, Figure 14

tion with the Houdry oil cracking process created the first field application for the newly emerging constant-pressure gas turbine. In many chemical processes industrial supercharging was impossible due to the high cost of compressing the air, but if gases from the process could be expanded in a turbine with enough power to drive the compressor, then supercharging would become economical.

Figure 4-64   *First Houdry turbocharger set with 20-stage axial compressor, $D_{1,Tip}$ = 0.82 m, and 5-stage reaction turbine* [223] *©ABB*

The sheer size of the required Houdry turbo-set was a challenge for BBC, with a volume flow of nearly 20 m^3/s at a pressure ratio PR = 4.2, Figure 4-64. The compressor unit with 20 stages was a considerable step forward in comparison to the previous Mondeville[224] and Zurich units, see point 6) in Figures 4-31 and 4-32. It was Henry Thomas (1881–1966), then chief engineer of Sun Oil Co. in Philadelphia, who realized the potential of expanding the BBC Velox axial turbo-technology to meet the demands of the revolutionary refinery process under development. Thirty years later Claude Seippel praised Thomas' dedication in his obituary, for favouring the BBC product without having any comparable technical reference.[225] His advance assessment did not disappoint: the excellent performance of the delivered turbomachinery meant the 4.4 MW compressor driving power was more than compensated by the 5.3 MW turbine output and a gearbox-coupled generator on the turboshaft produced electricity accordingly, Figure 4-64 and Figure 4-65. The turbomachinery was highly reliable from the start: a 1950 survey based on typically 3,000–4,000 possible operation days for the Houdry plants showed a turbomachinery-related downtime period of less

---

[223] Picture source: BBR, The Combustion Gas Turbine (A. Meyer), 1939, Figure 10

[224] See Hodge, Section 28. Industrial Process Applications

[225] See BBM, Mr. Henry Thomas, 1967. Thomas himself had filed a patent on a 'Process of (thermal) cracking of mineral oil' on 5 July 1928 which was granted as US 1938406 on 5 December 1933.

than 1.5 percent.[226] Mechanical erosion failures were identified as an isolated root cause of turbine blading defects, but were largely overcome after the installation of dust separators. The annual maintenance costs remained below 3 percent of the invested capital. The available information indicates that between 1936 and 1943, thirty-six such Houdry sets were built, six of these in Baden, while the remainder was licensed by BBC for Allis-Chalmers' US production. Standard turbo-set configurations ranged from 12 to 20 m^3/s, while one exceptionally large unit was installed at Tide Water, NJ with 30 m^3/s volume flow.[227]

Figure 4-65 illustrates this equipment in the final test set-up at Baden before shipment, where the refinery equipment was simulated by an auxiliary oil-burning combustion chamber. The elevated horizontal duct connects the compressor with the turbine section. The test revealed an 83.3 percent adiabatic compressor efficiency and a resulting excess turbine power of 1030 kW instead of the guaranteed 810 kW (with 510 ^{0}C turbine entry temperature). BBC had passed the initial learning curve in turbomachinery design.

Figure 4-65   *Testing of the first Houdry gas turbine set at Baden, CH 1936[228], from left: GT turbine component, GT axial compressor, gearing and generator, auxiliary combustion chamber on top*
©ABB

Although this test set-up at Baden[229] in August 1936 was sometimes designated as the true 'birth' of the gas turbine[230], the actual process sketch for the Houdry cracking plant in Fig-

---

[226] See BBM, Betriebserfahrungen, 1953, p. 144

[227] See Hodge, Section 28. Industrial Process Applications

[228] Picture source: Pfenninger, The evolution, Figure 15

[229] See BBC, Turbogebläsegruppe, 1936. The test report outlines also the complete first delivery from BBC for the new Markus Hook refinery of Sun Oil Co. in 1936: compressor/turbine turboset, 2.88:1 reduction gearbox and (1) auxiliary combustion chamber. The starter motor/generator came from Allis-Chalmers. In 2012, the fate of the traditional refinery was in danger.

[230] See Constant, The origins, p. 146

ure 4-66 illustrates that these facilities maintained more or less continuous compressor-combustor-turbine GT-type operation with a significant electric power output. The operational start of an industrial gas turbine can be pinpointed as 31 March 1937 when the Marcus Hook refinery was inaugurated, 30 km south-west of downtown Philadelphia on the banks of the Delaware River. During start up the rotor assembly was typically run for several hours by motor power M for equipment checks and a complete system warm up. During this period air was delivered via the bypass duct directly to the turbine. Combustor C1 in front of the turbine was then lit and the mass flow and pressure were increased by bringing the plant up to its design speed of 5,180 rpm within 8–10 minutes. The air was then directed to the various Catalytic Cracking Cases and combustor C2 was ignited to accelerate the warm up of the CCC units to prepare the regeneration of the catalysts. The regeneration of the carbonaceous deposits and the burning of the resulting vapour in the Combustion Case caused the gases from the crackers to supply the heat required and the auxiliary burners were shut down.

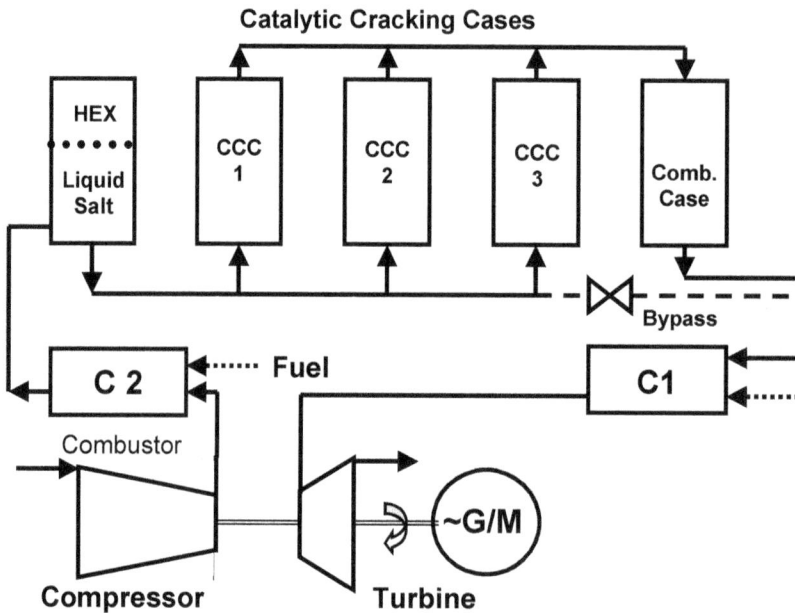

Figure 4-66   *Scheme of Houdry cracking plant during catalyst regeneration cycle* [231]

Within the CCC the packed bed of catalysts (C) is arranged between the cooling/heating tubes filled with liquid salt (S), Figure 4-67. The catalysts have an extraordinarily low thermal conductivity meaning operational switching from cracking (450 °C) to regeneration (510 °C) would require sustained time spans, as the adaptation to a new temperature level may not be supported by the salt-filled heat exchanger tubing which carries extra fin and extension surfaces to promote heat transfer. Some of the CCC's internal structure is also visible on the left hand side of Figure 4-69. It is remarkable that these process details

---

[231] See Hodge, Section 28. Industrial Process Applications

were part of a German publication in 1942, see also Footnote 232 of this Section, which confirms the later statement in Section 4.3.2.3: in view of the importance of the Houdry-based US aviation fuel production for war strategy, Germany had the expertise but no practical experience in catalytic cracking.

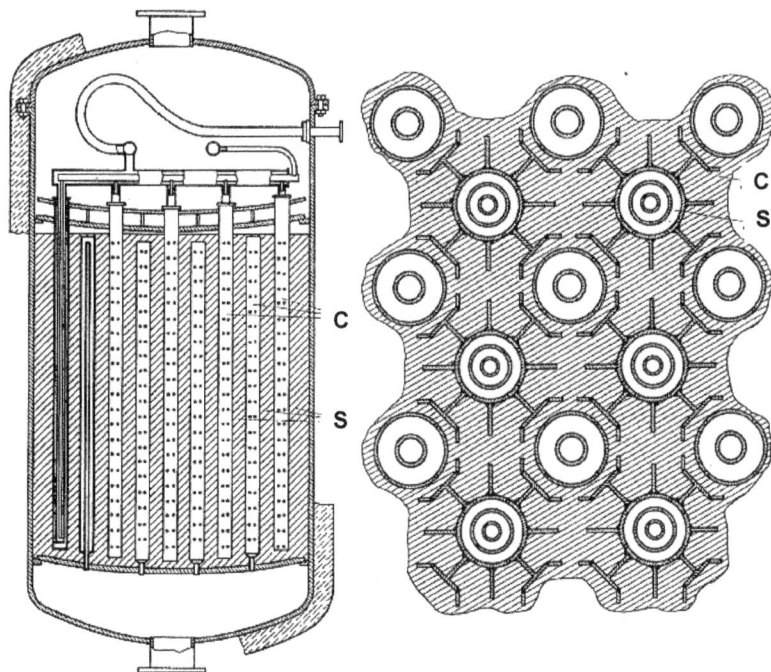

Figure 4-67   *Vertical (l) and horizontal (r) cross section of a Houdry catalytic cracking casing* [232]
© *1942 Springer*

4.3.2.2   Transforming Crude Oil into 100 Octane Petrol

Commercial petroleum production began in Titusville, Pa., USA in 1859 and the internal combustion engine was developed shortly afterwards. Crude oil is a mixture of hydrocarbon molecule compounds containing carbon and hydrogen atoms in different forms. Initially this mixture was separated by distillation into 'fractions' distinguished by their boiling points. Lower boiling fractions were used for lighting petroleum and the rest for machine lubrication, while little use was found for the actual petrol component. This situation changed considerably with the increasing number of automobiles, trucks and tractors at the beginning of the 20th century. A petrol shortage developed, until in 1913 Rockefeller's Standard Oil of Indiana (which later became Esso and Exxon) introduced a thermal crack-

---

[232]   Figure source: Marder, Motorkraftstoffe p. 308. The book has further details that the Houdry facility at Sun Oil's Marcus Hook refinery was erected by E.B. Badgers and Sons Co., Boston and that to Marder's knowledge, in 1941 in Germany – there were 12 Houdry plants in the United States, 2 in France (Raffinerie Vacuum Oil S.A.F., Gravenchon and Raffinerie de Berre, Berre l'Étang) and 1 in Italy (Raffinerie di Napoli, Naples); the actual number of installations at that time might have been twice that amount.

ing procedure that used high temperature pressure/steam to break down the larger, higher-boiling molecules into the smaller, lower-boiling fractions found in petrol. In due course it was recognised that more efficient engine performance could be achieved from a fuel that had a higher 'octane rating' as measured in a standard reference engine. This term describes the fuel's 'knocking resistance'[233], the advantageous quality of being able to withstand higher power-generating compression in a piston engine without exploding prematurely. The first significant increase in octane ratings was obtained in 1923 by adding TEL (Tetra-Ethyl Lead), an invention of General Motors.

Around that time the French engineer Eugène J. Houdry started his search for a catalyst to produce petrol from lignite.[234] Houdry, the son of a wealthy steel manufacturer, had visited the USA as a young man to attend the Indianapolis 500 where he competed in a Bugatti racing car. He was aware of the possibility that the sport was in danger of vanishing due to fuel shortages, especially in France where there were no oil reserves and they imported expensive, high-sulphur crude from Iraq. In 1922 a fellow racing driver showed Houdry a small bottle of petrol, said to be made from lignite, a low-grade brown coal plentifully available in France and Germany. A small laboratory operation was then set up near Paris. In 1929, with the promise of subsidies from France's Office Nationale des Combustibles Liquides and the backing of Houdry's family, a production plant (mine and refinery) to make petrol from lignite was built at St. Julien de Peyrolas, some 30 km north-west of Orange on the River Ardèche. Though the plant produced petrol, it was a financial disaster and the government withdrew further support.

But Houdry already had a second way of using crude oil instead of lignite for cracking. In a kind of trial-and-error method he tested hundreds of potential catalysts, with success, as verified by his Bugatti on a road near Versailles. The generated high-octane fuel accelerated the car easily and smoothly to 90 miles per hour! He had demonstrated that high quality fuel could be mass-produced without costly and technically limiting additives; he just needed the financial backing. In 1930 Houdry made his first contact with the US oil industry and moved to the US after successful demonstration tests. After some detours and excursions he came into contact with the management of the Sun Oil Company, a fortunate coincidence, since Sun was the only petrol-producing company which had resisted using TEL for the new high-compression engines from Detroit. Houdry came to Sun's Marcus Hook plant in Pennsylvania with the company offering nearly half a million USD to build a large-scale unit, Figure 4-68. In return Sun would receive the patent rights to any new technology developed there, of which there was a large amount. One of the first detections after testing thousands of potential catalyst materials was using Houdry's initial inven-

---

[233] Highly branched iso-octane, the octane which is least prone to knocking, is assigned a value of 100, while straight-chain n-heptane which knocks in practically any engine is assigned an octane number of 0. For example, a fuel with a knocking characteristic comparable to 70 percent iso-octane and 30 percent n-heptane has an octane number of 70.

[234] A 'catalyst' increases the rate of a chemical reaction solely by its presence, ideally without being changed or used up during the reaction itself. A catalyst would produce a considerable competitive advantage for the cracking of cheap, otherwise useless high-boiling fractions of petroleum into petrol. In the 1920s the science of catalysis was still in its early stage: most prominent at this time was a catalytic technique invented by the German Fritz Haber who had been using an iron oxide catalyst since 1909 to commercially fix nitrogen from the atmosphere to make nitro-fertilisers.

tion[235] regarding the suitability of Fuller's earth, a silica-alumina catalyst, in a derived form with added vanadium, chromium etc.

In addition, the engineers at Marcus Hook had two major problems to solve when expanding the laboratory process:

- regeneration of large volumes of catalysts by burning off tar and coal deposits which occurred after relatively short cracking operation times,
- efficiently balancing the heat requirements throughout the whole process of heat generation while burning off the 'poisonous' carbon from the catalysts and using heat to vapourise the crude oil feedstock.

The engineering team solved these problems one-by-one, led by Sun's research head Arthur Pew[236] and Henry Thomas. After determining a suitable airflow rate for regenerating the catalysts, they had to find a means of eliminating downtime during the regeneration cycle.

Figure 4-68   *The Marcus Hook refinery, Sunoco, Philadelphia, USA ~1990*   ©*Sunoco*

They did so by adding two (subsequently three) separate catalyst cases that could be automatically switched off- and on-line for batch operation. A typical catalyst case was approx. 10 m high and 3 m in diameter, filled with catalysts, then extruded hollow clay tube sec-

---

[235] Eugène J. Houdry applied for a patent entitled 'Catalytic materials and process of manufacturing' on 23 March 1932 which was granted as US 2 078 945 on 4 May 1937.

[236] See Pew, Operating Report

tions[237], later in crystalline form as in Figure 4-69. The optimum operation of a Houdry cracking plant was based on a 30 minute cycle: *'10 minutes on stream, 5 minutes to purge petroleum vapours with reduced pressure and steam, 10 minutes to regenerate the catalysts, and 5 minutes to purge combustion products from regeneration'.*[238] The optimum process heat control was achieved after finally abandoning a water-based cooling system to prevent catalyst overheating during regeneration for a molten salt substitute, indicated in Figure 4-66 by the Liquid Salt HEX Heat EXchanger on the left of the upper casing row and in Figure 4-67 by the tubing S. Using excess heat to preheat the crude oil feedstock and to provide steam for the rest of the refinery plant meant an excellent process heat balance and therefore much less energy consumption. A Houdry plant typically contained 400 t of molten salt at a continuous flow rate of 1 m^3/s.

Figure 4-69   *Sunoco Marcus Hook refinery: replacement of catalytic cracking units (l) and typical catalyst pellets (r), 1950* [239]

The timely axial turbomachinery development at BBC and the considerable extra risk taken when contacted by Sun's Henry Thomas in 1936 led to the unique convergence of two independent branches of mechanical and chemical engineering. The BBC contribution has not been explicitly noticed by the community of chemists yet, which rightfully praises the Houdry process and gave the Marcus Hook plant National Historic Chemical Landmark

---

[237] The best catalyst surface-to-volume ratios in chemical engineering have 'Raschig rings', hollow tube sections with equal length and diameter. Houdry's patented version came very close to this optimum shape.

[238] See Palucka, The Wizard of Octane

[239] Source: http://www.mindfully.org/Technology/2004/Eugene-Houdry-Octane1oct04.htm

status in 1966.[240] On 31 March 1937 Houdry unit Eleven Four came on line at Sun's Marcus Hook plant with brand new BBC equipment, Figure 4-70.

Figure 4-70   *The first BBC gas turbine for the Houdry cracking installations at the Marcus Hook oil refinery at the Sun Oil Co. Philadelphia, USA, 1936 [241] – with additional electricity output of ~900 kW [242]   ©ABB*

The plant was capable of cracking 15,000 barrels of residuum feedstock left over from a thermal cracking unit per day and it yielded 48 percent of 81 octane petrol from the start, twice as much as the previous thermal cracking output. Thick residuum from the fractionating tower was pumped into a still case where it was heated to 470 °C and vapourised. This vapour rose at a low pressure through the active catalyst cases where the long-chain hydrocarbons were cracked into smaller molecules. Another fractionation separated the resulting mixture into petrol and fractions with higher boiling points that could be put through the system again. In 1939 Sun Oil already had 10 Houdry plants in operation, based on further turbomachinery deliveries from Baden and the first, licence-produced equipment from Allis-Chalmers.

---

[240] Source: ACS American Chemical Society http://acswebcontent.acs.org/landmarks/landmarks/h:dr/index.html

[241] Picture source: BBR, The Combustion Gas Turbine (A. Meyer), 1939, Figure 9

[242] See BBM, Betriebserfahrungen, 1953 p. 144. The plant reached its 50th anniversary of operation and was scrapped afterwards, but BBC remained on the site with a newly installed GT8 gas turbine (personal correspondence with Sep van der Linden).

4.3.2.3   Strategic Importance and Side Effects

By catalytic 'reforming', i.e. passing the light petrol fractions from catalytic cracking through the Houdry units for a second time, Sun was able to raise the octane levels to aviation fuel quality. Units at Sun and other US licensees produced 90 percent of all catalytically cracked Allied aviation fuel in the first two years of American involvement in World War II. Starting in time to contribute essentially to the British success in the 'Battle of Britain'[243] (as some American sources claim), production rose to 40,000 barrels a day in 1941, climbing to 200,000 in 1943 and peaking in 1944 with 373,000 barrels per day, with a conceivable impact on the global war situation. In a contested article[244], Tim Palucka states: '*America's contribution was irreplaceable, because on the eve of the war, American companies had been extracting about 60 percent of world's petroleum, with the USSR accounting for 17 percent and Britain and the Netherlands most of the rest. The Axis powers extracted virtually no petroleum. The Germans had made great strides in producing liquid fuels from coal, and after their early territorial conquests, notably in Romania, they had ample oil supplies. But with almost no homegrown knowledge base in oil refining (and somewhat in contradiction to I.G. Farben's outstanding expertise in catalysis, also reflected in their corresponding cooperation with Standard Oil of New Jersey till the late 1920s), they were not able to catch up with the latest American advances, including catalytic cracking.*'

Houdry's invention generated wealth for him and added considerably to that of the Pew family, owners of the Sun Oil Co., founded in 1890 with headquarters in Philadelphia, run as a family-business since 1899. In 1937 he bought a 15 acre (5.3 ha) estate in the vicinity of the Pew family's land at Ardmore, Pa. which remained his home until his death in 1962. Today known as 'Clifton Wingates', Houdry called his estate 'Le Mesnil' ('Little Farm') Figure 4-71. After the war his neighbours brought zoning infringement charges against him for using his garage for manufacturing purposes. He had actually founded Oxy-Catalyst Inc. in Wayne Pa. to try and reduce the car emissions that he believed were causing an increase in lung cancer. He invented the first catalytic automobile converter for NOx and CO, ironically this was too early, due to the then widespread use of TEL which 'poisoned' his catalyst. His second great invention only became a success after his death when this fuel additive was banned in the late 20th century and 'unleaded fuel' became a worldwide standard.[245]

---

[243] The conflict cannot be solved with present information as will be outlined in further detail in the future book project on BBC's contributions to the early aero engine history: British sources acknowledge US deliveries came just in time to **compensate for** German air superiority in fuel quality, while US sources stipulate a **performance boost/superiority** for Spitfires etc. of 15–30 percent beginning at the time of the Battle of Britain in the second half of 1940 and based on Houdry-processed aviation fuel from the US. The latter position is convincingly supported by information on German synthetic aviation fuel production by coal hydrogenation.

[244] See Palucka, The Wizard of Octane

[245] This holds true today except for three countries: Yemen, Afghanistan and North Korea, and still for certain uses in aviation fuel.

Figure 4-71   *Eugene Houdry and his home, 1937–1962: 'Clifton Wingates'* [246] *at Lower Merion Township, 15 km north west of Philadelphia*   ©*Getty Images*

Sun Oil's success owed much to the war. It supplied most of the lubricating oils used by the Allies in WW I and as outlined before was a leading supplier of aviation fuel in WW II. Sun Shipbuilding was the Pews' second major commercial asset. Its first ship was built in 1917, just in time for America's entry into WW I and over the next 60 years they built 550 more ships[247], beginning in the 1920s with tankers for Rockefeller's Standard Oil Co. During WW II 40 percent of US oil tankers were said to have been built by Sun Co. In 2002 the Pew family foundation had assets of about USD 4 billion. Sunoco, now operating mostly in the Athabasca oil sands in Alberta and the North Sea, reported a net income of USD 312 million in 2003.[248]

---

[246] Source: http://www.flickr.com/photos/road_less_trvled/2570450819/ and E. Houdry portrait inset: http://chemgeneration.com/at/milestones/herstellung-von-benzin-aus-roh%C3%B6l.html After 1979 one of the later owners of 'Clifton Wingates' was Theodore 'Teddy' D. Pendergrass (1950–2010), a black R&B recording artist, part of the famous Philly Sound, who had six platinum records (1 million sales each) between 1977 and 1981. He even continued his career after becoming wheelchair-bound due to a tragic traffic accident with his Rolls-Royce car near Philadelphia in March 1982, singing together with artists such as Whitney Houston. According to a 1979 Ebony magazine article the house, originally furnished with Houdry's French antiques, was now filled with 300–400 stuffed teddy bears, all gifts from his devoted followers.

[247] An example with surprising BBC context is MS 'Andyk', as shown in Figure 5-82.

[248] http://coat.ncf.ca/our_magazine/links/53/pew.html.

BBC's engagement in the Houdry catalytic oil cracking project meant not only the final breakthrough for the industrial gas turbine, but also had further ramifications for solid fuel burning. Walter G. Noack's very last engineering activities before his early death in 1945 addressed problems with continuous turbocharged coal combustion, Figure 4-72. Coal was grinded in a beater mill and fed by means of a pressurised air stream to the combustion chamber. The figure illustrates (from left to right): the generator, the speed reduction gearbox and the axial compressor, above the compressor the air heater, followed by the gas turbine component and on its right side the combustion chamber. The two dark tubes above the combustor are the coal dust feed pipes from the beater mill to the combustor entry; the beater mill itself is located behind the stairways. The testing of the coal feeding process was hampered several times by severe coal dust explosions for which no cause could be identified immediately. Only after metallic friction and ignition (from metal pieces, screws etc.) was excluded, the danger of self-ignition by accelerated oxidation of the freshly ground coal under pressure was fully realised. As a counter-measure the entire coal ducting had to be designed thoroughly without dead ends and pockets to avoid the possibility of deposit building up.

Figure 4-72   *BBC coal combustion test facilities 1943*[249]: *left – coal-dust-fired gas turbine, 1.6 MW, coal consumption 1.3 t/h, GT speed 4,800 rpm, geno speed 3,000 rpm, right – coal dust-fired Velox boiler   ©ABB*

Coal dust combustion for direct application in gas turbines could not be brought to success at that time; despite satisfactory dust separation experiences in the Houdry project there were obviously still enough flying ash particles making their way to the turbine blading to

---

[249]   Picture source: Pfenninger, Vergangenheit, Gegenwart und Zukunft, Figure 30

cause severe erosion effects on both the stator and the rotor after just 250 operation hours. Similar negative experiences with flying coke deposits from early oil-fired GT plants have been successfully solved in the meantime. The coal ash problem was finally technically overcome by pressurised coal gasification; but even this approach was not economically convincing. In-depth studies of the gasification process revealed thermal losses of up to 10 percent of the theoretical coal heat due to ash removal, necessary cooling etc. The additional price of the pressurised coal gasifier hardware reduced the economic attractiveness of the whole set-up further in comparison to a conventional steam plant.[250] This brought the efforts at BBC to burn coal in gas turbines to a temporary halt. Future energy developments will show if this pioneering work will eventually gain importance again. Within a list of 15 BBC patents[251] about turbocharged gasification of solid fuels (coal), obviously collected in 1974 during BST separation talks, there is a hint at the ultimate coal gas turbine exploitation: Patent CH 250 740, dated 6 June 1946, describes a GT process to pressurise and burn out small, uneconomic seams in a coal mine and to direct the hot exhaust to a turbine for electricity production.

4.3.3 The World's First Utility Gas Turbine Set at Neuchâtel (1939)

4.3.3.1 A Consequential Engineering Harvest

After the development of Velox boilers, turbosets for the Houdry plants and numerous stand-alone axial compressor projects, the time was ripe for BBC to launch a gas turbine project solely for the generation of electricity. The Houdry turbomachinery experiences had demonstrated the excellent compressor efficiency, so that the turbines produced a continuous power surplus that could be used for electricity generation. The turbomachinery only had to be scaled in size from the Houdry design speed of 5,180 rpm to the European 50 Hz net frequency of 3,000 rpm, but the combustor represented an all-new design challenge. Initial cladding of the combustor walls used firebricks that crumbled over time. In 1947 this was changed to metallic, ribbed Martensite[252] bricks, but these could not survive the strong thermal shock stresses either and cracked and endangered the turbine. Only a cooled combustor steel jacket extended through the outer wall of the subsequent transition duct up to the turbine entry[253] brought lasting success.

Since 1902 the Service d'Électricité had operated a 300 kW steam power plant within the city of Neuchâtel[254] (then 30,000 inhabitants) for peak loads at their Centrale Thermique at Quai de Champs Bougin, sufficient for the city's electricity demand at that time. After a

---

[250] See Pfenninger, Vergangenheit, Gegenwart und Zukunft

[251] See BST, BBC Patente

[252] Martensite, named after the German metallurgist Adolf Martens (1850–1914), usually refers to a very hard crystalline steel structure, but can imply any crystal structure formed by diffusionless transformation. Martensite is formed by the rapid cooling of austenite. The martensitic transformation is the best known example of this kind of phase transformation. Shape memory alloys have recently gained significant interest based on the same effect.

[253] See Figure 4-79.

[254] See Footnote 2 in Section 1; Neuchâtel lies some 120 km south-west of Baden CH – BBC's engineering and production site.

necessary power rise to 1,100 kW in 1905, the next increase to 4,000 kW became necessary in the 1930s. First, two variants were alternatively investigated: diesel vs. Velox. The fear of war required a bomb-proof installation in a space-limited cavern. The whole installation environment was depicted in a meticulously precise model, Figure 4-73 and in reality on the right side of Figure 4-75. After lengthy negotiations and with considerable courage in promoting an all-new concept, Adolf Meyer won out with his proposal for a gas turbine stand-by unit, mainly based on the argument of lower capital costs. Advantages of the gas turbine were the low space requirements and lower operation and maintenance costs, making the disadvantages of lower thermal efficiency and therefore higher fuel costs economically acceptable. In 1937, BBC offered two design variants to the Services Industriels de la Ville de Neuchâtel, either a geared 3 MW gas turbine at 3,600 rpm or a gearless 4 MW gas turbine with a speed of 3,000 rpm. The 1938 order was for the latter version, which BBC developed and built in the record time of a year, finishing mid-1939.

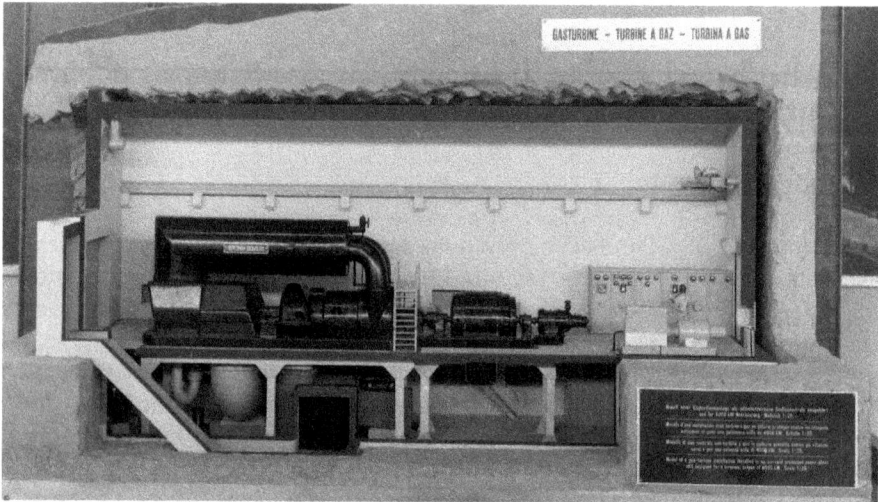

Figure 4-73   *Cavern installation of the GT Neuchâtel power plant (1939 model)* [255]

Figure 4-74 illustrates the principle set-up of this single stage gas turbine and typical temperatures of the air/gas stream. This power plant in its simplest form consisted of a compressor 1, a single-can combustion chamber 2, a drive turbine 3, a generator 4 and an electric starter motor 5 with exciter unit 6. The originally offered option to install a 110 kW diesel engine in case of a complete breakdown of the electricity network was abandoned. The group was started by an electric asynchronous motor with slip ring rotor and an adapted resistor timing to limit the current pulses. The synchronisation speed was approximately 30 percent of the gas turbine speed, so that the directly-coupled starter motor had to be designed for three times the synchronisation speed. A light fuel oil was burned which did not need to be pre-heated. The engine could be accelerated from idle to full speed within 8 minutes. It had a power output of 4 MW at the generator terminals. Rotating at

---

[255] Picture source: BBC, Collection of GT Neuchâtel original photographs, 1939–1940, ©ABB

3,000 rpm, the turbine, with an inlet temperature of 550 °C, provided 15.4 MW, of which the compressor at an air inlet temperature of 20 °C absorbed 11.4 MW.

In comparison to the previous Holzwarth GT design approach, BBC had completely omitted any kind of water cooling by using only surplus air for cooling purposes. Given the steel temperature limits of that time (around 550 to 600 °C), this meant a surplus airflow of four times that of theoretical combustion requirements.

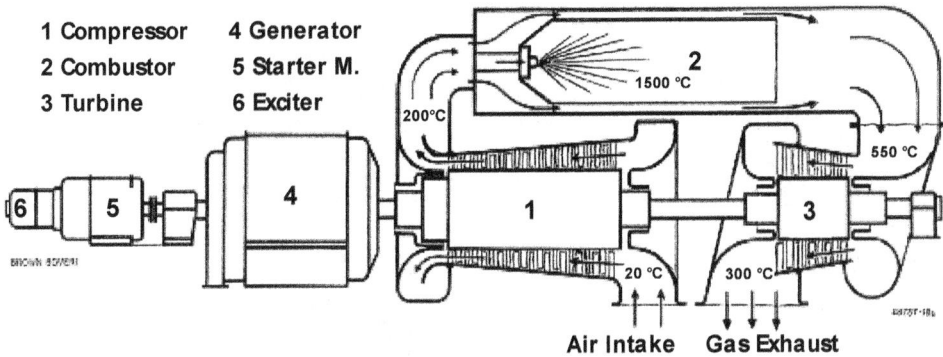

Figure 4-74   *Principle layout of the single-stage power generation gas turbine* [256]

The acceptance tests were commissioned to the nestor in the field of thermal machines, the 80 year old ETH professor Aurel Stodola. This decision certainly took Stodola's international reputation as an independent expert into account, not least in view of some unpleasant discussions which had accompanied similar performance assessments of Holzwarth's explosion gas turbine a few years ago[257].

4.3.3.2   Certified by Stodola and at the 'Landi 39'

The almost ceremonial spirit of these acceptance tests is captured in Figure 4-75 (left). Stodola with his frock coat and the traditional bowler hat of the senior engineer is framed by Jean von Freudenreich[258], head of the BBC TFVL 'Versuchslokal', who has a more 'modern', nevertheless class-distinctive attire with his light linen suit. The only blue collar

---

[256] Figure source: Pfenninger, The evolution, Figure 19

[257] Constant, The origins p. 97 has further details: '... *Holzwarth, then, stood accused of violating perhaps the cardinal norm of scientific engineering practice, that performance test results should be clearly and unambiguously reported. The invocation of such norms is clearly a means by which a community defends traditional practice. ...*'

[258] Jean von/de Freudenreich (1894?–1959), from Berne, joined BBC in 1914. Apart from his role in re-hiring Claude Seippel, see Footnote 19 in this Section, he was known for his fundamental research activities, mainly investigating/introducing wedge-type journal bearings which were beneficial for BBC. Seippel in his obituary: '... *Jean de Freudenreich fut un grand ingénieur, d'une originalité rare, une personalité telle que notre monde bouleversé en a si besoin. ...*' (JdF was a great engineer of rare originality, a personality of which our world, presently upside down, required more...)

worker and note taker, Hans Bellati[259], is on the right. According to Stodola's notes, the decisive full load test was carried out between 10:10 and 11:10 am on Friday 7 July 1939[260].

Figure 4-75   *GT certification by Stodola (80) at BBC Baden on 7 July 1939 (l) and GT cavern installation at Neuchâtel, CH 1940 (r)*[261]

Stodola's own words set the no-nonsense mood: '... *The load and governing tests were performed on 7 July 1939 at the maker's works in Baden under my personal supervision, with the assistance of the officially recognised Swiss Association of Steam Boiler Proprietors and the test laboratory staff of the Swiss Association of Electrical Engineers for the power measurements. The temperature and pressure measurements were taken by the firm's personnel. The calorific value of the fuel was determined by the fuels department of the Swiss Federal Material Testing Laboratories. ...*' A decisive sentence follows: '... *In spite of the simplicity of the arrangement the test showed the efficiency referred to the heat contained in the fuel, and the heat equivalent of the electrical output of the generator to be 17.38 %*[262], *representing an efficiency which, to-*

---

[259] Hans C. Bellati (1916–1988) started at BBC's test lab in 1937 with a starting salary of 240 SFr/month. In 1954 he moved to steam turbines where he led the design department between 1955 and 1981, his highlights being the design of the BBC 2-shaft 1,300 MW steam turbine and authoring a famous internal design handbook.

[260] Independent, if the start of the continuous, additional electricity production of the Houdry plants on 31 March 1937 at the Marcus Hook, Pa. refinery is taken as reference, or the 7 July 1939 at the BBC premises in Baden CH, the industrial gas turbine era has the lead in reference to the first flight of an GP-powered Heinkel HE 178 aircraft in the early hours of Sunday 27 August 1939 at Rostock-Marienehe. See also Footnote 3 in Section 1.

[261] Picture source: Pfenninger, The evolution, Figures 21 and 22

[262] This value has to be put into the perspective of today's 30–40 percent efficiency range for simple cycle gas turbines. From the beginning the gas turbine was characterised by performance deficits in comparison to con-

gether with the many constructional advantages of the set, renders it in many cases competitive. An especially noteworthy feature is the fact that the plant requires no cooling water. ...'

Summary of the documented data of the 'Neuchâtel Gasturbine' at the guarantee conditions of 20 °C inlet temperature and 3,000 rpm speed[263]:

– thermal efficiency	17.38	%
– fuel consumption	0.496	kg/kWh
– generator power	4	MW
– mass flow	62.2	kg/s
– volume flow	55	m³/s
– compression ratio	4.39	
– compressor efficiency ad.	84.9	%[264]
– expansion ratio	4.27	
– turbine efficiency ad.	88.4	%

Only two weeks after the date of the GT acceptance test, on Wed 19 July 1939, Stodola's planned publication of the test results also occupied the directorate of the Thermal Department, and the corresponding minutes[265] provide one of the very few insights into the political conception of the upper management at this critical time, six weeks before the outbreak of war: '... *Herr Faber informs that Herr Prof. Stodola intends to publish the article about the gas turbine in the "Zeitschrift des Vereins deutscher Ingenieure", which we do not really consider to be appropriate (...was uns nicht gerade zweckmässig erscheint)....'* It can be assumed that Stodola merely wanted to continue his proven path in this respect, and the VDI-Z had clearly been the top German technical journal in the decades before. In the end the conflict was resolved wisely, certainly with the sanction of BBC's board of directors. The article appeared in:

1. Engineering[266] UK	5 January 1940
2. Zeitschrift des Vereins Deutscher Ingenieure[267] D	6 January 1940
3. Schweizerische Bauzeitung CH	13 January 1940
4. Brown Boveri Mitteilungen[268] in German and Brown Boveri Review[269] in English	April 1940

At the end of the remarkable article Stodola gave a far-sighted hint regarding future BBC/ABB/Alstom gas turbine engineering. After addressing known standard methods to increase the plant efficiency by a) exhaust heat recuperation, b) adding a combined–cycle steam turbine, c) raising the combustion temperature, he continues with d) the introduc-

---

temporary steam turbines, which already achieved 25 percent at that time. This excluded the gas turbine from the base load power generation for nearly 50 years, and the situation only changed when both concepts were successfully merged into CCPP, Combined-Cycle Power Plants.

[263] See BBM, Leistungsversuche, 1940 and – see Stodola, Leistungsversuche

[264] The published adiabatic compressor efficiency values vary from 84.6–84.9–86.6.

[265] Direktionssitzungs-Protokoll, Vol. 49, Baden 19.07.1939, Traktandum 8, http://www.ethbib.ethz.ch/exhibit/stodola/stodola05.html, original document at the historic archive of ABB Switzerland, Baden CH.

[266] See BBR, Load tests...1940, Engineering

[267] See Stodola, Leistungsversuche

[268] See BBM, Leistungsversuche, 1940

[269] BBR, Load tests...1940

tion of a split-shaft turbine arrangement with separate compressor and generator drive and last but not least:

> '... e) Especially noteworthy are the possibilities of raising the efficiency by fractional-ised combustion, consisting of sub-dividing the gas turbine into two or more stages, and reheating the gases in an intermediate combustion chamber by means of additional in-jected fuel, coupled with inter-cooling of the air compressor. ...'

Stodola's early reference[270] to the two stage combustion concept found its first application in BBC's Beznau power station (1948–1950) and finally, 55 years after Stodola's recommen-dation, in the unique 'sequential combustion' configuration of ABB/Alstom's GT24/GT26 gas turbine family which proved all of Stodola's predictions.

Besides the official acceptance test on 7 July 1939, the final TFVL testing[271] at Baden took place between May and July 1939. However, the first phase (until mid-June 1939) was only in a limited speed range below 2,000 rpm, due to rotor cracks which were observed during manufacturing of the compressor section. A comprehensive reconstruction phase took place between 15 and 28 June 1939.

From mid-July 1939 the Neuchâtel gas turbine was displayed at the famous Swiss national expo, which took place between Saturday 6 May and Sunday 29 October 1939. Figure 4-76 illustrates the Neuchâtel gas turbine in the BBC section of the exposition environment. A highlight was the daily full-speed GT demonstration (without load), attracting large crowds, presumably from September until the end of the exposition period. The reason for the delayed start of the GT exposition demo runs was a cast combustor cover plate that was damaged during the full load testing at Baden in July. The visitors therefore saw the Neuchâtel gas turbine in the early weeks of its presentation only as a static display with a black-painted, wooden(!) combustor front plate[272].

The 1939 National Exhibition in Zurich, in German 'Landesausstellung' or in Swiss dialect 'Landi', was originally planned as an industrial fair displaying the latest 'high-tech' prod-ucts, like at similar preceding events in 1896 and 1914. However, the outbreak of war and the threat of Nazism added a new aspect and the fair is now remembered for promoting a 'spirit of national defence'. Visiting the 'Landi 1939' was considered a kind of national pil-grimage, expressed by the stunning fact that 10.5 million tickets were sold, measured against a total population of close to 4.2 million at that time.

An impression of the special spirit of this exhibition ('Landigeist') is shown in Figure 4-77, a unique combination of modern, transparent architecture, products of technical excellence and the polite cheerful attitude of the visitors, here shown by the 'Schifflibach', a tour of the expo in small boats.

---

[270] In his 1940 VDI-Z article Stodola actually gave the credit for proposing the 2-stage-combustion to G. Man-gold, VDI-Z, vol. 81, (1937), pp. 489–493. Before Stodola, Ad. Meyer had also discussed potential positive ef-fects of heat regeneration and two-stage-combustion in his presentation at IMechE London on 24 February 1939, see BBR, The Combustion Gas Turbine, 1939.

[271] See BBC, Gasturbine Neuchâtel, TFVL 1225, dated 1 April 1940, the timing of which indicates that the installa-tion on site at Neuchâtel, Figure 4-73 and Figure 4-75 (r), had happened in the meantime, presumably in March 1940, although without official commissioning.

[272] Private information from Olivier Seippel (1926–2012) on 3 August 2009 during a joint visit of the Birr Tech-nical Landmark GT pavillon.

Figure 4-76   *The world's first gas turbine at the Swiss National Expo, Zurich 1939, from left to right: starter motor, generator, lubrication pump, compressor, turbine and combustion chamber, horizontally on top* [273]

Figure 4-77   *The 'Schifflibach' – another Landi '39 attraction* [274]

---

[273] Picture source: BBC, Geschichte des Schweizer. Turbomaschinenbaus, 1982

[274] Picture source: http://archiv.ethlife.ethz.ch/articles/tages/ethistla39.html K. Angst/A. Cattani (Hsg.): Die Landi: vor 50 Jahren in Zürich. Erinnerungen, Dokumente, Betrachtungen. Stäfa 1989.

4.3.3.3   After 62 Years in Service – An ASME Technical Landmark

Commissioned in 1940, the GT Neuchâtel remained in service for 62 years. After accumulating 1,908 starts, 7,283 OH and 15,895 MWh of effective energy[275] produced, damage occurred to the generator and the plant was permanently closed on 18 August 2002.

In 1988 the ASME (American Society of Mechanical Engineers) had already awarded the GT Neuchâtel Historic Mechanical Landmark status, reserved for milestones of outstanding technical development. The ASME History and Heritage Landmark Programme began in 1971. The 4 MW gas turbine at the municipal power station in Neuchâtel, Switzerland was the 26th International Historic Mechanical Engineering Landmark and the 8th outside of the United States. In 1988 the 7 other landmarks outside of the US were in Britain (3), France, Germany, Australia and China. On 2 September 1988 the ceremony[276] took place on the original site in the cavern at Neuchâtel's Quai de Champ Bougin. At 11:47 the gas turbine was synchronised to the net to demonstrate its functionality and a memorial plate[277] was unveiled:

> **International Historic Mechanical Engineering Landmark**
> **4MW Gas Tubine, Municipal Power Stration**
> **Neuchatel, Switzerland**
>
> This gas turbine was the first successful electric power generation machine to go into commercial operation. It was designed and constructed by A.G. Brown Boveri, Baden Switzerland, and installed in 1939. It has since served continuously as a stand-by unit.
>
> **The American Society of Mechanical Endineers**
> **SIA Swiss Engineers and Architects Association**
> **1988**

In 2005 under the leadership of Walter Graenicher, then President of Power Service and of Alstom (Switzerland) Ltd., Alstom decided to relocate, restore and display the machine at Alstom's development and production facility in Birr, 14 km south-west of the Baden headquarters where the latest gas turbine technology is being validated. The ASME, in appreciation and acknowledgement of this preservation of a vital part of mechanical engineering history, re-designated this landmark in its new location on 4 June 2007. The following four pictures illustrate the GT Neuchâtel hardware at the new site.[278]

The axial compressor, Figure 4-78, has, except for the front stages, 23 stages of 50 percent reaction, still based on the early Goettingen profile design with untwisted 'repetitive blading', identical in the rear stages. At the 3,000 rpm design point it produces a total pressure ratio of PR = 4.39, with a mass flow of m = 62.2 kg/s and adiabatic efficiency of η_{ad} ~ 85 percent. The tip diameter at entry is $D_{1\,Tip}$ = 1.41 m, corresponding at nominal speed to a tip speed of u_{Tip} = 222 m/s and a relative tip Mach number $Ma_{1\,Tip,\,rel}$ = 0.72. The constant

---

[275] See BBC, Neuenburg – Hot Gas Path Inspection, 2002
[276] See Zimmermann, Vom Pionier zum Leader
[277] http://files.asme.org/asmeorg/Communities/History/Landmarks/12281.pdf
[278] The official ASME Landmarks list http://www.asme.org/about-asme/history/landmarks has the GT Neuchâtel as #135;. Visits and guided tours can be arranged via Alstom Birr, Portier 1, Tel +41 (0)56 466 6156, see also Lee Langston's informative travel report of 2010: Langston, Visiting the Museum

hub diameter has $D_{V\,Hub} = 1.07$ m. The relative position of the GT Neuchâtel compressor design in comparison to other projects is visualised again in Figures 4-31 and 4-32, where the Neuchâtel engine is shown as Point 8).

Figure 4-78   *Original GT Neuchâtel axial compressor*

The 7-stage uncooled axial reaction turbine (Figure 4-79) has an expansion ratio of 4.27 and an adiabatic efficiency of 88.4 percent. The turbine's outer diameter increases from $D_{5\,Tip} = 1.25$ m to $D_{6\,Tip} = 1.6$ m and the last three stages have damper wires at the outer rim. The hub diameter is again constant at $D_{T\,Hub} = 0.95$ m. This figure illustrates one of the early BBC design 'secrets', never publicly shown or discussed, but played an important role for the reliability of the whole configuration: The outer wall of the transition duct is actually double-walled from the combustor outlet to the turbine entry and hides a cooling air stream which is fed from the upstream combustor cooling jacket. It is very likely that this improvement was introduced together with the aforementioned combustor modifications in 1947.

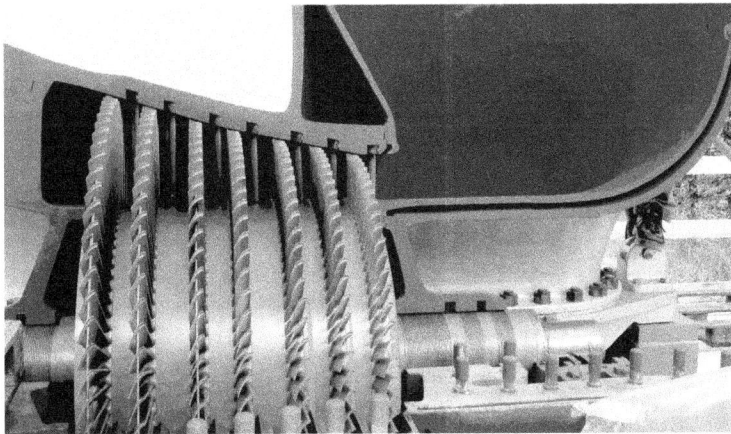

Figure 4-79   *Original GT Neuchâtel axial turbine with air-cooled gas duct to turbine*

Figure 4-80 shows the generator: on the cold side of the gas turbine, a design principle of all configurations from Baden. Referring back to the beginnings of axial turbomachinery development, more than 10 years before the realisation of GT Neuchâtel, the tapered blading of the single-stage generator cooling blower is clearly visible.

Figure 4-80   *The axial turbomachinery development 'starter': original GT Neuchâtel generator with axial blower wheel*

Finally, Figure 4-81 shows the ASME Historical Landmark showcase pavillion[279] with the GT Neuchâtel set-up at Alstom's rotor manufacturing facility in Birr.

Figure 4-81   *The ASME Technical Landmark, now relocated to Birr, CH 2007* [280]

---

[279] Designed by Lukas Zehnder, product designer from Ehrendingen, CH.

4.3.4   First BBC Power Generation Gas Turbines

The next milestone in gas turbine development was in 1946 when BBC put the first 10 MW power generating unit into operation at Filaret power station at Bucharest, Romania. This new power plant concept is depicted in Figure 4-82. Characteristic features are the 2-stage compression with intermediate cooling and the 2-stage expansion with intermediate re-heating. Both measures increased the efficiency and the power output so that the cost of installed kW power was considerably decreased. The ancestry of the BBC reheat cycle was especially addressed in the mid-1990s when the new GT24/GT26 engine generation came to market with the 'sequential combustion' concept.[281] The Beznau plant of 1948 was then taken as an early reference, but this closer view reveals that the foregoing 10 projects of Figure 4-84 could have founded even earlier origins.

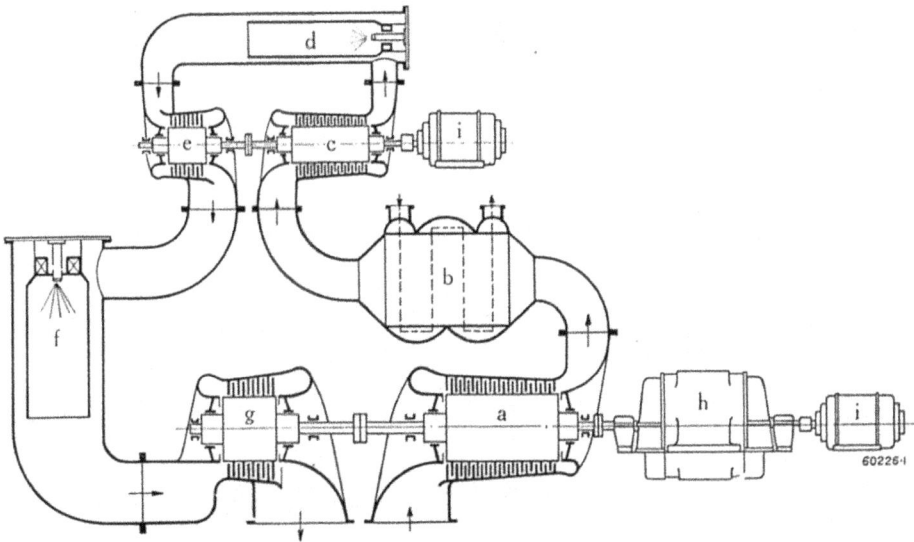

Figure 4-82   *Scheme of the 2-stage gas turbine plant for Filaret municipal electricity works at Bucharest, Romania, 1946[282] a – lp (low pressure) compressor, b – intermediate air cooler, c – hp (high pressure) compressor, d – hp combustor, e – hp turbine, f – lp combustor, g – lp turbine, h – generator, i– starter motor   ©ABB*

The risk of putting two axial compressors in a serial arrangement was taken for the first time in power generation, thus achieving an overall total pressure ratio of 10.[283] The idea to install an additional heat exchanger at Filaret for exhaust heat recovery was turned down, as this power plant was to be used primarily as a peak-load unit with low investment cost as a priori-

---

[280] See Alstom, The World's First Industrial Gas Turbine Set

[281] The CAES Compressed Air Energy Storage plant at Huntorf, Germany, Section 5.5.3, which started operation in 1977, is considered an important intermediate step of the application of this concept.

[282] Figure source: BBM, Betriebserfahrungen, 1953, Figure 4

[283] This statement applies for BBC Baden. As seen in Figure 4-45, before 1945 BBC Mannheim had already accomplished the risky combination of several axial/centrifugal compressor stages and a corresponding surge prevention concept at the WVA Kochel hypersonic wind tunnel test facility.

ty. The extensive (and costly) use of heat exchangers to raise the GT process efficiency characterised BBC designs (and that of their immediate Swiss competitors Escher Wyss and Sulzer) in the coming 25 years. There was nominal success with respect to efficiency values, but also sometimes formidable operational drawbacks, especially for low-BTU fuel. It was only in the 1970s when turbomachinery component improvements led back to the simple GT process.

Due to post-war difficulties the unit was only put into operation in 1951. The acceptance tests were carried out at Baden from March to September 1946[284] as part of BBC's first post-war customer conference, Figure 4-83. The guaranteed power output was actually surpassed by more than 2 MW. Taking into account the non-optimal installation conditions for these tests, with higher entry and exhaust duct losses, the corrected installed power was 12,020 MW at 20 °C and the corrected coupling efficiency was 23.72 percent, in comparison to a guarantee value of 21.6 percent.

The maximum turbine entry temperatures were between 566 °C (hpt) and 573 °C (lpt) and thus considerably below the material temperature limit of that time: 600 °C. The Filaret air mass flow at full load was 26.5 kg/kWh compared to 55.7 kg/kWh for the 4 MW Neuchâtel facility, which illustrates the increased compactness and consequent cost reduction for the 2-stage set-up.[285]

Figure 4-83   *BBC's first gas turbine with intercooling and sequential reheating – for 12 (10) MW power plant at Filaret, Bucharest, Romania, – work test demonstration at Baden, CH customer conference 1946[286]   ©ABB*

---

[284] See BBC, Gas turbine Filaret, 1946

[285] See Kruschik, Die Gasturbine, pp. 574–577

[286] Picture source: BBM, Ein neuer Schritt, 1946, Figure 8. The BBC Customer Conference 1946 claims to have had an international audience of 500 participants.

Average turbine and compressor efficiencies were determined to be in the order of 89 and 86.9 percent, respectively. The lp speed was intentionally chosen as 3,000 rpm so that the generator could be directly coupled with the lp spool. BBC carried out comprehensive studies on the optimum split of lp/hp compressor and turbine heads.[287] These studies revealed considerable operational advantages for a 2-stage compression/expansion configuration **with** exhaust heat recovery when the generator was coupled to the hp spool. This configuration, with considerably better part-load performance, was consequently selected for the Beznau plant a few years later, Section 5.1.5.2.

Figure 4-84 shows a comprehensive survey of the first 10 GT power generation projects after Neuchâtel, all ordered within the time frame until 1945. Different to the Neuchâtel configuration all of the following projects had separately standing, vertically arranged 'silo' combustors. As an example the hp combustor is visible in Figure 4-83 to the left of the lp combustor. Difficulties with the durability of the combustor internal cladding had already hampered the Neuchâtel unit. The BBC test reports for the Filaret and the Chimbote plants[288] documented ongoing problems in this area. Chimbote (ordered in 1945 and put into operation in 1949) was a single-stage gas turbine without exhaust heat recovery, very similar to the Neuchâtel design. In comparison to the guarantee values of 4 MW power and 18 percent efficiency, the actual values were 4.225 MW and 19.5 percent. The temperature distribution upstream of the turbine showed very pronounced distortions. Seen in the direction of flow, there were temperature differences of up to 180 °C between the left and right turbine sides. A direct consequence was a deformation of the gas turbine (casing) cylinder, with friction in the hp packing gland and two turbine blade rows. Similar observations were made for the Filaret hp turbine inlet flow, though the Filaret lp turbine showed a very uniform, circumferential temperature distribution. The part of the hp turbine entry casing in the immediate vicinity of the hp combustor again showed local temperature differences of up to 136 °C.

The Chimbote combustor had internal cladding with SiC stones from the Swiss porcellain factory Langenthal[289]. The brick lining started to deteriorate already after 12 Operation Hours (OH), the stones 'ballooned' and cracked. A test liner with 'fused magnesia' stones from Thermal Syndicate Ltd., Wallsend, Northumberland, UK could not convince either. Similar experiences overshadowed the Filaret test run, where the SiC stones of the hp combustor liner deteriorated after 30 OH. Only the introduction of metallic, ribbed Martensite bricks in 1947, Footnote 252 of this Section, brought a significant improvement.

For high thermal efficiencies and large power outputs Brown Boveri provided multi-stage gas turbines with regenerators at an early stage. The first two-shaft gas turbine installation was built for the Santa Rosa electricity works of the Lima Light and Power Co., Peru, Figure 4-84, No. 3. The generator was coupled to the hp shaft, thus allowing the speed of the lp set and the quantity of air delivered to be adjusted to the load conditions, making it possible to work at high temperatures even at partial loads, with acceptable efficiencies and correspondingly low fuel consumption. The installation was first set to work on 20 November 1949, but was put on hold between January and May 1950 in order to incorporate experience gained at Beznau, CH. From 15 May 1950 the GT set ran for 10–17 hours a day.

---

[287] See BBR, The determination, 1953

[288] See BBC, Gasturbine Filaret, 1946 and – see BBC, Gasturbine Chimbote, 1946

[289] Approx. 50 km south-west of Baden, half way between Baden and Berne.

No.	Order Year	Client	Plant	Power [MW]	Remarks fuel, speed turb./drive, type of drive
2	1943	Soc. Gen. de Gaz si Electrici-tate din Buchuresti, Bucharest, Romania	Filaret power station	(10) 12	natural gas, 3,000/3,000 rpm, generator, delivered in 1946, operational 1951
3	1945	Empresas Eléctricitas Aso-ciadas, Lima, Peru	Santa Rosa, Lima power station	10	oil, 3,000/3,600 rpm, genera-tor, with hex
4, 5	1945	Corporacion Peruana del Santa Lima, Peru	Chimbote power station, units II/III	2×4	oil, 3,600/3,600 rpm, genera-tor, with heat exchanger (hex)
6, 7	1945	C.A. Venezolana de Cementos, Caracas, Venezuela	Pertigalete power station, units I/II	2×1.65	natural gas/ oil, 5,350/1,800 rpm, generator, with hex
8	1945	Amalgamated Ice Factories and Cold Stores S.A.E., Alexandria, Egypt	House plant	1.2	oil, 6,650/1,500 rpm, genera-tor for refriger. compres-sors, with heat exchanger
9, 10	1945	AIOC Anglo-Iranian Oil Co., as of 1954: Iranian Oil Explo-ration & Producing Co., London, Great Britain	Tembi power station, Iran, units I/II	2×4	natural gas, 3,800/1,500 rpm, generator
11	1945	Altos Hornos de Vizcaya, Bil-bao, Spain	Barakaldo power station	2	blast furnace gas, 5,650 rpm, blower

Figure 4-84   *The next 10 GT projects after GT Neuchâtel #1 – up to 1945* [290]

The overall efficiency attained a very fair value of 24 percent, including the auxiliary ma-chines, at an average load of 73 percent over more than 8,000 OH.

BBC's increasing confidence in the new product became evident by the fact that GT units 6 and 7, Figure 4-84, for the Pertigalete power station in Venezuela were shipped without in-house testing in advance. Each of the sets had a terminal output of 1.65 W for 35 °C intake temperature. Both gas turbines were put into industrial operation immediately after their in-stallation in 1949 and have operated without disruption from the beginning. The acceptance tests were finally carried out after the first set had completed 1,100 OH and the second 2,000 OH, with an electricity production of 1.3 and 2.1 million kWh, respectively. As the operation of the attached cement factory depended entirely on the two gas turbines, it was not possible to carry out the tests at constant load, but average values had to be deduced instead.

At the end of the second year of service the cement production increased to such an extent that the energy consumption was almost doubled, corresponding to an almost full load for both machines. The fuel was a residual oil with a high ash content and high proportions of vanadium and sodium, meaning ash deposits on the turbine blades had to be washed out regularly after 900 OH. Hans Pfenninger proudly stated at the end of his 1953 operation review: '...*These machines have shown that gas turbines stand up well to industrial operation even when heavy fuel oil with a high vanadium content is used.* ...', a blow to the struggling Swiss competitor Sulzer and their Weinfelden GT project, which was severely hampered by problems with V_2O_5 deposits, Section 5.2.2. A short time after the described initial installa-

---

[290] See BST, Gas Turbines – Reference List

tions, two additional 5 MW BBC gas turbines of the robust single stage type without re-generators were constructed at the Pertigalete cement works near Lima, Peru.

Engine unit #8, a 1.2 MW single-stage gas turbine with regenerator, was put into service in the ice factory of the first owner Raymond Arcache & Cie., Alexandria, Egypt in the summer of 1950. The energy produced was used to drive refrigeration compressors and for that reason the plant was only run during the summer months.

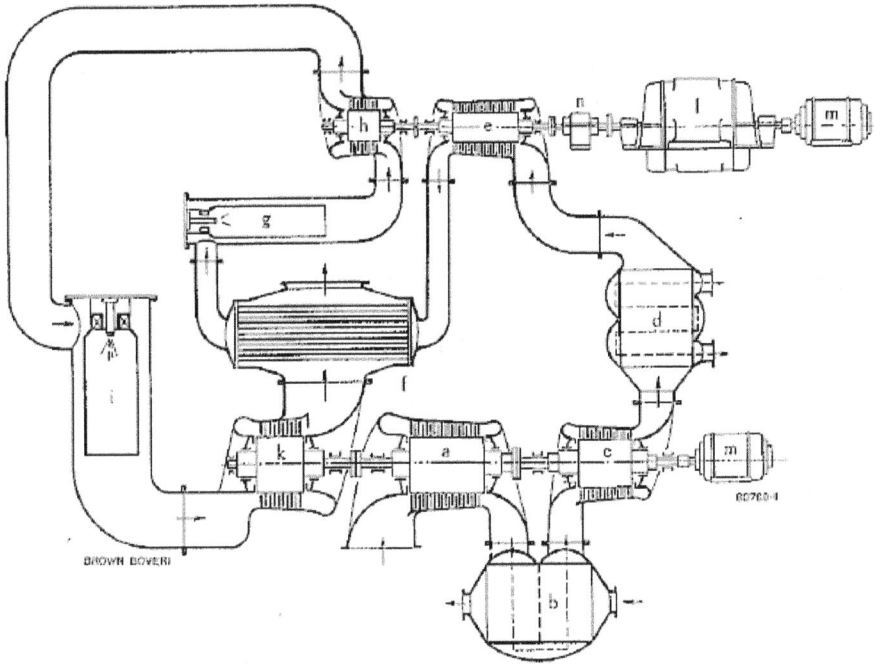

Figure 4-85   *First multi-stage gas turbine with regenerators installed in the Santa Rosa electricity works in Lima, Peru* [291] *a – lp compressor, b – intercooler I, c – med. pressure compressor, d – intercooler II, e – hp compressor, f – regenerator, g – hp combustor, h – hp turbine, i – lp combustor, k – lp turbine, l – generator, m – starter motor, n – gearing*

Another three gas turbines of the Chimbote type with a terminal output of 4 MW at 3,800 rpm were ordered by the Anglo-Iranian Oil Company. Two of these machines were installed as BBC GT units #9 and 10 at the Tembi Power Station, 200 km west of Esfahān, Iran, but due to the Anglo-Persian oil dispute[292] service had to be postponed after 1953. Due to frequent sand storms, special measures had to be taken to prevent sand from entering the compressors. Natural gas was used as fuel which was available at a pressure of about 28 bar. It was used for starting instead of an electric motor. It was expanded from 28 to 1 bar in an expansion turbine and then emitted into the atmosphere. During normal operation the natural gas was expanded in the turbine to the combustion pressure of 4.6 bar,

---

[291] Figure source: BBC, Brown Boveri Gasturbinen, 1963, Figure 35
[292] http://en.wikipedia.org/wiki/Anglo-Persian_Oil_Company

with an additional power gain of 150 kW. In order to prevent turbine icing a small fuel-gas preheater was built into the exhaust pipe of the gas turbine to heat the natural gas to 300 °C before expansion turbine entry. The control of this arrangement was patented.

Gas turbine #11 produced no electrical energy, only supplying compressed air for the Barakaldo[293] steelworks, 5 km north-west of Bilbao, Spain. The use of a combined burner for blast-furnace gas and fuel oil is interesting here. With a shortage of gas the speed of the set would drop, and lower the air pressure at the converter. This could have been danger-ous if it had fallen below the value necessary to prevent liquid iron from entering the blow-ing nozzle. With the combined burner, fuel oil could be immediately injected when the gas pressure fell, consequently avoiding a drop in speed.

To summarise, after a few years of operation, it could be stated that the early gas turbines were simple and reliable constructions which required little maintenance and had low ini-tial costs. However, the fuel consumption was relatively high; therefore, the units were particularly suitable for standby power stations, for peak-load stations with short annual operation times, or for plants in which fuel costs were only of secondary importance, for example in oil fields.

4.3.5   BBC Locomotive Gas Turbines

A short while after securing the order for the first utility gas turbine at Neuchâtel, in 1939 BBC's technical director Adolf Meyer managed to persuade the Swiss Federal Railways (SFR) to order a gas turbine locomotive with electrical transmission. It had a rated output of 2,200 hp and the concept comprised from the beginning an exhaust gas heat recuperator, also designated as an intake air pre-heater, Figure 4-86. In the electrical transmission only the gas turbine part was new, while generators, axle drive motors and switchgear were adapted without any alteration from the diesel-electric locomotive. Consequently, the gen-erator-driving gas turbine could always be operated at the most suitable compressor speed, resulting in high compressor efficiency at all speeds and loads. An inherent advantage of this drive concept is the simplicity of direction changes which requires considerable extra effort for other solutions. The entire GT machine set was built as a compact, integral unit which allowed for relatively easy assembly, Figure 4-87.

The GT-powered locomotive had to find its place between steam and diesel locomotives. For stationary plants the condensation steam turbines were hard to beat, but with atmos-pheric exhaust, as required for locomotives, the specific fuel consumption (sfc) was already in favour of the GT locomotive. On the other hand, though the sfc comparison was clearly pro diesel locomotive (with 180 g/PSh compared to 370 g/PSh for the GT configuration) from a Swiss perspective the considerable price differences between diesel fuel and heavy fuel oil (Masut) made the GT concept appear economically attractive at first.

---

[293] Barakaldo is a municipality in the Basque Country; the Spanish spelling is Baracaldo.

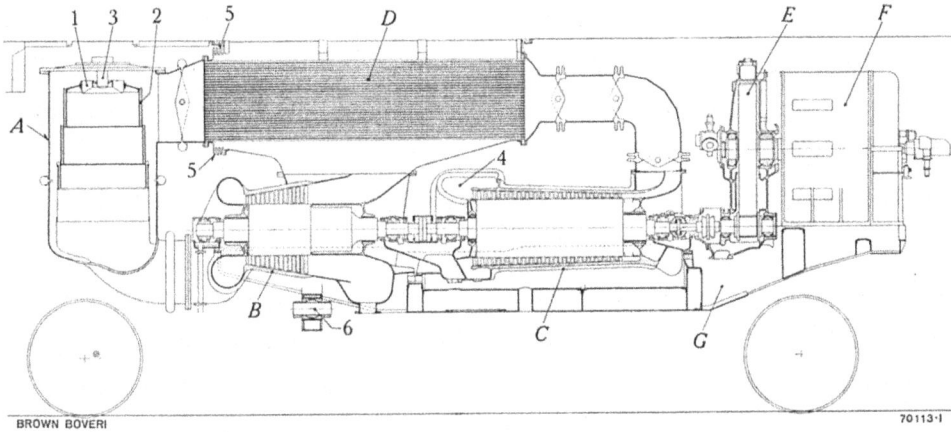

Figure 4-86   *Scheme of single-shaft SFR railway gas turbine installation with recuperator*[294]:
A – combustion chamber, B – axial turbine, C – axial compressor, D – heat exchanger
(recuperator), E – reduction gearbox, F – main generator, G – support frame structure
        ©*ABB*

A few characteristic parameters of the GT turboset were as follows for

		– best point (3/4 load)	– full load
– axial compressor speed	rpm	4,150	5,200
– air mass flow	kg/s	22.2	30.6
– compressor pressure ratio		3.06	4.21
– turbine inlet temperature	°C	550	577
– net power output at generator	kW	1,240	1,590
coupling	PSe	1,685	2,160

In an evaluation after the test campaign the test engineer in charge W. Thomann, Figure 4-11,[295] gave a few recommendations for improving the operational controls, which were all accomplished for the 2nd GT locomotive prototype delivered from BBC to British Rail in 1949. Some of his words are worth being repeated here:

'A gas turbine is a very simple engine. But its operational handling (starting, controlling, operation, stoppage, coupling etc.) is still too complicated and requires specially trained personnel. It should be our development target for the operational handling of the gas turbine to be as simple as the machine itself. ... This especially applies to a locomotive which after initial preparation should be able to be conducted by any average driver. Af-

---

[294] Figure source: BBM, Die Brown Boveri Gasturbinen-Lokomotive, 1945, Figure 9

[295] Werner Thomann (1907 – died after 1988) joined BBC TFVL in 1937 with subsequent responsibility for the commissioning of the first power generation gas turbine at Neuchâtel, the corresponding daily GT demonstrations at the 'Landi 39' and the test campaign of the first GT locomotive. After 1950, he led Dept. 1c for Thermal Power Plants – with the responsibility for 36 turn-key deliveries during eight years. After an intermezzo as president of BBC Canada (1958–1962), he returned to Switzerland as General Manager of General Atomics Europe till his retirement in1972.
His father Eduard Thomann (1869–1955) was Director of the BBC Railway Dept. from 1909 till 1933, – a career, highlighted by the realisation of the electric mountain railway to the highest European station at Jungfraujoch (3'454 m, 1912). Hauszeitung 1950/11, p. 176 and 1988/12, p. 9 (By Courtesy of N. Lang, Aug. 2013); http://de.wikipedia.org/wiki/Eduard_Thomann

ter the disappearance of steam locomotives, understanding and affection for the engine can no longer be taken for granted; he is the man who drives, who knows routes, signals and schedules, but who cannot be burdened with the right engine operation.'

The commercial arrangement between SFR and BBC demanded the fulfilment of the specified conditions for the GT locomotive (92 t total weight, 2,200 hp[296], 110 km/h max. speed) and a one-year operational test before SFR would accept the delivery, so BBC had to prefinance their own engine building, ordering mechanical railway equipment from SLM Schweizerische Lokomotiv- und Maschinenfabrik, Winterthur, CH. GT thermal group testing[297] took place at Baden between 25 November 1940 and 20 May 1941. Integrating the GT unit into the locomotive followed from 21 July 1941 until 1 September 1941 at the BBC Muenchenstein works on the southern outskirts of Basel, Figure 4-87. On the same day first test excursions were made successfully to Delsberg and Brugg. The official inauguration tour took place on 5 September 1941 from Basel via Koblenz – Eglisau – Schaffhausen – Stein a. Rh. – Romanshorn with a final transfer to SFR, Winterthur. For BBC, the new GT locomotive was the highlight of the company's 50th anniversary celebrations on 29/30 September 1941, touring between Zurich – Baden – Olten and back again.

Figure 4-87   *Assembly of the first GT locomotive at BBC Muenchenstein near Basel, August/September 1941, for SFR Swiss Federal Railways* [298]   ©ABB

---

[296] 2,200 hp was the maximum drive power for the selected 6 axle configuration with 4 driven axles.

[297] See BBC, Gasturbinen-Elektrolokomotive, 1940–1942. Testing in Halls 27/28 received a personal touch when acoustic emission values were determined amongst the surrounding work benches and the microphone (so-called 'Decibelmeters') positions were marked by the bench owners' names: '#1 – bei Huegli, #2 – bei Rotenfluh, ..., #6 – auf Meister Trimbach's Pult (desk), Distanz 80,6 m, ..., #8 – bei Stucki, #9 – bei Türe Turgi'. Though some found the high-frequency GT sound painful, the absolute maximum value of 85 dB remained at the level of diesel locomotives.

[298] Picture source: Pfenninger, The evolution, Figure 23

Figure 4-88   *The 'Maker': Director Meyer on the footplate of the GT locomotive Am 4/6 1101 – under a heavy smoke cloud*[299]   ©ABB

The designation Ae/m 4/6 1101 for the new locomotive followed the Swiss classification system, used for more than a century to classify the rolling stock operated on the railways of Switzerland[300] :

- A   means a standard gauge locomotive with v_{max} over 80 km/h,
- e or m   designates the traction type, here 'electric' or 'diesel, gas turbine'[301],
- 4/6   number of drive/total axles,
- 1101   110 km/h class, first of its kind (visible on the front of the locomotive in Figure 4-88).

From the beginning fuel shortages caused by the war overshadowed the development of this project. Diesel fuel was extraordinarily expensive, so meaning the stationary efficiency measurements had already had to be limited to the ¾ load best performance point (~18 percent) and full load. One of the first modifications was the adaptation to (cheaper) fuel oil so that diesel was only required for starting. The negative consequences were dark smoke emissions, previously suppressed by diesel burning, especially with a part load or when idle. Consequently, a

---

[299] Picture source: Pfenninger, The evolution, Figure 24

[300] http://en.wikipedia.org/wiki/Swiss_locomotive_and_railcar_classification

[301] The designation Ae 4/6 1101 corresponds to early BBC usage, see for example BBM, Die Brown Boveri Gasturbinen-Lokomotive (H. Pfenninger), 1945, which changed to the common terminology **Am** 4/6 1101 a few years later; at present, most internet and literature sources such as the excellent website http://www.voepelm.de/t_b/sbb5/SBB Am 4-6 1101/SBB Am 4-6 1101.htm refer to the BBC GT locomotive as **Am** 4/6 1101 which in this context becomes an 'Ae 4/6 III 10851' after a final reconstruction as a multi-current AC/DC test locomotive between 1958 and 1961.

thick cloud of smoke billowed over Adolf Meyer's head when the official picture was taken with him at the footplate, Figure 4-88, which for official BBC publications had to be accordingly cropped. Nevertheless, the regular daily GT train service was made possible by an extraordinary fuel contingent from KIAA[302]. After the equipment to burn fuel oil had been installed in December 1942, tests trips took place on the route Basel – Zurich – Chur with trailing loads of up to 500 t on the outward journey and 300 t on the way back. The demonstrated maximum power was 2,800 hp, the peak velocity reached 128 km/h. In March 1943 a lengthy measurement tour led from Baden to Olten – Solothurn – Lausanne – Fribourg – Berne – Olten – Sissach in which all requirements of the product specification were fulfilled. Pull/acceleration tests were carried out on the 'old Hauenstein line'[303]: A velocity of 40 km/h was reached within 56 s with a 200 t load on an ascending slope of 21 °/oo, clearly superior to the specification demands. Since 23 May 1943 the GT locomotive had been in regular service on non-electrified lines, and from 3 June 1943 until 18 July 1944 there was a continuous daily service for 297 days on the 150 km long, secondary line from Winterthur to Stein/Bad Saeckingen, a mixed passenger/cargo traffic to gather valuable operational experience in which more than 50,000 km had been accumulated. After accomplishing all specified duties the locomotive was officially transferred to the SFR inventory on 1 October 1944. On the basis of a lease contract between SFR and SNCF Société Nationale des Chemins de fer Français the GT locomotive was in regular passenger service between Basel and Strasbourg from 30 October until the end of 1945, interrupted by a widely noticed stay at Gare de l'Est in Paris from 29 November to 1 December 1945.[304] Between 15 March and the end of July 1946 there was a regular service on the 250 km line between Basel and Chaumont, France, again with excursions to Paris. Between January and July 1947 the GT locomotive clocked up 50,000 km in express train services between Basel and Zurich, to Lucerne and Berne. Finally the GT locomotive was operated from 20 June until 2 November 1950 by DB Deutsche Bundesbahn (German Federal Railway) in scheduled service mainly on the 140 km long, important north-south section between Treuchtlingen and Wuerzburg, where the Am 4/6 1101 was said to have shown its superior performance in comparison to the legendary steam locomotive DRG – BR 01, at best on the ascending slopes of the Altmuehl Valley.

The rapid electrification of the Swiss rail network led to a certain disinterest in the GT propulsion concept in Switzerland, so BBC paid more attention to the development of the British rail market. In 1949 a more robust version of the first GT locomotive was built for British Railways, originally ordered by the Great Western Railway in 1946. The main differences between the SFR and BR locomotives were the higher total weight 92 vs. 121 t, the max. velocity 110 vs. 145 km/h, overall length 17 vs. 19.2 m, GT power 2,200 vs. 2,500 hp and GT speed 5,200 vs. 5,800 rpm. This mainline gas turbine-electric locomotive prototype spent its working

---

[302] KIAA stands for 'Eidgenössisches Kriegs-, Industrie- und ArbeitsAmt' which organised the economic planning in Switzerland during WW II. The head of KIAA between 1941 and 1946 was Ernst Speiser (1889–1962), a Swiss politician representing the Canton Aargau in both the 'Nationalrat' and 'Ständerat'. He gained professional business experience in the 1920s in France, Great Britain, USA and the Far East. In 1931, he joined BBC Baden as Vice Director, becoming Commercial General Director in 1938, in which function he also directed KIAA activities. From 1955 until his death he was a member of the BBC Verwaltungsrat (supervisory board). http://de.wikipedia.org/wiki/Ernst_Speiser

[303] http://de.wikipedia.org/wiki/Hauensteinstrecke The north-south route Basel – Olten across the Swiss Jura Mountains was completed between 1853 and 1858 with a 'blood toll' of 63 workers as one of the earliest highlights of the great technical history of Swiss rail. It was originally planned by the English engineers Robert Stephenson (1803–1859) and Henry Swinburne, dating back to an 1850 request from the Swiss Bundesrat.

[304] See BBM, Rückblick... im Jahre 1942, 1943, p. 58; BBM, Rückblick... im Jahre 1943, 1944, p. 85 and – see BBM, Die Brown Boveri Gasturbinen-Lokomotive, 1945, p. 355

life under the designation BR 18000 in the western region of British Rail, operating express passenger services from London's Paddington station. Like its Swiss forerunner, the locomotive had a small diesel engine for GT starting and carrying out slow speed locomotive movements. The GT was started only a few minutes before the train was due to leave the station, saving fuel and minimising annoyance from noise and exhaust fumes.

The machine proved troublesome during operation; ash from the low-grade fuel oil damaged the turbine blading and the combustion chamber liner turned out again as the Achilles' heel of BBC's early GT designs, requiring frequent replacements. It also suffered from the destruction of its heat exchanger in a fire at Temple Meads station, Bristol, when combustion deposits in the hex exhaust side caught fire a considerable time after the machine had been shut off. Relatively early in service it became clear that even demanding express passenger schedules implied inevitably extended periods of part-load operation and correspondingly high fuel consumption. Unfortunately, the BR 18000 never achieved an acceptable level of reliability nor was it possible to operate it under reasonably economical fuel conditions.[305]

The successor machine BR 18100, built in 1951 by Metropolitan-Vickers, Manchester is also part of Alstom's technical heritage.[306] The MV design was based on aircraft practice and had six horizontal single-can combustors spaced around the turbine shaft and no heat exchanger. In this respect, the emphasis was on power while the fuel consumption was high. It was designed to use aviation kerosene (nickname 'Kerosene Castle') and was much more expensive to run than the BR 18000 from BBC which used heavy fuel oil. The major differences of the BR 18000/18100 versions were the locomotive weight, 121 vs. 130 t, the output horsepower, 2,500 vs. 3,000 hp and the number of traction motors, 4 vs. 6.

The BR 18100 had its last run in 1954 and shared the fate of its Swiss ancestor Am 4/6 1101 in December 1957, when it was converted into a 25 kV AC electric locomotive E1000 (which became E2001 in 1959).

A further, completely new machine element which was applied for the first time during this peak of BBC's engineering creativity was the 'Comprex' pressure wave exchanger, combining compression and expansion in a single impeller. In the years 1944 to 1946 a gas-turbine locomotive with a power of 4,000 hp at the coupling was built with a Comprex high-pressure 'topping' stage. Figure 4-89 shows the gas turbine frame with its extraordinarily compact arrangement of components at the test site in Baden. After accomplishing the usual test programme the whole equipment must have been ready for installation. Later explanations which reduce the whole effort to a Comprex test set-up only are not very convincing. Given BBC's very cautious investment policy at that time, one must come to the conclusion that a similar locomotive project was likely near to completion at that time. As a writer close to BBC, Kruschik has many details[307], but the actual background of the project cannot be unveiled. Since it is mentioned as suitable for mountainous areas, it was

---

[305] http://en.wikipedia.org/wiki/British_Rail_18000

[306] As visualised in Figure 2-1, MV was merged with AEI in 1959, becoming GEC in 1970, GEC Alsthom in 1989 and finally Alstom in 1998.

[307] See Kruschik, Die Gasturbine, Chapter XIV 'Gasturbinenlokomotiven und fahrbare Kraftstationen', Section 5 '4000 PS Gasturbinenlokomotiv-Projekt von BBC', pp. 713–715 and – see Jenny, The BBC turbocharger, p. 200; here, Jenny at least provides an argument for the project cancellation: the Comprex development revealed unforeseen difficulties in achieving high pressure levels which in the early 1950s could be overcome more easily with conventional compressor design advancement.

most likely a cancelled SFR project, possibly coinciding with Adolf Meyer's retirement (and a consequent lack of motivation?) on 31 March 1946.

Figure 4-89   *4,000 hp gas turbine set ready for installation in an (unknown) locomotive project, 1946*[308]
a – axial compressor, b – recuperator, c – axial turbine, d – Comprex unit, e – combustion chamber, f – exhaust duct   ©ABB

The set had the same turbo-components as the 2,500 hp GT unit; the principle of pressure exchange is illustrated in Figure 4-90. After the axial compressor exit the air is directed to the Comprex unit, with a resulting pressure rise from 4 to 10 bar. From the Comprex wheel the air is sent through a small centrifugal blower to compensate for losses in cycle pressure and to alleviate the charge exchange, before a rise in temperature to 800 °C takes place in the combustion chamber. Afterwards, the combustor hot gas stream is split in the mainstream to the pressure wave exchanger, which handles equal volumes of gas and air. The resulting hot gas 'overflow' is therefore directed to an auxiliary turbine, with the precondition that the gas temperature must be reduced to an acceptable level by the addition of cooling air. This cooling air is diverted from the high pressure air flow to the combustor. The air/gas mixing ratio and consequently, the temperature upstream of the auxiliary turbine is determined so that the temperatures of the gas mainstream and after the auxiliary turbine are equal. Comprex development began as early as 1910: Jenny describes the complicated process of an alternative turbocharging concept until BBC's trials in the 1970s.[309] Interest grew due to the expectation to omit the 'torque hole' in conventional turbochargers. Claude Seippel filed the decisive patents (with Swiss priority as early as 7 De-

---

[308] Picture source: Pfenninger, The evolution, Figure 26
[309] See Jenny, The BBC turbocharger pp. 198–203. A recent summary of Comprex is given in Sané, Wave rotor.

cember 1940) but later gave[310] some credit to Georges Darrieus for valuable suggestions, such as a Comprex variant with fixed, non-rotating cells, but rotating feeding and taking neighboring components. In his fundamental patent he showed the calculations and how the engine should be dimensioned to profit from the pressure waves. Since the rotor of the pressure exchanger is alternately in contact with hot gases and cold air, its temperature is lower than the maximum process temperature. Consequently, it was possible to select a higher maximum temperature for the gas turbine and increase efficiency. Seippel also coined the term 'Comprex' for the innovative machine concept. The practical applications of the Comprex such as for a locomotive gas turbine's 'topping' stage were part of Seippel's second patent, illustrated by the original sketches of US 2461186, Figure 4-90, with Swiss priority date 20 February 1942.

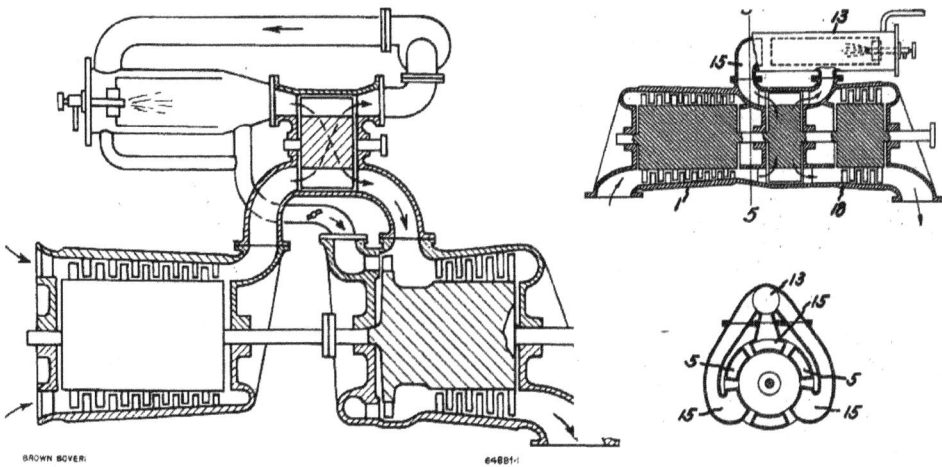

Figure 4-90   *Comprex pressure wave exchanger, 'topping' stage for 4,000 hp gas turbine locomotive project (l), compact lightweight configuration (r) 1942–1946; figures show variants of Seippel's US patent 2461186*

Free-wheeling Comprex rotor hardware with curved flow channels is shown on the left hand side of Figure 4-91, while the operation principle can be deduced from the scheme on the right, which applies to the simpler, actively driven concept of a Comprex car turbo-charger with straight channels:

The Comprex acts as an additional compression stage to shift the air pressure level from 4 to 10 bar, simply by pushing a 'piston' of hot gas against the cold air column.

Before and during WW II, the history of the BBC power generation gas turbines has particularly widespread implications in other fields of application, mostly aeronautical and military, which deserve a separate in-depth discussion in the future. Three of these side developments with obvious reference to BBC will be addressed in short.

---

[310] DE724998 'Druckaustauscher, z.B. für Kaeltemaschinen' with priority 7. Dec. 1940, see Section 4.1.2 and CH229280 = US2461186 'Gasturbinen-Anlage' with priority 20. Feb. 1942. For the credit to Darrieus – see BBM, Hommage à Monsieur Georges Darrieus, 1968

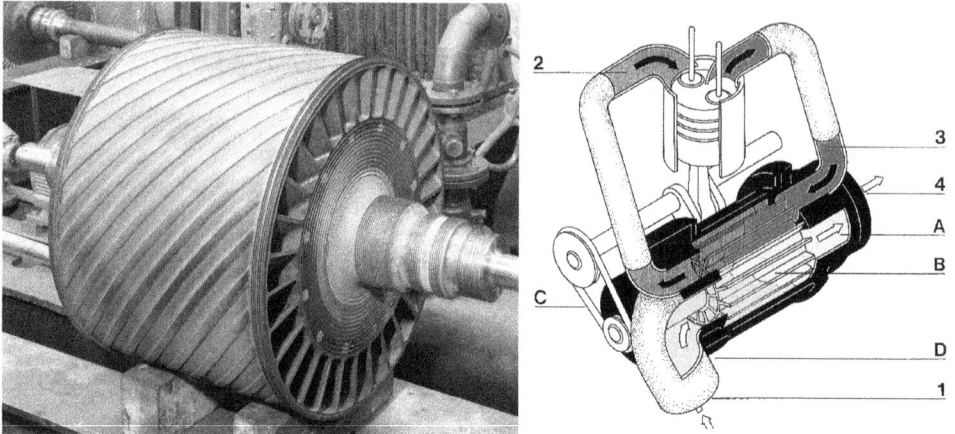

Figure 4-91   *Rotor of Comprex pressure-wave exchanger (l) [311] and operating principle for piston engine pressure wave supercharger (r) [312] 1 – fresh air, 2 – charge air, 3 – cylinder exhaust, 4 – exhaust gas, A – exhaust gas housing, B – Comprex wheel, C – belt drive, D – air housing © ABB (l)*

4.3.5.1   Von Ohain's 'Tuttlingen Engine'

In Figure 4-90 basic Comprex applications were illustrated with reference to Claude Seippel's second Comprex patent; surprisingly and unexpectedly Comprex was also part of Hans-Joachim Pabst von Ohain's work between 1943 and 1945 at Heinkel-Hirth in Germany, documented comprehensively in Antony L. Kay's book.[313] Figure 4-92 shows Fig. 2.33 from the book, clearly related to the patented Comprex configuration on the right of Figure 4-90. Tuttlingen, today a small town of 35,000 inhabitants, approximately half way between Stuttgart and Baden, CH, some 50 km away from the Swiss border, gave its name to the new aero engine concept as the location of a dispersed Heinkel-Hirth site where the R&D work was performed.

For those familiar with the BBC Comprex, the operation principle of the 'Tuttlingen engine' is straightforward, except the pressure wave exchanger wheel called the 'compressor/turbine rotor element', Figure 4-92. The Comprex rotor drives a pre-compression diagonal fan at engine entry, typical of the Heinkel-Hirth 109-011 turbojet. After passing this fan, the air flows into the Comprex's bottom section and is drawn up with the wheel rotation into the combustion chamber at the top. At the same time, a fuel/air mixture is burned there and the hot gases exhaust down through the other side of the bladed rotor into the free-wheeling power turbine. The planned applications of this engine are unknown; according to Kay, von Ohain stated that a small test model of 60 hp had been built and run at Tuttlingen, presumably in BIOS[314] reports immediately after the end of the war. However, there is not the slightest hint in von Ohain's indirectly quoted statement that the basic ideas for this concept might have had origins other than von Ohain himself. In Kay's words

---

[311]   Picture source: BBM, Forschung im Dienste unseres Maschinenbaus, 1953, Figure 6
[312]   Figure source: Jenny, The BBC turbocharger, p. 198
[313]   See Kay, German Jet Engine and Gas Turbine Development pp. 52–54.
[314]   BIOS British Intelligence Objectives Sub-Committee

'*Von Ohain proposed that a portion of the temperature drop available from the hot compressed gas should be used to compress the cold air to the higher pressure. For this purpose, the Tuttlingen engine had the essential feature of a single rotating element which performed the function of a compressor and a turbine*'..., **But** this is the BBC Comprex principle, patented and published before the Tuttlingen experiments. Of course, the reference in Kay's book is no clear proof that von Ohain actually omitted the necessary hint to the foregoing Swiss activities, perhaps Kay simply forgot to mention it.[315] Nevertheless, for some sceptics, especially in the UK, this small, likewise unimportant issue may be particularly annoying, especially for those who believed that the argument concerning Frank Whittle's and Hans Pabst von Ohain's 'parallel' turbojet inventions looked somewhat fabricated, in addition overshadowed by early 'secret patent' issues on the German side.

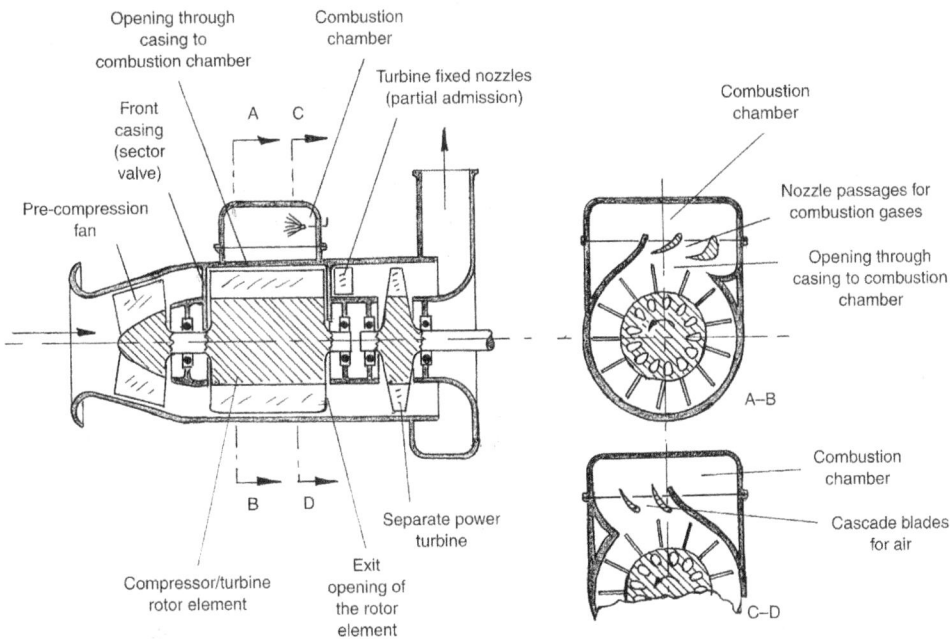

Figure 4-92   *Hans-J. Pabst von Ohain: basic arrangement of the 'Tuttlingen engine', 1943–1945*
          © *The Crowood Press*

The Tuttlingen engine concept has another unexplained link to BBC engineering. Again according to Kay, '*the aim of the* (Heinkel-Hirth) *designers was to provide an improvement of the Holzwarth constant-volume combustion turbine by avoiding the intermittent combustion of the cycle with its associated mechanical difficulties.*' In fact, from 1940 to 1941 the Heinkel engine project inventory had a constant-volume turbojet project under the designation HeS 40 (TL). Before a more thorough investigation of BBC's participation in

---

[315] Max Adolf Mueller received a patent DE913116 for a 'Jet engine or gas turbine plant with cell wheel – pressure exchanger for larger powers' since 20. April 1952 and one month later DE916607 for a 'Cell wheel gating system for jet engines'; in both applications there is reference made to DE724998 (Seippel 1940) and CH229280 (Seippel 1942).

aero engine developments during WW II is carried out, it can be speculated if the necessary Holzwarth concept expertise, Section 3.3, was achieved

- by own Heinkel-Hirth R&D activities on the basis of earlier Holzwarth and BBC publications,
- through direct contact with Hans Holzwarth, who had filed his own aero engine patent DE 923,337 straight after the end of the war with priority date 2 March 1945, as described already in Section 3, Footnote 72. This is most likely as Schramberg and Tuttlingen are only about 50 km apart,
- through conscious or unconscious 'information leaks' from BBC; the following episode means this latter alternative is more probable.

4.3.5.2   German Tank Gas Turbine Developments

As previously mentioned, BBC Mannheim had maintained an independent design office TLUK/Ve (mainly) for aero engine development since April 1941, led by Hermann Reuter.[316] Significant contributions to German military gas turbine developments were carried out here. Some of these activities addressed tank GT developments at the very end of WW II in 1944/1945. In the present chapter about the locomotive gas turbines it is worth reflecting on certain relationships between these two topics... and again remarks from Antony Kay fuel speculation. In his words[317] *'During 1943, some serious thought was given in Germany concerning the possibilities of using the gas turbine for overland traction purposes. At that time, the only land vehicle which had been driven by a gas turbine was an experimental locomotive which the Swiss Federal Railways had ordered at the beginning of 1939. The power unit built by Brown Boveri of Baden (Switzerland), delivered 2,200 hp to drive the locomotive through an electrical transmission system. Rail trials began after acceptance tests on the power unit in September 1941, but the trials were limited through a wartime shortage of fuel. Back in Germany it was not until 1944, with the Second World War grinding towards its conclusion, that an official interest was shown in developing the gas turbine for land traction purposes'.* We remember, the Swiss actively paraded the successfully completed GT locomotive in front of the Germans: Between 3 June 1943 and 18 July 1944, they serviced the line Winterthur – Stein/ Bad Saeckingen daily, where the elevated railway tracks follow the Rhine river for more than 100 km, the border with the neighbouring German Reich.

A prime reason for late German interest in the gas turbine for tank propulsion was its ability to run on a much lower grade fuel than the internal combustion engines then in service, as well as the 700 hp power limit of these engines: the tank gas turbine projects had a typical output of 1,150 hp. As an example, Figure 4-93 illustrates the installation of the BBC GT 102 in a German 'Panther' tank.

An urgent need for information about the Swiss GT locomotive project can be deduced by the location and organization of the tank project. Overall responsibility for the GT project was in the hands of Dipl-Ing Otto Zadnik (1887– died after 1956)[318] of Porsche KG. He was primarily

---

[316] See Kay, German Jet Engine ... p. 289
[317] See Kay, German Jet Engine ... Section 3, Gas Turbines for Land Traction, pp. 156–173.
[318] http://www.komenda.at/Zadnik_d.htm

an electrical transmission specialist and together with his small team, had the task of designing the transmission gear and overlooking the GT installation in the vehicle. The electric
drive of the Swiss locomotive and its compact GT installation must have looked attractive for
the German engineers. Even more intriguing is the location of the small design office at
Rheinau(weg) in Bregenz-Hoechst, literally **on** the Austrian-Swiss border. The place[319] is only 100 m north of the border, represented here by the 'Old Rhine' some 10 km before its con-
fluence into Lake Constance. Since Zadnik grew up in nearby Bregenz, he may simply have
favoured his home town in his choice of office location. However the location in an open
space in the immediate neighbourhood of the Swiss border is surprising, especially since
similar German facilities were usually underground at that time. The only explanation in the
moment is the assumption that the vicinity to the Swiss border either created advantages
against Allied air raids or alleviated cross border 'communications' or espionage.

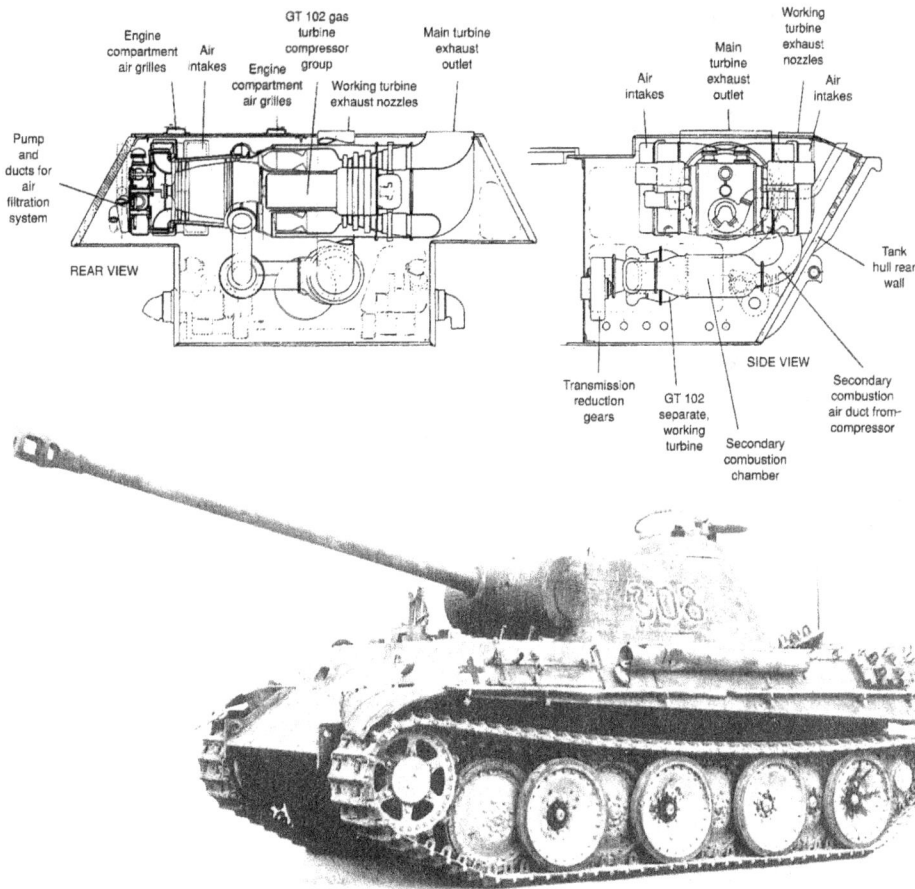

Figure 4-93   *BBC GT102, 1,150 hp engine project, installation drawing (top) for German 'Panther' tank
(bottom)* [320]

---

[319] Search for <Bregenz Rheinau Austria> in Google Maps.

[320] See Kay, German Jet Engine... (top) Fig. 3.7, p. 165 © The Crowood Press and (bottom) Fig. 3.4, p. 161, © Royal
Armoured Corps Centre, Bovington UK; http://en.wikipedia.org/wiki/Panther_tank

4.3.5.3   Separate Gas Generators for Large Transport Aircraft

G.G. Smith[321] was apparently the first (in 1942) to describe the possibility of using the 'stand-alone' feature of the BBC locomotive GT combustor, Figure 4-86, also as a separate gas generator, preferably for large transportation aircraft. Interestingly, in 1941 Professor Ackeret had already suggested a similar aircraft propulsion unit independently of Smith, Figure 4-94 (right), with a central axial compressor in the aircraft fuselage with the propeller propulsion units, relatively light-weight gas generators and axial turbine drives distributed on the wing. The further development of this propulsion system was part of the war-time activities of BBC's gas turbine department at Baden.[322] The piston engine in the aircraft fuselage was planned to have 1,000 hp; with the extra combustion power the propulsive power was meant to double, so that each propeller drive would deliver 1,000 hp. The central mass concentration of the compressor and the heavy piston engine in the fuselage was to be used advantageously to absorb the recoil effect of a heavy, high calibre gun, an idea which only materialised after WW II in various ground battle aircraft and 'gunship'[323] configurations. Figure 4-94 (left and middle) shows two alternative combustor designs from BBC. The so-called 'louvre-type' configuration on the left, actually used for the locomotive gas turbine, found wide-spread application in many post-war aero engines with co-axial, multi-can combustor arrangements. The inner liner is formed by a series of nesting conical sections of sheet metal; the design required an inner casing able to withstand temperatures of about 980 °C without deformation and an outer shell to support the pressure of the compressed air and the combustion gases. Approximately 20 percent of the air flow supported the swirling fuel spray and combustion in the central region, while the remainder of the air flow was used to lower the temperature of the combustion gases to 500–550 °C prior to admission to the turbine. At each peripheral slot by the nesting sections some of the excess air flows into the inner casing to form a boundary layer, protecting the casing from heat and eventually mixing with the gases. The second combustor design, as seen in the middle of Figure 4-94, is a plain cylindrical combustion unit with special screening features to reduce radiation losses.

---

[321] See Smith, Gas Turbines..., 3rd ed. pp. 79–82
[322] See Pfenninger, Die Gasturbinenabteilung, p. 685
[323] http://en.wikipedia.org/wiki/gunship

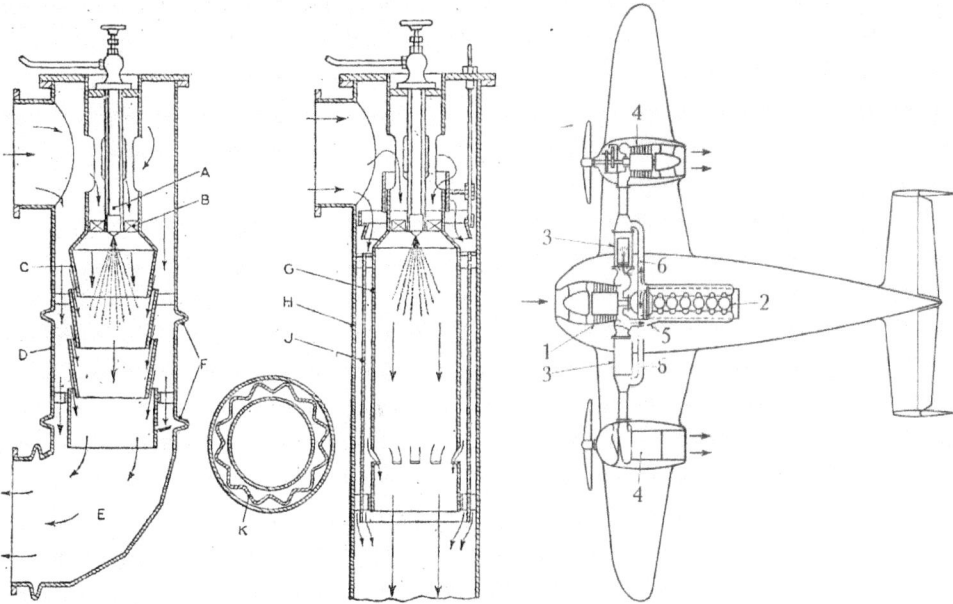

Figure 4-94   *BBC gas generators* [324] *and their aircraft applications (BBC patent CH 221503 and Ackeret 1941)* [325] *Left: Louvre-type combustion chamber (locomotive GT) A – fuel nozzle, B – air guide vanes, C – inner casing, D – outer shell, E – mixing space, F – expansion corrugations; Middle: Screen-type combustion chamber G – inner casing, H – outer shell, J – intermediate screen, K – alternative corrugated screen; Right: Aircraft propulsion unit as suggested by Prof. Ackeret, 1- central axial compressor, 2 – piston engine drive, 3 – combustion chambers, 4 – distributed axial turbines, 5 – turbocharged air duct to piston engine, 6 – piston engine exhaust ducting*

---

[324] Combustor, Figure 4-94 left, – see Patent US2,268,464 by Claude Seippel with Swiss priority starting 29 September1939. See also Smith, Gas Turbines, p. 79, Fig. 55.

[325] No corresponding publication by Ackeret has been found yet, but the illustration was included in a picture collection belonging to Hans Pfenninger in BBC, Geschichte des Schweizer Turbomaschinenbaus, 1982, with reference to a proposal from Prof. Ackeret. Independently (see Munzinger, Duesentriebwerke) this illustration refers to an early BBC patent CH 221503, patent grant dated 31 May 1942 and priority date 9 December 1940.

5 Gas Turbine Technology Development 1945–1988

The following chapter covers more than forty years of BBC's power generation gas turbine history, more than twice the length of the period of early beginnings, covering decisive technological breakthroughs towards initial product accomplishment. For the company, these were forty years of technical and economic hardships and repeated organisational restarts in order to adapt to a rapidly changing environment. In most problem cases, gas turbine power generation was one of the causes of concern, as well as an area where solutions were tested. In the end all efforts were in vain: BBC lost its status as an independent company and merged with Asea to ABB Asea Brown Boveri Ltd. The corporate side of this development change has already been addressed in the introductory Sections 2.1.1 and 2.1.2. From a technical point of view it is possible to determine certain questionable development decisions, but none of these had a critical or permanent impact on the outcome.

The sub-sections of this Section first comprise a chronological survey of both business and technical developments in Brown Boveri gas turbine engines[1], followed by a few impressions of the most prominent engineers of that period. The next chapter first deals briefly with the Swiss competitors Sulzer and Escher Wyss, highlighting the interesting fact that the network of these three neighbouring Swiss companies was self-sufficient before international competition came into focus in the 1960s with competitors who still dominate the field of power generation today. Following the basic structure of the book, which combines each of the three time phases with one of the major gas turbine components, the middle Phase II from 1945–1988 focuses on combustor design as a central GT component. In addition, two of the remaining sub-sections will deal with some important mechanical design features of the engines under consideration, and peculiarities of the gas turbine manufacturing and supply network. Finally, the last sub-section will present a number of special GT projects in the context of power generation.

---

[1] Technical descriptions of the GT11-GT11N2, GT13-GT13E2 and GT9/GT8 engine development lines reach partly beyond this Phase II upper time limit of 1988.

5.1   Development Survey at BBC

5.1.1   'Die Muehen der Ebene'

'We left the exertions of the mountains behind ... and see the exertions of the plains ahead'. These lines from Bertolt Brecht's 1949 poem 'Perception'[2] sound like a symbolic prophecy of the difficulties approaching BBC at that time. Up until the mid-1950s, the first ten years were very successful for Brown Boveri[3], but business conditions were still determined by the war and Switzerland's comparatively privileged starting position. The war created extraordinarily high market demands all over Europe, especially for power generation equipment, which BBC was ready to fulfil based on all kinds of imaginable resources. This situation may have prevented thorough analysis at times, but in hindsight a broad mixture of factors inhibiting sustainable business development can be identified:

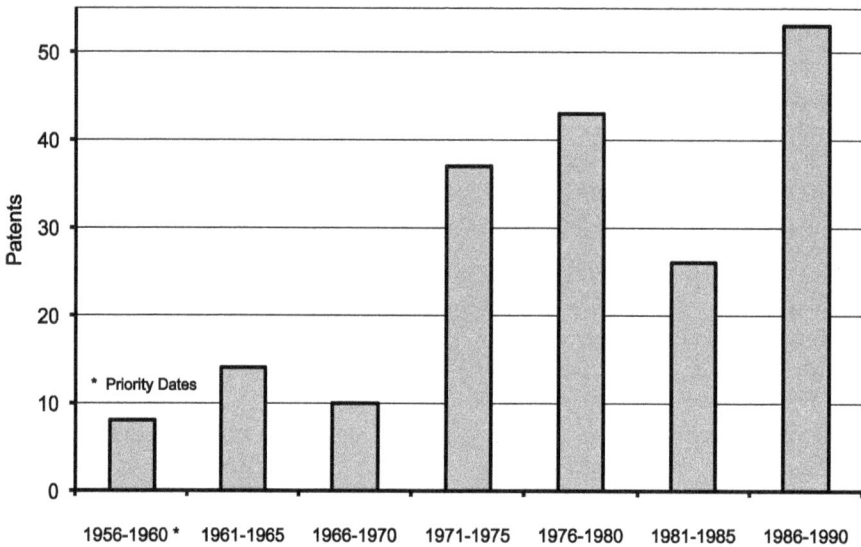

Figure 5-1   *BBC GT patent development between 1956 and 1990 from the 10 most productive engineers of that period* [4]

Innovations Levelled Off The slow restart in innovative engineering development has already been addressed in Figure 2-5, and even after some delayed but visible recovery of the patenting activities, the best figures for Phase II in the 1970–1980 period reach only 50 percent of the high marks for Phase I and Phase III at best. This general assessment is quite naturally complemented by the dwindling creativity of the top engineers as they aged, Fig-

---

2   See Brecht, Wahrnehmung. The original reads 'Die Mühen der Gebirge liegen hinter uns ... Vor uns liegen die Mühen der Ebenen'.

3   Figure 4-29 illustrates the rise of axial compressor sales in that period.

4   The data may contain some uncertainties, but the approximate ranking in view of filed inventions for the time frame under consideration is: J.J. Keller, H. Bellati, R. Fried, H.U. Frutschi, W. Endres, H. Pfenninger, Cl. Seippel, D. Mukherjee, R. Kehlhofer, W. Burger.

ure 4-6, obviously not compensated for by the hired new engineering recruits.[5] The patents filed between 1956 and 1990 of the 10 most active engineers are illustrated in Figure 5-1 and underline the relative lull up until the pronounced plateau of the 1970s, until the intermediate decline of the last BBC years becomes visible and is ultimately stopped by the ABB upturn at the beginning of the 1990s.

A categorisation of the same data into engineering disciplines, Figure 5-2, reveals a certain emphasis on system solutions (including heat exchangers) collected here under the term 'other', but definitely not on the GT main components and with a disturbing lack of continuing attention to the compressor.[6] As shown by the actual product developments (e.g. for the Beznau plant), the main focus of GT engineering developments until the mid-1960s was directed towards performance improvements by means of extensively used heat exchangers (for intercooling and heat recuperation), while the technology of the GT core components (compressor, combustor, turbine) remained widely unchanged[7]. This course was set out at a relatively early point by Ackeret in a 1942 paper[8] '...The thermodynamic principles of the constant pressure process are simple. The Carnot process with two isotherms and two adiabatic lines creates large compression ratios. However, it is possible to define a fully equivalent process with two isotherms and two lines of constant density (isopycnes) with heat exchange between these, that works at moderate pressures and technically acceptable temperatures. The sole task is to operate engines and plants within these set limits with the lowest losses. ... Without heat exchange, relatively high compression ratios with many stages are needed and heat exchange determines rather modest compression ratios instead. Best conditions are achieved for a pressure rise of 2- to 4-times that of the ambient.'

---

[5]   Figure 2-2 illustrates the continuing personnel growth worldwide, and Figure 2-3 shows percentage-wise even better numbers for BBC Switzerland for the period under consideration. A speciality of BBC's attitude towards patenting is the time after 1980, Figure 2-5, where filings of innovative ideas in thermal machinery were even intentionally reduced for cost reasons.

[6]   This surprising observation stands in strong contrast to the general attitude of BBC's senior management towards technology, as expressed in a BBM editorial (see BBM, Rueckblick... in den Jahren 1959 und 1960, 1961, p. 4) by Theodor Boveri, jr. (1892–1977, son of Walter Boveri): '...in the interest of the future of our company, we shall not neglect the development of all-new areas (in power generation). This implies not only to address research problems of nuclear fission and fusion, but also fuel cells, thermo-electric elements, thermo-ionic processes and magneto-hydrodynamic machinery for the direct transformation of heat into electric energy. Even though it cannot be judged today, if only one of these areas will gain industrial importance, we cannot afford to ignore such developments even in a research organization of primarily industrial character.'

[7]   This approach was somewhat surprising in comparison to the previous Velox design principles, for example, which stressed simplicity and compactness as primary design targets.

[8]   Ackeret gave a fundamental description of the closed cycle GT, also known as AK Ackeret-Keller Process, see Section 5.2.1, in a presentation to the Deutsche Akademie der Luftfahrtforschung, Berlin on 4 December 1942, see Ackeret, Auf dem Weg zur Gasturbine.

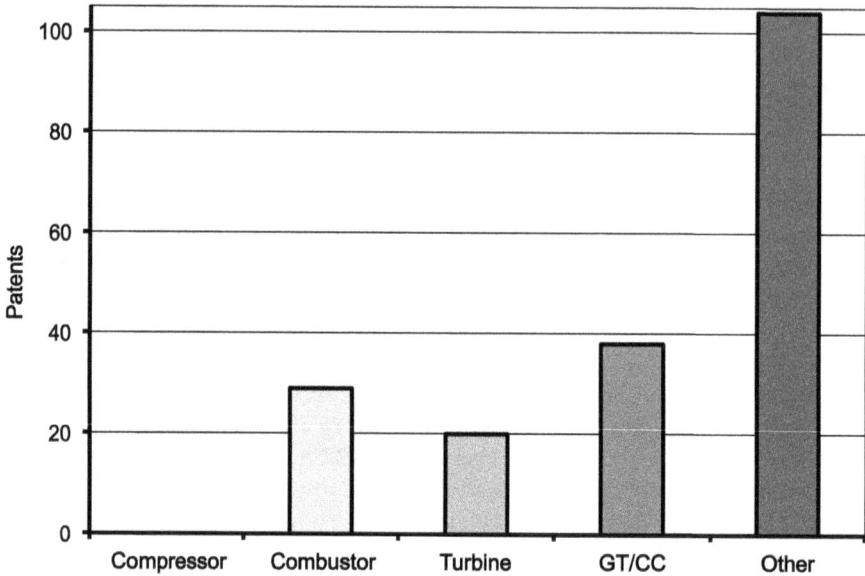

Figure 5-2   *BBC GT patent contributions between 1956 and 1990 from the 10 most productive engineers of that period – by engineering disciplines*

These design peculiarities were not limited to BBC, but were typical for all major players of that time. Besides BBC, this was mainly GE Energy and with a proper distance Sulzer, Westinghouse etc. In view of the final breakthroughs in gas turbine technology during the 1990s it would have been better to prioritise the CC Combined-Cycle mode (GT + ST), also by addressing still existing technological hurdles **and** to switch to gaseous fuels much earlier. As indicated in Figure 5-2 and as will be discussed in detail in Section 6.3, BBC/ABB/Alstom belonged to both the early CC pioneers and promoted the widespread use of CC technology in the 1990s.

Business As Usual? After the post-war boom was over, it appears that the GT development moved along rather slowly. The major achievements in BBC's Thermal Engineering happened in the steam turbine division instead, especially in the rise of turbocharger sales, Figure 3-11. With respect to efficiency and fuel consumption, gas turbine usage appeared to be limited to emergency applications, oil-rich countries like Saudi Arabia, where difficulties with water supplies prevented the use of steam turbines for power generation, or the expensive and unreliable complexities of multi-shaft GT arrangements with heat exchanging (Section 4.3.4.).

Measurable economic growth only occurred again at the end of the 1960s, partly reflected in the previous patent history. Until then the organisational ties between Baden headquarters and the various subsidiaries were rather loose and direction was controlled almost exclusively by licence agreements.

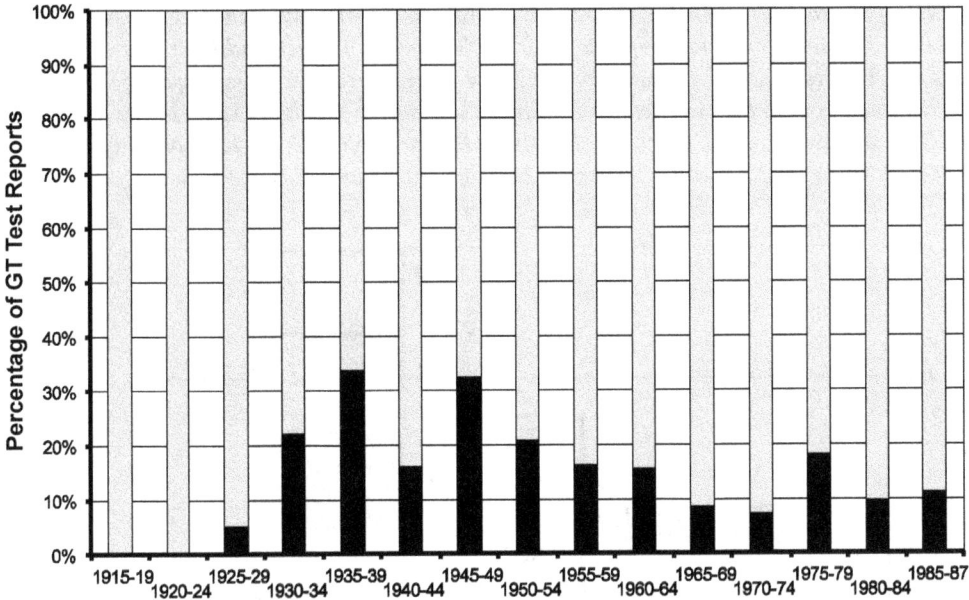

Figure 5-3   *Share of GT test reports (black) as part of BBC's total reporting output between 1915 and 1987*

The relatively reduced position of power generation gas turbines combined with a rather loose control of in-house engineering and development activities can be indirectly deduced from Figure 5-3. The TFVL/ZX test report records over the impressive time span between 1915 and 1987 have been evaluated from this perspective. The figure illustrates the relative share of GT-related test reports over time, culminating in two peak phases from 1935–1939 and 1945–1949, when GT activities comprised nearly 1/3 of the total. The second peak was partly a wrap-up period for undocumented activities during WW II. Compared to the steep ascent until the first peak, the decline is rather shallow and expands across more than 20 years, up until the early 1970s. However, the importance of gas turbine development never returns to peak levels. Rather unexpectedly, the period between 1975 and 1979 interrupted the long-term downward trend. Figure 5-4 illustrates this effect in detail: for the category 'GE Generators and E-Motors' and also 'GT Gas Turbines' the numbers of issued test reports increase significantly, while for 'DT Steam Turbine' and 'TL Turbochargers' the reporting activities stagnate at high level. Georg Gyarmathy[9], who was closely attached to the TFVL/TX/ZX test department at that time, commented on the situation in a typically humorous way in an e-mail in 2009, only a few weeks before his premature death: '...*Another large reorganisation took place. Dr. E. Jenny*[10] *became Director of TL, for electro*

---

[9] Prof. Dr.-Ing. Georg Gyarmathy (1933–2009) received a degree in mechanical engineering from the Technical University Budapest in 1956. After the Hungarian Revolution in the same year he came to Switzerland as a refugee, where he defended in 1962 his doctoral thesis on the subject of condensation in steam turbines under the supervision of Prof. W. Traupel. After a research stay at USAF Dayton Oh., he came to BBC Baden in 1967 where he was engaged in various design, research and management tasks before returning to the ETH Zurich in 1983 as successor to W. Traupel at the Institute for Thermal Turbomachinery, a position he kept until 1998. In the following years he ended his professional career as the General Consul for Hungary in Bavaria.

[10] Dr.-Ing. Ernst Jenny (1923–2004), see Jenny, The BBC turbocharger

motors, gas-bearing spinning machines, all central laboratories and technical service units. He was shocked about enormous reporting delays ("Restanzen") which had built up in the previous era. The result was – I would say somewhat in retrospect – a large output of old reports, which could explain the observed Dirac impulse. ... Another possibility could be the redirecting of GT test activities away from the short-lived BST and joint BBC-Sulzer company structure (1965–1970), i.e. away from the Sulzer test labs at Winterthur and back to TX. ...'

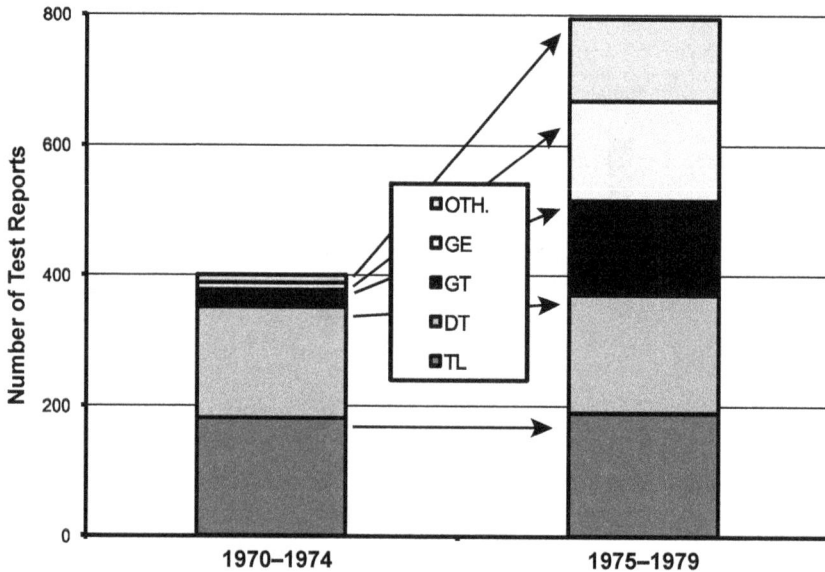

Figure 5-4   *Test reporting close up : the transition from 1970–1975*

Another alternative interpretation of this data is to assume a pull effect and not a push behind these developments, especially on the GT side. Substantial organisational changes happened in the wake of the BST dissolution, the short-lived joint venture between BBC and Sulzer from 1969 to 1974. Effective from 1 July 1974, BBC regained sole responsibility for gas turbines, with Hans Baumann as head of TM 'Thermal Mid-Size Units' and Wilhelm Endres of TMA 'Open Gas Turbines'. Independent of the actual cause, a kind of 'out with the old, in with the new' effect is apparent here. It may be also indicative of the somewhat daring management style of the newly implemented GT organisation, which will be more closely reviewed in Section 5.1.2.

Which Type of Fuel? Besides the uncertainties of markets and technologies which have already been addressed, even the fundamental question of the best fuel for power generation was not settled. The late W.G. Noack's insistent trials to make coal processing and burning acceptable for GT purposes before 1945 ended in frustration. Diesel fuel and even heavy crude oil were the name of the game for the next 30–40 years. As will be outlined in the 'Combustor' Section 5.3, BBC had some success in this area, but the combination of low grade fuels with complex heat-recuperating GT configurations was a deadly mixture. Vanadium traces in diesel fuels present a permanent corrosion hazard and it is the main fuel component influencing high temperature corrosion. During combustion, it oxidises and reacts with sodium and sulphur, yielding vanadate compounds such as the notorious V_2O_5 vanadium pentoxide, with melting points as low as 530 °C, meaning it attacks the passivation layer on steel and makes it susceptible to corrosion. In addition, the solid vanadium compounds also cause abrasion of engine components.

One of the advantageous consequences of the 'fuel crisis' in GT development for Switzerland and further afield occurred when Walter Traupel[11], Head of Turbomachinery and GT Development at Sulzer Bros. Winterthur until 1954, left enervatedly Sulzer's V_2O_5-haunted semi-closed gas turbine project (Section 5.2.2) to continue as a renowned Professor for Thermal Turbomachinery at the ETH Zurich.

Over time the fuel oil quality improved, together with the multi-fuel capability of the GT combustors. The two oil crises in the 1970s destroyed all assumed market certainties, and nobody was more alienated than BBC by the absence of new power plant orders. The subsequent demand in electricity had been hopelessly over-estimated. An expected move to get rid of oil consumption by raising electricity production as a form of alternative(!)[12] energy was a complete misunderstanding and the market reacted to the oil price shocks basically by saving energy only. Finally, the special market situation of the newly rich, oil-producing Saudi Arabia alone delayed BBC's fate for another decade.[13] On 23 July 1977 the contract for a turnkey GT powerplant worth 330 million SFr was signed at Riyadh. This order for 'Riyadh 5'[14] comprised 8 gas turbines with 550 MW total power and the corresponding service equipment; this order was the largest ever issued in the Middle East. Riyadh's population of 500,000 was then served with electricity exclusively provided by 22 BBC gas turbines, and the whole of Saudi Arabia hosted 48 BBC powerplants in total, the first one put into service in 1954.[15] Before the 1980s, when natural gas completed the portfolio of gas turbine fuels, primarily in CCPP Combined Cycle Power Plant (steam/gas turbine) applications, it was nuclear energy which indirectly determined the development of conventional turbomachinery equipment. Between 1950 and 1980 BBC participated in all major nuclear power generation programmes, ending with excessive financial losses[16] that otherwise could have been used for projects such as an earlier introduction of the highly efficient and environmentally friendly CCPP technology. A BBM editorial[17] of January 1948 was clear in

---

[11] Prof. Dr. W. Traupel (1914–1998) started his scientific career in turbomachinery in 1942 with his doctoral thesis 'Neue allgemeine Theorie der mehrstufigen axialen Turbomaschine' which handled steam and gas turbine, compressor and turbine component design within a uniform dimensionless methodology for the first time. After 16 years at Sulzer Bros. Winterthur he came to the ETH Zürich in 1954, where he led the Institute for Thermal Turbomachinery until 1983. His book in two volumes (see Traupel, Thermische Turbomaschinen) became a technology classic and appeared in 3 editions.

[12] If limited to hydro power generation, this statement was justified. Nevertheless, this terminology sometimes had considerably different meanings, and it is therefore not really surprising to learn about an early form of environmental conscience, in which troublesome emissions from burning crude oil caused massive public demand for more 'clean' nuclear energy.

[13] Even Saudi Arabia only represented short-term advantages for BBC. A few years after 'Riyadh 5' a possible order extension was withdrawn due to insufficient funds.

[14] See also Figure 2-15 and corresponding text on 'Riyadh 8'.

[15] See Wildi, Organisation und Innovation bei BBC, p. 46

[16] See Catrina, BBC Glanz-Krise-Fusion 1891–1991, p. 178 ff. The majority of the losses from nuclear experiments at BBC were generated on the German side. It is not known if the management responsible for GT tried to change the conventional BBC preferences for steam turbines thereafter.

[17] See BBM, Rueckblick... im Jahre 1947, 1948

this respect, an opinion which apparently deteriorated later amongst BBC's senior management: '*Today it is almost impossible to talk about power generation without taking nuclear energy into account. ...The probability of supplementing or replacing our conventional power plants with nuclear facilities is not predicted in the short-term. Given the present state of nuclear technology, the costs of such power plants would be very high, operations expensive and dangerous and disposal of the highly radioactive waste products would be very difficult in inhabited areas. ...*' This statement has not lost its inherent truth more than 60 years later.

Competitive Pressures Besides objectively unpredictable market developments it was the mighty GE Energy which put considerable pressure on BBC and its established market position in the 1960s, but this competitive pressure also had a beneficial effect. In comparison to its biggest competitor, BBC must have realised the painful consequences of missing the unique business opportunity at the end of WW II: to participate in the military turbojet market. Boosted by the Korean War (1950–1953) and in subsequent arms build-up GE increased its personnel at the Evendale, Ohio plant from 1,200 to 12,000 over just a few years, producing military turbojets like the 35,000 type J47 or the 17,000 type J 79 engines[18] in record breaking quantities, and thus laid a solid technology base for the related power generation GT business at Schenectady, NY.

Both leading OEMs suffered the same technological problems, for example burning 'Bunker C' fuel. While BBC managed to tackle the V_2O_5 problems using superior combustion technology and fuel additives (see Section 5.3)[19], GE developed a de-salting scheme using DeLaval centrifuging equipment that removed sodium and added magnesium to inhibit the vanadium corrosion. This resulted in an ash that removed itself at shut-down, which proved to be a satisfactory solution, provided that the turbine was only operated for intermittent service periods.

However, for both competitors, the gas turbines were costing more than the market price[20] and, in the early 1960s various concepts were adopted to reduce the total cost: i) all kinds of direct cost cutting measures, ii) modular standardisation of building blocks for the 'power island',[21] iii) enclosing the turbine in a packaged power plant, Section 5.5.1 and iv) advance orders to considerably shorten delivery times.

The fact that the cost pressures were relative is underlined by one of several neatly handcrafted power plant models for a 1959 GT sales project in Colombia, Figure 5-5.[22] As shown

---

[18] The GE J47 with 4,800–7,200 lbf first flew in May 1948 and was the first axial-flow turbojet approved for commercial use in the United States. Some major military applications were the B-47 Stratojet, the Convair B-36 and the F-86 Sabre. The engine was produced until 1956 but stayed in service until 1978. The GE J79 in the 14,000–18,000 lbf thrust class first flew in May 1955, in a variety of fighter and bomber aircraft such as the F-104 Starfighter and F-4 Phantom. Over time, GE introduced a number of aero-derivative turboshafts as cost-effective IGTs, such as the LM 1500 (J79), LM 1600 (F404), LM 2500 (CF6) and LM 6000 (CF6-80C2).

[19] This BBC special department was outsourced in 1978 to become Turbotect Ltd. in Baden. Their website provides details on the various chemical and mechanical treatment options for GT operation http://www.turbotect.com/product1.shtml

[20] http://edisontechcenter.org/gasturbines.html

[21] BBM, Rueckblick... in den Jahren 1963 und 1964, 1965, p. 15

[22] Picture source: BBM, Rueckblick... in den Jahren 1959 und 1960, 1961, tables IV and V (shown). The picture illustrates the interior of a GT plant model for Tibu, Colombia with a planned 5×6 MW construction; of course, this model was supplemented by corresponding outside miniatures and the whole series of images were distributed in glossy colour print.

by Th. Boveri's extensive (and costly) power generation technology requirements, Footnote 6 of this Section, there was still enough budget available for 'nice-to-have' expenses.

Figure 5-5 *In conflict: cost cutting needs and technology, 1960 Handcrafted model of a 5 × 6 MW GT power plant, Tibu, Colombia*

In 1965 the stagnating GT market received an unexpected push forward. The big blackout of the New York area occurred on 9 Nov. 1965, 5.16 pm and one of the Long Island Lighting Company's gas turbines restarted the system with a 'black start'. Combined with solving the technical problems with the Frame 5 design, this event was the impetus needed to turn the struggling GE business around and can be even considered as one of several turning points in the gas turbine business in general. Though the tools were the same, the initiation of counter-measures from BBC was somewhat delayed, and issues of cost and throughput time[23] absorbed the attention of the BBC management until the very end.

Besides the aforementioned commercial- and production-related effects, BBC felt increasing technological pressures after the mid-1960s. GE's Belle Isle 3.5 MW power generation gas turbine from 1949 serves as a reference case for evaluating axial compressor technology. The GE stage pressure ratio $PR_{Stage} = 1.127$ and the stage pressure coefficient $\Psi_{St} = 0.1598$ for 1949 correspond with the BBC reference marks in Figures 4-31 and 4-32.

The engine (see Section 5.1.5) represents BBC's first 'standard' GT product, in service since 1953. According to the internal Alstom 'Poseidon' database there were 71 units sold, of which 26 have been dismantled in the meantime, 12 are on stand-by, and 33 are still in service (as of 2009).

[23] In the early 1970s BBC was struggling with throughput times of 22 months for GT11D4, while GE were investigating reducing the time from 12 to 6 months.

The technological 'status quo' between the two major competitors remained largely unchanged until the mid 1960s, all in view of modest, sometimes not even measurable GT market activities. GE also exploited the advantages of their superior aero engine technology for industrial gas turbine applications, at best reflected by the short-lived popularity of aero jet expander projects[24], a combination of stationary turbojet gas generators driving a power generating turbine. GE only moved into the heavy gas turbine market slowly, with significant difficulties.[25] BBC felt relatively prepared for the difficulties of crude oil combustion, due to their superior burner/single combustor concept, and problems such as a lack of sophisticated cooling technology were not really felt regarding the potential dangers of cooling hole clogging. In addition, there is another point worth mentioning in order to understand the relative complacency regarding GT engineering at BBC.

In the period under consideration (1950–1980) BBC was an internationally successful supplier of steam turbines. Until the beginning of the 1970s, gas turbines were not competitive in comparison and it was obviously assumed they would remain as they had been for the past 20 years. Traditionally, BBC saw the gas turbine as confined to the areas of emergency power supply and locations where water shortages prevented the use of superior steam turbine machinery. Besides this, GT orders were tailor-made and Hans Pfenninger (head of the GT department between 1948 and 1968, Section 4.1.2) was heard to state that 2–3 projects per year were considered sufficient to preserve the 'status quo'. Otherwise, the marginally existing gas turbine would have occupied too much precious production space, needed for the continuously booming steam turbine business. Half-jokingly, gas turbines were relegated internally from a superior steam technology point of view to the category of 'entropy augmentors'.

The situation changed with the consistent global energy demand from the 1960s onwards, which correspondingly provoked reactions from the power generation gas turbine market. Figure 5-6 illustrates the annual global oil production, which at that time underwent its largest increase ever recorded (only interrupted by the first and second world energy crises in 1973 and 1979).[26]

As seen by the dashed line in Figure 5-6, industrial GT sales reacted to this strong and sustained energy demand with some delay, as shown by the late 1960s, but then experienced an unprecedented surge, meaning the installed GT power grew nearly 10 times within 10 years, untouched by the energy crisis in the given time frame. This engineering success was predominantly caused and experienced by GE, who had prepared the ground in time with attractive GT performance levels by providing acceptable solutions to their crude oil problems[27], but more importantly with a significant rise in unit power, from 20–30 MW up to the 50–80 MW range. After first offering turbo expander solutions for power generation,

---

[24] This short-term boom also reached BBC, see Section 5.5.2.

[25] The GE Frame 5 type gas turbine, a 26–30 MW range turbine, represents the GE heavy-duty approach of that time, with approx. 2,800 units installed worldwide in the meantime.

[26] The crisis in 1973 was caused by the OPEC oil embargo and the 1979 energy crisis occurred as a result of the Iranian Revolution, http://en.wikipedia.org/wiki/1970s_Energy_Crisis

[27] In this context one can certainly assume that this strong rise in oil production went hand in hand with a broad expansion of refinery capacity, so that fuel distillate became available in considerably higher quantities. For the mid-1970s a price difference between residual and distillate oil was reported in the order of 40 percent, but both were still rather low.

i.e. several turbojet gas generators working on a turbine, GE introduced their heavy-duty Frame 5 (MS5001, 26–28 MW) in significant quantities (> 100 units/a) from 1968 onwards. The strong demand in power capacity required larger power units and even with a high degree of modular design and standardisation, the strong upturn in the mid-1970s must have been restricted by production capacity limitations more than once.

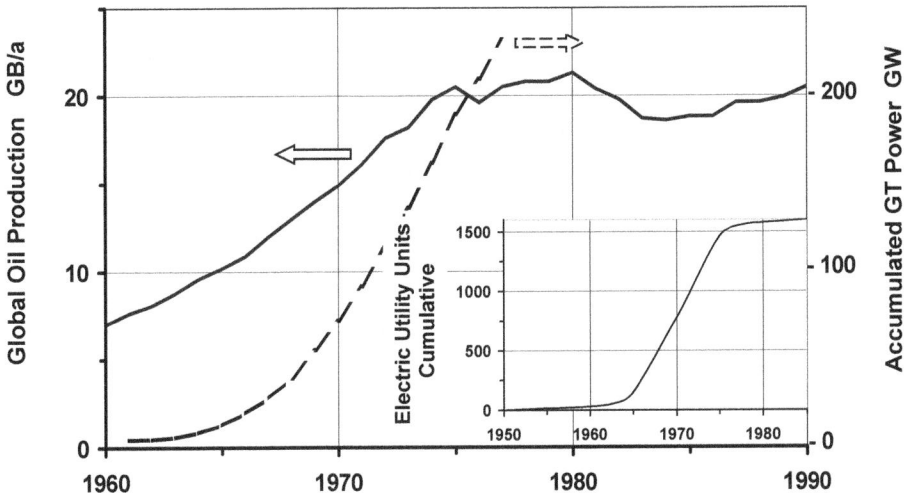

Figure 5-6   *Annual global oil production in GB/a, giga barrels per year(l)* [28]*, accumulated industrial GT sales worldwide up until 1977(r)* [29] *and US electric utility GT units (inset)* [30]

5.1.2   A Rare Top-Down View of the BBC Gas Turbine Business (1978)

During this period, a BBC reaction was often not immediately recognisable by the documented materials available, and even then not very clearly. In the early 1970s BBC became aware of its weak position in the important US market for power generating gas turbines. Not only did they fear commercial protections there, but expressed concern about falling behind and being partially excluded from the newest developments in high temperature materials. One exception to this somewhat disturbing silence is a thorough comparative performance analysis[31] from 1972, which came to the alarming conclusion that *'... compared to our D-rating, the GE PG 7841*[32] *has an operation temperature ('Mischtemperatur vor Turbine') of 1,080 °C, i.e. 120 deg. C higher. In combination with the relatively high compressor and turbine efficiency values this advantage cannot be compensated with the existing 5-stage BST turbine concept. A new 2–3 stage turbine... should replace the present 5-stage design.'* This remark from the company's mechanical design quarters reveals the growing necessity

[28]   http://en.wikipedia.org/wiki/Peak_oil

[29]   See Sawyer, Sawyer's Gas Turbine Catalog 1976, p. 4

[30]   See ASME, 3500 kW Gas Turbine

[31]   See Zaba, Vergleich der BST-Gasturbinen

[32]   The referred GE PG 7841 engine has been marked in Figure 5-8.

for fundamental design improvements, which led to the all new GT15/GT17 project with 11-stage transonic compressor and 3-stage turbine in 1976/7, Section 5.1.5.4.

Since in-house strategy documents from that period did not survive (if there were any), it is a rare and unique opportunity to observe BBC's activities of that time through the eyes of GE. There is a 'white paper'[33] from General Electric in our historic files dating from December 1978. It starts with a list of BBC's strengths:

> 'BBC has established a worldwide reputation in the power industry that suggests substance, capability and quality. BBC is a worldwide presence in fact rather than in a marketing sense only. Its worldwide network of manufacturing subsidiaries and sales organizations provide BBC with local identity and knowledge of local business practices, alternate manufacturing sites to obtain local or advantageous sourcing. BBC benefits from Swiss and West German aids such as low interest customer financing, favorable tax treatments and export incentives and protected home markets. BBC has a high contributed value, manufacturing many of the thermal and electrical products required for power projects and particularly turnkey projects which represent a rapidly increasing share of worldwide gas turbine business. BBC is a capable participant in extended scope gas turbine projects. It is organized to provide power plant design and the integration of subcontracted components. When partners are required such as for steam generators or for the construction of civil works, BBC is experienced in forming appropriate consortia.

> BBC has exhibited a strong capability for fuel flexibility. A considerable range of experience extends from very low-BTU gas such as blast furnace gas to residual and crude oils. BBC's single combustor design has supported this flexibility in the lower firing temperature ranges.

> BBC's strong worldwide sales coverage is centralized in BBC International which also allocates material sourcing by corporate component. BBC has established a strong US and 60 Hertz manufacturing site at St. Cloud, Minnesota. BBC has a joint gas turbine R&D program with Sulzer which broadens their available base.[34]

The subsequent list of BBC weaknesses reads as follows:

> 'The strong Swiss and West German currencies have placed BBC at a pricing disadvantage in the worldwide gas turbine marketplace for products sourced in those countries.

> BBC's gas turbine history has reflected an extremely conservative level of firing temperatures during the period of substantial market size. The energy crises have now placed BBC in strong competition for a reduced market with more efficient, higher temperature gas turbine designs. BBC's recent attempts to rapidly escalate temperatures have yielded unsatisfactory results in the form of unit damage and missed performance.'

---

[33]   See General Electric, A Brief Critical Assessment

[34]   The latter point is remarkable for 1979: the BST Brown Boveri-Sulzer Turbomachinery phase had officially ended in 1974, see Section 5.1.3, presumably without too much publicity, meaning GE intelligence missed the information.

The latter argument is accompanied by an illustration, Figure 5-7, which shows considerable turbine inlet temperature[35] differences of more than 150 $^0C^{36}$ between both companies from the 1960s onwards. It could therefore be argued that BBC had already lost the GT rating race of the mid-1970s (Figure 5-8) in 1960, long before it actually began. There is no direct proof that the BBC management realised the approaching calamity at the time. In all likelihood the situation was not considered too dangerous: a) the company occupied an unchallenged, leading position in the steam business, b) after more than 20 years of a small-scale, tailor-made GT market, gas turbines were still not considered for large-scale power generation, c) given the continuing difficulties of crude oil burning, BBC must have felt well-positioned. This was due to their very conservative GT cycle parameters, their rugged single burner/ combustor technology and their uncooled turbine operation. It was clearly not even considered that all of these circumstances could change so quickly.

Some forty years later, it is difficult to understand the decisions (or non-decisions) and motivation of that time, all the more so because there is a noticeable lack of thoughtful and reflective writing on the situation, both at the time and later on. Even a prolific author like Claude Seippel leaves today's reader somewhat puzzled. In 1966 he published a keynote speech for the 150th anniversary of the Technical University Vienna, entitled 'Development trends in steam and gas turbines'[37] – what an opportunity for a comprehensive technological 'tour d'horizon'! Although he was in his mid-sixties by then, he was still a leading engineer with many scientific aspirations, as proven by his nearly simultaneous seminal publications on combined cycle power plants.[38] Seippel addressed a number of current research topics, amongst those most prominently the 3D flow distribution in the rear lp stages of steam turbines and water analogy issues of supersonic turbine cascade flow. For gas turbines he addressed all the significant trends: '*As for the steam turbine, development activities in recent years have focused on increasing unit size (for peak load operation). While the first gas turbine for the city of Neuchâtel in 1936 (sic!) produced 4000 kW at 3000 rpm, today 34,000 kW can be achieved on just one shaft and with a dual-shaft configuration approximately twice as that.*' and '*The combination of gas and steam turbine is better suited for base load operation rather than peak.*' But, alas, he continues... '*The first realisation of this concept took place in the Velox boiler.*'! This was in 1966, five years after the first combined cycle power plant had been commissioned by BBC: the 75 MW, natural gas-fired plant 'Korneuburg A', only 15 km north of the festive convention. Instead of referring to what would become the high-efficiency showcase concept for the whole industry, he reflects upon '*alternatives to the high temperature gas turbine*'. In Claude Seippel's opinion of 1966 these are '*The Comprex*' and the '*MHD Magneto-HydroDynamic Generator*'.[39] Nothing comparable is said about either sophisticated gas turbine component technology or aerodynamically im-

---

[35] GE terminology 'turbine operation temperature' refers to the temperature after the first turbine stator row, in general slightly higher than BBC's reference 'Mischtemperatur T_{el} or Tmix (mixed out temperature)'.

[36] It appears as though GE had not taken notice of BBC's efforts to achieve higher turbine inlet temperatures, expressed by the '750 °C GT family' from the 1950s onward, the best example of which was the St. Dizier plant. Although several of these engines had been delivered to neighbouring countries like Canada and Mexico, none went directly to the USA, and there was a temperature limit which only allowed 730–750 °C for natural gas operation, while heavy oil operations remained restricted to 650 °C.

[37] See Seippel, Entwicklungstendenzen

[38] See BBR, The theory of combined steam and gas turbine installations, 1960

[39] In principle the MHD generator transforms thermal energy into electricity without moving parts. Its development was surpassed by the cheaper natural gas-fired CCPP and fuel cells, http://en.wikipedia.org/wiki/MHD_generator

proved compressors, as one might have expected from the 'father of the axial compressor'. He does not mention turbines either, including the increasingly important area of turbine cooling. Strangely, this same kind of technological self-sufficient attitude prevailed during discussions between Theodore von Kármán and BBC 20 years earlier, Section 4.2.4.6. *'Stagnation is regression'* was certainly as well known then as it is today, but the lack of specific compressor or turbine issues creates doubt whether the necessity of regular turbomachinery testing and corresponding, relevant R&D activities were acknowledged at all during that period. A similar concern was expressed in the context of Figure 5-2 and the observed, disturbing absence of any compressor patenting activities.

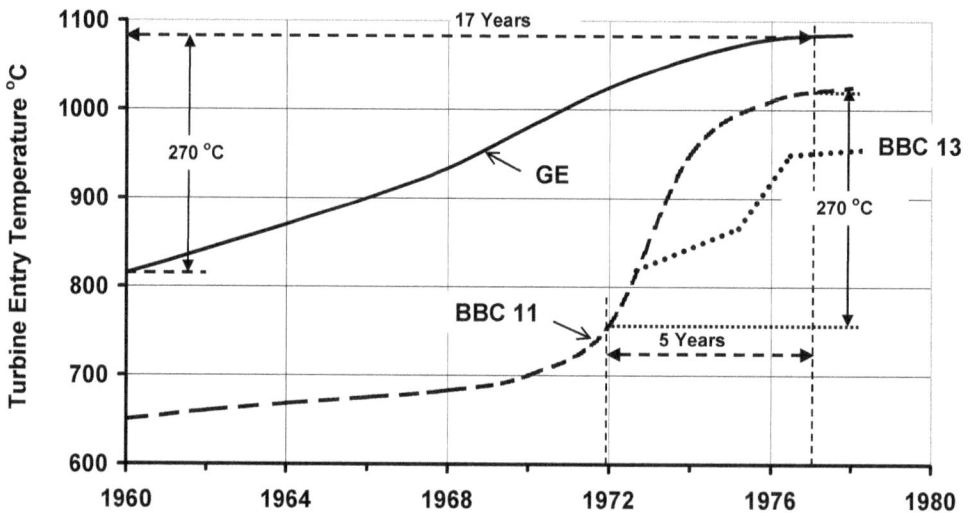

Figure 5-7   *GE competitor analysis (1979; dotted line 'BBC 13' for GT13 added)* [40]

Recognising the strong market upturn after 1970 and the corresponding GE strategy to develop larger GT power units, BBC started to announce a similar move for their related new GT11 (60 Hz) and GT13 (50 Hz) families. While in the early 1970s the base GT11-0 had started in the 35 MW class, aggressive rating increases of up to 78.2 MW were marketed for the GT13D and up to 70 MW for the GT11D4/D5. The aim was to achieve this by the late 1970s. The GE chart shows that the corresponding temperature rise of 270 °C required 17 years of steady development progress on the GE side, so that BBC's intended time frame of just 5 years was observed with some astonishment and, as was proven, justified scepticism.

This risky approach is also visible in the following engine rating comparison of that time, Figure 5-8. The upper curve represents the power output of a variety of GE gas turbines from approx. 50–80 MW, in a time period from roughly 1968 to 1988. For 1973 the GE PG7841 (60 MW) is highlighted, which was addressed by BBC in the aforementioned comparative discussion. The BBC GT11 engine family is illustrated by a MIN/MAX power bandwidth, deduced from a total of 134 correspondingly ordered engines. It appears that around the time of the large temperature rise in 1975 the bandwidth spreads to incomprehensibly large differences

---

[40]   Part of the data from – see General Electric, A brief critical assessment

in power, apparently confirming the presumed risk that the target setting was sometimes too aggressive and could not be accomplished on a continuous basis, meaning a subsequent de-rating was the predictable next step. Extreme excursions can be observed for the GT11D2 in 1976 and the GT11D4 in 1978, and a significant consolidation of the situation took place until the mid-1980s in the context of the GT11D5 development, Section 5.1.5.3.

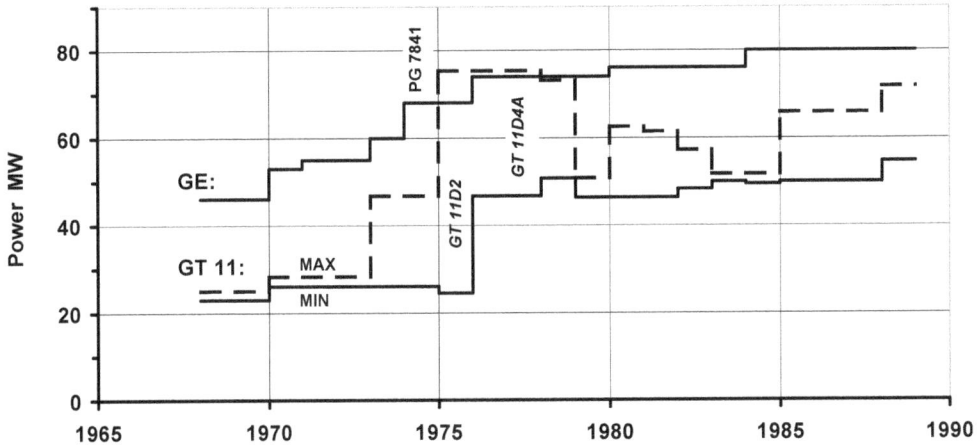

Figure 5-8   *The GT rating race 1968–1988: GE*[41] *vs. BBC GT 11$_{MIN}$/GT 11$_{MAX}$*

Half-hearted support for gas turbines may still have been acceptable at low market figures, partially even compensated for by special long-term customer relations, but it became an immediate problem when the technological competition increased at the end of the 1960s. Obviously unprepared, BBC saw itself confronted with a huge gap in technology in general, but especially in the hot gas section. As always, the 'steamroller approach' was not advisable. Over-rated engines predictably caused problems in the field. Uneven temperature distributions in the silo combustor combined with manufacturing/ welding problems of the U transition tube and on combustor tiles caused material failures which resulted in several complete engine losses. All of a sudden, limited engineering resources were tied up in short term recovery solutions, rather than being used to compensate for existing deficits in expertise through urgently required fundamental R&D activities.[42]

The available GE documentation from 1979 about the GT11 performance in the US contains detailed failure records for various projects, Figure 5-9, which clearly document that the aggressive market entry of BBC's US partner company, Section 5.1.3, was paid for with considerable problems and material losses, but also with BBC's reputation.

Further implications of these failure records for Turbodyne and their GT11 developments will be discussed in Section 5.1.5.3. For BBC, besides the striking incapability to control and influence a partly incompetent, definitely over-ambitious licencee, there are several rea-

---

41   GE data from – see Endres, Erfahrungen mit Gasturbinen –, GT 11 data deduced from order listing.

42   The development organisation of the time, without any division between long-term product development and short-term project engineering, contributed to this effect. Nevertheless this had quite positive effects on multi-challenged engineers' acquisition of experience.

sons for this negative development. The missed opportunity to become a substantial player in the technology-driven turbojet field at the end of the war must have resurfaced unpleasantly. Besides deficits in high-temperature materials and corresponding mechanical integrity analysis, turbine film-cooling technology (see Section 6.2.2) is the most striking example of how BBC did not do themselves any favours with their technical idleness during the decades in between, despite having a clear lead in 1940.

Comm. Date Rating	Customer Project	# of Units	P [MW] Iso	Turb. Inlet [°C]	Operat. Hours to 6/78	Comments
1974 C	Northern States Blue Lake	4	50	885	Average 1'640	4th stage turbine blade failure after 200 h, units have high forced outage rates.
1977 D	South. Cal. Edison Long Beach	7	56	945	Average 2'290	Cracking of combustion mantle. Loose piece damaged turb. blading. Do not meet emission.
1975 D	Public Service Oklahoma Weleetka	3	57-68	945	Average 350	Has combust. and control problems, also with unattend. start. Emission levels exceed limits.
1977 D	City of Braintree Braintree	1	59-68	985	990	Had turb. damaged by internal object after 900 h. Experienced anoth. failure 100 h later, when a large section of U-gas duct wiped out all 5 turbine stages. Does not meet performance.
1976 D2	Houston Light & Power Houston	6	59 - with water inject.	945	1'000-2'000 each	Missed guarant. output by 5 MW. Burned one unit trying. Combustor U-duct cracking & buckling. Customer pays for ext. modifications to solve probl. and increase output to 64 MW. Had severe vibr. w. water injection.
1976 D2	Sasketchewan Power Saskatch Landis	1	68-75	985	8'000-9'000	Turb. blades damaged due to internal objects after 8'000 h of operation.
1976 D2	Chugach Electric Beluga	1	60-68	985	11'000	Restricted to part load operation (840 °C) due to combust. duct problems.
1977 D2	DOW Chemical Plaquemine	1	65-68	985	4'700	Internal object damage, 1st stage turbine blades replaced twice.

Figure 5-9   *GE listing of BBC/Turbodyne GT 11 field service experience* [43] – *comments in Italics added*

---

[43]   See General Electric, A brief critical assessment

For a short while, the company may have considered leaving the challenging GT product market completely, as Sulzer did as a result of BST developments, see next section.[44] On the contrary, BBC decided to tackle the situation, but the counter-measures were exclusively commercially and strategically defined by a reworked product portfolio and new business partnerships, especially in the decisive US market. The only wise decision would have been to patiently stabilise their market position and then begin a comprehensive technology recovery programme, if not alone, then at least with a competent partner. This move would have required a lower market presence for at least five years, but in realising the market upturn, the situation became somewhat desperate for the management (if analysed in depth at all). Organisational chaos, objective irresponsibility combined with a muddle-through attitude cannot be excluded.

The 'BBC 13' curve indicates that a different approach was possible, as added to the original GE chart of Figure 5-7 to represent the turbine entry temperature development for the GT13 family, mainly manufactured by BBC Mannheim. The convergence with GE target values happened cautiously here, with a less pronounced gradient[45] in comparison with the Turbodyne-influenced GT11 development, marked in Figure 5-7 by the 'BBC 11' curve. It appears as if these different strategies had been determined at Baden[46] (in the case of the GT13 approximately 3 years before its actual implementation) and that comparable Baden data existed for the GT11 family as well. But apparently, Turbodyne decided to be independent for the higher 'BBC 11' turbine entry temperature values, leaving Baden with the task of adapting their own data sheets approximately 2 years after the different GT11 types had been implemented in the US.

5.1.3 Strategic Failures: The BST and Turbodyne Interlude

At the end of the 1960s a process took place in the Swiss machinery industry that resulted in a 50:50 merger of the power generation and turbomachinery activities of BBC and Sulzer Bros. AG, becoming BST Brown Boveri-Sulzer Turbomachinery Ltd. The official press announcement read: *'On 6th February 1968, the two Swiss company groups Brown Boveri/ Oerlikon Engineering Works and Sulzer Brothers/ Escher Wyss announced their decision to*

---

[44] Such interpretations can also be deduced from the Introduction of the internal GT Design Handbook by Wilhelm Endres, (see Endres, Erfahrungen mit Gasturbinen) who was put in charge of BBC's open GTs in mid-1974, when certain critical strategic decisions must have already been in effect. Besides the very valuable design experiences collected in the handbook, it can also be read between the lines. He states *'High temperature technology had been developed for aero jet engines at a cost of many billions (of dollars). It saw a fast-paced introduction to industrial gas turbines from 1970 (precision castings, blade cooling) with consequences such as a doubled power output for the GT11 within 5 years. The (BBC) development costs soared, with costs for 'design errors and losses' rising from 30,000 SFr annually to 3,000,000 SFr, i.e. by a factor of 100. Other manufacturers had similar problems and some disappeared from the market for years due to an excessive number of technical problems. The market volume fluctuated greatly and at times, almost every manufacturer considered stopping their gas turbine activities completely.'*

[45] It is interesting to see how well the development of the Siemens V94, 50 Hz family confirmed the BBC 13 approach in comparison: V94.0/1974/Tmix=850 °C, V94.1/1977/920 °C, V94.2/1979/939 °C, V94.4/1982/1040 °C. See BBC, GT11N early development, 1985–1990, an internal memo collection which refers to an unidentified KWU publication from 1985.

[46] Several TGT3 documents exist from 1969/1970 which describe the GT13 performance approx. 3 years before it was actually put into practice. Similar TGT3 documents existed with corresponding dates for the GT11, but these data sheets were cancelled and re-issued after the very engine realisation in the US and with data corresponding to the GT11 curve in Figure 5-7.

join forces in the gasturbine and turbocompressor fields, in order to rationalize their hitherto parallel activities and to strengthen their effectiveness in these branches of production. The main point of this important coordination was outlined as follows: A new, jointly-owned company "Brown Boveri-Sulzer Turbomachinery Ltd., Zurich – BST" was to be set up for the purpose of the development, design and sale of gasturbines and turbocompressors. These products, which had been previously built by both company groups with world-wide success, would in (the) future only be developed, designed and sold in standardized type series by the joint undertaking.

The parent companies provided the newly-founded firm with a staff of experienced employees for sales, development, design and commissioning. Furthermore, a considerable amount of technical data in the form of ideas(!), developments, knowledge, experience, tradition(!) and documentation was transferred.'

These words are repetitive, redundant and hollow, hiding principle defects of the construct from the very beginning. In fact the joint venture only survived for five years, the transaction reversed on 1 July 1974.[47]

The total effect of this undertaking for BBC was presumably rather negative, mainly due to inevitable organisational disturbances and diversions at a time when the long-dormant GT market exploded and should have received the complete attention of a focused engineering team. From a GT perspective alone, Figure 5-10 illustrates the somewhat disproportionate elements that were merged. The merger had been attempted more or less continuously throughout the 1960s and had previously been cancelled more than once due to management incompatibility. A hand-written document[48] by Hans Pfenninger, dated 14 December 1970, Section 4.1.2, reveals that negotiations about simplifications and the removal of overlapping elements from product portfolios had already lasted 8 years (and consequently also accompanied the first two years following the foundation of BST). The negative spirit which characterises this document (Pfenninger: '*Typical Sulzer negotiation tactics – take everything, give nothing*') must also have overshadowed daily cooperation. Other contemporary witnesses remember an aggressive, almost hostile competition between the later 'partners' during the 1960s: an attitude that apparently continued throughout the BST years with hefty disputes about the best design solutions. Finally, recognition must have dawned that the whole undertaking was a mistake and the phase of dissolution began. This ended with BBC keeping large gas turbines, including closed cycle GTs from the Escher Wyss legacy, while Sulzer took responsibility for small gas turbines and industrial compressors.

On paper, BST could at least record a considerable product portfolio reduction for both parties. While the reference publication 'Sawyer's Gas Turbine Catalog' lists for 1969[49]

---

[47]   The BBC Hauszeitung from 06/1979 states that in contrast to the formal split of BST in 06/1974 '*...1978 was the last year, when GT Plant and Equipment were at different locations – TCV-3 and -4 and TCG at Escher-Wyss-Platz, Zurich, the TCT14 Group at Oerlikon, while the rest were in Baden. Fewer communication and coordination problems have been reported, especially in the sales department, after all activities were concentrated in the "Pavillons" on Roemerstrasse under Dr. Wilhelm Endres.*' The mentioned 'pavillons' are still a core site of GT development today.

[48]   See Pfenninger, Einige Gesichtspunkte

[49]   See Sawyer, Sawyer's Gas Turbine Catalog 1969, pp. S-4/5

- for BBC: 13 different gas turbine types in the 8–30 MW power range and
- for Sulzer: 10 'Open Cycle Turbines' between 5–30 MW and 6 ex-Escher Wyss
              'Closed Cycle Turbines' for 2–25 MW,

'Sawyer's Gas Turbine Catalog 1974'[50] has 'in production' still

- from BBC: 3 new GT9, 11, 13 engines, Section 5.1.5, for 27–80 MW (normal rating –
              on paper)
- from Sulzer: 2 'open cycle' GT versions for 3 MW, Type 1 and for 11 MW, Type 7 (both
              engines as single and split shaft versions)[51] and for both companies in the
              'design stage'
              2 'closed cycle' GT configurations for 300 and 1,000 MW nuclear
              He gas-cooled reactors with integrated helium turbine, which BBC's PR
              department introduced in the same publication one year later under the
              striking title 'Now – Science Fiction becomes Fact' !

Design activities for BBC's new GT family concept dated back to the mid-1960s from BBC's development department of 70–80 employees. For a more comprehensive picture of the workload of BBC's gas turbine development department, it is worth remembering that groundbreaking activities for example in the field of combined gas/steam cycle power plants took place at the same time, which led to another far-reaching 'first' with the commissioning of Korneuburg A, Austria in January 1961: the first utility CCPP (Combined Cycle Power Plant) with 75 MW for exclusive power generation use, Section 6.3. From the mid-1960s to the mid-1970s the number of personnel in the GT department in Baden only rose from 55 to 80.[52] This number was clearly far too small in view of the challenging engineering tasks, even under the assumption that there might have been some additional support from other BST sources, later from Turbodyne, and eventually continuously from the steam department for tasks such as drawing work. The first performance and design calculations began for GT11B in June 1966, followed by GT9B in February 1968, GT13B in January 1969 and finally GT11C in March 1971. Engine prototype testing at BBC's main test facility in Muenchenstein on the southern outskirts of Basel was carried out in the same order between May 1970 and the end of 1973.[53] Although any direct contribution from Sulzer to these engine designs has not been documented, the example of existing Sulzer engines, Section 5.2.2, may have influenced decisive BBC design changes during the BST era and afterwards. Potential candidates[54] in this context are

---

[50] See Sawyer, Sawyer's Gas Turbine Catalog 1974,

[51] There was also a short-lived Type 3 with 4 single combustion chambers, which was investigated for a power plant in Saarbruecken, D in the 1980s, with a GT-turbocharged fluidised bed boiler.

[52] There is a breakdown available for 1977, see BBC, BBC/TCT Kapazitätsbelastung, 1977, addressed to T2 H. Baumann, TC W. Endres, with reference to the group heads TCT-T. Zaba, TCT-11 P. Zaugg, TCT-12 A. von Rappard, TCT-13 F. Farkas, TCT-14 A. Wicki, TCT-15 D. Mukherjee, TCT-36 A. Pesendorfer. At that time the development team had grown to 114. The project breakdown for 1977 mentions GT8-21MJ (man/year), GT9-6MJ, GT15-10MJ, GT17-2MJ, GT11D4A-6MJ, GT11E-8MJ, CAES 60Hz-3MJ, 'Claudia'-3MJ.

[53] See BBC, Entwicklungs-Dauer, 1985, also for the quoted numbers of personnel in the GT department. The GT11C prototype was tested at the Blue Lake plant in March 1974; this engine, first manufactured at Turbodyne, was a direct down-scaling of the predecessor GT13B/C. The relatively short development times of 3–5 years imply that the compressor and turbine blading design were accomplished in advance.

[54] Input from Franz Farkas, Section 5.1.6, based on experiences during and after the BST period.

- the single shaft engine set-up, supported by just two journal bearings
- variable compressor IGV (Inlet Guide Vanes). Though used by BBC for all stages in several previous projects[55], the focus on front row IGVs for increased mass flow flexibility happened during transonic compressor developments, implemented in the GT8 and later in the GT13E
- introduction of convection-cooled, precision cast turbine blading[56]
- first considerations about premix combustion technology.

A positive achievement from the BST period was the first demonstration of transonic compressor technology, Section 4.2.5. The first feasibility studies of this high mass flow/ high pressure ratio concept dated from the 1940s in Germany and from corresponding NACA activities during the 1950s, while Swiss investigations began with the seminal doctoral thesis by A.J. Wennerstroem, at Professor Traupel's Institute for Thermal Turbomachinery at the ETH Zurich in 1965.[57] The first company studies then followed, by R. Flatt[58], Sulzer in 1967 and Franz Farkas[59], BBC in 1969. After BST designs of isolated transonic compressor stages, Farkas proposed a universal compressor design for a new, scalable 150/180 MW GT family (GT15–60 Hz, GT17–50 Hz) and a comprehensive rig test programme was set up accordingly for test rigs PALUE 1 with 4 stages, PALUE 2A (6 stages) and BERNINA 2A' (9 stages) and successfully carried out at the BST test facilities in Winterthur. While GT15/GT17 projects were cancelled in the 1980s, both the Sulzer GT10 engine (with 10 compressor stages) and BBC's GT8 (12 stages), Section 5.1.5.3, represented practical implementations of transonic compressors in industrial gas turbines.[60]

As outlined already in Section 2.1.1, Brown Boveri began extensive restructuring in 1970. The company's subsidiaries were divided into five groups: German, French, Swiss, 'medium-sized' (seven manufacturing bases in Europe and Latin America), and Brown Boveri International (the remaining facilities). Each of these groups was further broken down into five product divisions: power generation, electronics, power distribution, traction equipment, and industrial equipment.

Throughout the 1970s, Brown Boveri struggled to expand into the US market. The company negotiated a joint venture with Rockwell, the American manufacturer of high-tech military and aerospace applications, but the deal fell through when the two companies could not agree on financial terms. While Brown Boveri counted a handful of major US customers as its clients, among them large utilities such as the Tennessee Valley Authority and American Electric, Brown Boveri's American market share was dismal considering the

---

[55] See for example the Cornigliano blast furnace gas compressor, Figure 4-21.

[56] Though internal turbine cooling had already spread from turbojet engine design to heavy-duty gas turbines during the 1960s, BBC waited until the BST period when the first design guidelines were set up and patents were filed together with Sulzer experts. The first cooled vane was introduced for the GT9C prototype and corresponding tests took place in 1973. The first cooled blade was implemented in the GT13D version at Donge/Geertruidenberg, NL in 1975.

[57] See Wennerstroem, Simplified design theory. His thesis was published in 1965, although the actual work was completed in 1962, (from a recent private communication with Art Wennerstroem) to a large extent at Sulzer Winterthur where he worked part-time due to an agreement between Prof. Traupel and Pierre de Haller, Sulzer's head of research at the time.

[58] See Flatt, Entwicklung einer transsonischen Axialverdichter-Stufe

[59] See Farkas, Die Berechnung der transsonischen Beschaufelung

[60] See Stoff, Warum die stationären Gasturbinen-Axialverdichter mit Ueberschall laufen

company's international standing (North American sales accounted for only 3.5 percent of total sales in 1974 and 1975, Figure 5-11), and the company continued to search for a means of effectively entering US markets.[61]

In 1970 the Studebaker Worthington subsidiaries Worthington Turbine International and the Electric Machinery Manufacturing Company were combined to form the Turbodyne Corporation with three major operations: 'Electric Machinery', 'Worthington Steam Turbine ' and 'Turbodyne Gas Turbines'. Finally, Colt Industries Inc. in West Hartford, Connecticut, which had been a BST licence holder, was sold and became part of Turbodyne Gas Turbine with a central production site based at St. Cloud, Minnesota, USA. The Colt licence from BBC was renewed for Turbodyne in the early 1970s, including a technology transfer from BBC.

As already illustrated in Figure 5-6, participation in the rapidly growing GT power generation market in the US must have seemed mandatory for BBC so that its already depleted market share would not fall into complete insignificance, Figure 5-11. Details of the decision process to start the 'American adventure' cannot be reconstructed. It was under somewhat shaky conditions, with an all new, largely untested engine concept (GT11) and under extraordinary technological challenges, with organisational weaknesses in a new partnership (Turbodyne[62]). However, it is very likely that it was a joint BST board decision. Beginning in 1971, Turbodyne started to manufacture the GT11C for the US market, at first quite successfully, with 37 of the 134 GT11 in total coming from the St. Cloud plant, operating to a large extent in the USA itself. But as expected, fundamental design and quality deficits became apparent after a while and were noted by the competition, Figure 5-9. The shift to the USA, in combination with the highly demanding 'rating race', also put BBC's traditional 'reference plant' concept into question, where at best a nearby plant in Switzerland had been used for testing new technology. Without a sufficient home market and at the same time lacking any meaningful test facilities directly at St. Cloud, the reference plant concept was unavoidably rolled out to 'unsuitable customers', with the predictable effect that difficulties and failures gained the attention of the competition and other potential customers after a while.

While corrective design actions for both GT11 and GT13 took place, leading to a certain process stabilisation in Baden, the US partnership with Turbodyne deteriorated under the enormous strain, which was also due to the loss of key US personnel. In 1978, BBC tried to correct Turbodyne's obvious manufacturing deficits by integrating the St. Cloud plant into BBI Brown Boveri International.[63] The remainder of Turbodyne Corp. Operations continued as

---

[61] http://www.fundinguniverse.com/company-histories/ABB-Ltd-Company-History.html

[62] According to the accounts of some participants, Turbodyne directly felt pressure from the GE competition, meaning that on their initiative the GT11 ratings were repeatedly raised no less than 9 times in a row in the critical years after 1975. At the same time the unconditional principle to make design modifications that could be retrofitted in reverse considerably limited the potential solutions. A rather spectacular result of bad coordination between Turbodyne and Baden occurred when Turbodyne decided to change their blade stagger angles with the intention of improving turbomachinery efficiencies. This was done without realising that this inadvertently opened up the possibility of severe hot gas ingress, with 'exploding rotor heatshields' and locally critical rotor over-temperatures. In the end it was these kind of experiences which forced Baden to regain full authority for designs and to end the 'US adventure'.

[63] This was less than two years after a new Consortial Agreement for a successful joint future had been signed between BBC & Turbodyne in August 1976: the GT15 project start, unimpressed by the GT11 'fiasco', was accordingly planned for 1 November 1976. (see BBC, Termine GT Prototyp 15, 1976)

part of Studebaker-Worthington. 10 further GT11D units were sold for BBI, but none of these to the US market. In the mid-1970s, the growing demand for large power generating facilities in the Middle East distracted BBC from its push into the North American market. Oil-rich Arab and African nations, such as Saudi Arabia and Nigeria, created new markets for Brown Boveri's heavy electrical engineering expertise in their attempts to diversify their manufacturing capabilities. But this situation in the Middle East ended in 1979, when a possible order for a large GT project was not accepted due to insufficient profit margins.[64] In view of a still challenging US market BBI dissolved its US subsidiary in 1983.

This was a somewhat questionable decision by BBC in the context of an enforced US engagement, but it should not be forgotten that the difficulties in the power generation market in the late 1970s were objectively so serious that seemingly bigger and better prepared companies also struggled and disappeared as independent firms. The prime example of this is Westinghouse, in the meantime then the clear No. 2 behind General Electric.

In comparison to BBC's shortage of adequate test and development facilities on site, Westinghouse's CTSD (Combustion Turbine Systems Division) completed their GT Development Centre[65] at Concordville, Pa. in 1976, capable of full-scale testing of compressor, combustor, turbine, and auxiliary system components over the entire range of operating conditions (exhaust system designs were developed on a reduced scale). This lab included a high-bay area that could accommodate a full-size gas turbine for testing and development purposes, as well as a large conference room and offices for the managers, engineers and technicians who operated the facility. It was big enough for full-scale combustion testing, requiring a large, motor-driven air compressor. It also required a gas-fired heater to simulate combustor inlet conditions: this facility was vital for the development of the W251 (40 MW) and the W501 (100 MW) series of gas turbines.

By 1984, more than 1,200 Westinghouse-designed gas turbines had been put into operation in 57 countries. However, by the mid 1980s, due to high labour and manufacturing costs and low demand for new utility-scale turbines, Westinghouse's gas and steam turbine divisions laid off a large percentage of their personnel in the Philadelphia area and moved their engineering and marketing functions to a new facility in Orlando, Florida. Turbine manufacturing was outsourced under a licensing agreement to Mitsubishi Heavy Industries (MHI). Shortly afterwards, when efforts to operate the gas turbine development lab on a contract basis for the military proved fruitless, the CTSD Concordville site was also closed and buildings were subsequently destroyed. In 1997, a decade later, Westinghouse sold off its non-broadcasting operations as part of its merger with CBS, with much of the power generation business becoming part of Siemens.

5.1.4   Business Facts

Previous discussions revealed the period at the end of the 1960s to be a decisive turning point in the gas turbine market. Before then tailor-made design solutions prevailed for more than 20

---

[64] See Wildi, Organisation und Innovation bei BBC, p. 65
[65] http://en.wikipedia.org/wiki/Westinghouse_Combustion_Turbine_Systems_Division

years, and it was only recently that turbojet expanders[66], a combination of cheap, mass-produced and therefore easily available turbojet gas generators with a specific project-designed expansion turbine, broke up the established market scenario. Another decisive step forward was the prefabrication of GT modules, shipped and erected in pre-installed compartments so the period from order to installation was decreased from years to several months. The development was spearheaded by the first standardised heavy-duty configuration, GE's MS5001[67], better known as 'Frame 5'. Significant quantities of 20 units/year were produced in 1958, peaking in the 1970s with several years where 100–150 units were sold. A snapshot for the year 1967, focussing on GE vs. BBC is illustrated in Figure 5-10. The total installed GT heavy-duty power was still within a rather comprehensive magnitude of 5 GW with less than 400 units. BBC's deliveries so far totalled to 60 units, approximately 1/3 of the corresponding GE value.

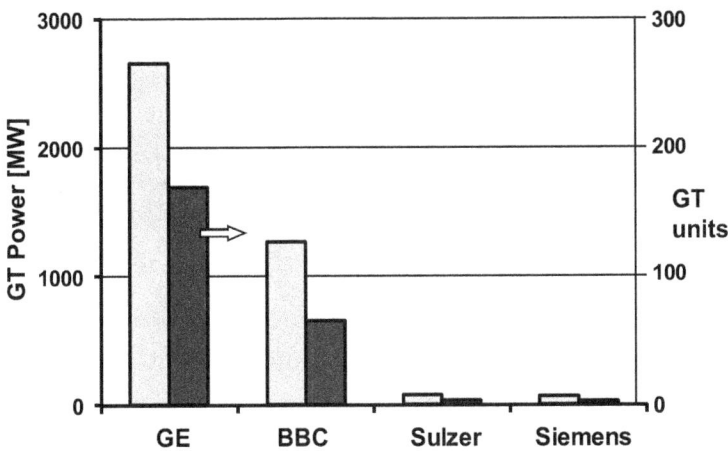

Figure 5-10   *Gas turbine power generation status 1967: accumulated GT power and sold units per selected OEM* [68]

The next new feature was the sustainable trend towards bigger power units, a direct consequence of the strong power demand at the end of the 1960s. While in the early 1960s the share of gas turbine installations with 15 MW or more was close to 40 percent, the same category had already increased to 75 percent by 1970. Figure 5-11 illustrates that the oil crisis of 1974 and 1978/9, with the result of a fourfold increase in fuel prices, predictably had no immediate impact on the strong market rise during the 1970s, but immediately after-

---

[66] Interestingly a turbojet expander solution was also considered as a kind of qualification certificate from Turbodyne in the eyes of BBC, when the company had successfully implemented an impressive combination of 6 turbojets driving a turbine wheel with a 4 m diameter.

[67] The MS5001 came to market in the late 1950s. It is still produced by GE's subsidiary Nuovo Pignone, Florence, I. With approximately 2,800 units installed worldwide, it is by far the most successful type of its class. Available in compact and reliable, single and two-shaft versions, it is designed for long life and ease of maintenance. The gas turbine has only two main bearings, and this and other features (e.g. compartment design) make it formative for later BBC design decisions. It consists of a 17-stage, high efficiency axial-flow compressor coupled to a 2-stage turbine and a combustion system with 10 chambers, able to burn a wide range of fuels. The electrical output is between 26.3–27.3 MW base/peak load.

[68] Data source: Sawyer, Sawyer's Gas Turbine Catalog 1967, also for Figure 5-11 (l)

wards the power demand peak was interrupted, only returning to similar levels 20 years later during the 'US bubble', when market deregulation and newly appearing IPP (Independent Power Producer) caused a record new demand.

Figure 5-11 also puts BBC's market participation in this development boom into perspective: BBC's market share drops continuously from an almost monopoly position in the early 1950s (in overall terms not so impressive) to less than 5 percent during the turbulent 1970s.

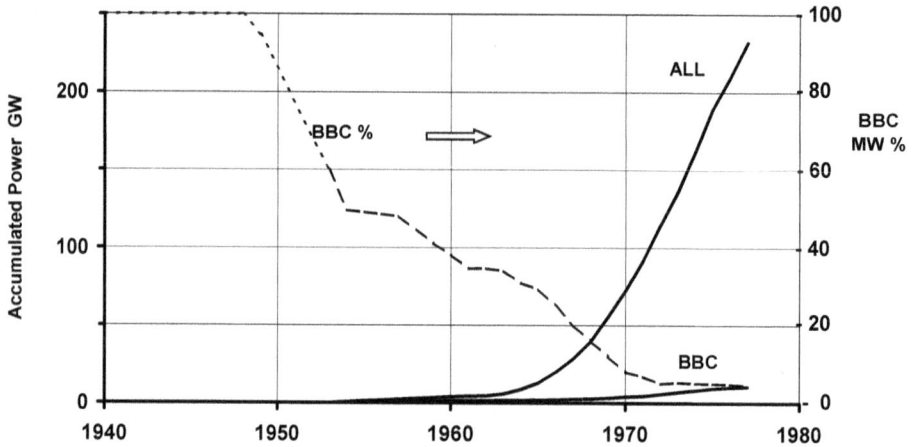

Figure 5-11   *Accumulated industrial GT sales worldwide 1950–1980: both total and BBC (l) – and development of BBC's relative power share (r)*

5.1.5   Early Plants and the Key Development Projects GT8/GT9/GT11/GT13

5.1.5.1   BBC Gas Turbine Development Roadmap 1945–1988

The first 10 GT projects after 'Plant #1', the world's first utility gas turbine at Neuchâtel, Section 4.3.3, had been ordered still in 1945, i.e. during Phase I of this historical review, and have been dealt with in Section 4.3.4.

The urgent need for larger power units at elevated efficiencies led to the introduction of two-shaft concepts with compressor intercooling and heat recovery at turbine exhaust. This was immediately after the war, under the impression that fuel costs were relatively high. Targeted plant efficiencies of 31 percent and the possibility of cheap crude oil combustion should have brought these gas turbines into immediate competition with the steam turbines of that time, which reached comparable efficiency levels in 1947. The development of the showcase plant with 40 MW record power at Beznau, AG, Switzerland, practically within sight of the BBC Baden facilities, is described in the following Section 5.1.5.2. Tailor-made two-shaft plants of this kind which had been started with great enthusiasm were implemented in 30 projects up until the mid-1950s, 14 of which used heavy crude oil. A typical example was illustrated in Figure 4-85 for the 10 MW St. Rosa, Lima power station, mainly identical with the later Beznau I facility. These overall sales numbers were somewhat disappointing, given the large development efforts in combination with extreme difficulties in finding acceptable solutions for the combustion problems of low-BTU fuels. The root cause for this negative experience was the same as in several comparable cases in the future: the 'moving target', the steam turbine. And, as experienced repeatedly, it was advancement in BBC steam turbines which set the pace, as in the years 1954 –1965 when the introduction of intermediate steam reheat and significantly higher operation temperatures considerably raised ST efficiencies and unit power.

In principle, this move towards complex, what were known as ICR Inter-Cooled, Recuperated 2-shaft GT plants with a group of heat exchangers in the form of compressor air intercoolers and turbine exhaust heat recuperators[69] should have exploited the inherent, high performance potential of this set-up, but in reality it often revealed many times the devil hiding in unforeseen details. The most prominent of these, vanadium corrosion, has already been addressed in Section 5.1.1.

BBC's standard GT plant concept was radically simplified again. The principal sketches in Figure 5-12 illustrate the move from a simple GT Neuchâtel type arrangement, via the complex ICR-GT configuration, back to the simple GTs in use until 1966.

The latter type, produced in growing numbers, had a large similarity with established steam turbine arrangements (as seen on the right-hand side of Figure 5-12): an elevated mounting platform with air feed and compressor-combustor connections below, so that the upper halves of the compressor and turbine casings were free from interconnecting tubing and easily accessible for engine maintenance. A single silo-type combustion chamber was placed vertically beside the turbine. The operation was extraordinarily simple and automatisation could be realised with low extra effort. Approximately 100 engines were built in the power range from 6–17 MW according to this mounting concept, acquiring a reputation for robustness and reliability.

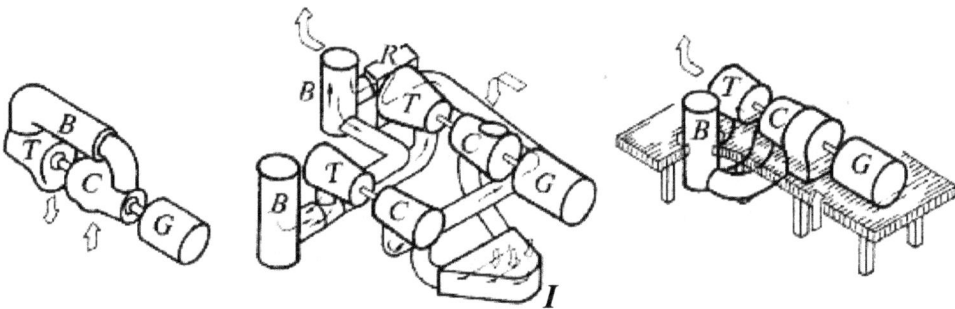

Figure 5-12   *Development of basic BBC GT arrangements I:[70] type GT Neuchâtel 1939 (l), type ICR-GT till 1960 (m), simple GT types up to 1966 (r), C – compressor, B – combustor, T – turbine, G – generator, I – intercooler, R – recuperator*

---

[69]   The exhaust heat exchanger transfers typically approx. 75 percent of the available temperature difference between turbine and engine exit to the air between compressor exit and combustor intake. Besides the designation 'recuperator', the term 'regenerator' is also in use for power generation. Special applications like automotive gas turbines or the long-range 'regenerated turboprop' from BBC Mannheim, 1943–1945, used rotating 'regenerator' drums, so that stationary heat exchangers were called 'recuperators' instead.

[70]   Figure source: Endres, 40 Jahre Brown Boveri Gasturbinen, Figure 2

A comparably small but steady trend towards increased power output became visible in engine mass flows, increasing to approximately 160 kg/s by the mid-1960s, i.e. 2.5 times that of the Neuchâtel value. Similarly, the thermodynamic performance parameters of the BBC gas turbines grew rather moderately: turbine entry temperatures only rose by 100 °C in the 25 years after the Neuchâtel GT to 650 °C and the standard compression ratio roughly doubled compared to the Neuchâtel starting value of PR = 4.4, accomplished in 17 stages. It was in response to specific customer demands for high thermal efficiencies that peak values were achieved, like the (guaranteed) 750 °C at turbine entry demanded by EDF Électricité de France in 1953, put into practice as a so called GT08* at St. Dizier (6.7 MW, 21.8 % efficiency, 4,750 rpm) in a special version with increased compressor inlet flow, co-alloyed turbine blading and rotor heat shields and the peak pressure ratio PR = 18 which was already in use in 1955 in a standard GT12/8 reheat engine, ordered by Società Elettrica Selt-Valdarno (1906–1970) for their power plant at Livorno/ Leghorn[71], I. (26 MW, 24.6 % efficiency). After an identical Livorno GT-2 group had been installed, for a short while in 1957 this GT site represented with 52 MW the largest GT power capacity in the world.

The annual delivery rates for these engines remained almost constant until 1965, at a level of 3–7 units, with the exception of three subsequent years (1955–1957), when sales rose to 25 units. The sales volume spread across 8 GT types, of which by far the largest share was made up by the GT10* (3,600 rpm, 6–9 MW)[72] with more than 80 units produced. Geographically, GT plants from BBC were erected in approximately 40 countries world-wide up until 1970, with certain local cumulation – with more than 15 installations in Canada, Germany, Saudi Arabia and Venezuela. In contrast, weaknesses in the important US market manifested quite early, with only 2 units sold.

As discussed already in Section 5.1.4, the period of tailor-made GT fabrication also ended for BBC almost 5–7 years after a comparable move of the market leader General Electric, due to the introduction of pre-fabricated and container-packaged GT models in large quantities. Although BBC could claim another 'technical first' here, developing the first mobile GT-powered generator set worldwide as early as 1955, as a direct consequence of BBC's unique experience in gas turbine locomotives more than 10 years before, BBC obviously missed the inherent opportunity for larger sales numbers by means of rapid product standardisation. From 1965 onwards it was predominantly the GT11L with 15 MW output power which belonged to this category, which with 79 sold units represented a relative success story. But while GE decided to maintain a competitive status for the comparable Frame 5 engine by continuous design adaptations – for more than 50 years and until today justified by approximately 2,800 sold units – for BBC/BST the GT11L only represented a short-lived episode. These 'packaged energy stations' will be dealt with explicitly in Section 5.5.1 as part of several BBC 'special projects'.

In the mid-1960s BBC started serious efforts to increase the GT unit power size considerably and at the same time to introduce a family design concept, so that variations of the

---

[71] Leghorn is the English name for Livorno, a port on the Tyrrhenian Sea on the western edge of Tuscany. http://en.wikipedia.org/wiki/Livorno

[72] Originally, due to the lack of standardisation, BBC had no specific type designations and used the project names instead. In the 1970s the 'old engines' were numbered according to speed, with designations GT05 (6,630 rpm), GT07, GT08*, GT09*, GT10*, GT12 (3,000 rpm) for simple cycle GTs, in parts 'asterixed' to differentiate from later GT developments with the same numbering.

output power were achieved by simple geometric up- or down-scaling.[73] In addition, the family members had either 3,000 or 3,600 rpm so that the power generators could be directly coupled. A major precondition was the development of (still sub-sonic) compressors with optimised flow capacity.[74] While BBC's traditional design preference for a degree of reaction R = 0.5, Section 4.2.3, was confirmed in principle, considerable improvements were achieved by fine tuning radial loading and airfoil thickness variations for both rotor blades and stator vanes. The resulting compressor design adaptations were characterised by inlet diameter increases, an increase of compressor entry tip speeds from 270 to 330 m/s with nevertheless, thoroughly observed critical Mach number limitations and typical degree of reaction values of R ~0.68 for the compressor tip sections and R ~0.4 correspondingly at the hub. In comparison to the previous GT11L design standard the compressor intake volume was more than doubled, even though the number of axial compressor stages had been reduced from 19 to 15. Similarly, by increased turbine expansion ratios, in parallel by rotor and turbine vane carrier cooling, the number of required turbine stages was reduced from 7 to 5.[75] Both design changes combined then allowed for a substantial shortening of bearing distances and in due course, a design change from 4- to 2-bearing arrangements.

These design changes were tested in two prototype projects:

- *GT13S* was the first gas turbine with the new compressor blading, but still carried out as a 4-bearing-configuration. Two 42 MW units were ordered in 1967 by EDF for a large combined cycle plant in Vitry, F. The units were built by BBC's French subsidiary C.E.M. and put into operation during 1971.
- *GT11S* represented the corresponding development for the 60 Hz market. The design of this gas turbine, beginning in 1967, used the same turbomachinery blading and rotor cooling system as the GT13S, but for the first time had a two-bearing-rotor. The combustion chamber was an up-scaled version of the GT11L, mounted vertically beside the turbo-set. In comparison to the previous GT11L, the new GT11S produced twice the power with very little increase in size.[76] The prototype was thoroughly tested at BBC's Muenchenstein test facility up until May 1970[77], when the engine was shipped (and at the same time renamed GT11B) to Rainbow Lake, Alberta, Ca. as the first 60 Hz customer of the new GT family.[78] In fact, this GT11B represented the nucleus of GT family development over the following 30 years.

---

[73] It is worth remembering the basic scaling rules: 1. Linear dimensions scale directly with the scale factor 60/50 Hz = 1.2, 2. rotor speed scales inversely with the scale factor, 3. mass flow like power output scales with the square of the scale factor, 4. mass and volume scale with the cube of the scale factor.

[74] See Pfenninger, Axialverdichterbeschaufelung für grosse Foerdermengen

[75] See Seippel, The evolution of compressor and turbine bladings

[76] See Bankoul, A New 30-MW Packaged Gas Turbine. The 30 MW GT11S concept design started with a firing temperature of 780 °C which was raised to 850 °C to provide the renamed, now 33 MW GT11B with better market opportunities in North America.

[77] First ignition of the new engine family at Muenchenstein was on 11 November 1965. Based on 330 starts the total test time was more than 200 h – with special attention to engine starting and compressor rotating stall/surge. See Endres, Three Gas Turbine Types

[78] Entry into service at Rainbow Lake was on 12 October 1970.

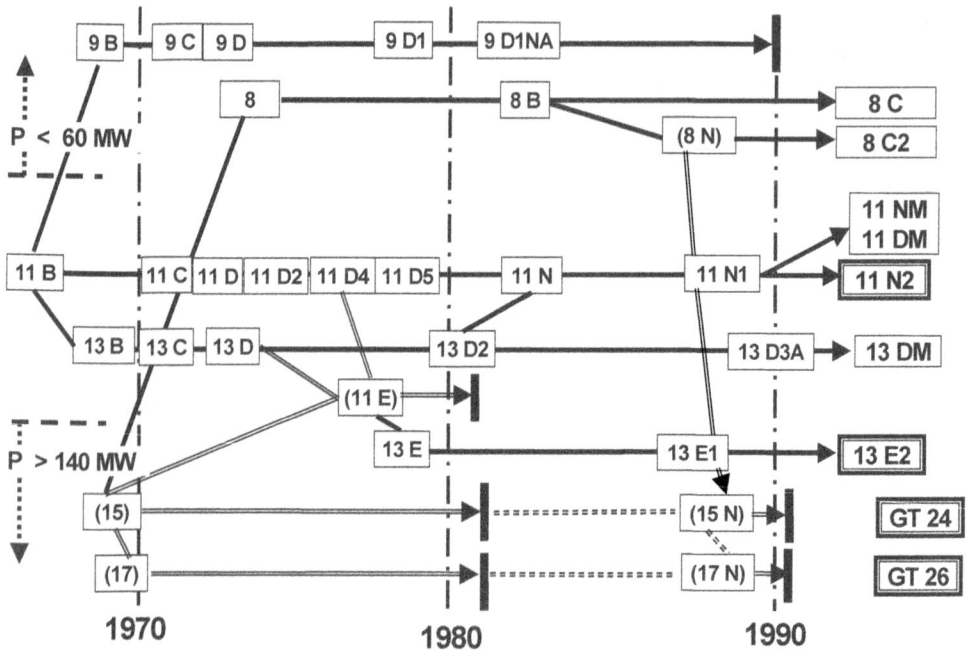

Figure 5-13   *Basic BBC/ABB/Alstom GT family roadmap [79], rating vs. order date – since late 1960s*

The GTxxS/GTxxB designation referred to an uncooled turbine blading which limited the firing temperature to 780 °C. These two GT prototypes with integrated design changes prepared the ground for the introduction of the GT9 – GT11 – GT13 family, which is described in full in Section 5.1.5.3 and on the basis of Figure 5-13. The figure illustrates the GT developments emerging from the first GT11B engine, typically with reference to the first order date. The roadmap is schematically divided into three power ranges

- up to 60 MW in the upper section – with GT9 and GT8
- an intermediate range from 60 to 140 MW – with GT11 and GT13
- and at the bottom a range beyond 140 MW – with GT13E → GT13E2, GT15/GT17 twice started and stopped
- the new GT24, GT26 family members, which will be described separately in Section 6.1.3.

This power classification also indirectly reflects the various responsibilities for the development strings, especially in the 1970s. While the family concept GT9-GT11-GT13 had been planned and put into practice up to the B versions more or less exclusively in Baden, this technology standard soon proved insufficient in the competitive reality and upgrades had to be prepared under pressure. For this broad approach the Baden engineering resources would have been overstretched[80], so that more and more design tasks were delegated, beginning

---

with the C versions, to the licensee Turbodyne for the GT11 line, 60 Hz in the USA and to BBC Mannheim for the GT13 configurations, 50 Hz. Especially in the first half of the 1970s,

Turbodyne tried to cope with the extraordinary pressure on GE's home market alone. Several serious development mistakes occurred, a few are listed in detail in Figure 5-9. But instead of correcting and maturing the design standard at the intended technology level, a new and higher rating was aimed for salvation. As an example, for the short period from 1974–1977 GE observed three ratings: GT11C, D and D2 which, besides the development troubles, see Figure 5-9, were characterised by sales figures that were far too low – GT11C (4), GT11D (11) and GT11D2 (12). In hindsight it is difficult to decide whether Turbodyne pushed forward impetuously or if the licence agreement would have allowed a limit to the image damage which Turbodyne caused for Baden. On the other hand there are indications that the slower, more moderate and apparently more cooperative Mannheim-Baden development of the GT13 line profitted more than once from Turbodyne's 'maverick approach'. Beginning in 1975, the Saudi GT11 market came to the rescue, covered exclusively by Baden which implied that Turbodyne's solo attempts were reined in anyway.

The reconstruction of a GT type naming convention at BBC at the end of the 1960s is somewhat shaky, mainly due to an apparent lack of relevant documentation. The desire for a rough type categorisation apparently only came up after 20 years of GT production without such a remedy. Then, at the end of the 1960s, potential type designations '11' and 1.2-times scaled '13' had not been used, but coincidentally the constant hub diameters of the new family components (of the compressor $D_{Hub, C}$ = 118 cm and for the turbine $D_{Hub, T}$ = 112 cm) were close to exactly 110 cm for the intermediate shaft section in the combustor range. This characteristic measure reappears within a certain tolerance for the major production families and can be extended at least up to the GT13. For the GT24 and GT26 developments[81], these designations were fixed in 1992. 'GT2x' is interpreted as indicator for a sequential, 2-stage combustor arrangement, while the second numbers stand for the design mass flow regimes of 400 kg/s and 600 kg/s, respectively:

- GT 8   ~ 80 cm rotor diameter in combustor region
- GT11   ~110 cm rotor diameter
- GT13   ~130 cm rotor diameter[82]
- GT24   2 combustors and 400 kg/s
- GT26   2 combustors and 600 kg/s.

General development targets of the various GT families over time were

- a steady power and performance increase
- improved fuel flexibility at reduced emissions
- reliable and cost-effective operation.

---

prominently the technologically challenging, high powered developments of GT15/GT17, for real or claimed economic and market-related reasons.

[81] Actually, the 2-stage combustor design and naming convention was applied after early single-stage, annular combustor design solutions GT14 and GT16 of these mass flows had been defined on the drawing board. For GT15 and GT17, projects that were twice started, but not realised in 1981 and 1991, the designation was related rather to the intended power class of 150/170 MW.

[82] GT13B/C/D/D2/D3/D3A have exactly, and the GT11N2 approximately 140 cm rotor diameter in that area (input Chr. Jacobi), which brings the rule of thumb above into doubt, and a designation as GT14 would have been equally justified, already then.

The various GT ratings of Figure 5-13 can be partially differentiated according to the applied turbine cooling scheme:

- B without vane and blade cooling (but with cooled rotor and vane carrier)
- C as B, in addition only vane 1 cooled
- D as C, in addition blade 1 cooled
- D2 and higher as D, in addition stage 2 cooled.

In the 1980s the GT11D6 rating was renamed to the simpler GT11N with N = 'New performance', due to sales considerations.

5.1.5.2   Beznau and Other Tailor-Made Gas Turbines

In 1946 NOK[83] decided to build a thermal power plant to cover peak energy demands and to bridge the lower power supply of hydro plants in dry years. Based upon thorough studies[84], a gas turbine plant proposal from BBC was finally selected for erection at Beznau, site of NOK's traditional running water power station[85] on the River Aare. The order was given to BBC Baden; construction relied on already built units to reduce design and erection time; consequently, a 13 MW unit (Beznau I) and a 27 MW unit (Beznau II) were combined, both in two-shaft reheat/recuperator configuration, which in the same arrangement had already been sold before (e.g. for Lima, Peru – Figure 4-85[86]). Nevertheless, the units were the first reheat units in commercial service, Beznau I in January 1948 (1.5 years after ordering), followed by Beznau II one year later.[87]

From the first firing of Beznau I there is a rare document showing the dark smoke and soot clouds during the crude oil start-up phase, Figure 5-14, and at the same time a variation of corresponding management faces, depending on the professional discipline ranging from amused astonishment to sceptical reservation. As was to be expected, the complex plant created numerous problems which had to be tackled by systematic testing in 1948 and

---

[83] NOK Nord-Ostschweizerische Kraftwerke AG (1914–2009), today Axpo AG in Baden CH, produces and distributes electricity in the north-east part of Switzerland for 3 million inhabitants, owned by the cantons in that area, http://de.wikipedia.org/wiki/Axpo_AG

[84] See Pfenninger, Zur historischen Entwicklung der Gasturbine; he mentions in this context a positive expert study from Prof. Bauer, ETH Zurich in favour of BBC's economic proposal, based on cheap crude oil use and a GT reheat concept with more than 30 percent thermal efficiency.

[85] At present, 11 turbines are fed typically by 418 m^3/s to produce 19.5 MW. http://de.wikipedia.org/wiki/Aarekraftwerk_(Beznau)

[86] Power duplication of the Beznau II turbo-set was provided by dual-flow configurations for the lp compressor and lp turbine, see Kruschik, Die Gasturbine p. 577. Like for the first launched 'reheat with recuperator (regenerator)' plant at Santa Rosa electricity works in Lima, Peru, the hp shaft was chosen for electricity production to secure high part load efficiencies. For the smaller Beznau I engine a gearbox is put between the hp compressor and the generator. Later reheat engines were configured without a recuperator and the generator was coupled to the lp shaft; examples are Livorno/Leghorn, Korneuburg A and 20 other following engines. Both Beznau engines were equipped with a fully automatic turning unit which kept the rotors slowly spinning for several hours after shut-down to prevent thermal rotor deformations.

[87] Earlier reheat engine orders of 1945, Figure 4-84, became operational shortly after Beznau: Lima 1949, Filaret 1951. Besides the GT plant, BBC delivered the turbo-generators, starter and auxiliary electric motors and the control and switch gear for Beznau. Building construction works were carried out by Ed. Zueblin & Cie. AG Zurich, Jaeggi Brugg and AG Conrad Zschokke, Doettingen. Sulzer Bros. Winterthur delivered the fuel tanks, an auxiliary steam boiler plant and air conditioning. The total plant cost was 17.4 MSFr, equivalent to 435 SFr/kW.

1949, parallel to the ongoing construction of Beznau II. The following Figure 5-15 shows a view of the Beznau powerhouse after completion and Figure 5-16 provides a corresponding cross-section, with a marker of the plant's absolute altitude, 340 m ASL (above sea level).

Figure 5-14   *First firing of Beznau I GT plant in January 1948 ©ABB (top) with BBC management*[88] *attending (see roof cut-outs on the left, – for follow-up Beznau II chimney installations)*

In the first weeks after firing of Beznau I the crude oil burning process was optimised, with the end result that smokeless operation was achievable in a wide operation range, except for low loads with visible 'white smoke' and full loads with inevitable 'black smoke' emissions. Black smoking could be shown as a direct function of the fuel nozzle spraying angle and of fuel atomisation quality; nevertheless, it appeared to be impossible to achieve smokeless combustion throughout the whole operation range. Due to environmental problems the initial fuel 'bunker oil C' was replaced over time by ultra-light fuel oil.[89] On the 27 April 1948,

---

[88]   Picture source: BBC, Geschichte des Schweizer. Turbomaschinenbaus, 1982. Out of the 7 men in the picture, the following have been identified – from left to right: Mr. Hoffmann, Head Thermal Machinery Section 1C; 2. Mr. Riollo(?); 3. Mr. Blum or Felix Huber; 4. Mr. Kerez, Head Sales Dept. VA1; 5. Hans C. Bellati, TFVL; 6. Mr. Frick (deputy of 4.), 7. (?)

[89]   See Zaugg, Ein doppeltes Gasturbinen-Jubilaeum

after only 284 operation hours, three blades of the hp turbine rotor failed, causing complete rotor damage. Inter-crystalline cracking in the blade root section and operation temperatures of 50 K above past experience were determined as the main causes. After similar problems had occurred in the lp turbine rotor after 2,350 h, the notch effect was reduced by introducing larger radii in the blade root sections and improving the turbine assembly process.

Figure 5-15   *#1 worldwide: gas turbine power plant Beznau 1949,*[90] *engine hall with 27 MW GT group II in the foreground, 13 MW GT group I in the rear, combustors are visible as black vertical cylinders on the left, air pre-heaters in the open space to the right   ©ABB*

After first performance measurements had been evaluated, the measured power output of 9–11 MW showed considerable discrepancies compared to the guaranteed value of 13 MW for Beznau I. Instead of a calculated thermal efficiency of 30.6 percent, the measured values appeared to be 6–8 efficiency points below target. After a while turbomachinery fouling came into focus. First ash deposits on the turbine blading directed the root cause search towards the crude oil combustion, but then even bigger performance defects were analysed, especially for the lp compressor sections. A cumbersome process of performance assessments for individual turbomachinery components had to be carried out, interrupted by intermediate cleaning phases, so that the negative deposition effects for the performance of every component could be evaluated. The intense measurement/cleaning phases lasted from 7 November 1948 until 22 January 1949 for the Beznau I GT[91], while Beznau II was investigated immediately

---

[90] Picture source: BBC, Brown Boveri Gasturbinen, 1963, Figure 34

[91] More than 60 years later, Olivier Seippel (1926–2012) remembered that he accompanied his father several Sundays in a row on his 'pilgrimage' to Beznau, where special teams were regularly employed to scrape off tar depositions from the compressor blading. This is true: every action is meticulously recorded in the TFVL test notes: For Beznau I, 'scraping days' in 1948 were on 7 November (SUN), 13 November (SAT), 28 November (SUN), 12 December (SUN) and in 1949 on 16 January (SUN) and 22 January (SAT). These actions continued

after operation start-up between 29 January and 11 April 1949. After collecting some operational experience, it became evident that especially the lp compressor fouling was the result of a fundamental design fault of the Beznau construction. With prevailing westerly winds the plant was in the wake of a nearby forest tree line, and the ramp-type roof of the power house aggravated leeward recirculation tendencies. The relatively short smoke stacks in the immediate vicinity of the fresh air entry opening on the southern, narrow side of the powerhouse led to close-coupled air/ exhaust gas mixtures, and finally to loss-generating tar depositions on the compressor blading[92], Figures 5-16 and 5-14.

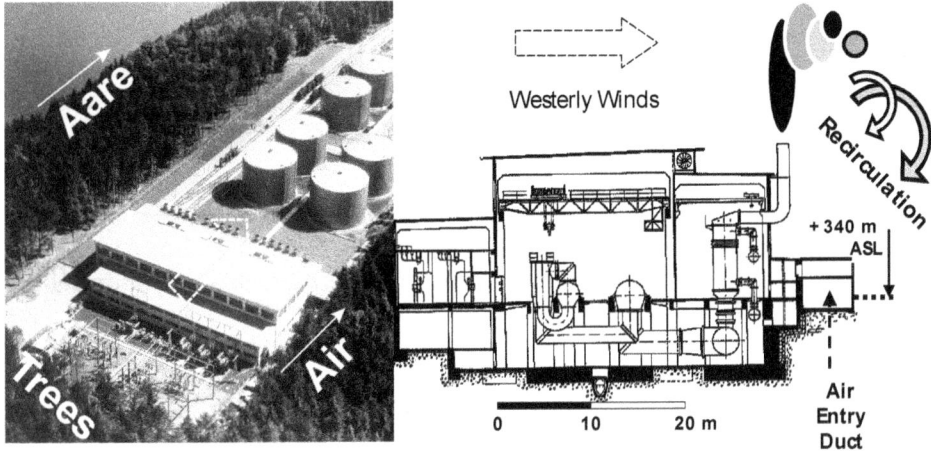

Figure 5-16   *Beznau power station, aerial view (l) from south-west:* [93] *(l) River Aare and fuel tanks (top), air intake (bottom right) and west-east cross-section of power house (r) with exhaust gas recirculation   ©ABB*

In the end it became clear that even with cleaned turbomachinery components the guaranteed thermal efficiency values were missed – for Beznau I by 28 percent instead of the intended 30.6 percent, and for the larger Beznau II plant by 27 percent in comparison to the intended target of 28 percent.[94] These problems were finally solved by raising the turbine hot gas temperature to 650 °C[95], for Beznau I in 1951 and subsequently for Beznau II in

---

[92] for Beznau II until April 1949. Later the crude oil operation demanded regular thorough turbomachinery washing for 2 h after 300–400 operation hours, after the tar had been liquefied by the previous steam injection.
This is a simplified explanation. In fact, the most obstinate deposits were found in ip and hp compressors. High air humidity led to condensation in the intercoolers, which in combination with the sulphur-containing, recirculated exhaust gases led to the formation of sulphurous and sulphuric acids. The acids attacked the thin copper fins of the coolers to form water-soluble copper sulphate, which was carried on with the condensate to the subsequent compressor, where the carrying water evaporated and encrusted the compressor blade surfaces. These problems were overcome by adding water-separating drain valves to the intercoolers.

[93] Picture source: BBM, Betriebserfahrungen, 1953, Figure 23. Storage capacity was 30,000 m³, sufficient for 3,000 operation hours at full power, or more than a winter period.

[94] To put the measured efficiencies into the right perspective, it has to be noted that the specified winter application of the plant implied a standard intake air (and water) temperature of +5 °C only.

[95] 650 °C was a new peak value for BBC, but GE's Belle Isle plant had already achieved 760 °C at around the same time and Westinghouse started commercial offers in 1954 with 730 °C. So the US companies already had

1953, so that after this date the plant represented both the world power and best performance record of its time in its own right, with 40 MW and nearly 30 percent efficiency[96].

The Beznau plant was operated mostly at peak load for 1,000 to 2,500 hours per year. As it became increasingly difficult to operate the plant within environmentally permissible limits, the plant was finally shut down in 1988 and replaced with a 54 MW GT8C reserve plant from ABB at Doettingen, further downstream from the old location on River Aare. The GT exchange visualised the GT development process over time remarkably well. The new, typical simple-cycle GT product status of the mid-1990s required only one fourth of the ground floor space of the old plant.

The Beznau two-shaft, reheat process of 1948 with two combustors was in a certain way also a forerunner of Alstom's later success models GT24/GT26 of the mid-1990s with sequential combustion, Section 6.1.3. In 1978 the 2-stage combustion on the basis of one turbine rotor had been successfully tested in a CAES Compressed Air Energy Storage plant at Huntorf, D, which will be outlined in Section 5.5.3.

A considerable part of post-war GT projects still dealt with furnace gas burning in steelwork operations. The previous Velox applications in steelworks were already addressed in Section 4.3.1.3. The classic wind converter process generated substantial quantities of blast furnace gas with considerable heating values. Since 1950 more and more gas turbines were being used in this process, where besides the generator for electricity production a blast gas compressor is additionally driven. Moreover, the atmospheric blast furnace gas had to be compressed to combustion pressure level. BBC produced 15 of these plants with approximately 13.5 MW electric power and an additional 4-5 MW for the wind compressor drive. After the increasing success of the basic oxygen steelmaking process[97] there was no longer waste blast furnace gas available and consequently, the gas turbines steadily disappeared from steel industries worldwide.

On 25 September 1959[98] the inauguration of a 100 MW GT plant represented another world-record high power installation for BBC[99] at Port Mann, on the left bank of Fraser River, 25 km east of downtown Vancouver, Ca. Four 25 MW 2-shaft gas turbines, comparable to Beznau II, were combined here, practically a last highlight of the concluding period of tailor-made gas turbine plants. Though there were precautions for crude oil operation,

---

[96] an advantage of approx. 100 deg C, which rose to 160 deg C by 1960, in comparison to standard BBC plants, Figure 5-7.

These high performance values were achieved for simple-cycle gas turbine only 25 years later, though at that time based on a much simpler GT configuration.

[97] The process, also known as the 'Linz-Donawitz Process' after the two Austrian towns where the process had been first commercialised (at VOEST Linz 1952 and at OEAMG Donawitz 1953) was invented by the Swiss engineer Robert Durrer (1890–1978). http://en.wikipedia.org/wiki/Robert_Durrer

[98] The commissioning ceremony was led by the Surrey 'reeve' Robert Nesbitt. In this part of Canada a 'reeve' is the elected president of a town council. http://www.translink.ca/~/media/Documents/rider_info/Buzzer %20Vault/1950s/1959/Buzzer_1959_09_25.ashx. Port Mann and previous BBC applications of the GT intercooling concept recently returned as the focus of discussion when Sep van der Linden, a former Alstom employee, challenged an ASME statement about GE's LMS 100 as 'the world's first intercooled industrial gas turbine'. The resulting CCJ editorial http://www.ccj-online.com/1q-20012/gas-turbine-historical-society/ is worth reading.

[99] See BBM, Die groesste Gasturbinen-Kraftzentrale, 1960. Picture source: BBC, Brown Boveri Gasturbinen, 1963, Figure 38

the actual operation was with natural gas which was directly taken from a nearby pipeline with sufficient pressure. The four gas turbine groups were arranged one after another in an impressive powerhouse of 140 m length, 25 m width including control rooms and 18 m height, Figure 5-17.

The Port Mann facility represented another pioneering 'first' with respect to the first-time installation of a remote GT operation and control system[100] for such a big thermal power generation assembly in a highly-populated area, from the headquarters of the owning British Colombia Electricity Co., at that time at Carrall St., downtown Vancouver. The first section of the communication path was via a directional radio link to the Ingledow power sub-station and from there by means of HF impulses on a high-voltage transmission line to the Port Mann power station, Figure 5-18.

Figure 5-17   *Two-shaft gas turbine power plant 4 × 25 MW, British Columbia Electric Co. at Port Mann,*
                    Vancouver Ca., 1959   ©ABB

Since the whole electric system of B.C. Electric Co. up to the erection of the Port Mann plant was based on hydraulic power, long distance remote control was an established principle, not least as a direct consequence of the high Canadian salaries. Saving labour costs also meant an immediate request for a thermal peak load and emergency power operational profile, so that remote controlled, automatic operation also belonged to early customer demands for the 'Largest GT Power Plant in the World at 100 MW', as BBC had proudly announced this milestone in the company's technical history.

---

[100] In 1958 BBC also delivered the remote control system for the smaller, though also automatically operated GT plant El Convento of C.A. La Electricidad de Caracas, VE, which generated valuable experience for the Port Mann application.

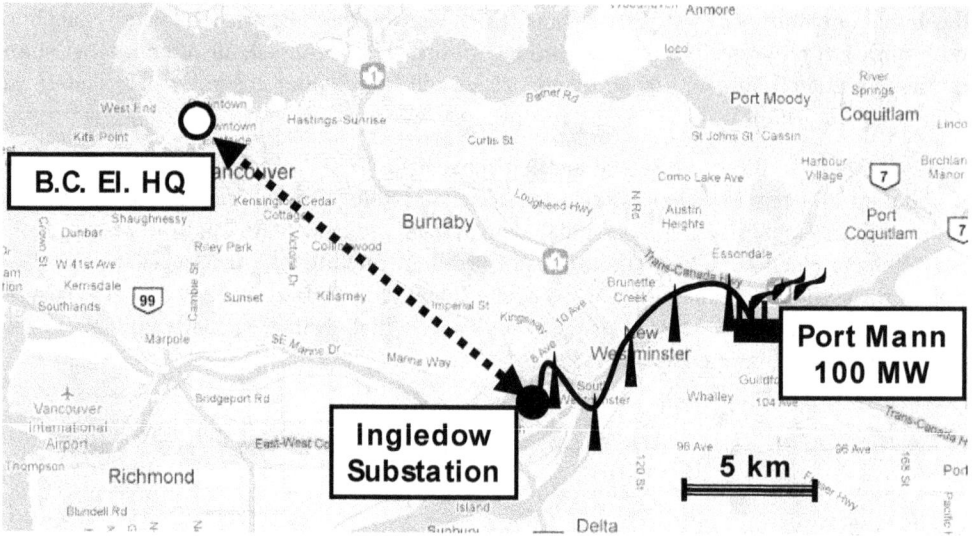

Figure 5-18   *First high-power station remote operation & control 1959 – between Port Mann 100 MW GT plant and B.C. Electric headquarters, Vancouver Ca.,*   © 2012 Google

5.1.5.3   The Development of Ready-Made GT Families after 1970

The transition to cheaper, series-produced GT designs, simultaneously with considerably shorter erection times, had already started in the mid-1950s with the introduction of a) mobile and packaged power stations and b) 'turbojet expanders', i.e. a combination of a mass-produced aero GT engine as gas generator and a subsequent, tailor-made drive turbine. These developments will be described later in a group of special projects, Section 5.5.

In 1955 the mobile power plants with a rated output of 6 MW were brought to market as another BBC 'first'. The next step was the development of packaged gas turbine units, mainly to reduce erection time in barren areas. These were assembled ready for service in the factory and dispatched as separate 'blocks' or packages. These sets (GT11S and GT13S) which were already addressed in short in Section 5.1.5.1, operated at gas inlet temperatures of 750 °C and a pressure ratio of 7:1. The relevant thermal efficiency was 25 percent, a value without the application of a recuperator so that the engine configuration was considerably simplified compared to previous developments. Typical single-stage GT packages achieved a power output of 42 MW, while two-stage arrangements even reached up to 72 MW. An essential precondition for these developments was new turbomachinery blading for both compressor and turbine, which allowed the engine mass flow to be raised at 3,000 rpm to the order of 300 kg/s.

In continuation of BBC's GT long-time development roadmap, which was partly described in Section 5.1.5.1, Figure 5-19 visualises the considerable benefit of both the design switch from a 4- to 2-bearing arrangement and of improved turbomachinery blading (compressor 19 → 15, turbine 7 → 5 stages), which took place in the late 1960s. These design changes are illustrated here by comparing the GT11L with the subsequent GT11S. The envelope of the latter 30 MW 2-bearing configuration fits nearly into the envelope of the 15 MW GT11L.

The standardised BBC GT family concept started with the 33 MW GT11B for the 60 Hz market, as illustrated already in Figure 5-13. In Figure 5-20 the mentioned design characteristics are highlighted again. The rotor's critical speed was checked and carefully tuned out of the operation range. As another design element of proven steam turbine design standard, the 'Turbine Vane Carrier' (TVC) was introduced[101], a separate casing inset in the gas turbine casing for the fixation of the turbine vanes, of the intermediate 'stator heatshields' and to distribute the cooling air in the turbine entry casing.

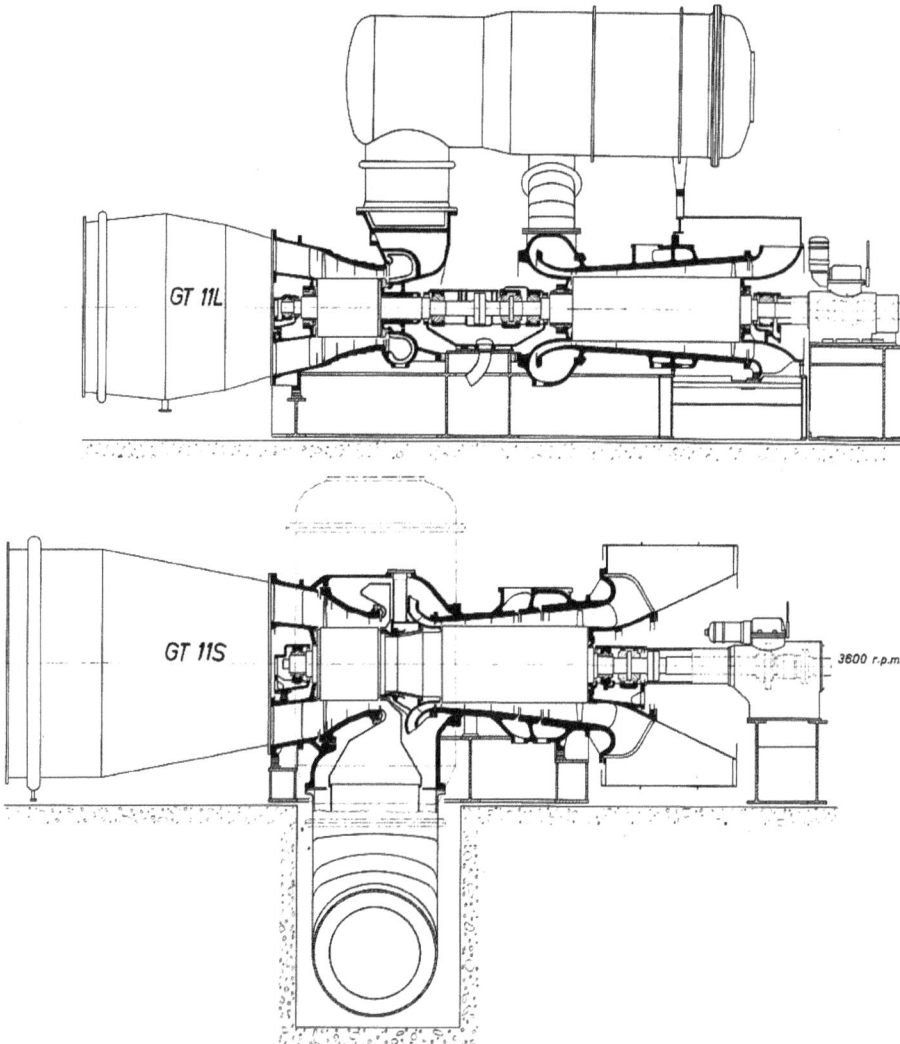

Figure 5-19   *Size comparison of GT11L vs. GT11S and transition from 4- to 2-bearing arrangement* [102]

---

[101] First TVC introduction for GT13S and GT11B.
[102] Figure source: Bankoul, A new 30-MW packaged gas turbine power plant, Figure 4

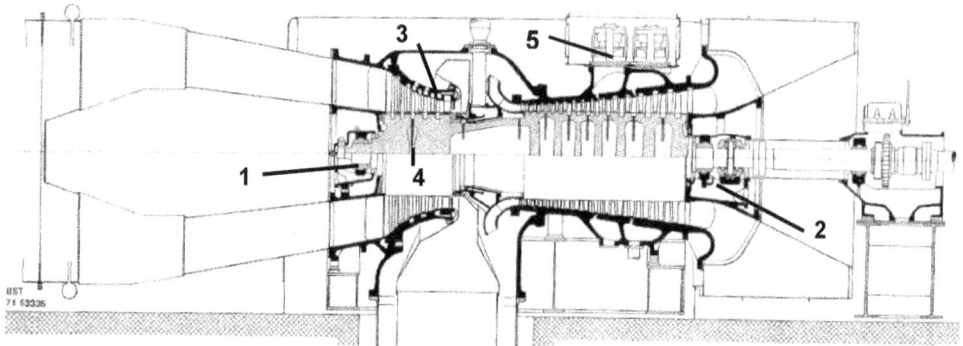

Figure 5-20   *Cross section of GT11B [103]: 1 – Turbine journal bearing, 2 – compressor journal and thrust bearing, 3 – turbine shroud cooling, 4 – turbine rotor cooling [104], 5 – compressor bleed valves*

The higher GT11B rating with gas temperatures at the turbine inlet of up to 850 °C required the first-time introduction of a complex air-cooling system for the hot turbine rotor and the turbine vane carrier. A principle flow scheme is given in Figure 5-21: high pressure air, tapped from the compressor bleed valve(s), Figure 5-20, is fed to the turbine rotor and casing sections, and depending on the GT thermal rating, distributed to the cooled objects in that area:

- rating B   no vane and blade cooling[105]
- rating C   B standard plus vane 1 cooling
- rating D   C standard plus blade 1 cooling
- ratings D2 & D3   D standard plus blade 2 cooling.

This compressed air is partially lost for power generation and is therefore carefully dealt with.[106] In fact these design moves of the late 1960s opened up a completely new aero-thermodynamic design field, the 'SAS Secondary Air System', which besides cooling tasks mainly comprises the careful tuning of the sealing of the sensitive engine section against hot gas intrusion by a pressure-balanced SAS.

Whilst at the compressor shaft section the temperatures hardly exceeded 350 °C without causing problems, the turbine part was exposed to higher temperatures and had to be protected by an adequate cooling system. The design philosophy behind the selected cooling solution was to limit the heat penetration and to remove it immediately under the rotor skin. This system was so effective that as a maximum the rotor temperatures could be kept below 420 °C, so that ferritic Cr-Mo-V steel was still applicable. Between the blade rows, the rotor surface was protected by rotor heat shields, small cast rectangular sections with

---

[103] Figure source: Endres, Three gas turbine types, Figure 3

[104] This Figure shows a later, improved GT11B configuration, for details see Figure 5-21.

[105] Besides the two-bearing rotors, the GT11B prototype testing also saw cooling air fed from and through the rear rotor end for the first time, as indicated in Figure 5-21. A more detailed description of the cooling system is also given in Section 6.2.2.2.

[106] In addition to the outlined scheme of Figure 5-21, rotor bore cooling from the rear/hot end was practically used for GT11N2/GT13E2 for the first time (in the meantime for GT24/GT26 as well) to prevent the need for an external cooler. 'Cheap' low-pressure compressor bleed air is directed by the new rotor end cooling to the mid-turbine blades.

ribbed[107] inner surfaces to guide the cooling air, which had to be designed and assembled safely to withstand considerable centrifugal loads at elevated temperatures. The rotor-protecting function of the heat shields worked not only in continuous similar heat shield protection, a somewhat alleviated design task due to the absence of centrifugal forces. In addition, the turbine blades were protected by a chromium coating against corrosive (sulphur) and abrasive attacks, which expanded the operation times beyond 20,000 h, 4–5 times more than without such a protective layer.

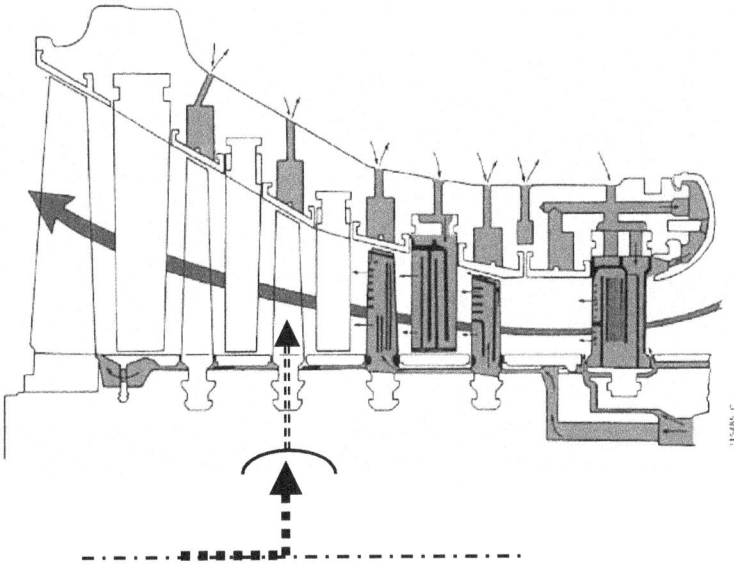

Figure 5-21   *Details of turbine rotor and vane carrier cooling* [108]

The common starting point of the GT9-GT11-GT13 family concept is illustrated in the comparative Figure 5-23, where cross-sectional outlines of the corresponding B versions are presented; engines with simpler and clearer conceptions, greater reliability and operational safety were the common design target. The main performance data is listed in the following tabulation of Figure 5-22. The three new gas turbine types cover an output range from 20 up to 55 MW for base load operation.

GT Type	Start	Speed [rpm]	P [MW]	η [–]	m [kg/s]	$PR_{Comp.}$	Compr. Stages	Tmix [°C]	Turb. Stages
GT9B	1973	4482	21.0	0.246	146	7.75	15	800	4
GT11B	1970	3600	33.0	0.255	225	7.75	15	800	5
GT13B	1973	3000	57.3	0.274	363	9.3	17	820	5

Figure 5-22   *The initial family concept: Data for GT9B, GT11B and GT13B, ISO rating (ambient air, 15 °C, 1.013 bar, no inlet and exhaust losses)*

---

[107] Rotor heat shields for the GT9/GT11/GT13 family were introduced with smooth inner surfaces, the improvement of ribbed surfaces followed from GT13E, GT15/GT17 onwards.

[108] Figure Source: BBC, Brown Boveri Gas Turbines, 1987, Figure 6

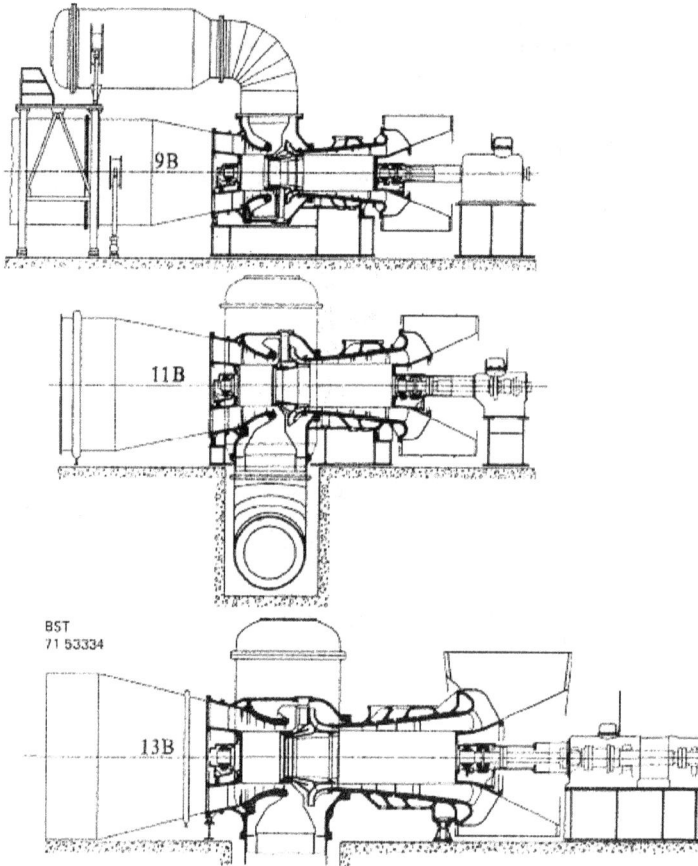

Figure 5-23   *The initial family concept: cross sections of GT9B, GT11B and GT13B* [109]

Besides the primary performance-influencing parameters entry mass flow m and turbine gas entry temperature Tmix, the improved turbomachinery design standard of the new GT9-GT11-GT13 family [110] is reflected in the GT efficiency values η, which are 4-5 points better than the non-recuperator engines of the two previous decades. The further principal GT configuration development is illustrated in Figure 5-24, a continuation of Figure 5-12. The important differentiation criterion is the single-burner combustor arrangement:

– (l) horizontally, for the packaged GT11L and the GT9B and C versions
– (m) vertically, as 'silo combustor' aside of the GT axis with the characteristic transition U tube, from combustor outlet to turbine inlet, which caused many documented problems [111], especially for the GT11B – GT11D4A variants

---

[109] Figure source: Endres, Three gas turbine types, Figure 2
[110] For the standard reference for all performance data in this context, see ABB, GT Liste. In any case P- and η-values refer to the generator output power. Endres, Three Gas Turbine Types, quotes in his survey somewhat higher values in reference to the GT/generator coupling.
[111] Figure 5-9 has several references to damage causing 'internal objects, combustion duct, combustion gas U-duct' etc. The duct design is described in short in the silo combustor Section 5.3.1.

–   (r) vertically 'top-mounted', in line with the GT axis (without U tube), as used for all
    follow-on configurations from 1974 up to the introduction of annular combustion
    chambers, first for GT8B[112] after 1987.

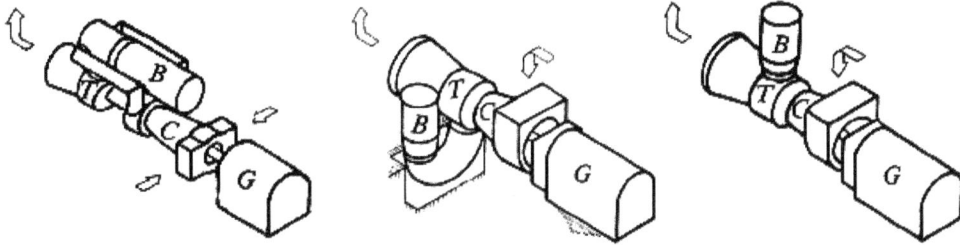

Figure 5-24   *Development of basic BBC GT arrangements II:[113] GT11L – GT9C 1963–1972 (l); GT11B –*
                   GT11D5, GT13B – GT13C 1969–1978 (m); GT9D ff., GT11N ff., GT13D ff., GT15 – GT8 1974–
                   1992 (r); C – compressor, B – combustor, T – turbine, G – generator

The GT11B rating was still sufficiently low, meaning there was no need to cool the turbine
blading, but over time, especially for the GT Type 11 in an exceptionally high number (9) of
rating adaptions[114], the turbine hot gas entry temperature climbed up to 1,100 °C in the late
1970s and 1,230 °C another decade later. For the first time the new GT family was equipped
with a lifetime counter which multiplied the running time by a factor enabling the lifetime
consumption of the turbine front blade rows to be monitored, differentiated by emergency,
peak or base load operation.

Originally, it was planned to adapt the three family members wherever required to the
50/60 Hz demands by a load gearbox[115], but except for the lower power GT9 and later GT8
configurations, this principle was not implemented: GT11, with its different ratings, was
exclusively reserved for 60 Hz markets, while GT13 represented the 50 Hz applications.

---

[112] Four out of 24 sold GT8B units were equipped with an annular combustor, RBK 'RingBrennKammer'. One of
the strong advocates of the annular combustion concept was Ferdinand Zerlauth (1916–2001), head of Sulzer's
design department until the early 1980s, in which time he designed the advanced 21 MW GT10 engine with
the same transonic compressor as GT8, an annular combustor with design input from Lucas, UK, but with integral (no horizontal split) compressor casings. This was against the traditional BBC design heritage and related to turbojet designs. In a kind of competitive race with the GT8B, Sulzer announced the first GT10 sold to
an English customer in 1986. However, shortly afterwards, in 1989, this engine was transferred from Sulzer to
ABB, which put the Swedish ABB Finspång in charge of further developments. In 2003 along with other small
power IGTs (Industrial Gas Turbines), the GT10 went from Alstom to Siemens. The Austrian F. Zerlauth was
hired after his retirement to work as a design consultant for BBC between 1983 and 1985, where he came up
with several advanced concepts for the NGT New Gas Turbine project, see Alstom, New Gas Turbine Concepts. It can be assumed that the BBC/ABB readiness to switch to the new annular combustor concept was
considerably accelerated by Zerlauth's influence and argumentation towards higher power density and cost
reducing measures.

[113] Figure source: Endres, 40 Jahre Brown Boveri Gasturbinen, Figure 2

[114] This high number reflects the competitive pressure especially for the Type 11 gas turbine on the North American market at that time, as already addressed in Section 5.1.2. It must be kept in mind that each new rating
meant considerable extra effort, beginning with design and drawing preparations and adaptations of the
complex supply chain, up until the bookkeeping for the necessary exchangeability of all new parts.

[115] See Endres, Three Gas Turbine Types

Bleed-air valves eased the flow during the compressor start-up. Because of its higher compression ratio, GT13B had three bleed-air valves, whilst both the other types were provided with only two, as shown in Figure 5-20 for GT11B.

Immediately after the first GT11B had been shipped to the first customer (Rainbow Lake[116], Canada) design and manufacturing of the derivative engines GT9B and GT13B began. The first two GT9B units were ignited in January 1973 in Monrovia, Nigeria as local units #III and IV, while the first two GT13B followed in May 1973 in Kindby, Denmark.

After this initial family survey, the remaining highlights of the subsequent 20 years will be addressed individually for each family line, always based on the family roadmap of Figure 5-13, starting with the GT11, for which the essential development events (especially in the 1970s) have already been covered in previous sections.

GT11 Development Line (60 Hz)   As outlined previously, the development history of the GT11 and its excessive number of ratings over time can only be viewed in the context of a strong inter-relationship between the BBC Baden headquarters and the US partner and later subsidiary, Turbodyne, St. Cloud, Minn., between 1971 and 1982, Section 5.1.3.

The sequence of rating increases initiated from Turbodyne started immediately after GT11B delivery with a power step-up to 50 MW for GT11C, at turbine hot gas entry temperature Thg = 955 °C[117]. Two additional compressor front stages were introduced to raise the total compressor pressure ratio to PR = 9.9. However, substantial changes happened for the turbine and it is very likely that this was the origin of further troubles. The first turbine vane received a simple convection cooling and a shrouded hub at the inner annulus; the distance between the front vane and the first blade was considerably expanded. Compared to GT11B, the turbine blade numbers were reduced by 10 percent and those of the vanes by nearly 27 percent overall; both measures lifting the engine efficiency to 28.4 percent. The compressor entry mass flow grew from 225 to 267 kg/s. The corresponding strive for increased turbine capacity became most significant for the fifth stage, where blade numbers changed from 115 to 59, and correspondingly the vane count from 122 to 60. This massive design change had to be compensated by considerably increased airfoil chord lengths for vane 5 and blade 5, so that the rotor length and its cooled surface also grew accordingly.

The power gap with GE appeared to be substantially narrowed by the new GT11C, all the more so because at the same time the GT thermal efficiency grew in comparison to GT11B by 2.6 points to 28.1 percent[118]. It can be assumed that when the four GT11C were commissioned after extended testing at the customer site Northern States Power's Blue Lake station at Minnesota, USA, there must have been high hopes for a sustaining market success. But as indicated already, these units showed up early on GE's deficiency records, Figure 5-9, after an av-

---

[116] First GT11B ignition at Rainbow Lake was in 1970.

[117] In view of a complex, new and widely untested cooling system this single temperature step of 105 deg C between GT11B and GT11C still appears overambitious and one of the key causes for the inadequate reliability of this GT design.

[118] GT11C was intended as a strict downsizing of GT13B/GT13C, see BBC, Entwicklungs-Dauer, 1985. The engine efficiencies were measured on the Blue Lake customer site between 25 March and 11 June 1974, with critical BBC attendance, see BST, Leistungs- und Wirkungsgradmessungen; an abnormal scatter of performance results was observed amongst the four identical engines – with efficiency discrepancies to target up to five points, presumably an early indicator of the poor manufacturing/assembly quality of Turbodyne/BBT.

erage of only 1,640 OH[119], with the miserable assessment: '... *Units have high forced outage rates.*' Consequently, it was not surprising that only these four units were sold. But instead of carefully analysing the cooperation's false start, necessary conclusions were put aside in favour of an ongoing rating 'race'. While BBC initially had high hopes to gain sustainable access to the important US power market via Turbodyne, the licence relationship rather deteriorated into a struggle for design authority and leadership, to Baden's unexpected surprise when confronted by a demanding and overly self-confident Turbodyne.[120]

From 134 GT11 engines sold in total up to 1989, of which approximately 1/3 were manufactured at Turbodyne's St. Cloud plant, 50 percent or 67 units had the D5 rating in production after 1978 when Turbodyne was fully owned and controlled by BBC Baden[121]. Consequently, it is worth briefly discussing this GT11 version here.

The nominal generator power of the GT11D5 was 70.5 MW at a thermal efficiency of 30.8 percent and the corresponding nominal turbine inlet temperature rose to 1,002 °C. The compressor had a design flow capacity of 284 kg/s at PR = 10.8. The majority of these engines were manufactured in Baden and shipped directly to the newly emerging BBC market of Saudi Arabia. Figure 2-15 has already been presented as a typical example of such a 800 MW gas turbine power station on the basis of the GT11D5. For this application the engines were only individually loaded up to 50 MW, so that in total 16 diesel-fuelled engines were ordered in 1981/1982 and put into operation between January 1983 and February 1984. At the time of the last reading in December 2005, each of these engines had clocked min. 91 to max. 118 kOH.

Only two years after BBC's disappointing farewell to the sluggish US market in 1983, the US business picked up again and BBC jumped on an unexpected opportunity to return. Before, a hefty in-house discussion took place on the best GT configuration for continuation in the 60 Hz market. The GT11D5 with its ill-fated U-tube concept was no longer competitive, the GT11E, originally targeted as new approach for the 100 MW class, had been put on hold (Figure 5-13) as too costly and would have required too long for market readiness. Not too surprising after 15 years of drawback and frustration, the US competition was ranked in an overpowering position, '*five to eight years ahead of us*', which was not completely justified. So a low cost, short-term design modification towards a 80 MW GT11D6 was the

---

[119] Independent of the remarkable fact that such detailed information reached the competition, it can be concluded that these figures had been correctly communicated at the time of GE's analysis in Dec. 1978. The 'last readings' for the four GT11C engines at Blue Lake as listed in the in-house database were taken between Aug. 1992 and July 1993 with an average of only 2,368 OH and 508 starts.

[120] An example can be found in an ASME paper (see Hoppe, Construction and Initial Operating Experience) of the presentation of the new GT11C upgrade (for the Blue Lake plant, together with the GT13B for Kindby, DK) at the ASME Gas Turbine Conference & Products Show between 30 March and 4 April 1974 in Zurich. The engine was introduced under the Turbodyne designation '...*model GT-55 (type 11) gas turbines built and installed on a turnkey basis by BST's American licensee, Turbodyne Corporation of Minneapolis, Minnesota...: The turbines are upgraded versions of the original type 11 design* (a reference to a Turbodyne design paper follows "Lehman, B.G.: A New Gas Turbine for the Utility Market, JPGC Boston 1972")' and in the following, comparative plant description '...*The basic plant design concept, independently pioneered by Brown Boveri-Sulzer Turbomachinery (BST) and Turbodyne Corporation (formerly Worthington Turbine International, Inc.) is being followed in both plants; ...*' Predictably, BST engineering management was not amused.

[121] BBT Brown Boveri Turbomachinery, Inc., part of BBI Brown Boveri International, consolidated Turbodyne in 1977 and continued BBC's North American GT business in St. Cloud, Minn. until 1983. After the take-over only 10 additional engines were sold outside the USA (Trinidad, Mexico), while US activities concentrated mostly on keeping the existing fleet going.

name of the game, good enough for a few more years against GE's MS7001E (unchanged since 1977), but threatened by GE's new 100 MW F-class which was expected in 1986: *'just around the next corner'*.[122] The first, long-overdue design modification of the GT11N, as the 'New performance' GT11D6 was now called, was the introduction of the top-mounted silo combustor, 10 years after GT9D and GT13D. In addition, the compressor airflow was increased by approx. 10 percent, the stator front row was built as VIGV Variable Inlet Guide Vanes and a compressor stage was added, while on the other side the turbine entry temperature only rose by a small extent.[123] In contrast to the hesitant start, the GT11N with the following GT11N2 version (since 1994) became a successful engine programme with more than 150 sold units, which have now demonstrated the beneficial combination of rugged, flexible and reliable design elements for more than twenty years: advanced turbomachinery with a variety of combustor options[124]. A short survey of the GT11B/GT11N/GT11N2 performance data development is given in Figure 5-25.

GT Type	Start	Speed [rpm]	P [MW]	η [-]	m [kg/s]	PR$_{Compr.}$	Compr. Stages	Tmix [°C]	Turb. Stages
GT11B	1970	3600	33.3	0.255	225	7.75	15	800	5
GT11N	1989	3600	80	0.320	311	12.4	18	1027	5
GT11N2	1994	3600	110.0	0.342	368	14.6	16	1085	4
GT11N2	2000	3600	115.4	0.339	394	15.9	14	1085	4

Figure 5-25   *40 years of GT11B/GT11N/GT11N2 development* [125]

Several GT11N2, **50** Hz installations also represented a world-record in turbo-geared power transmission[126], with 115 MW shaft power. The progress of the technology is clearly visible in the steadily increasing compressor pressure ratios and turbine entry temperatures, while at the same time a decreasing stage count. In the meantime the newest GT11N2 is equipped with 3 rows of VIGV which provide a 40 percent airflow turndown capability. Low electricity costs are achieved due to high availability values and predictable maintenance costs.[127]

---

[122] In fact GE's first F-technology unit only entered commercial service in 1990, and struggled with considerable reliability and availability problems. http://www.netl.doe.gov/technologies/coalpower/turbines/refshelf/GE Overview of 7FB Turbines.pdf

[123] See ABB, ABB Gas Turbines

[124] There are three combustion alternatives, a Low Calorific Syngas Combustor (LBTU) with process gas heating values as low as 2–5 MJ/kg, a 'SB' Single Burner combustor for oil dry operation (18–35 MJ/kg) and an 'EV' dry low-NOx combustor (35–50 MJ/kg) for < 25 ppm NOx emissions.

[125] See ABB, ABB Gas Turbines; Jacobi et al., Modernization...: ABB's GT11N2 gas turbine, and – see http://www.alstom.com/power/fossil/gas/gas-turbines/gt11n2/ In 2000, first at the Tung Hsiao 6, TW, combined-cycle power plant, the first 14 stages of the GT24 compressor were introduced to the GT11N2 as part of the family concept considerations, increasing the power output to 115 MW. The new 3 VGV compressor additionally increased the mass flow flexibility for Low BTU applications.

[126] See Section 7.1, **1999**

[127] Typically, GT11N2 combined-cycle power plants demonstrated 99.1 percent reliability and 96.8 percent availability.

The breakthrough for this, now Alstom's longest running engine family, came with a substantial order of 12 GT11N from MCV (Midland Cogeneration Venture) in 1987.[128] This was a win-win situation for both the customer and BBC. After the failed nuclear installation the plant required a power generation alternative quickly and at an environmentally acceptable NOx emission level. This was an option which could be implemented with a 1,500 MW natural gas fired electrical and steam co-generation plant on the basis of 12 of BBC's newly developed GT11N gas turbine units to provide base load power as well as steam and electricity at Dow's neighbouring facilities. When it began operation in 1991, it was the largest gas-fired steam recovery power plant in the world.[129]

At times in 1986/1987, BBC Corporate was in critical financial conditions which had an immediate impact on the gas turbine development plans. Dr. Anton Roeder, KWG General Manager of combined-cycle and gas turbine power plants subdivision at the time and obviously not a strong supporter of the GT11D6, wrote a Meeting Report[130] on 12 January 1987 under the revealing reference 'GT11D6 (alias GT11N) (sic)' with the following statements:

- Development cost up to date – 13 million SFr, GT delivery time 24 months, US orders expected in 1987
- KWG development budget cut by 1/3, focus on high priority strategic projects
- GT11D6 development continuation and finalisation is not planned
- Development freeze until the first firm order is received, in this case GT8, GT13E and RBK (Ring-BrennKammer, annular combustor) priorities remain unchanged, in the worst case external personnel will have to be hired'.

---

[128] Midland, Michigan, lies 25 km west of Saginaw Bay on Lake Huron and has 40,000 inhabitants. The plant was to be sited adjacent to Dow Chemical's Midland plant, the world headquarters of the chemical giant that was founded there in 1897, and near Dow Corning, a Dow Subsidiary. Originally designed as the Midland Nuclear Power Plant with twin pressurised water reactors, the project was abandoned in 1984 (when 85% complete), citing numerous construction problems. These problems included sinking and cracking of some buildings on the site due to poor soil compaction prior to construction, as well as shifting regulatory requirements following the 1979 accident at Three Mile Island. Conversion of the plant began in 1986 and was completed at a cost of 500 million USD, almost twice the original estimate of the nuclear facility. When the plant finally opened in 1991, 17 years after the project's start, a total of more than 4 billion USD had been spent.
Over time, construction was also opposed by environmentalists, led by Midland resident Mary P. Sinclair (1918–2011). She worked as a technical researcher at Dow, Midland, before she became 'one of the nation's foremost lay authorities on nuclear energy and its impact on the natural and human environment' during the decade-long struggle. She followed the idea of life-long learning; in 1988 she entered a doctorate program at University of Michigan and six years later, at age 75, she was named a Doctor of Philosophy in the field of Environmental Communications. Interestingly, Midland remained in the press when a 750 MW coal plant project was cancelled in 2009, again after substantial community protest, this time in favour of the gas-fired MCV plant which had been underused over the past decade due to the high price of natural gas. http://en.wikipedia.org/wiki/ – cite_note-HOF-2http://en.wikipedia.org/wiki/Mary_P._Sinclair

[129] See IPG, MCV-Midland

[130] Other attendees KW1 – Sonnenmoser, KWGT – Wicki, KWGV – Kehlhofer, KWG-ST – Dr. Endres, see BBC, GT11N early development, 1985–1990. Dr. Roeder's general mood during this critical period also comes across in two preserved documents: on 11 August 1988 KWG launched the 'PGT Strategic Plan' with the commitment 'With our top-quality industrial capability, we – the designer and supplier of the power plant – will become the overall world leader in the field of gas-turbine-based power plants', followed by the target-setting statement 'PGT's business target for 1990 is to increase our overall profit rate on operating capital (ROOC) from 2.5 % to 22.5 %.' See ABB, PGT Strategic Plan, while, fitting, in the Minutes (see ABB, Seminar KWG Entwicklungsstrategie) of a KWG Development Strategy session nine weeks later on 20 October 1988, E. Borinelli daringly noted "...The term of '...proven BBC mechanical design...' in the previous discussion tempted Dr. Roeder to reply 'The GT13 is too heavy, the GT8 too expensive, the GT9 is old and the GT11 not yet operative!'" Not to mention (DE) '... and the GT15 is ready to be scrapped.'

There was an immediate threat that highly qualified development key personnel would have been laid off. Further GT11N activities were restricted to the absolute minimum, so-called 'consolidation tasks' were specified during 1987 to remain within a budget of max. two man years. Further in-house writing carefully differentiated between GT11D6/ GT11N's 'cancellation (Annullierung)' and 'suspension (Sistierung)' depending on the originating source, and somehow miraculously the programme survived. After the initial installation was equipped with standard silo-combustors and steam/water injection, Figure 5-26, environmentally demanding US regulations on NOx emissions meant that in 1991 a retrofit of the unique EV burners had to be carried out, Section 5.3.4, already representing the second generation of lean premixed, DLN Dry, Low-NOx combustion with (at that time) record low NOx values of < 25 vppm for natural gas combustion.

Figure 5-26   *GT11N 85 MW gas turbine configuration (model, status 1989)* [131] *with initial water/steam injection at top of silo combustor, with DLN since 1991*

Ironically, the decisive years 1986 and 1987 hid a few meaningful lessons:

-   BBC's financial troubles prior to the merger with Asea had been caused mainly by several, multibillion misinvestments in nuclear power generation technology projects in Germany, so that the final survival of the gas turbine business from orders following a cancelled nuclear project appeared to have an inherent logic
-   in the end it was the 'cancelled/suspended' GT11N programme with its mini-budget which commercially succeeded in comparison to a well-cared development programme for the GT8, for example, with 5-times higher expenditure.

Figure 5-27 shows the GT11N Midland reference plant in full operation. Besides ABB who were responsible for the gas turbines, another major subcontractor was Combustion Engi-

---

[131]   Picture source: ABB, GT11N – The 82 MW gas turbine, front page

neering[132] for the 12 HRSG Heat Recovery Steam Generators to supply steam to the existing nuclear unit No. 1 steam turbine, which became an integral element of the combined cycle co-generation plant. Six of the HRSGs were equipped with duct burners to generate additional steam to provide process steam to Dow Chemical at all times and for required operation conditions. Requirements from the market place for more unit power in hotter ambient conditions, as well as improved combined cycle efficiencies, led ABB engineering to a further uprating of the already improved GT11N1 (83.8 MW, 32.9 percent efficiency) in 1994, in the wake of the initial GT11N production units starting with the Midland Cogeneration project. The introduction of the DLN EV combustor had limited the GT11N to ISO rating, as the benefit of power augmentation by steam injection had become obsolete.

Figure 5-27   *1,500 MW (12 × GT11N) plant at Midland, Michigan, in 1991 largest gas-fired steam recovery power plant in the world, former nuclear vessels at the rear on the right* [133]

GT11N2 basic design considerations started from the given turbine enthalpy drop, for which a 4-stage turbine configuration was selected. A slight increase in hub speed relative to the GT11N and a corresponding increase in shaft diameter made it possible to reduce the number of compressor stages from 18 to 16 while maintaining the aerodynamic stage loading. Developing a common compressor family concept together with the GT24, the compressor was re-

---

[132]   Combustion Engineering (C-E), headquartered at Stamford, Ct, USA, was an American engineering and power systems firm with approximately 30,000 employees. The company was acquired by Asea Brown Boveri as part of Percy Barnevick's 'global player' strategy in the late 1980s. C-E's financial debt and lingering asbestos liabilities brought ABB to the brink of bankruptcy in the early 2000s. The boiler and fossil fuel businesses were purchased by Alstom in 2000, without the asbestos risks. ABB was able to resolve the remaining asbestos claims filed against Combustion Engineering and Lummus Global in 2006 with a billion-plus dollar settlement agreement. http://en.wikipedia.org/wiki/Combustion_Engineering

[133]   © http://www.sourcewatch.org/images/2/2d/Mcg.jpg and http://www.sourcewatch.org/index.php?title= Midland_Power_Plant

placed and the stage count of the GT11N2 reduced to 14 stages as of 2000, while at the same time increasing the pressure ratio to PR ~16.[134] The 4-stage turbine is built from materials with reconditioning advantages. From its introduction in 1994 the design offered the possibility of reducing turbine cooling to only two stages, a considerable cost and performance advantage in comparison to the three cooled stages of the previous 5-stage turbine configuration. As seen in Figure 5-28 for the newest GT11N2, the first rotor blading is additionally coated by a thermal barrier coating. The figure illustrates also the blade damping in the outboard section of the last row. These damper bolts caused problems with wear and failure for the early Midland 5-stage turbine versions. As early as 1990 this problem had been fully solved by redesigning the bolt and hole combination and by introducing additional hole coating.

Figure 5-28   *GT11N2 4-stage turbine rotor blading, 2011* [135]

Different from the naming convention which put the GT11N2 into the GT11 series, it represented an all-new design, except for the silo combustors where a higher pressure ratio compensated for the effect of higher mass flows and kept the specific combustor loading almost unchanged. The main development target of the GT11N2 project team was to create a 'bridging product' until the introduction of the planned GT15 (later GT24), at a competitive cost level with short delivery times, mainly as a 'workhorse' for the US market. The main design principles were:

- a moderate increase in speed to reduce the turbine stage count from 5 → 4
- IN738 LC as standard material for turbine blading and heat shields
- a significant reduction in the number of parts, especially the number of airfoils per row
- casings cast to near-net-shape to reduce manufacturing costs
- a welded rotor consisting of only four forged sections
- equal number of turbine blades per row (60) with the same axial fir roots, so that two lines of the blade attachment grooves (180 deg apart) could be milled in the rotor in one continuous working operation

---

[134] See also Footnote 125 of this Section.
[135] http://www.alstom.com/power/fossil/gas/gas-turbines/gt11n2/

- provisions to exchange the first row of turbine vanes and/or the last blade row without opening the turbine casing
- a turbine vane carrier with cast-in feed pipes for cooling air, a patented feature[136]
- outer casings from nodular graphite iron with considerable cost savings compared to steel casings.

The GT11N2 development from the project's launch in March 1990 up until erection at the first engine customer's site was accomplished in fewer than 1,000 working days on the basis of an integrated international project team at Baden/CH, Finspång/S, Berlin/D, Stellenbosch/SA and Richmond/USA. In addition to this, the 12,000 manufacturing drawings were completely CAD-generated for the first time, due to a lack of available working stations in two-shift sessions. The fleet of 60 GT11N2 engines represents a considerable market success far beyond the originally intended 'bridging' function.

Relative to the previous GT11N, there were two main compressor-related modifications: a) the hub speed/ rotor diameter was increased along with some re-staggering of the profiles and b) the first two stages were redesigned to handle the increased intake mass flow, using DCA Double Circular Arc profiles, which had been developed and tested separately, Section 4.2.5. This selection was to improve the efficiency of the front stages, in view of the relative tip Mach numbers (slightly above Ma = 1).

The present 14-stage compressor, originating from the GT24, uses advanced CDA Controlled-Diffusion Airfoils; the three rows of VIGV provide the wide operational flexibility to maintain process steam conditions from 100 percent down to 40 percent gas turbine power. Summarising a decade of development, the GT11N2 combines:

- fuel flexibility on the basis of the three different combustion systems available
- operational flexibility across a wide part load range and changing load demands
- reliability and availability figures with 99 percent start reliability in peaker units
- low environmental emissions, with NOx < 25 vppm for natural gas baseload operation.

A view of 40 years of remarkable ups and downs of the GT11 development line is best supplemented by an illustration of the present fleet distribution, Figure 5-29. Here, the nearly equivalent sales figures for the GT11 (and its many ratings) and the subsequent GT11N/ GT11N2 family are compared, based on an imaginary turning point in 1985. The evaluation is focused on the key market shares in the USA, Saudi Arabia and for the rest of the world. The figure contains two basic messages:

- the world-wide accepted success of an environmentally-friendly product like GT11N/ GT11N2 had to be paid for with the loss of extraordinary loyal BBC customers in Saudi Arabia and their preference for crude oil burning
- the desperate struggle to access the important US market in the second half of the 1970s accompanied by a number of spectacular failures of GT11 projects was finally won in an impressive fashion.

The complete GT11 development line has been sold over a period of nearly 4 decades in more than 280 cases, and remains part of Alstom's GT portfolio as a fuel-versatile and reliable option for medium-size power generation applications.

---

[136] DE19546722 B4, inventors Chr. Jacobi and E. Primoschitz

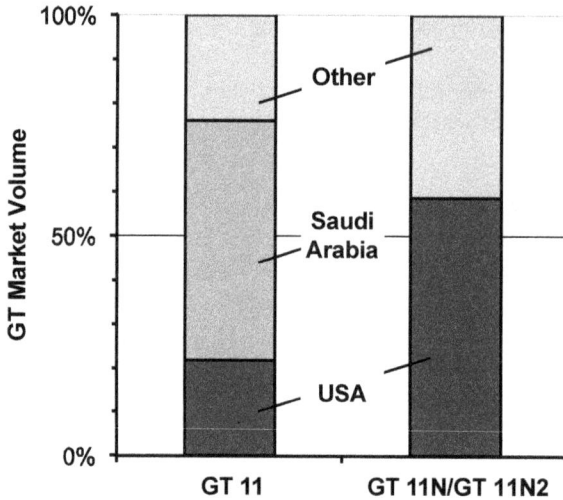

Figure 5-29 *Fleet distribution GT11 vs. GT11N/GT11N2*

GT13 Development Line (50 Hz) In comparison to the GT11 development history, the 50 Hz GT13 line developed in a much steadier fashion, nevertheless with considerable dynamics. The whole GT13 family covers a power range from 57.3 MW to 184.5 MW within nearly 40 years and the simple cycle GT efficiency rose from 27.4 percent to an excellent 37.8 percent.[137] The series started with the GT13B, Figures 5-22 and 5-23, which was fired for the first time on 19 May 1973 in Kindby, Denmark, where the first two units were installed and extensively tested before handover to the customer. The ISO turbine entry temperature was 820 °C, the compressor mass flow measured 363 kg/s at a compression ratio of 9.3. Test analysis revealed certain performance discrepancies which were corrected for the following C/D ratings by adapting turbine blading/ degree of reaction.[138] A continuous rise in the turbine entry temperature, first to 880 °C (GT13C), then to 945 °C allowed for a power increase to 77.5 MW, equivalent to the D rating. The first GT13D went to testing on 23 October 1975 at Donge, Geertruidenberg, NL in a combined-cycle application with natural gas firing. This was at the same time the first GT13 configuration with top-mounted combustor[139] according to Figure 5-24 (r), with the first turbine stage cooled. For all these engines the mass flow was kept constant and the pressure ratios grew only slightly due to the higher turbine entry temperatures.

The introduction of the rating standard D2 in 1983 was accompanied by an additional end stage for the original 17-stage compressor and an expansion of turbine cooling to the first

---

[137] For further information, see ABB, 20 Jahre ABB-Gasturbinen Typ 13

[138] See BBC, Gasturbine Typ 13B, C und D, 1974. In particular the fifth blade row was opened up at the hub and closed at the tip so that the velocity profile at the GT diffuser entry was improved, with inproved pressure recovery and consequently increased total turbine head (drawback: the blade root had to be moved away from the preferred axial direction).

[139] Apparently this mounting arrangement did not only have advantages. W. Endres reported (see BBC, Besprechung mit Motor Columbus, 1977) that the new arrangement caused thermal deformation of the Hot Gas Casing (the transition duct from combustion chamber to turbine inlet) due to direct radiation from the combustor, and Motor Columbus expressed customer concerns about the 'combustion tower' under earthquake impact.

two stages.[140] In the meantime the power had reached 98 MW and the thermal efficiency 32.3 percent, as the combined effect of increased turbine entry temperature, mass flow and total pressure ratio. The inherent advantages of a compact design had already been demonstrated: in comparison to the base version GT13B the power increase of 66 percent was accomplished within an unchanged bearing distance.

In 1983/1984 GT13E development started, which represented the largest power and efficiency jump of the whole GT13 series, at times when BBC was already struggling with economic difficulties. Plans for a radical new approach for a large engine which had been followed as GT15/GT17, Figure 5-13, since the late 1960s regarding high mass flow and transonic compressor applications had to be stopped, Section 5.1.5.4. Considerations for a medium size engine dated back to the late 1970s in the 60Hz market, when Turbodyne promoted a successor of the GT11D5 which emerged as an all-new GT11E concept. While this approach did not materialise after the described, intermediate retreat from the US market, the concept was transferred to the 50 Hz variant GT13E, Figure 5-30. In addition, designers from BBC Mannheim introduced numerous detail improvements.[141]

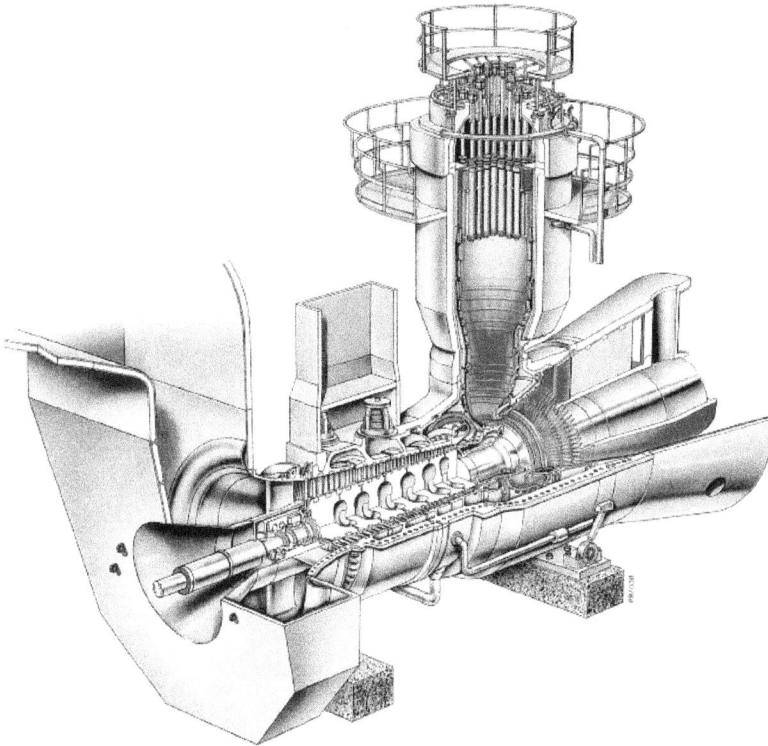

Figure 5-30   *3D view of GT13E with low NOx combustion chamber*[142]

---

[140] Compared to D2, the D3A rating had additional film cooling holes on turbine vane 1 and blade 1, a different re-staggering of vane 1 and new airfoils for blades 3 and 4.

[141] See BBR, Type 13E, 1985

[142] Picture source: ABB, Development and technology of a 145 MW gas turbine, Figure 5

Improved component efficiencies[143] and accordingly adapted process parameters drove the power output up to 147 MW and the thermal efficiency to 34.6 percent. This technical data qualified the GT13E to become the largest and best gas turbine of its class worldwide, as predicted by a thorough comparative market outlook[144] from as early as 1983, Figure 5-31. The basic precondition for this achievement was to increase the compressor stage count to 21 by adding 3 front stages (of which the first row was built as adjustable VIGVs) and one exit stage. The turbine was kept at 5 stages and the same stage loading, where the higher turbine work was compensated for by a slight increase in shaft diameter and thus circumferential speed. The number of cooled turbine stages was increased from two to three. Here, the vane cooling had been put into practice for the first time at BBC as 'impingement cooling', Section 6.2.2.2. As shown in Figure 5-30, the turbine rotor 1 carried a rather unusual tip shroud to improve overall performance. The GT13E turbine annulus was increased to limit the axial flow velocities and thus maintain a comparatively low loss level. Another milestone for the GT13E was the first time achievement of a combined-cycle efficiency: > 50 percent in 1988 at Hemweg 7, NL and in due course 53 percent, for example at Killingholme, UK in 1993.

Turbine blading materials were Ni-based precision castings for vane rows 1–3 and blade rows 1–4. The castings fully exploited the emerging state of the art of precision investment casting technology at the time, with respect to part size and complexity (internal cooling). For cost reasons, the fourth vane row was cast in three airfoil segments together with the single vanes of the fifth row, both in newly developed cobalt base alloy. Finally, the fifth blade was forged from a nickel based wrought alloy. Stages 1 and 2 were protected by high-temperature corrosion coating. In 1997, the new GT13E2[145] intermediately achieved 164.3 MW at 35.7 percent efficiency, values which rose for the same type steadily to more than 200 MW and 38 percent efficiency at present. Figure 5-32 illustrates the impressive process parameter development of the 50 Hz GT13 line over nearly 4 decades in compact form.

GT Type	Power [MW]	Thermal Efficiency
KWU V 94.2	127	0.317
GE Alsthom Frame 9E	107	0.315
Mitsubishi 701 D	119	0.320
Westinghouse 501 D	95.2	0.316
BBC Type 13E	140.0	0.330

Figure 5-31   *BBC market outlook 1983[146]: GT13E vs. competitor engines, ISO conditions, base load, oil fired, generator power*

---

[143] One characteristic design feature of the GT13E was the comparably low axial turbine exit velocities (150 m/s vs. 200–250 m/s) with relatively long (and costly) blading for the 5-stage configuration. The higher stage count was selected for performance reasons, with the negative consequence of relatively low turning (and rigidity) in the fourth stage which could only be compensated by rather thick airfoils.

[144] See BBC, Markt- und Wettbewerbssituation, 1983

[145] Officially the GT13E2 was launched at the end of 1991 with two 164 MW engine orders for a CCPP for National Power, Deeside, UK. See Turbomachinery International, ABB launches GT13E2

[146] For 1983 these were conservative estimates for the coming GT 13E; as can be seen in the text before, the achieved performance data in 1987 was considerably higher: P = 147 MW, η = 0.346.

GT Type	Start	Speed [rpm]	P [MW]	η [-]	m [kg/s]	PR$_{Compr.}$	Compr. Stages	Tmix [°C]	Turb. Stages
GT13 B	1973	3000	57.3	0.274	363	9.3	17	820	5
GT13 E	1989	3000	147.2	0.346	491	13.9	21	1070	5
GT13E2	2012	3000	202.7	0.380	614	16.9	21	>1120	5

Figure 5-32   *Steady development over decades: from GT13 to GT13E2*

The GT13E2 has an extraordinary high open-cycle efficiency for a conventional class gas turbine with a high exhaust energy level, an extra benefit for combined cycle applications, Section 6.3.2. This implies significant fuel cost savings over competing technologies for a wide variety of applications. It is a result of carefully optimised, existing proven technology without the need for expensive and high-maintenance technologies. The GT13E2 can operate in either a performance optimised mode or a lifetime-focused mode without a change of hardware. Steadily introduced GT13E2 improvements use validated engineering tools which have been tested and confirmed at the Alstom Power Plant, Birr, CH. The GT13E2 servicing concept achieves maintenance intervals of 36,000 equivalent operating hours (EOH) between hot gas path inspections. The main design features of the present GT13E2 configuration are addressed in Figure 5-33.[147]

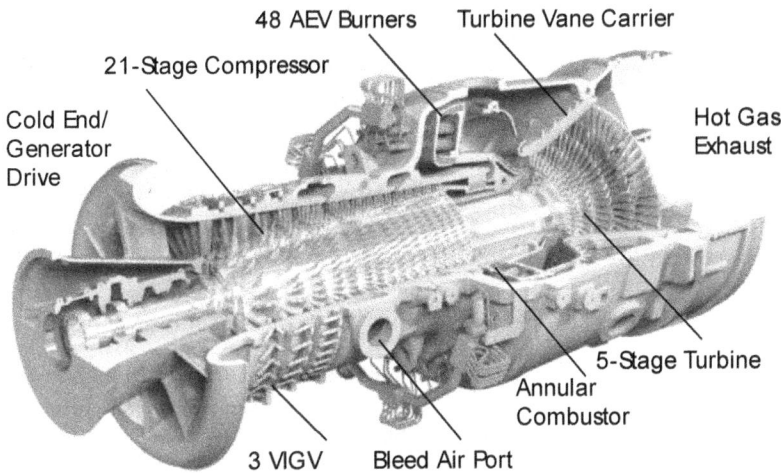

Figure 5-33   *GT13E2 design features (2012)*

Following BBC's classic design concept since the 1960s, the *Generator Drive* is placed at the cold GT end. The 10 percent increase in the power output of the 2012 upgrade was obtained by a corresponding rise in air flow. The old GT13E2 compressor was replaced by the first 16 stages of the GT26 for (unchanged) 21 stages in total. Variable guide vanes, once provided only on the first stage, are now attached to the first three stages (*3 VIGV*) to increase flexibility and turndown capabilities at high compressor efficiency. The compressor

---

[147] Picture source: Chellini, Upgrade of the GT13E2 gas turbine

Bleed Air Port matches the turbine cooling requirements; blow-off during start-up is injected into the exhaust system, thus eliminating the previously used silencers. The *Annular Combustor* was upgraded to further enhance the emission behaviour and the operational flexibility by implementing the AEV (Advanced EnVironmental burner), Figure 5-67. The main difference between the AEV and the previous EV burners is the dedicated mixing section in the burner body downstream of the swirler. Since the AEV specific mass flow capacity was increased, the number of burners has been reduced from 72 EV to *48 AEV Burners* currently. An important feature of the AEV burner is the front-stage gas injection through multiple nozzles which are arranged circumferentially around the burner exit plane. This solution increases the stability of the flame generated by the internal fuel stage against flame extinction and pulsations. Hence, the AEV burner system can be permanently operated over the entire load range with low NOx emissions, which have been reduced from 25 to 15 vppm for natural gas at full load. In addition to allowing quick and wide variation in fuel gas compositions, the AEV burners provide the ability to switch over from gas to oil operation, ensuring uninterrupted stable power supply for capacities between 40 percent and full load. The cooled *Turbine Vane Carrier* is inserted into the outer casing with sufficient tolerances to allow for the required thermal expansion. The axial *Hot Gas Exhaust* system facilitates a straight arrangement of gas turbine and heat recovery steam generator (HRSG) and thus compact combined-cycle plant layouts. The elimination of the turning elbow reduces back-pressure and turbulence at the HRSG inlet with advantageous unit performances in simple and combined-cycle mode. The welded rotor eliminates maintenance work such as restacking, disk replacement and factory rotor overhaul. The rotor shaft runs in two journal bearings and has an additional axial thrust bearing on the cold end of the gas turbine to compensate for the differential forces of compressor and turbine. The last row of the low aspect ratio rotor blading of the *5-Stage Turbine* carries a tip shroud damper ring, eliminating the loss-generating damping bolts. Advanced turbine aerodynamics and multi-convective cooling schemes contribute significantly to the GT13E2's outstanding efficiency. The design of the first turbine stage, combined with Thermal Barrier Coating (TBC) and conservative firing temperatures allows for extended inspection intervals of up to 36,000 EOH. All Inconel 738 conventionally cast turbine airfoils and heat shields ensure the parts have a long life and allow for full, cost-effective refurbishment.

Supplementary to this principal sketch, Figure 5-34 illustrates the turbomachinery/casing set-up of the GT13E2 hardware for the past 'workhorse' configuration 2010. The blow-off cavities after compressor stages 4, 8 and 12 for the blow-off valves and for high-pressure cooling air supply are clearly visible along the compressor casing annulus.

Over the past 4 decades the GT13 line was sold in total for approximately 220 units, up until the most recent GT13E2 version. Backed by more than 7 millions of fired hours, the GT13E2 gas turbine engine has proven itself to be the ideal core of a power plant, producing reliable, competitively priced electricity. In summary, the GT13E2 provides:

– improved flexibility through 2 operating modes – switches online between maximising performance or operation time between inspection intervals
– grid frequency support with high response and no extra lifetime factors
– adaptability for a range of industrial applications including aluminium smelters and desalination plants.

Figure 5-34   *GT13E2 thermal block 2010: view with upper casings removed, compressor entry in the foreground and axial turbine exhaust in the rear* [148]

GT11/GT13 Development Lines Upgrades As indicated in Figure 5-13 some of the roadmap side branches found numerous upgrades, mostly in response to corresponding customer requests. The most important ones have been listed in Figure 5-35 for

- GT11DM (as an upgrade from the intermediate GT11D5)
- GT11NM (as an upgrade from GT11N1), whereas
- GT13DM was deduced from GT13D3A until 1997.

GT Type	Start	Speed [rpm]	P [MW]	η [-]	m [kg/s]	PR$_{Compr.}$	Compr. Stages	Tmix [°C]	Turb. Stages
GT11DMC	2003	3'600	79.4	0.330	295	11.4	17	1'000	5
GT11NMC	2003	3'600	96.9	0.347	345	14.0	17	1'030	5
GT13DM	1997	3'000	109.7	0.341	396	13.2	18	1'020	5

Figure 5-35   *Examples of currently available GT11D/GT11N/GT13D upgrades*

---

[148]   Picture source: ABB, GT13E2 – the most efficient low-emission gas turbine, Figure 2

GT9/GT8 Development Line[149] The GT9 started in 1970 as a replacement for the short-lived GT11L, as the smallest member of the new GT9-GT11-GT13 family in the 20 MW range. Its end came in the mid-1980s after a fairly excessive number of 8 ratings in a relatively short lifetime of just 10–15 years. During the ABB period and the first years of Alstom the lower power class was represented by the GT8 family with a final output close to 60 MW. Basic milestones from 30 years of development history are highlighted in Figure 5-36.

The GT9B represented a downscaled GT11B, especially in the turbomachinery section, Figure 5-20. The main differences have already been addressed in Figures 5-22 and 5-23. These were the horizontally arranged combustion chamber, still a relic from the GT11L, and the reduced number of turbine stages (4 instead of 5). Like the following GT8, these engines ran at speeds of > 4,000 rpm (GT8 > 6,000 rpm), so that 50/60Hz power generation applications were covered by corresponding turbogear sets[150].

GT Type	Start	Speed [rpm]	P [MW]	η [-]	m [kg/s]	PR$_{Compr.}$	Compr. Stages	Tmix [°C]	Turb. Stages
GT 9 B	1973	4482	21.0	0.246	146	7.75	15	800	4
GT9D1NA	1982	4486	32.2	0.266	158	9.0	15	945	4
GT 8 B	1984	6340	48.0	0.315	170	15.7	12	1085	3
GT 8 C2	2000	6210	57.2	0.344	195	17.6	12	1100	3

Figure 5-36   *Major performance data of the lower power classes GT9/GT8*

The GT9 rating progress was accompanied by the detailed features:

- C (1975) introduced the cooled and shrouded vanes 1, which allowed a Tmix increase of 60 deg C; to accelerate the development pace and especially to gain experiences under long-term full load operation, a collaboration agreement was signed with the first GT9C customer at Bochum, D, where this engine ran for the first time on 14 June 1975
- D (1976) saw the top-mounted silo combustor also become part of the GT9, Figure 5-24 (r), with cooling expanded to the complete stage 1
- D2 (1978) with turbine up-flowing (vane opening by 4 deg) and improved vane outboard sealing
- D1 (1980) with + 6 percent in compressor mass flow
- D1NA (1982) finally saw a rotating speed increase of 2.3 percent and flow improvement in the exhaust diffuser area.

---

[149] From 1990 to 2003 the low power class portfolio also comprised the following gas turbines for 50/60 Hz applications, under the responsibility of ABB Finspång, S, and Lincoln, UK: GT10–21 MW, see Footnote 112 of this Section, GT35-Jupiter – 16.9 MW, Mars – 8.8 MW and a bunch of industrial gas turbines from former Ruston origin (TA, TB, TD, Typhoon, Tornado, Tempest, Cyclone). Siemens bought the small GTs (3–15 MW) from Alstom in May 2003, followed by the medium class (15–50 MW) in August 2003; due to their short residence time at ABB/Alstom these GTs are not explicitly presented here.

[150] In the early 1990s these turbo-gears caused considerable safety concerns in the engineering/ purchasing departments, after a severe accident with 4 fatalities of a geared Siemens Gas Turbine V 64.3 at the Leipzig cogeneration power plant on 1 November 1994. http://www.udo-leuschner.de/energie-chronik/941116.htm
The crack propagation behaviour of standard gearing Cr-Ni-Mo steels (with low Ni content of ~1 percent) and failure-prone ultrasonic material inspections came into focus in general. In 1998, ABB decided to switch to steels with 3 percent nickel content for GT8C/GT8C2 gearboxes.

The sales records reveal the sales of 106 GT9 units in total, over a time span of nearly two decades. Important sales events took place in 1974 with the introduction of the GT9C and correspondingly 17 new orders and in 1981 when the annual sales record peaked again with 16 units of D1NA, D1N and D1A rated engines. The Type 9 life cycle ended in the late 1980s due to the newly emerging, apparently superior GT8 concept, Figure 5-36. Especially after the company had strategically detected the combined-cycle market as an opportunity for further growth, the GT9 without VIGV mass flow adaptation capability at part load no longer had a productive future (as predicted from the beginning for the GT8).

The GT8 is a low power offshoot of the ambitious, high power GT15 (60Hz)/GT17 (50Hz) programme from the early 1970s, Figure 5-13. After the main development programme had to be cancelled due to influences which will be discussed in the following Section 5.1.5.4, the GT8[151] became the sole remaining carrier of innovative ideas and technologies from this BBC era into the present.

After the development of all three engines was almost accomplished in the late 1970s, the GT8 received the lead role for the de-facto in-house prototype testing at Doettingen/ Beznau, Aargau, CH from 1 June 1981 until 1 March 1982. The GT15 testing should have happened with the GT8 lessons-learned, digested and already implemented at the first customer between 1 April 1982 and 31 August 1982.

The disaster[152] began on 12 January 1982 with medium-scale damage to the first, only lightly instrumented GT8 with 'Nimonic' as a base rotor material. The damage occurred at nominal speed (6,340 rpm) during the very first loading. Mistakenly the failure was first blamed on the 'weak' gearbox segment bearings and the engine was rebuilt in steel[153] with only minor modifications and prepared for testing, this time extensively instrumented, on 5 January 1983. The second failure occurred during an overspeed run (7,015 rpm) with no load. This time the failure was more impressive: the rotor was broken into 5 parts, all indicating clearly forced ruptures with strong plastic deformation. Subsequent in-depth investigations revealed the common failure initiated from the growth of fatigue cracks on the last turbine blade row. However, this did not explain the severity of the damage occurred, which this time was far beyond past experiences. The extent of the unexpected sensitivity to unbalance (blade-off) was finally explained by a rotor configuration that was too slender[154] and the equally slender intermediate shaft which was operated above its critical speed. This enforced the introduction of completely new rotordynamic design criteria with 'operation below the first free-free mode frequency'. Some time afterwards this hidden knowledge became official, for example as part of ISO standard 10814. After digesting this lesson, the cure for the new GT8B was rather straightforward: a stiffer rotor was designed,

---

[151] See BBR, Design and testing of a 45 MW Gas Turbine, 1985

[152] See Alstom Power, The Test Failures of the GT8A. The term 'disaster' is not exaggerated, as the most serious result of these tests was the consequence that the all-new GT15/GT17 projects (which contained BBC's well-funded aspirations in the upper power class, not least by the ground-breaking and successfully developed transonic compressor technology) were halted and finally had to be scrapped as it was simply not possible to correct them, see BBC, GT15/ FPL, 1987.

[153] 'Nimonic' – the first GT8 rotor was St 460TS/Nim901/St 580S – a registered trademark of Special Metals Corp. designating a family of nickel-based superalloys. First developed in the 1940s at Wiggin Works, Hereford, England, in support of the development of the Whittle turbojet engine. The second rotor steel was St 13TNiE/St 572S.

[154] Rotor slenderness ratio L/D – rotor length/rotor diameter.

an additional bearing for the intermediate shaft implemented, improved journal bearings used and the flutter-prone last turbine blade row first damped by lacing pins and then later more effectively addressed by the completely shrouded blades 2 and 3. The first corrected GT8B was installed at Shell Pernis[155], NL, where it ran successfully for more than 100,000 OH until it was replaced with a new GT8C.

Besides necessary improvements of the mechanical integrity, the GT8 development had the target of developing an annular combustion chamber, primarily as a means of cost reduction, while at the same time more demanding emission regulations had to be fulfilled. The first ideas for an annular combustor were vented in the early 1980s, but tests of 'RBK-W' configurations (Ring-BrennKammer – W 'Wabenbrenner'/honeycomb burner) at Doettingen revealed considerable difficulties in terms of flame stability, non-uniformities, pulsations and the achievement of emission targets. Consequently, a new concept, 'RBK-D' with swirler burners (D – 'Drallkörperbrenner') was implemented (Section 5.3.2) with the unique result that out of the sold 24 GT8B units in the field, four of these were equipped with an annular combustor. The first of these 'multi injection burner' configurations was started in 1987 for Purmerend, NL, 12 km north of Amsterdam, and ran successfully until 2009 when it was retrofitted with EV burners. The following GT8C[156] rating was already equipped with EV burners, representing the second generation of dry low-NOx burners for lean, pre-mixed combustion. Figure 5-37 illustrates the different steps of combustor evolution for GT8B/C.

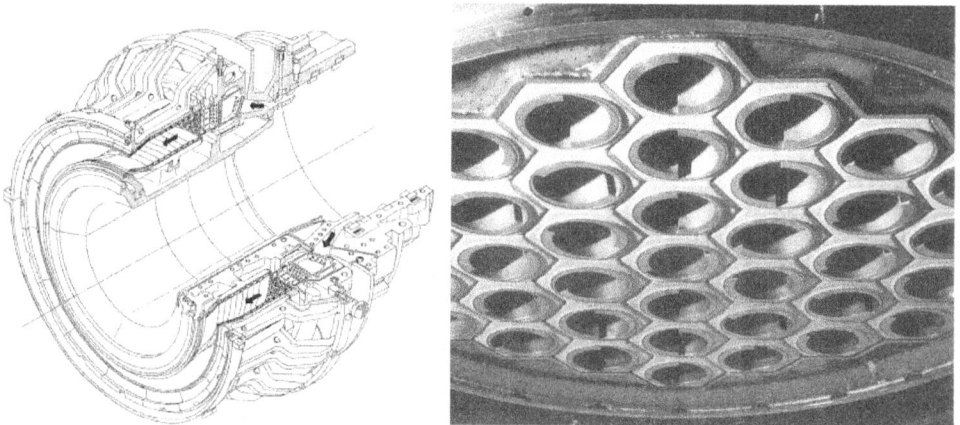

Figure 5-37   *GT8 low emission combustion:* [157] *GT8B annular combustor with 'multi injection burner' 1986 (l), new EV-Silo Combustor 1990 (r) for the GT8C*

For a short time in the late 1980s the GT8 played a key role in the NGT (New Gas Turbine) project with the prime targets:

---

[155] Shell Pernis, close to Rotterdam, is the largest refinery in Europe.

[156] For a comprehensive description of the new GT8C features, see ABB, The GT8C: An Improved, Low-Emission Gas Turbine

[157] Picture source: (l) Jeffs, Asea Brown Boveri introducing... and (r) ABB, New EV-silo combustor, front page

- to deduce the cost reduction potential on the basis of a new family concept GT8 – GT15N – GT17N
- to consider future GT upgrades by using RR technology[158]
- to establish the GT8 as technology lead carrier for the planned 'family'.

This phase of GT8 development history is marked in Figure 5-13 with the box '(8N)'; however, early on during ABB/RR design cooperation work it became evident that upscaling of the smallest family member was not the right basis for further studies. In due course (December 1988) the activities focused on GT15N and even a GT11F configuration, principally bigger engines with better potential for improvement, but as indicated without any sustainable results in the end, which were urgently required for a promising restart of ABB's power generation gas turbine developments.

Finally, the long-term development target of an emission-, loss- and cost-effective solution materialised in the form of a combined annular combustion chamber and DLN EV burners for the GT8C2, Figure 5-38.[159] In 1983 the GT8 family was the carrier of the GKR (Gasturbinen Kraftwerks-Rationalisierung) project[160] to achieve cost reduction targets of 12–14 percent.[161] A basic achievement was the introduction of a universal modularisation concept[162], bearing fruit in the GT13E and in following projects.

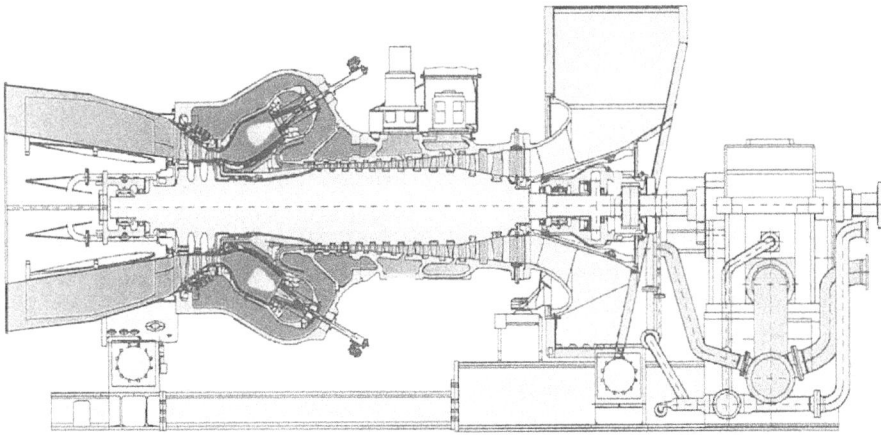

Figure 5-38   *GT8C2 cross section of thermal block*

The GT8C2 provided a high degree of operational flexibility for combined-cycle application, with its beneficial mass flow turn down capability combined with an easy to handle dual fuel

---

[158] Since 1987 the BBC management had established a power generation GT development cooperation with Rolls-Royce (RR), which was put on hold after 4 years due to 'irreconcilable cultural differences'. Prime technical support had originally been expected in the areas of turbomachinery aerodynamics, Computational Fluid Dynamics (CFD, the buzz word of the time), turbine cooling technology and cost reduction processes.

[159] The first such combination was introduced in the GT13E2 and the GT10.

[160] See BBC, Projektauftrag GKR, 1983

[161] An initial analysis revealed a cost deficit of 22 percent compared to competitive market prices, on the basis of 10 engines sold annually.

[162] See Von Rappard, Modular Concept

provision. The gas turbine set-up is convincing due to its clear and optimised structure, combining a 12-stage high mass flow density, transonic compressor with an annular low-emission combustor and a high efficiency 3-stage turbine with cooled stages 1 and 2. The combustor contains 18 single EV burners. In dry gas operation the inherent lean premix principle with vortex breakdown stabilisation provides extraordinarily low NOx values of < 25 vppm. During oil operation the NOx emissions are controlled by water injection. The compressor is equipped with a single front row of variable inlet guide vanes. The rotor-supporting journal and thrust bearing arrangements were mounted externally within the compressor intake housing and the exhaust diffuser for ease of maintenance. The compressor intake, compressor/ turbine housing and exhaust diffuser were bolted together to form one rigid unit, giving easy access to compressor and/or turbine components after removal of the top half housings.

In summary the GT8 family demonstrated the final GT8C2 version as a reliable, heavy-duty workhorse for industrial, co-generation and utility applications. In total the GT8 fleet grew to 89 units from 1984 until today. Of these, the split between early GT8 and the improved GT8C/ GT8C2 was nearly 50/50, with an approximate share of 50 percent for European installations.

5.1.5.4   GT Designs on the Brink – and Beyond

After the gas turbine development roadmap, Figure 5-13, has been discussed in terms of the major power classes before the arrival of the all-new GT24/ GT26, Section 6.1.3:

- P < 60 MW   GT9/GT8
- 60 MW < P < 140 MW   GT11
- P > 140 MW   GT13,

it is worthwhile to briefly address a few projects which were not brought to market even after the development had been accomplished, or which were cancelled prematurely. The timeline in Figure 5-13 reveals a certain cumulation of these events at critical business times for BBC in the early 1980s (marked by vertical 'stopper bars') and again at the beginning of the 1990s when ABB had taken over:

- P < 60 MW   GT9 (officially replaced by GT8)
- 60 MW < P < 140 MW   GT11E (1981)
- P > 140 MW   GT15/GT17 (1981) and GT15N/GT17N (1991).

The GT11E became a relatively short episode between 1976 and 1980, while the GT15/ GT17 project tied up considerable capacity and attention at Baden from the late 1960s until 1981, and for a second time between 1987 and 1991 before the projects were ended without resulting in a product.

The initiative for the **GT11E** came in response to the US market and was communicated back to Baden by Turbodyne as early as 1973: to keep the GT11 competitive until around 1985, it was thought that the developing power gap between the standard GT11 line and the newly planned GT15 in the 150 MW class should be closed by the GT11E, with approx. 100 MW and a thermal efficiency of > 35.8 percent. A feasibility study clarified that a modified GT11 was clearly more advantageous than a geared, downscaled GT15.[163] The decisive

---

[163] See BBC, Studie GT-Typ 11E, 1976

process parameters were set to a compressor pressure ratio PR = 14.1 and a turbine entry temperature Tmix = 1,080 °C.[164] The compressor relied on the established N blading, Section 4.2.5, while hub loading was unchanged according to existing experience. Two additional compressor front stages were introduced for a higher design mass flow of 284 kg/s, and two stages added at the backend to raise the pressure ratio from a GT11D5 level of 10.8 to >14. Special features were put into the 5-stage turbine re-design: the hub diameter and correspondingly the circumferential turbine speed were increased by 18 percent, all of the vanes had inner duct shrouds and the first three turbine blade rows were shrouded. The degree of reaction of the turbine blading had to be reduced to ~0.35 to limit the axial thrust for an acceptable bearing life. GT11E for 60 Hz and GT13E for 50 Hz applications were planned as identical designs with the speed-related scaling factor 1.2. The GT11E was originally planned to go ex-works in 1981; this date, finally delayed to 1983, brought these engine design activities into direct conflict with the closing of the BBT and BBC's intermediate retreat from the US market. Although a release of the first engine and tooling for long-lead turbine blading castings had already been issued in 1979, the GT11E remained a paper engine. Later on the concept was briefly revived in the late 1980s and even transferred into an improved GT11F version, again without success.

The first engineering considerations for a top-end gas turbine in the power range of 150–210 MW, GT15/60 Hz and GT17/50 Hz dated back to the late 1960s, when the transonic compressor was discovered to be a powerful tool for increasing the front face mass flow density[165] of a gas turbine and consequently, its power output. Correspondingly, the other process parameters rose substantially, for example: from a typical GT11D5 level to the new GT15, the turbine entry temperature rose from Tmix = 1,000 °C to 1,100–1,200 °C and the compressor ratio from PR = 11 to 16–17. Figure 5-39 contains the major performance data for these engine concepts, both for the first approach with a planned market introduction in the early 1980s, which was cancelled in the wake of the general BBC crisis, and for the renewed effort with moderately increased target parameters, this time in collaboration with Rolls-Royce engineering ten years later, which ended prematurely again.

GT Type	Start	Speed [rpm]	P [MW]	η [-]	m [kg/s]	PR$_{Compr.}$	Compr. Stages	Tmix [°C]	Turb. Stages
GT 15 B	1976	3'600	168	0.358	573	16.5	11	1'100	3
GT 17 B	1976	3'000	238	0.351	825	16.5	11	1'100	3
GT15 N	1988	3'600	190	0.354	565	15.9	11	1'160	3
GT17 N	1988	3'000	274	0.354	815	15.9	11	1'160	3

Figure 5-39   *Major performance data of the (cancelled) top-end engines of the 1980s: GT15/GT17[169] and ~1990: GT15N/GT17N*

All this was accompanied by an impressive increase in the engine dimensions, Figure 5-40. The turbine exit diameter went up from 2.2 m (GT11D5) to 3.2 m (GT15) and the bearing

---

[164] See BBC, Pflichtenheft GT-Typ 11E/13E, 1977

[165] Between 1960 and 1980 the relative mass flow density of a transonic compressor in relation to a conventional subsonic design increased by a factor of 2–3.

distance from 6 to 9.2 m. In hindsight this was a step too far and was closely related to the final failure of the project after the disastrous test experiences with the similarly configured GT8 rotor became known.

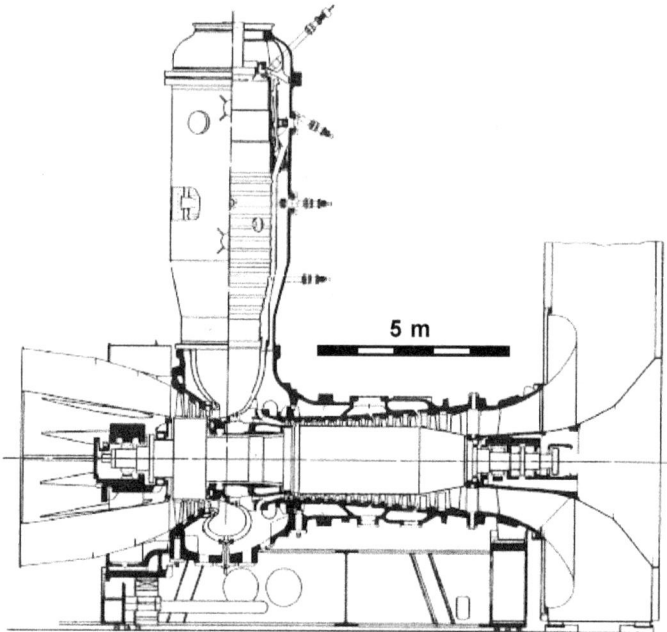

Figure 5-40   *GT15 general layout* [166]: *BBC project of a 60 Hz, >150 MW gas turbine with approx. 32 percent thermal efficiency (guarantee values)*

During a period of nearly 10 years a comprehensive component development programme was carried out. Transonic compressor flow up to a relative tip Mach number of 1.3 was tested in 4-, 6- and 9-stage rig set-ups, Section 4.2.5, and evaluated with newly developed, analytical tools. Figure 5-41 shows the non-dimensional geometry of the previous GT11 compressor front stage in the high subsonic regime, in comparison to the GT15 transonic entry stage, both plotted for the same shaft speed. The highly subsonic GT11 stage diffuses the flow according to the principles of a subsonic diffuser. The GT15 transonic stage is of significantly greater size and has correspondingly higher relative Mach numbers in the outboard inlet sections.

The calculated flow field of Figure 5-42 illustrates the working principle of a multicompression system with a supersonic inlet portion and a subsequent subsonic diffuser.[167] This advanced concept allowed increases in both mass flow and stage pressure ratio in comparison to the subsonic solution. As seen in the resulting flow field calculation, the BBC concept of a transonic compressor was of the 'shock-in-rotor' type. These advanced

---

[166] Figure source: BBC, Transonic compressor development, 1984

[167] The relative Mach number iso-lines show an oblique shock wave in front of the rotor, extending into a normal shock at passage entrance. At the front part of the airfoil suction surface a deceleration due to an oblique shock wave can be seen, followed by a strong expansion resulting in high Mach numbers in front of the normal shock, which separates the supersonic flow zone from the subsonic passage area downstream.

calculations were achieved by newly available numerical 'time-marching methods' in the early 1980s.[168] In addition, the impact of variable inlet guide vanes were extensively tested together with blow-off valve operation for part load optimisation. The first 7 of the 11 vane rows were shrouded for improved sealing against the rotor surface. The welded rotor had a disk for each of the 11 blade rows, whereas the last 4 stages were designed with rotor cooling.

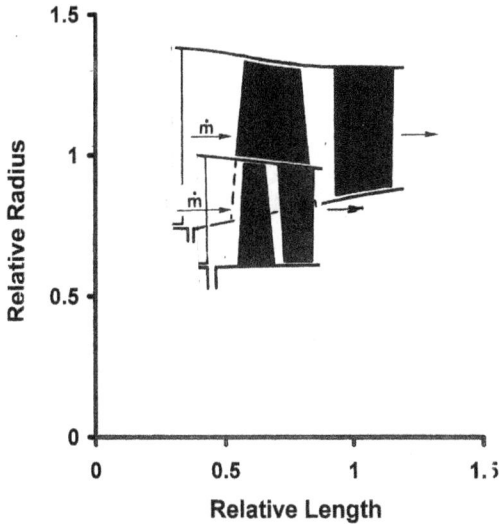

Figure 5-41   *Relative size comparison of conventional (GT11, at smaller hub radius) vs. transonic (GT15) compressor front stage* [166]

The top-mounted combustor was based on the previous GT13 development, but still had a quarter-scale larger volume. Special care was taken to improve and optimise the cooled hot gas casing design. The turbine testing up until 1976 addressed the performance, especially the two cooled front stages where the blade cooling was based on re-cooled air. In contrast to the three vane rows, all the blade rows were unshrouded and required no damping wires. Turbine blading was based on IN738LC precision castings and an upgrade to IN792 with a 20–40 deg C higher material temperature potential was planned for a GT15C[169] rating.

GT15 prototype manufacturing was launched in April 1976 with the promise of 20–25 percent lower specific plant costs. The spent development cost for all three types GT17 – GT15 – GT8 are quoted at that time as 26 million SFr, with an expected cost rise up to a total of 67 million SFr[170] (including costs for prototype testing) until the end of 1981. The first four

---

[168] See Farkas, The Development of a Multi-Stage Heavy-Duty Transonic Compressor

[169] The corresponding Tmix was targeted at 1,210 °C, while the values in Figure 5-39 referred to the original GT15B rating. See BBC, Aufgabe des Prototyps der Gasturbine 15, 1976

[170] Comparable to the extraordinary low figures for the involved development personnel, these cost figures may certainly have been correlated with BBC experiences of that time, but nevertheless appear to be an order of magnitude below the realistic needs. This pattern traditionally reappears in the responsible management circles: 20 years later (1998) the development of an all-new engine family GT24/GT26 was again announced as officially finished after spending 300 million SFr, a number considerably away from the real development cost (as learned in hindsight).

units were planned to be manufactured at Baden. Generator and engine assembly should have been carried out by Turbodyne, which was also foreseen for the manufacturing of the complete GT15 thermal block after unit #4.[171]

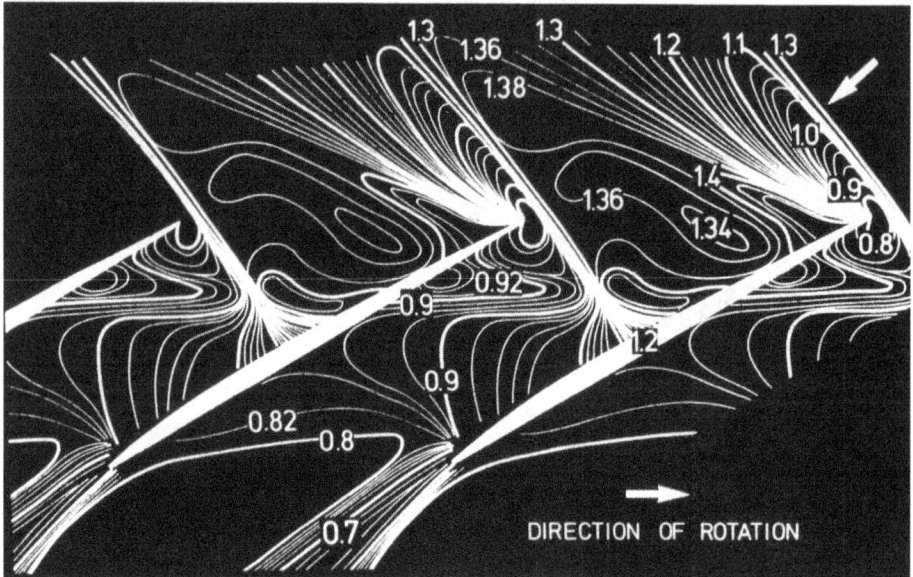

Figure 5-42   *Typical transonic compressor flow field at rotor 1 blade tip: Calculated lines of constant Ma* $_{w, rel}$[172]
              ©ABB

The prototype manufacturing proceeded to new impressive dimensions, Figure 5-43, and with Saudi National Company a launch customer could even be presented for two GT15 engines for a Jeddah plant, planned to be in operation in 1982/3. After 1978 the long-lead items for the GT17 had also reached manufacturing state.

Several publications had been prepared in time – for the GT15 at Baden and correspondingly for an immediately following GT17/GT13E sales campaign out of Mannheim.[173] The marketing buzzwords came from the strikingly simple basic design features:

–   'ONE welded ROTOR,
    For compressor and gasturbine with two bearings only, made from single forged discs of oldfashioned, well established ferritic steel. ...

–   ONE BURNER,

---

[171] Again this assessment is difficult to accept in view of the aforementioned serious quality problems of Turbodyne production for the GT11 series at the same time.

[172] Figure source: BBC, Transonic compressor development, 1984

[173] See BBC, Typ15 (June 1980), BBC, BBC Type 15 (~1980) and – see BBN, Neue BBC-Gasturbinen grosser Leistung (July 1980) with reference to GT17/GT13E.

...with the capacity of burning liquid and gaseous fuels simultaneously and the possibility to change over ... under load.

Figure 5-43   *GT15 prototype manufacturing at Baden, CH:*[174] *10 m rotor in comparison to a 400 MW ST rotor in the rear* [175]   ©ABB

–   *ONE COMBUSTION CHAMBER,*
     for a proven fuel flexibility, long combustor residence time, resulting in a high combustor efficiency. ... Direct and easy access to the first stage blades for inspection and maintenance.'

On Thursday 12 January 1982, the day of the first GT8 test failure, a period of uncertainty and doubt began, which grew after the second more serious GT8 failure on Friday 5 January 1983. Very shortly thereafter it became clear that the then existing GT15/GT17 design concept and the produced hardware could only be cured (if at all) by excessive additional spending; the rotor design was too closely related to the GT8. Thorough design investigations up until 1987 finally revealed that a GT15 with remedial design modifications was not qualified as a sound concept for future developments. The compressor was assessed as too long and too costly. In

---

[174]   Picture source: BBC, BBC Type 15 (~1980)

[175]   The picture played a special role in calming possible customer concerns with respect to the engine size. After mentioning the similar sized GT13, of which already 35 units were in operation, and the even bigger Huntorf plant, Section 5.5.3, the PR text (see BBC, BBC Type 15, ~1980) continued: *'The turbine/ compressor rotor as the center piece of the machine is still considerably smaller than a steam turbine rotor also manufactured by BBC.'* The truth can be seen in the picture, but as learned from the GT8 rotor damage 5 years later, not the 'rotor size' that was the critical issue, but the 'rotor slenderness ratio L/D'.

the meantime, the turbine cooling concept had become outdated and it had been realised that there was still no strategically settled combustor concept available for further developments. Therefore, not completely unexpectedly, the decision was made to scrap the last relics of more than 20 years of ambitious GT development on Saturday(!) 7 August 1987[176], just in time to be included in the conclusion of 96 years of Brown Boveri as a company at the end of the year.

In the following years the GT15/GT17 experienced a short revival within concept investigations for a strategic project, 'NGT – New Gas Turbine', with the target of laying the base for a new GT family concept. The new approach was predominantly to find convincing answers to overcome identified weaknesses in the following areas:

– advanced turbomachinery design and calculation methods
– improvements in design/manufacturing of effective turbine cooling and corresponding material issues
– all aspects of cutting costs.

A development partner was sought, preferably experienced in the aero engine business. This was found with the industrial gas turbine division of Rolls-Royce, UK.[177] In 1988/9 the main NGT targets were to assess the cost reduction potential of the complete GT product portfolio (GT8, GT11N, GT13D and GT13E) and to investigate possible engine upgrades on the basis of superior aero engine technology.

Prime development priorities in 1989 were:

– a low-cost design freeze for GT8, GT11 and GT13
– a generally applicable dry, low-NOx annular combustor design solution
– definition progress in the NGT field
– urgent improvements to CCPP cost (- 15 percent) and throughput time (< 21 months).

In the first round of the NGT design effort, the GT8 had been selected as a basis for development, but very soon the larger GT15 power class came into consideration, at first as a scaled version of an advanced GT8N concept, Figure 5-13.[178] After a thorough competitor analysis, the process parameters for a new GT15N/GT17N[179] base concept were selected, close to the data of Figure 5-39, under the assumption that a four-times scale of the already existing IAE[180] V2500 hp compressor would be used and a corresponding 3-stage turbine

---

[176] See BBC, GT15/ FPL, Sitzung vom 87-08-07.

[177] The gas turbine collaboration effort with Rolls-Royce began in the 1980s, when initial study work concentrated on GT8 improvement, by a 100 deg C higher temperature capability and + 2 percent efficiency.

[178] The saying goes that these preferences for GT15N (instead of GT8), for an intensified aero engine partner support, and the hiring of former aero engine key personnel for project management, can be traced back to Goeran Lundberg, who took on the GT management lead post around that time from A. Roeder. A pure scaling of the GT8 led to a gas turbine with 550 kg/s airflow. However, this machine was too big with respect to power output, the last turbine blade was too large to be cast and the rotor stresses were high.

[179] The designations GT15N/GT17N, as part of the NGT project, are used here and in the following to differentiate these activities from similar approaches 10 years earlier.

[180] IAE, International Aero Engines AG, is a Zurich-registered joint venture manufacturing company formed in 1983; current shareholders are Pratt & Whitney (65 %), MTU Aero Engines (12 %) and the Japanese Aero Engine Corp. (23 %). Fiat Avio withdrew as a shareholder of the programme early on but the now-renamed Avio S.p.A. remains a supplier. The Roman 'V' product nomenclature remains as a legacy of the five original shareholders. In October 2011, Rolls-Royce plc, which originally had contributed the 10-stage hp compressor with PR = 16–18, agreed to sell its 32.5% stake in the company to Pratt & Whitney, giving them a 65% stake.

designed in a joint team effort. In addition, the engine was to use an annular EV combustor from ABB which had been under development since 1986, Figure 5-37, and which was later applied for the GT13E2 at a comparable mass flow level as well. These boundary conditions were transferred into a GT15N concept drawing which is shown in Figure 5-44.

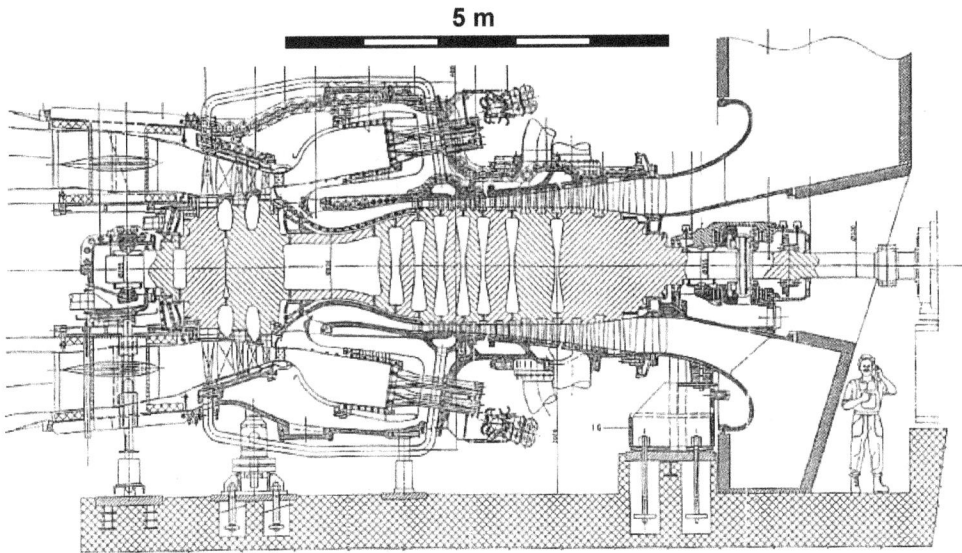

5 m

Figure 5-44   *GT15 N, ABB/RR 190 MW conceptual design (cancelled), 1991* [181]

Relatively early in the design process the ABB design team set new priorities for the scope of application for the forthcoming GT15N/GT17N engines:

1. Combined cycles with separate or coupled steam turbine and natural gas firing, a decision which had been overdue for 20 years
2. Oil firing with wet emission control and alternative dual fuel
3. MBTU gas firing and dual fuel operation.

The basic assumption of the entire effort was that ABB and RR could work together in a close and efficient manner. According to a statement of the then responsible project manager Allen Pfeffer in the 'GT15 Turbine, Final Report'[182] '...*this assumption was correct on the working level. But teamwork was hamstrung by proprietary data rights problems.*' RR cooling technology should have become the basis for increased turbine temperature capability. However, many of the proposed schemes were considered as too complicated to be

---

IAE's purpose is the development, production and aftermarket services of the IAE V2500 turbofan family with 23,000–32,000 lbf take-off thrust, which powers the Airbus A320 family and McDonnell Douglas MD-90 aircraft. EIS Entry-Into-Service of the first V2500-A1 was in May 1989.

[181] Figure source: ABB, GT15 turbine cross section and WBZ, 1979

[182] Hans Wettstein had recovered the report electronically in 2002, the original paper version of the report was issued at the end of 1992, but carried no information on date, author(s) and context, presumably for secrecy reasons.

cast economically on a large scale.[183] A great deal of time and effort had to be spent developing castable and effective cooling schemes.

On the material side, advanced aerospace superalloys like CMSX-4 promised stress and temperature benefits, but were again considered far too expensive to be cost effective. The compromise alloy choice, CM247, was commercially available (and had not required RR involvement). Neither RR nor ABB could provide the required lifing data for the superalloys; for RR the GT15 part life targets of several 10,000 OH were beyond their experienced time horizon, while ABB lacked information on superalloys in general. Surprisingly, several unexpected mechanical and aerodynamic design problems emerged during the cooperation. ABB's standard rotor design and the RR airfoil designs were largely incompatible, meaning a hybrid design had to be developed in order to provide a high temperature root.

A major part of the RR technology package was the 10-stage compressor, developed from the V2500 turbofan programme. Figure 5-45 illustrates the compressor's similarity by a split cross section of GT15N (top) and V2500 turbofan (bottom). In this illustration the compressor dimensions were intentionally made equal, while in reality the planned GT15N compressor was 4 times bigger than the V2500 ancestor. The design details of the V2500 master compressor were considered absolutely 'sacrosanct'/untouchable so that even a simple contour straightening of the compressor flow annulus ('swan neck') as suggested by ABB for manufacturing cost reasons was out of the question, given the risk of unpredictable consequences for the delicate compressor tuning. This design philosophy was somewhat typical for turbomachinery products of the time, which emerged only through lengthy, empirically based development cycles.[184]

In contrast the design philosophy of the turbine was conservative, in that the machine had to have ample growth and to be able to run with conventionally cast material if necessary. The RR first stage vane and blade design did not require much optimisation in order to achieve the required life. As a result ABB probably did not receive the best RR had to offer in terms of engineering experience and cooling technology levels.

The GT15N development programme came to a dead end during 1991[185], when it became more and more obvious that the selected configuration would not achieve the targets with respect to manufacturability at a reasonable cost level. The contract with RR was put on hold and later cancelled. At the same time the idea of sequential combustion, which was already one of Stodola's recommendations for a 'fractionalized reheat combustion' after testing the Neuchâtel engine, Section 4.3.3.2, was rediscovered as a possibility for limiting the technological exposure. The new concept, which later resulted in the GT24/GT26 engines, Section 6.1.3, was pushed forward in strict secrecy, together with MTU Munich and NPO Saturn as new aero technology and design partners.

---

[183] The final report from 1992 has interesting details which reflect somewhat the disappointing project status (and spirit): *'The largest DS (Directionally Solidified) blade to date is that of the (GE)7F, which is 2/3 the height of the GT15 stage 2 blade. For the 7F part there are many production problems reported with (ceramic mould) shell cracking, with 25 percent of all moulds scrapped.'* But in all fairness, one has to concede that the casting houses were still improving their capabilities (especially on large scale casting of turbine blading in superalloys) 10 years later, so that the assessment for 1992 is realistic.

[184] The final GT15N project versions had the 11-stage compressor design as illustrated in Figure 5-44.

[185] The official GT15N stop was announced internally during the GT24/GT26 project kick-off meeting at 'Stadtcasino' Baden, CH in October 1991.

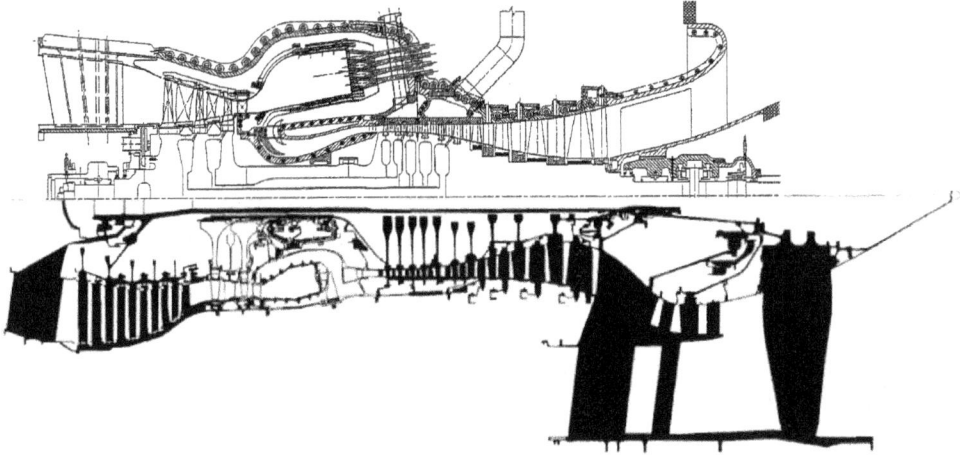

Figure 5-45   *Related 10-stage compressors:* [186] *GT15N (top) vs. IAE V2500 turbofan (bottom), presented here in equal compressor size, real scale factor ~4 : 1*

5.1.6   Prominent Engineers

The description of the long 'Phase II' GT development period between 1945–1988 revealed rather inconsistent and contradictory tendencies, which lacked also the dominant and continuous handwriting of prominent engineers, especially in comparison to the previous Phase I. Leading figures from Phase I, Section 4.1.2, such as Claude Seippel and Hans Pfenninger, also partly influenced the first half of Phase II. As engineering representatives of the remaining period in the 1970/1980s two engineers have been selected who considerably contributed to a smooth in-house engineering cooperation between the Mannheim and Baden teams for the first time: Albert Eiermann and Wilhelm Endres. While the late Claude Seippel, for example, still earned lasting engineering merits in the 1960s for laying the theoretical ground for an effective usage of combined GT/ST power plants, he and others had not managed to exploit the advanced turbomachinery design expertise immediately after WW II, then allocated at Mannheim, for a general, beneficial use within the whole of BBC. This drawback might have only become fully visible after his period of management responsibility, when his successors sometimes appeared to be helplessly exposed to the superior turbojet technology applications of the main competitor GE.

The third prominent engineering personality who will be addressed in this context is Franz Farkas, at best known for his lasting merits in developing and applying transonic compressor technology for first time power generation applications.

5.1.6.1   Albert Eiermann (1924–2013)

Albert Eiermann had lifelong ties with the Kurpfalz[187]/ Rhein-Neckar area, where he was born in 1924, in Dielheim, 15 km south of Heidelberg, D. He remained just 30 km from his birth-

---

[186] Figure source: Top – ABB, GT24/GT26 weekly reports, Bottom – http://www.scribd.com/doc/84519527/IAE-V2500

place during his entire professional career as a development engineer for fluid machinery with Brown Boveri at Mannheim-Käfertal. After receiving his diploma in mechanical engineering, he began on 1 January 1950 and stayed until the very last day of BBC's existence on 31 July 1987. His chronological project record is also an excellent representation of BBC Mannheim's turbomachinery activities over almost four decades, from 1950–1990:

Figure 5-46   *Dipl.-Ing. Albert Eiermann (1924–2013, picture ~1982)*

- topping steam turbine with 300 bar for GKM 'Grosskraftwerk Mannheim', (today 1.7 GW with additional 0.9 GW until 2013/14) in the early 1950s, with first realisation of double-reheat and a record efficiency for condensation steam plants at the time of 38.2 percent
- optimum design of various turbomachinery equipment for steel works in the Ruhr area, blast furnace gas compressors and turbines, oxygen compressors with detailed real gas compression studies, 300 bar high-pressure compressors for BASF Ludwigshafen, D in the 1960s
- participation in key development projects for nuclear plants in the 1970/80s:

  a   dual-flow centrifugal compressor with PR < 4 for separation nozzle technology (^{235}U enrichment facility, see below),

  b   He compressor (50 bar) and turbine for the helium-cooled HHV high temperature helium test facility, built at Mannheim with test operations at nuclear research centre FZ Juelich, D, and finally put into practice as Thorium High Temperature Reactor THTR-300, 300 MW, at Hamm-Uentrop, D.[188] Related to this: patent DE3141841 for a 'Centering device and safety gear for a rotor shaft in non-contact magnetic bearings', filed on 22.10.1981

---

[187] Electoral Palatinate or County Palatine of the Rhine (German: *Kurpfalz*) is a historic state of the Holy Roman Empire.

[188] Operations started on the plant in 1983, and it was shut down on 1 September 1989. The THTR was synchronised to the grid for the first time in 1985 and started full power operation in February 1987. The THTR-300 cost € 2.05 billion and was predicted to cost an additional € 425 million up until December 2009 for decommissioning and other associated costs. http://en.wikipedia.org/wiki/THTR-300

–   involved in various GT development projects over the years, amongst these:

c   the Aero Jet Expanders, Section 5.5.2

d   the 290 MW Huntorf air storage power plant, the first of these CAES plants (be-
     coming attractive again today) in 1978, Section 5.5.3

e   the 150 MW GT13E, Section 5.1.5.3, for Hemweg, NL in 1988. This became Albert
     Eiermann's final masterpiece as head of the BBC 'Gas Turbine Technology' de-
     partment, with substantial support from Wolfgang Keppel as project manager.
     This development milestone was acknowledged by an entry into the Guinness
     Book of Records 1988 as the most powerful gas turbine in the world.[189]

The aforementioned separation nozzle process a) for the enrichment of ^{235}U was developed
in the early 1970s at the Karlsruhe Nuclear Research Centre as an alternative to gaseous
diffusion or ultra-high-speed centrifuging processes. Highly stressed rotating machinery
was avoided, as the separating centrifugal forces were generated in the separation nozzle
method by deflecting a high-speed jet consisting of uranium hexafluoride (5 percent of UF_6)
and helium or hydrogen as a light auxiliary carrier gas.

Figure 5-47 illustrates a prototype test facility at Karlsruhe, containing about 80, approx.
2 m long tubes as 'separation elements'. An enlarged inset shows the cross section. Here,
the feed gas UF_6/H_2 mixture enters the 0.1 mm wide (and 2 m long) nozzle slot from the left
under pressure, and the lighter ^{235}U fraction follows the inner flow path through the mid
section while the depleted heavier fraction exits to the right. The best separation efficiency
was achieved at Reynolds numbers Re ~100.

The shown test stage consisted of the tank containing the separation elements, the cross
piece for gas distribution, the two-stage cooler and the high-speed dual-flow centrifugal
compressor with a firmly coupled e-drive motor from BBC.[190]

---

[189] The GBR entry was initiated by Communications/Marketing at BBC Mannheim, additionally supplemented
by an entry titled '*The first 4 MW utility gas turbine at Neuchâtel, 1939*'. Recently, Siemens took the opportunity to file the new SGT5-8000H at Irsching in the same GBR category, as announced by Uriel J. Sharif on 21
December 2007.

[190] Between 1969 and 1977 the German Industrial Research Organisation for Combustion Engines (FVV Frank-
furt/M.) organized a working group for gas turbines and sponsored a comprehensive experimental (DLR Cologne and TU Hannover) and theoretical (Prof. Walter Traupel, ETHZ, see Traupel, Thermische
Turbomaschinen I, p. 374 and 402) research project on centrifugal compressor flows, in which the author
(then at DLR, Institute for Propulsion Technology, Cologne, Profs. Gert Winterfeld and Heinrich Weyer) had
the privilege to contribute detailed research results, like the first time application of Laser-2Focus
Velocimetry (Richard Schodl) for 3D flow investigations within a high-speed centrifugal impeller. Albert
Eiermann participated in this working group, which was chaired by Erwin Schnell, KHD Oberursel, D, as one
of the experienced industrial key advisors. Schnell was chairman of a parallel FVV working group on 'turbine
flows' for six years. Although the uranium enrichment development activities at BBC were not publically
known at that time, it is likely that results of the FVV research programme immediately influenced this challenging centrifugal compressor development. http://www.fvv-net.de/cms/upload/FVV_R555_Sonderheft-
40JahreRV_2Seiten_web.pdf

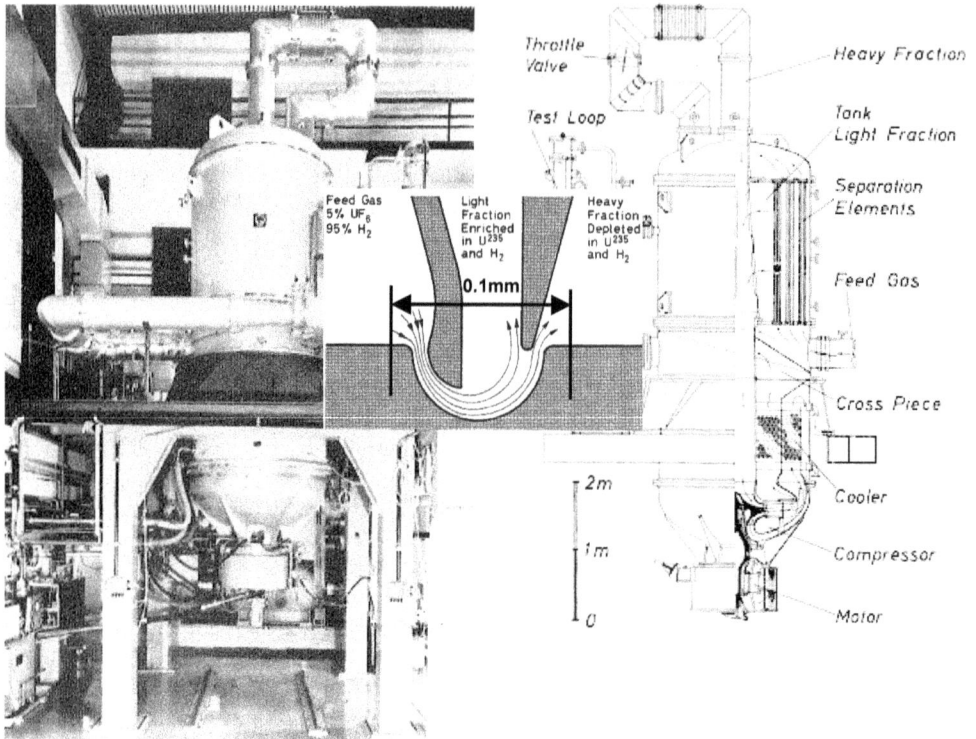

Figure 5-47   *Nozzle separation prototype test facility, FZK Karlsruhe, D ~1972,* [191] *A. Eiermann's dual-flow centrifugal compressor in the bottom casing*

Albert Eiermann's UF_6/H_2[192] compressor provided a feed pressure of approx. 0.1 bar (PR < 4) at a volume flow of 28 m^3/s; a three times bigger full scale unit was planned for 85 m^3/s.[193]

---

[191] See Becker, Present state and development potential of separation nozzle process. To achieve the required degree of enrichment for light water reactor operation, several hundreds of such units would have to be arranged in a cascaded row, corresponding to 30–50 km slot length of the shown width (typically 0.1 mm). The corresponding manufacturing technology was originally developed at FZK for the fabrication of separation nozzles and then found widespread industrial application for the manufacturing of high-aspect-ratio microstructures under the name LIGA technology, a German acronym for LIthographie, Galvanoformung, Abformung (lithography, electroplating and moulding). LIGA is known for the manufacturing of high-pressure diesel engine injection nozzles; Mimotec of Sion, VS, Switzerland supplies the Swiss watch market with LIGA-made metal parts of nickel and nickel-phosphorus materials.

[192] The engineering design challenges are illustrated by the properties of the gas mixture. Uranium hexafluoride (UF_6), referred to as 'hex' in the nuclear industry, is a compound used in the uranium enrichment process that produces fuel for nuclear reactors and nuclear weapons. It forms solid grey crystals at a standard temperature and pressure, is highly toxic, radioactive, reacts violently with water and is corrosive to most metals. As well as this, with 95 percent H_2 as a carrier gas, it would be explosive when exposed to O_2, making the encapsulated design necessary. However, it reacts mildly with aluminium, forming a thin surface layer of AlF_3 that resists further reaction. Consequently, it is likely that forged Al alloy and Al coating has been widely used for that BBC turbomachinery design. http://en.wikipedia.org/wiki/Uranium_hexafluoride

[193] See Becker, Present State and Development Potential of Separation Nozzle Process.

After his official retirement in 1987 Albert Eiermann continued to work as an ABB consultant in the GT13E2 development programme until 1991, focusing especially on annular combustor integration. He died in August 2013 at Munich, D.

5.1.6.2   Wilhelm Endres (1927–2007)

Wilhelm Endres had a broad international training before he came to BBC Baden in 1955. Born in 1927 in Augsburg, Germany (the site of MAN Maschinenfabrik Augsburg-Nürnberg) the family followed the professional career of father Wilhelm[194], a combustion engine designer, first to Danzig[195], then to Chemnitz and finally to Munich, where Wilhelm Jr. studied Mechanical Engineering at the THM from 1946–1952, finishing as Dr.-Ing. in Mechanical Integrity with Prof. Ludwig Foeppl[196].

In 1955 Wilhelm Endres came to the steam turbine department of BBC Baden, first receiving responsibility for the hp turbine design group in 1959 and classically moving on as assistant to the Technical Directorate TD-Th in 1964. In 1967 he was transferred to the gas turbine business, where he finally became head of the GT Department on 1 September 1968 as successor to Hans Pfenninger, Section 4.1.2. In 1969 he was promoted to Deputy Director and head of the BST Gas Turbine Section. His close cooperation with Albert Eiermann from BBC Mannheim began, with whom he maintained a trusting and very cooperative relationship throughout his professional career. Wilhelm Endres had a lasting impact on GT mechanical design methodology, as outlined typically in Section 5.4.1. He was highly appreciated by employees reporting directly to him for his human character, friendliness and true personal interest. Looking back, he associated the BST period with an *'extraordinary high personal workload'*. After the official end of the BST BBC-Sulzer joint venture and the corresponding BST department TGT-3, Section 5.1.3, the 'Open Gas Turbine Technology' > 20 MW came back to Baden. From 1 July 1974 this department was led by Endres again, now as Dept. TMA and after a reorganisation, as

---

[194] Wilhem Endres sr. (1893~1970) left MAN Augsburg around 1930 to work at the International Shipbuilding and Engineering Company Ltd. (Danziger Werft), which was owned by Great Britain (30%), France (30%), Poland (20&) and the Free City of Danzig after WW I. He continued as head of piston engine development at 'nomen est omen' DKW (Dampf-Kraft-Wagen – steam powered vehicles) Chemnitz and became a well-known author on combustion engineering (Sammlung Goeschen) and professor for combustion engines at the TH Munich from 1946–1962. His Institute of Combustion Engines and Motor Vehicles (LVK Lehrstuhl für Verbrennungskraftmaschinen und Kraftfahrzeuge) was founded in 1936 with the original name 'Aircraft Engines and Engine Theory'. The grandfather of our Wilhelm Endres jr. was Max Endres (1860–1940), a renowned forestry scientist, whose wife Sascha Endres was the daughter of Franz Wilhem Nokk (1832–1903), the Prime Minister (Praesident des Staatsministeriums) of the Grand Duchy of Baden, Germany, from 1893–1901.

[195] Wilhelm Endres jr. attended primary school in Danzig, then between 1937 and 1944 in Chemnitz he attended the Humanistische Gymnasium (secondary school focusing on Latin and Greek). At 17, he was called to the German Army and luckily became a US prisoner. As a student he came to Switzerland for the first time within his 'Freiwilliges Landjahr', a voluntary work programme. Received by Swiss students at Basel, the hungry young Germans were stunned by the bending breakfast buffet. In 1953/54 he was a Fulbright scholar at the Northwestern University, Evanston, Ill. on the northern outskirts of Chicago, specialising in the theory of elasticity and photoelasticity.

[196] Karl Ludwig Foeppl (1887–1976) was an engineer and university professor, best known for introducing the method of studying stresses by photoelasticity in 1928. He was brother-in-law to Ludwig Prandtl. During WW I he worked as a code-breaker, who in 1915 managed to successfully decipher the Gronsfeld Code used by the Royal Navy for the first time. In 1920 he became a professor at the TH Dresden, before following his equally renowned father August Foeppl (1854–1924) as professor and head of the 'Mechanical-Technical Laboratory' at the TH Munich from 1922–1955.

TC from 1 January 1976, with Hans Baumann as line manager of the newly founded area 'Thermische Mittelgrosse Anlagen', which had been called 'Energieerzeugung Mittlere Anlagen' (thermal mid-size units) since 1976. On 6 April 1981 another reorganisation took place, which left Wilhem Endres in charge of his Deputy Director post for TC 'Gas Turbine Power Plants'.

Figure 5-48   *Dr.-Ing. Wilhelm Endres (1927–2007, picture ~1975)*

Not much is known about how Wilhelm Endres lived through the critical years of the second half of the 1970s, but as was indicated in Section 5.1.4 already, thoughts of cancelling BBC's struggling gas turbine business completely must have existed. Enthusiastic engineering freshmen from the mid-1970s remember Wilhelm Endres in a somewhat fatalistic mood, when asked about the gas turbine product strategy, he replied "*We have no need for an own strategy, we get this from GE for free*".

With a number of in-house and external international collaboration programmes, the 1970s also meant the beginning of a period of intensified business travel.[197] Figure 5-49 documents a discussion amongst combustor specialists at NASA, Cleveland Oh., USA.

---

[197] John Corrigan, currently Vice President for Technology, Engineering, and Research, since 1967 with Howmet Corp., one of the leading manufacturers of precision investment castings, has reminiscences about Wilhelm Endres in the late 1970s. Endres came to Hampton, Va., USA, presumably during the phase of intensive GT11E/GT13E preparations, Figure 5-13, with a pack of turbine drawings – alone. This scene addresses the decisive moment, when an OEM starts first discussions with a key supplier, both to clarify manufacturability as well as the cost of a new product. After more than thirty years of mutually trustworthy cooperation in this field, these kinds of meetings have become rather routine; typically, a joint team approach of 10–15 man-years is required for an all-new turbine design, before an optimum design compromise among aspects of turbine castability, cooling performance and target costs can be settled. The fact that Wilhelm Endres showed up to this drawing check alone could have had several simple explanations, such as travelling cost considerations or '*the boss is the best expert*' (which was certainly true for him), but could also reflect a certain fear that possible negative evaluation results might spread easily afterwards if some in-house 'confidants' were in attendance.

Figure 5-49   *BBC expert meeting at NASA Cleveland, Oh., ~late 1970s, from left: Dr. W. Endres, Dr. P. Felix [198] and H. Koch [199]*

Wilhelm Endres left BBC like Albert Eiermann and many other qualified engineers during the transition from BBC to ABB on 31 July 1987, but afterwards continued for another six years as a Professor for Mechanical Engineering, Design and Machine Parts at the University of Stellenbosch, RSA, 50 km east of Cape Town, and as a part time consultant for ABB. In 2000, he summarised his dedication to the gas turbine in the comprehensive GT Design Handbook[200]: long-term experience with challenging engineering design tasks and consequential lessons learned; a lasting and valuable heritage of company-internal studies for generations of engineers to come.

5.1.6.3   Franz Farkas (*1935)

Franz Farkas contributed to BBC/ABB/Alstom gas turbine developments for more than 40 years, predominantly in the area of advanced turbomachinery. Born in 1935 in Hungary, he went to school there and studied Mechanical Engineering. After the Hungarian uprising in 1956 he left the country and found a new home in Zurich, Switzerland. He continued his studies at the ETH Zurich, finishing with a diploma thesis supervised by Professor Traupel on the '*High-Temperature Helium Gas Turbine for Nuclear Power Plants*'.

In 1961 he commenced working at BBC, as a mechanical designer until 1963, before he moved on to numerical design and at the end of the 1960s found the subject of his life's work as head of the BST compressor group, the industrial realisation of the transonic compressor. The inherently high mass flow capability of this new compression approach opened the door for the so-called 'Grenzleistungsmaschinen' (power at the limit), the first generation of GT17 (220 MW, 50 Hz) and GT15 (150 MW, 60 Hz). In the 1970s he became

---

[198]   After studying chemical engineering , Dr. sc. techn. ETH Peter Felix (*1944) started at Sulzer Winterthur in 1970 as a research engineer for new material development. After engineering breaks at BST and Turbodyne, he moved on to BBC's Sales & Projects Dept. in 1981. After various regional sales positions for ABB in Saudi Arabia, Middle East and North Africa, he finished his professional career as Senior VP, Process Mgmt. & Finance in 2003 at Alstom Paris, F.

[199]   See Section 6.1.4

[200]   See Endres, Erfahrungen mit Gasturbinen

project leader of this GT family, of which finally only the small GT8 and the GT10 for the departing Sulzer side (after the break-up of the BST joint venture) became launching vehicles for the transonic compressor, Sections 5.1.5.3 and 5.1.5.4.

Figure 5-50   *Dipl.-Ing. Franz Farkas (*1935, picture ~1985)*

In 1978 he was promoted to head of all numerical design (comprising compressor, turbine, combustor, the 'blade office' and sales support), at which time the GT13E was also realised, together with BBC Mannheim. After 1985, during the second round of design efforts for the GT15/GT17 together with Rolls-Royce, Franz Farkas was named head of the aero design team, with the responsibility for compressor and turbine aero design as well as turbine cooling calculations.

Finally, at the end of the 1980s, a project team was brought together to realise for the first time, what would later become the GT26 (260 MW, 50 Hz) and GT24 (180 MW, 60 Hz): a new single shaft, two bearing gas turbine concept with sequential combustion, i.e. two low-emission combustors in a row, separated by a hp turbine. The main advantages were lower combustion temperatures, improved part-load performance (especially beneficial for combined cycle applications) and environmentally attractive, low emission values. An extraordinary challenge in this context was the design of a single-shaft compressor with a previously unheard of total pressure ratio of PR > 30 and with special design optimisation requirements for starting, part-load operation and shaft dynamics. Franz Farkas coordinated the ABB-internal aero team again, Section 6.1.3. Before he applied for early retirement in 1997 to spend more time with his family, especially his two grandchildren, he systematically documented his broad experiences in various internal documents. However, afterwards he continued part-time as a technical consultant and experienced design reviewer for another eight years. During this period Franz Farkas contributed his vast experience to numerous transonic compressor upgrade programmes, to turbine aerodynamic and cooling evaluations, as well as mechanical integrity assessments.

5.2   Competitive Designs and Developments

References to technical issues of competitors and competitive designs are generally mentioned throughout this text. Nevertheless, it is worth addressing a few of the competitors specifically, amongst these the earliest local competitors in Switzerland and their typical products.

5.2.1   Escher Wyss – The Closed Cycle AK Engine

In 1805 Hans Caspar Escher founded mechanical spinning works in Zurich together with the banker Salomon von Wyss, to which they attached a mechanical engineering department primarily producing textile machinery, and in support of these activities water wheels and turbines as well as power transmission units.[201] From 1835 complete ships with boiler and steam engine were built there, followed by stationary steam plants and locomotives. In 1840 the business expanded to Loosdorf, Lower Austria and in 1856 to Ravensburg, Germany. In 1860 their own spinnery was closed and led by Heinrich Zoelly (1862–1937), the company focused on the production of water/ steam power engines and refrigeration plants. Between 1904 and 1929 the Zoelly-Syndicate produced steam turbines for power generation worldwide, as well as ship propulsion and locomotives; at the same time Escher Wyss AG (since 1935) gained a leading position in hydropower stations. After 1945 E-W acted globally with Zurich as headquarters for research, development and the manufacturing of precision parts, with at best, a Swiss workforce of 2,300 in 1963. At the end of the 1950s the MFO MaschinenFabrik Oerlikon (very traditional for BBC's early history) was integrated into E-W. This merger was only the prelude to a further concentration process of the Swiss turbomachinery industries, when as described in Section 5.1.3 BBC, Escher Wyss and Sulzer decided to join forces to form BST Brown-Boveri-Sulzer Turbomaschinen AG. The Escher Wyss steam turbines came under BBC responsibility. Walter Traupel called this a '*harrowing tragedy, life-long achievements of Zoelly and others had been destroyed...*'[202]

Jakob Ackeret worked for Escher Wyss between 1927 and 1931, after his stay in Goettingen and before he was called up as Professor for Aerodynamics at the ETH Zurich. At Escher Wyss he worked mostly as a specialist in hydraulics, but with considerable influence on fluid machinery and steam turbine design. In 1936, rumours spread about the gas turbine-related activities at BBC. Ackeret and Curt Keller, head of the Escher Wyss research department at the time, decided on a different approach. Their basic idea was to operate the machine in a closed air cycle, the Ackeret-Keller process, to which the heat is added from outside the closed loop by means of an air heater (heat exchanger). A first test set-up became operable at Escher Wyss during 1939[203] and the acceptance tests, carried out by Prof.

---

[201] The 'Continental System'(1806–1814) of Napoleon I supported the initial success of Escher Wyss.

[202] See Traupel, Marksteine. Traupel has also remarkable details on the BST break-up. In 1974 the fifth BST anniversary had been celebrated by a dinner in great harmony, but only two weeks later the unsuspecting employees were informed about the break-up decision as a result of a conflict within/between the management board(s), with everybody free to leave either for BBC or Sulzer. The ensuing, still ongoing, regular '*engineering experience exchange conferences*' obviously considered Traupel as a kind of justified protest of the neglected personnel and as a strong sign of engineering solidarity.

[203] See Ackeret and Keller, Eine Aerodynamische Waermekraftanlage and – see Traupel, Marksteine. The aforementioned first AK publication appeared on 13 May 1939 in the Schweizerische Bauzeitung in dedication to

Quiby as the successor to Prof. Stodola, showed a promising efficiency of 32 percent in January 1945. As known, in thermal power plants energy in the form of transferred heat is transformed from the combustion gases into the working fluid, which undergoes a cyclic process. The AK process consists of the following four changes of state:

- isothermal compression
- isobaric heat addition
- isothermal expansion
- isobaric heat removal.

Because this process was originally proposed by the Swedish engineer John Ericsson (1803–1899), it is sometimes also called the Ericsson cycle. Since isobaric heating is made equal to isobaric cooling, the thermal process efficiency is equal to the Carnot cycle in principle, however, the technical realisation of this process is difficult because isothermal compression and expansion are hardly achievable due to the fact that they only can be approximated by multi-stage adiabatic compression with intermediate cooling.

Particular attention was paid to the closed cycle during wartime at AEG Allgemeine Elektrizitäts-Gesellschaft, Germany (translation: General Electric Co.), though nothing is known about a licence agreement with Escher Wyss. The process appeared to have certain advantages for solid fuels, economically essential for power generation. A principle scheme of a closed cycle gas turbine is shown in Figure 5-51 with air or other gas as the working medium. The air was heated indirectly in the air heater, an external combustion system and a second time upstream at the heat exchanger for recuperating turbine exhaust energy. Before the compressor was re-entered the air was cooled back. A small compressor was used to pressurise the system, and a valve to discharge it for power adaptation when required. Good part load efficiency was possible since both the working pressure and the fuel input could be adjusted. While the turbomachinery and the ducting could be small as a result of the high working pressure, the air heater had to be very large owing to the comparatively low combustion pressure. The control of the system had been assessed at AEG as difficult, and the cycle efficiency also appeared to be over-estimated because of the expected susceptibility to pressure losses. Correspondingly, pioneering work on the closed cycle gas turbine was left to the Swiss Escher Wyss.

In general, air was used as a working fluid, but other gases like helium or nitrogen would be possible as well. The AK power output was easily adjustable by cycle pressure/density variation. The frightening fouling effect of open GT crude oil operation would have been prevented by selecting suitable operation gases. A drawback in comparison to open plants were the higher energy costs because a cooler was required, and high-quality, high temperature steels were needed for the (air) heater.

Prof. Stodola's 80th birthday three days before, under the motto 'Quidquid agis prudenter agas et respice finem', from the Aesopian fable #78 about the two frogs: 'Whatever you do, do it well and think about the end'. It is interesting to note, however, that the related patent US2172910 'Power plant' was only in Curt Keller's name, with priority in Switzerland on 12 July 1935 and assigned to 'Aktiengesellschaft fuer Technische Studien Zurich', obviously C. Keller's private venture, independent of Escher Wyss.

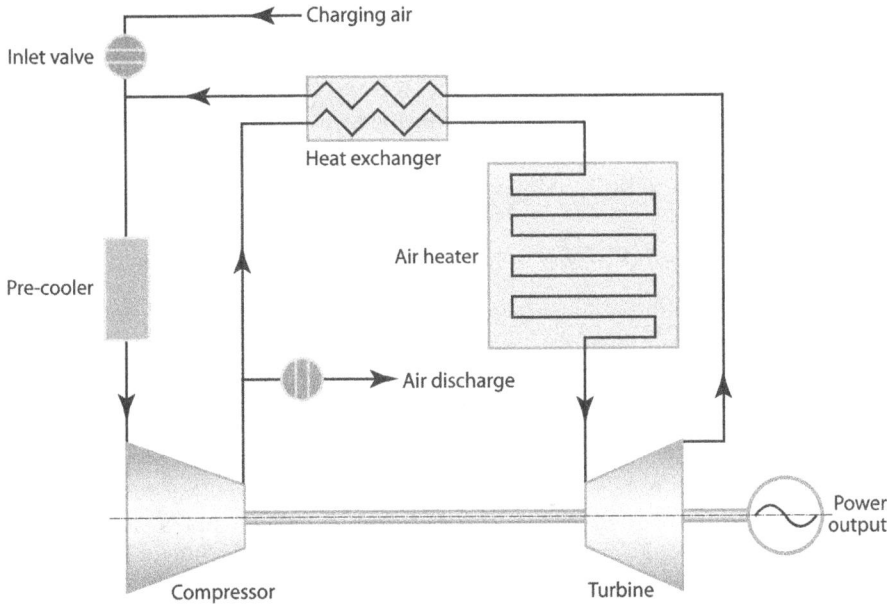

Figure 5-51 *Scheme of the closed cycle gas turbine* [204]   © *The Crowood Press*

After WW II Escher Wyss GmbH of Ravensburg, Germany and Gutehoffnungshuette (GHH) of Oberhausen-Sterkrade, Germany built the world's first commercial closed cycle gas turbine unit, which drove a 2 MW generator and ran entirely successfully on bituminous coal from 1956.[205] Escher Wyss' work on the closed cycle gas turbine is described in a paper from 1966[206] with seven reference plants in detail: 2.3 MW at Ravensburg, Germany, 2 MW at Toyatomi, Japan, 6.6 MW at Coburg, Germany, 12 MW at Kashira, Russia, 12 MW at Nippon Kokan, Japan, 14.3 MW at Oberhausen-Sterkrade, Germany and 6.4 MW at Haus Aden, Germany. These plants were fired on coal, natural gas, blast furnace gas and mine gas at turbine inlet temperatures between 660 and 720 °C. The first closed cycle helium gas turbine was produced in 1966. Though theoretically a sound idea, the AK process did not succeed in the competitive reality of the 1950s against its major contender, the ever-improving steam turbine cycle. Originally, the AK engine was planned at the upper performance end of the steam process, but in the reality of 40 years of competitive development this target was never achieved and one of the very last options for the high temperature reactors, with the reactor replacing the conventional AK air/gas heater, finally disappeared with the whole application concept.[207]

---

[204] Figure source: Kay, German jet engine and gas turbine development, p. 207.

[205] See Kay, German jet engine and gas turbine development, p. 208.

[206] See Taygun, Die Gasturbine mit geschlossenem Kreislauf als Mehrzweck-Anlage

[207] Interestingly the early AK projects also addressed nuclear reactors, e.g. the Ackeret manuscript collection at the ETH Zurich library & archive contains within section Hs 552, item 156 the original drawing of an AK-related reactor design dated 13 November 1945. In his presentation, see Ackeret, Auf dem Wege zur Gasturbine (Towards the Gas Turbine), at the Deutsche Akademie fuer Luftfahrtforschung (German Academy for Aeronautical Research) on 4 December 1942 in Berlin (L. Prandtl attending), J. Ackeret addressed three subjects:

In total 24 closed cycle plants were built by Escher Wyss and the associated companies John Brown Engineering UK, GHH Germany, Fuji Electric Japan and La Fleur Corp., USA. The Escher Wyss technology went to Sulzer Ltd. in 1969 and to BBC after the BST breakup in 1974. There were no new activities after 1981. Independent of the indicated drawbacks, closed cycle gas turbines experience a resurgence of interest once in a while and still remain a promising technology going into the 21st century.[208]

5.2.2   Sulzer – Gas Turbine with Semi-Closed Process

The first gas turbine plant with a conventional open cycle was tested at Sulzer Ltd. Winterthur, Switzerland in 1942. Since Sulzer had a special interest in ship propulsion units[209], the development of a gas turbine with a semi-closed cycle had started as early as 1940. The advantages of this cycle were seen in the high peak and part-load efficiencies and in the high power concentration.

In 1947 the Nordostschweizerischen Kraftwerke AG, Baden, CH ordered such a plant with 20 MW output for power generation at the Weinfelden power plant, approx. 10 km south of Constance/ Kreuzlingen on Lake Constance, starting operation in 1949 with a measured thermal efficiency of 32 percent. The basic process advantage is the combination of a small open cycle lp ducting for intake and exhaust, and a closed hp cycle; the integral cooling water demand is one quarter of that of a steam turbine plant in the same power class.

Without discussing further technical details here[210], the complexity of the semi-closed GT working principle becomes immediately obvious in Figure 5-52. The basic problems of the plant operation, fouling and corrosion by ashes and sulphurous fuels was described in de-

---

The simple (in his words) open GT concept of Brown-Boveri Baden CH with 18 percent efficiency, and the achieved operational reliability at 550 °C, encouraged the company (BBC) to build a gas turbine locomotive. Due to the extraordinarily compact design this would also only have 19 percent efficiency, but a small extent of heat exchange (for the association of the BBC GT locomotive of 1941 and German GT tank projects beginning in 1943, see Section 4.3.5). He also suggested a gas turbine with closed cycle operation 'which the author together with Herr Keller developed at Escher Wyss. The plant operates safely at temperatures of up to 650 °C and with efficiencies, surpassing comparable steam plants. Since there is in principle no hindrance in using fuels other than liquid (sic!), a high economic performance is achievable.' Is it too speculative to assume that 'other than liquid fuels' could have meant (besides coal) also uranium, in 1942, in Germany? After reading Wildi, Der Traum vom eigenen Reaktor, the similarity of German reactor projects before 1945 and Swiss reactor concepts afterwards is most striking, as both were of the rather rare 'heavy water moderated' type. Wildi's book has also details about the catastrophic failure with partial core melting of the 9 MW test reactor at Lucens, Canton Vaud, Switzerland on Tuesday 21 January 1969. http://de.wikipedia.org/wiki/Reaktor_Lucens
The failure was obviously caused by leaking 'liquid ring' seals of the cooling gas blowers within the responsibility of Escher Wyss, and a subsequent oxyhydrogen gas explosion in the reactor core. Curt Keller was named as co-inventor for the corresponding 'sealing device', US2965398 with priority in Switzerland from 1 September 1956. The sealing liquid was described here as a 'suspension of an adsorption substance in a solvent mixture'; as an example, active carbon dust was to float in a mixture of alcohol and water. Afterwards a committee was set-up with C. Keller as a member, which investigated the failure for ten(!) years, predictably without naming Escher Wyss in the final report at all. See Wildi, Der Traum vom eigenen Reaktor

[208] The most comprehensive presentation of the subject is found in Frutschi, The closed cycle gas turbine. Hans-Ulrich Frutschi started his engineering career under Curt Keller of Escher Wyss in Zurich, before he continued at BBC-ABB-Alstom. Closed cycle gas turbines based on helium or supercritical carbon dioxide hold promise for use with future high temperature solar and nuclear power generation.

[209] A characteristic advantage of the semi-closed GT cycle is the relatively small intake and exhaust volume flows, especially beneficial for space-limited ship installation.

[210] See Kruschik, Die Gasturbine, p. 559 ff.

tail by Walter Traupel[211] as one of his last industrial tasks before he became a professor at the ETH Zurich.

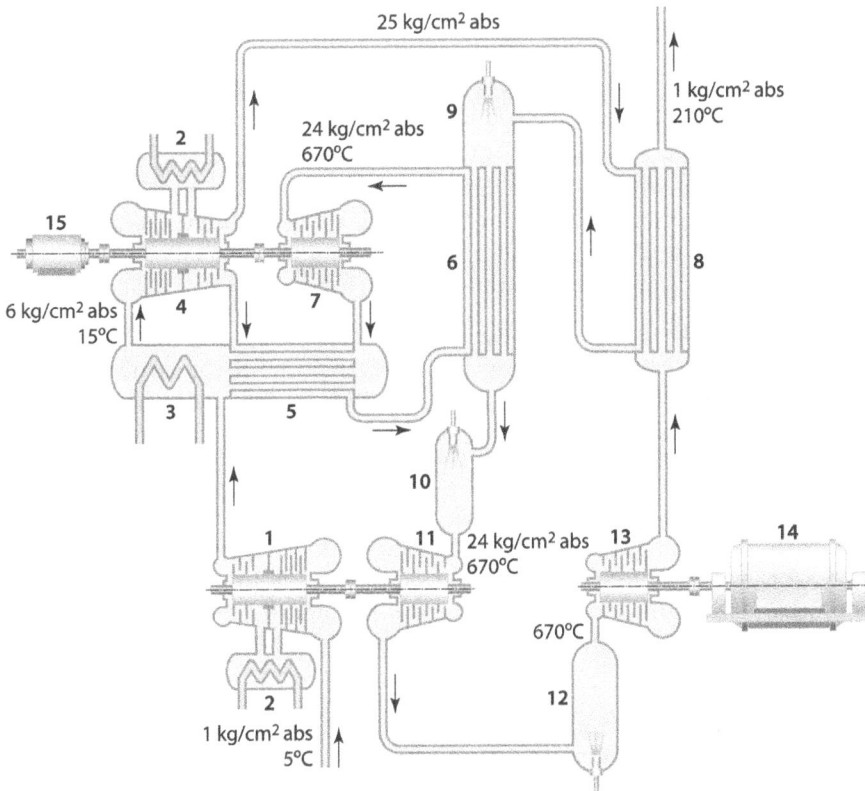

Figure 5-52   *Working scheme of the Sulzer 20 MW Weinfelden power plant with semi-closed GT cycle[212]*
1 – lp compressor, 2 – intercooler, 3 – heat exchanger, 4 – hp compressor, 5 – air heat
exchanger, 6 – air heater, 7 – air turbine, 8 – gas heat exchanger, 9 – main combustion
chamber, 10 – secondary combustion chamber, 11 – compressor drive turbine, 12 – intermediate
combustion chamber, 13 – generator drive turbine, 14 – generator, 15 – starter motor
© 1960 Springer

The intermediate Brown Boveri – Sulzer joint venture BST between 1969 and 1974 was already described in Section 5.1.3. In the 1970/80s less profitable areas were under scrutiny and were to be sold off if they were not profitable within a time limit. Gas turbines, knitting machines, and the historic diesel engines came into focus and were among the areas in which disinvestment took place. After Sulzer Ltd. had integrated the turbomachinery activities of Escher Wyss in 1969, Sulzer Turbo itself became part of MAN Turbomachinen AG in 2000, which was renamed to the traditional designation MAN Turbo in October 2004. In addition, in January 2006 steam turbine activities were bought from the Thyssen-Krupp

[211] See Traupel, Die Entwicklung der Gasturbine in der Schweiz, p. 3776 ff.
[212] Figure source: Kruschik, Die Gasturbine, p. 562

daughter company B+V Industrietechnik, former Blohm und Voss, Hamburg. Finally, effective as of 1 January 2010, after a merger with MAN Diesel, a new company with traditional roots was launched: MAN Diesel & Turbo SE.

5.2.3   Other Companies

General Electric Corporation The competition between Brown Boveri and General Electric started long before the introduction of the industrial gas turbine. As early as 1901 BBC had signed its first licence agreement with a US company and it was as early as 1904 that the dynamic Swiss enterprise caught the direct attention of General Electric, who placed a request for information from BBC to learn more about the licence conditions for the manufacturing of BBC machinery.[213] Before these first formal company contacts were established there was already an information exchange on a 'family level'.[214]

The first practical business step into the 'New World', launched by the BBC subsidiary Scintilla, came to a somewhat disappointing end after Lindbergh's spectacular Atlantic crossing in 1927, and the 'Spirit of St. Louis', equipped with Scintilla magnetos. BBC sold this company after a long period of economic slump, just before a Lindbergh-triggered sales bonanza started in the US.

Several BBC business operations to gain ground on the US market finally failed after the stock market crash on 'Black Thursday', 24 October 1929[215], and in the subsequent deep economic recession, when protectionist laws prevented imports from Switzerland completely. Immediately afterwards the mighty arch rival threatened Brown Boveri directly: General Electric bought extensive shares of their European competitors, more than 20 per-

---

[213] The special relationship between Switzerland and General Electric dates back much earlier. Since 1871 John (Johann Heinrich) Kruesi (1843–1899) had been a member of the staff of Thomas A. Edison. Kruesi, born at Heiden, AR (Appenzell), Switzerland as the 'illegitimate' son of the local apothecary's daughter, grew up in an orphanage, finished his apprenticeship as a mechanic in 1864 and after a few years as a travelling journeyman came to the US in 1870. According to one of his co-workers, Kruesi *'understood work in the drafting room and could decipher Edison's sketches no matter how crude they were...'.* In 1877, one year after the move to Menlo Park, he made the first phonograph following Edison's design and in 1879, he was part of the team of inventors responsible for the electric light bulb and was the principal mechanic on countless other patented innovations, including 'Kruesi tubes', isolated power distribution cables. With the development of Edison's system of electric lighting Kruesi moved to higher management positions. In 1881, Edison put him in charge of Edison Electric Tube Co., responsible for the installation of underground power distribution cables from the central generating stations. He still kept close contact with his workers who called him 'Honest John'. When General Electric was founded in 1892, one year after Brown Boveri & Cie., as a merger of several companies, Kruesi was promoted to General Manager and then to Chief Mechanical Engineer at the Schenectady site in 1896. However, Kruesi made a severe mistake in the very last stages of his career: he clearly underestimated the potential of the Curtis steam turbine under development at GE Schenectady. Despite his merits, Kruesi was replaced as head of development in 1896 by W.L.R. Emmet (1859–1941), in charge of GE's Lighting Department at the time, who built the first 500 kW Curtis vertical steam turbine, which was in commercial operation from February 1903 at Newport, RI, USA and since 23 July 1990 the 144th ASME International Historic Mechanical Engineering Landmark on exhibition at IPL E.W.Stout Generating Station, Indianapolis, In., USA. See ASME, The First 500-Kilowatt Curtis Vertical Steam Turbine.

[214] Julius Pfau, the cousin of BBC founder Charles E.L. Brown, worked in GE's Testing Department at Schenectady, NY, USA in the early 1900s and obviously had a broad insight into GE's development activities at that time. A letter from 1903 proves a technical data exchange between the two relatives on turbines and dynamos, but without any further possibility of identifying who gained more from this. Nevertheless, C.E.L. Brown indicated clearly that he advised his cousin to stay at GE as long as possible in order to build up further valuable experience. See Novaretti, Die Rolle sozialer Netzwerke in der BBC-Fruehphase

[215] In Europe known as 'Black Friday', 25 October 1929, but the actual catastrophic crash was only on the following Tuesday, 29 October 1929.

cent of BBC and a substantial percentage of the German AEG. Even the somewhat uninterested Boveri family sold a considerable part of their shares to the aggressive Americans due to financial troubles. The Swiss stock law and the corresponding introduction of name registered stock shares finally brought Swiss banks to the rescue and prevented unwanted GE dominance. A last attempt of General Electric to gain a seat on the BBC supervisory board based on a certain level of owned shares was definitively rejected in 1933.

The GE heavy duty gas turbine division, now GE Power Generation, as part of GE Energy, Atlanta, is based in Schenectady, New York, USA. As mentioned earlier in Section 5.1.1, the first US utility use of a gas turbine was in 1949 at the 3.5 MW Belle Isle plant[216] of Oklahoma Gas & Electric. This gas turbine design was established immediately after the WW II aircraft engine experience of the TG180 turbojet project.[217] A similar 3.5 MW gas turbine was installed in El Paso in 1953 and was still in operation 50 years later. Between 1966 and 1976 there were over 1,400 GE gas turbine units installed in the USA, rated more than the initial 3.5 MW of the Belle Isle plant. In the 1950s GE had introduced their 'frame' gas turbines as direct scales with power output between 16 and 23.2 MW, by 1965 with considerably increasing firing temperatures and pressure ratios. The first ever GE combined cycle plants were the City of Ottawa, On., Canada, 11 MW FS3 and the Wolverine Electric Ottawa, On., Canada, 21 MW FS5 installed in 1967[218].

In 1970 the aluminium smelter in Bahrain (ALBA) became the first ever GT powered aluminium smelter in the world. The 24 MW Frame 5 was installed with 25 machines in one row. Smelters had previously always used hydro power and were located accordingly.

The GE Frame 7 gas turbine appeared in 1970 with a rating of 47.2 MW and a turbine inlet temperature of 900 °C, quickly followed in 1972 by the 51.8 MW Frame 7B. In the early 1970s GE entered into a joint venture with Alsthom to develop the Frame 9, 80.7 MW single shaft engine for 50 Hz applications, the first installed by EDF, Paris for peaker duty. Five additional models (frame 9B) were built in 1979 for the Dubai Aluminium Smelter.

Model E gas turbines started production in 1980 with ratings passing the 100 MW mark. The range of machines continued to develop and were widely used in the 1990s with inlet temperatures at the time of 1120 °C. In 1988 the 147 MW 'F7F' arrived with a pressure ratio of 13.5 and turbine inlet temperature of 1260 °C. A further step was then taken in 1991 with the 'FA', in the range of 175 MW and inlet temperatures reaching 1316 °C; the 50Hz version 9FA produced 240 MW.

The older GE B and E class gas turbines had been designed with simple cycle duty in mind, with a hot end drive which required an exhaust collector vertically upwards or sideways.

---

[216] See ASME, The 3500 kW Gas Turbine at the Schenectady Plant. The Belle Isle plant became the ASME 73rd National Historic Mechanical Engineering Landmark on 8 Nov. 1984.

[217] See Hendrickson, Gas Turbine Electric Powerplants. The TG180 had been flight tested as early as 1942.

[218] This date '1967' applies for a standard utility combined cycle power plant by GE, which ranks as no. 4 in the list of the first CCPP plants: No.1 1961 Korneuburg A, Austria 75 MW (BBC), no. 2 1963 Oklahoma Gas & Electric Co. Horseshoe Lake Powerplant OK, USA, no. 3 1965 Hohe Wand, Austria 12.8 MW (Siemens), see Hunt, The History of Industrial Gas Turbine, Part I, p. 16.
'Combined' or 'binary' cycle power generation installations had been constructed prior to Korneuburg A, mainly in the United States, notably GE's mercury/steam systems dating back to 1928 (the best known being the 40 MW Schiller plant in Portsmouth NH, USA, 1950) and there had also been conventional power plants using gas turbine exhaust for preheating boiler feedwater. See Eckardt, Celebrating 50 years of combined cycle

The following F class versions were already developed for combined cycle applications from the beginning, with a cold end drive to allow for an axial exhaust to the immediately coupled HRSG (Heat Recovery Steam Generator).

The largest combined cycle gas turbines in operation are the H class machines, manufactured by General Electric, Mitsubishi Heavy Industries and Siemens, typically with firing temperatures of 1,430–1,480 °C (2,600–2,700 F). Each increase in firing temperature yields a dual benefit of additional efficiency and increased specific work to the overall life cycle cost of the power plant.

At the forefront of H class gas turbine development, GE Energy has three installations worldwide (2010) at CCPP sites. GE's first H class installation began service in 2003[219] at Baglan Bay power station in Wales, UK. This 520 MW, 50 Hz 9H system has compiled in the meantime more than 36,000 OH and served as a technology validation site. Three other 50 Hz 9H gas turbines work at Tokyo Electric Power Co.'s (TEPCO) Futtsu[220] thermal power station 4. The first system went there into service in 2008. In the meantime the third unit in operation has brought the total H technology output at this TEPCO site to 1,520 MW. GE's first 60 Hz 7H system went online in 2008 at Inland Empire Energy Center in Riverside County, Ca., USA, accumulating more than 8,000 OH up until 2010. With the start-up of a second turbine the Inland CCPP has a maximum net rated electric output of 775 MW.

In general, the H systems are designed to achieve 60 and more percent net plant combined cycle efficiency. The three key components of the GE H-technology gas turbines are a) the closed-loop steam cooling of stages 1 and 2 of its four-stage turbine, b) an 18-stage, 560 kg/s axial compressor, deduced from the GE CF6-80C2 turbofan engine with an optimized PR=23 and c) the GE low emission DLN Drx-Low-NOx system adapted to the H conditions.

Siemens AG Due to early wrong management decisions towards the 'wet GT' concept (Humphrey, Stauber, Section 4.1.2.2), due to the economic situation in Germany during WW II and the special lack of high-temperature materials for non-military projects, Siemens was not in a position to realise planned gas turbine activities at the time. After further delays due to Allied restrictions after the war, it was only on 11 June 1956 that the first Siemens 1.5 MW gas turbine VM1 (Verbrennungs-Maschine – combustion engine) made its first successful test run.[221] In 1952 Friedrich refreshed his experiences with water-cooled

---

[219] At times, there had been speculation that a BBC patent on turbine steam cooling, as used in GE's H class for the first time in industrial production, could have delayed the market introduction of this concept during the 20 years of patent validity. US4424668, inventor Dilip Mukherjee, had Swiss priority beginning on 3 April 1981. The essential claim was '*In order to operate a gas turbine and steam turbine plant, the parts located in the hot-gas stream of the gas turbine being steam cooled, and the cooling steam, thereby raised to a higher temperature, being fed to the steam turbine for further expansion.*'

[220] Futtsu 4 is part of a larger TEPCO complex that produces more than 5 GW of power, making it one of the world's largest GT combined cycle installations. Besides the three H systems, the site employs 18 smaller GE gas turbines.

[221] See Leiste, Development of the Siemens Gas Turbine. This author also stresses the connection of Rudolf Friedrich (1909–1998), sometimes called the '*Father of the Siemens Gas Turbine*', to the German wartime turbojet development of the Junkers Motorenwerke Jumo 004. In fact Friedrich had belonged to Herbert Wagner's development team, which left Junkers AG as early as summer 1939, while the Jumo 004 compressor was clearly designed by W. Encke, AVA Goettingen, with a degree of reaction R=1. Wagner's R=0.5 compressor design of 1934, later further propagated by R. Friedrich, presumably followed the example of (BBC's) stationary gas turbine. Friedrich continued to work after the leave from Junkers, for two years at Heinkel, Rostock, before joining Brueckner-Kanis GmbH, Dresden, D, where he came in contact with developments of a 5,000 hp 'Drehkessel'

turbines under a contract from BASF Ludwigshafen, Germany, where the turbine rotor exhaust steam was used in the local chemical plant. This unit, later designated as VM2, had an output of 1.3 MW, with compressor inter-cooling and a 7-stage axial turbine. Originally equipped with ceramic vanes for a 1,000 °C operation temperature, these failed after only 8 rapid temperature cycles and had to be replaced by air-cooled metal vanes. The next reported project, VM3, was similar to the VM 1 and applied an uncooled turbine design again, for firing temperatures up to 650 °C. Due to an integrated exhaust gas recuperator, the 2.8 MW output power was achieved with 26 percent thermal efficiency. After some activities with 5.6 MW blast furnace gas turbines between 1956 and 1960, the evolution of the first advanced gas turbines began with the types VM80 and VM51.

The VM80 was the first large GT unit (mass flow 184 kg/s, PR=6, TIT=720 °C) for Siemens and with 23.4 MW output and 32 percent efficiency the world's largest single shaft gas turbine for a short while. The first plant went into operation in September 1962 as a peaker unit and for district heating at Stadtwerke (Municipal Power) Munich, Germany. This gas turbine still had a two shaft, 4 bearing concept, but immediately afterwards the 13 MW VM51 came along: the first single shaft, two bearing arrangement for the Siemens Hohe Wand CCPP, Austria in 1965.

In 1969, the high demand for development expenditure required a concentration of resources in Germany. AEG and Siemens merged their activities by founding KWU Kraft-Werk Union and Transformatoren AG, a business set-up which only briefly outlived the parallel BST activities in Switzerland. In 1977 Siemens bought out AEG and integrated KWU into its Power Engineering Group. GT production was concentrated at the former AEG factory at Berlin-Moabit, erected by Peter Behrens in 1909 as a masterpiece of advanced industrial design, while engineering R&D continued at Muelheim/Ruhr and Erlangen, Germany.

In the 1970s Siemens introduced their 50Hz V94.2 gas turbine in continuation of the VM51 concept, followed by the scaled 60Hz V84.2. After renaming, these models were designated SGT5-2000E and SGT6-2000E; a total of 240 units have been produced to date. The development steps were: in 1970: V93, 51 MW and V94, 86 MW; in 1977: V93.2, 148 MW; in 1985: V94.2, 148 MW; in 1994: V94.3, 213 MW and in 1996: V94.3A (annular combustor), 232 MW. The design was characterised by a 16-stage compressor, two large, side-mounted silo-combustors and a 4-stage turbine, of which the first two stages were air-cooled. The rotor design relies on a built-up disk structure, incorporating radial Hirth serrations and a central tie bolt. Over time the V94.2 output was raised to 163 MW and the simple cycle performance grew to 34.5 percent efficiency.

In the 1990s SGT5-4000F followed (ex-V94.3, 292 MW, 50 Hz), as well as SGT6-4000F (ex-V84.3, 187MW, 60 Hz, on customer request only) and SGT-1000F (ex-V64.3, 70 MW geared – for 50/60 Hz). In the meantime more than 120 SGT-1000F units have been produced. The

---

(rotating boiler) gas turbine unit according to proposals of Vorkauf and Huettner with a 13-stage, water-cooled axial turbine and 1,800 °C turbine inlet temperature. See Kay, German jet engine and gas turbine development, p. 176. The results were obviously very contradictory, so that in his book of 1949 (see Friedrich, Gasturbinen mit Gleichdruckverbrennung, p. 73) he came to the surprising conclusion: '*Works on water- and air-cooled turbines were not without some success in the past. Nevertheless, there is the impression that the water or air-cooling of combustion turbines will find no general application in future stationary units or in railway traffic.*' Friedrich worked for Siemens-Schuckertwerke AG, Muelheim/Ruhr, Germany between 1948 and 1964 and afterwards was a Professor for Thermal Turbomachinery at the TH Karlsruhe, Germany until 1976.

design of this GT family originates from a short-term collaboration between the aero engine manufacturer Pratt & Whitney and Siemens, thus representing state-of-the art aerodynamic, cooling and materials technologies of that time. The 15-stage axial compressor with PR=18.2 has CDA Controlled Diffusion Airfoils and endwall corrections on all stages. Optimised vane staggering allowed the number of VIGV to be reduced to just one row. The annular combustor incorporates 24 hybrid burners with either air-cooled metallic or ceramic tiles. The four stage turbine consists of PW1483 single crystal alloy for the first two blade rows. As a special feature a hydraulic tip clearance adaptation mechanism has been introduced here by moving the rotor upstream relative to the casing (also during operation) to increase the performance at base load operation.

In 1997 Siemens acquired the Westinghouse power generation business and moved the GT development center to Orlando, FL, USA. This merger survived two 60 Hz engine types:

- Of the former Westinghouse W501F, now SGT6-5000F, 200 units have been sold in the meantime. The latest model of this engine generates 208 MW at 38.7 percent simple cycle efficiency.
- The last pure Westinghouse development was the W501G in 1997, now SGT6-6000G; various improvements have increased its simple cycle power output from originally 230 MW to currently more than 260 MW.

As the first and only all-new GT project after the Westinghouse acquisition, Siemens introduced the SGT5-8000H into commercial service in 2011. The gas turbine produces 375 MW in the simple cycle mode at 40 percent efficiency, making it the most powerful gas turbine in the world at present.[222] The 13-stage axial compressor has an inlet mass flow of 800+ kg/s at PR = 19.2. It has an air-cooled 'can annular type', ULN Ultra-Low NOx combustor and the 4-stage turbine, in contrast to the more complex, steam-cooled GE H-configuration, is again exclusively air-cooled. The GT prototype was successfully tested at the E.ON power plant Irsching 4 near Ingolstadt, D. At first the gas turbine and its components were tested alone, and the CCPP installation only completed thereafter. A corresponding combined cycle test operation in May 2011 resulted in the new CCPP record value of 60.75 percent total efficiency, on the basis of an expected electric power output of more than 570 MW for the so-called SCC5-8000H combi plant. The completed CCPP unit Irsching 4 started commercial operation on 15 September 2011.[223]

The order book for the scaled 60 Hz, 274 MW version SGT6-8000H at the end of 2012 comprised a further 19 firmly ordered units, nine gas turbines for three FPL Florida Power & Light combined-cycle power plants, while the rest will be distributed worldwide, all these expected to become operable in 2013–2015.

The Japanese Companies Japan's first power generation gas turbine was installed in 1949 at the Maruzen Oil Co. Ltd. in Chiba, 40 km east of Tokyo. It was a 1.6 MW single shaft machine from Tokyo Shibaura Turbine Co., a predecessor of the current Toshiba

---

[222] The impressive dimensions are 13 m length, 5 m height and 440 t weight. Siemens invested approx. € 500 million in the SGT5-8000. See http://www.siemens.com/press/pool/de/events/media_summit_2006/krueger_speech_1388741.pdf and http://www.siemens.com/press/pool/en/events/media_summit_2006/mediasummit_2006_energy_krueger_1387413.pdf

[223] See http://de.wikipedia.org/wiki/SGT5-8000H

Corp. In the 1950s several Japanese turbine companies started prototype GT developments, among these Hitachi Ltd., IHI Ishikawa-Harima Industries, MHI Mitsubishi Heavy Industries, Mitsui Shipbuilding and Engineering Co. and Toshiba Corp. However, between the late 1950s and the beginning of the 1960s, more and more domestic gas turbine suppliers partnered with US and European manufacturers to provide simple-cycle industrial gas turbines for the home market. Technically, it was the transition period from first-generation non-cooled turbine blading to the second generation with forced air cooling. Most of the present high performance gas turbines now operating are improved and refined versions of these second generation products. The 'Moonlight Project' started in 1978 as Japan's approach to energy saving technologies and was maintained until 1993, with a comprehensive development programme for energy conservation technologies including gas turbines, fuel cells and heat pump systems. The Engineering Research Association of Advanced Gas Turbines was formed, comprising 6 national research institutes along with 14 accompanying companies, with the development target of a 100 MW gas turbine with 55 percent combined cycle efficiency.[224] This joint effort laid the ground for Japan's unique third generation of gas turbines.

The combined cycle era in Japan began around 1980, with immediate success due to the superior efficiency in comparison to conventional gas and steam plants. A range of new technologies (superalloys with improved material strength, the adoption of crystal formation control such as directional solidification and single crystal castings, advanced cooling technology in combination with blade coatings, continuing improvement of dry low-NOx combustion) paved the way for fourth generation gas turbines with significantly better performance values based on firing temperatures in the range of 1,300–1,500 °C.

A great deal of the Japanese progress in industrial GT technology was achieved by MHI and its partner companies. Traditionally, MHI had cooperated with Westinghouse Corp. since 1923, when it first entered into a licence agreement for electrical equipment. From 1965 (until Westinghouse were acquired by Siemens) a cross licencing, technology exchange agreement specifically for gas turbine technology was in place. Up until the mid-2000s MHI had supplied approx. 430 gas turbines worldwide. MHI is developing a new 'J series' gas turbine for the high-performance CCPP market. In May 2011 Mitsubishi Heavy Industries achieved a turbine inlet temperature of 1,600 °C on a 320 MW 'J class' gas turbine, 460 MW in gas turbine combined-cycle power generation applications in which gross thermal efficiency is claimed beyond 60+ percent.[225] The company even announced plans for increasing turbine inlet temperature up to 1,700 °C, which would bring CCPP efficiencies of 62–64 percent into reach, at CCPP powers of ~700 MW.[226]

---

[224] The development stages were: first, the development of a reheat gas turbine for a pilot plant (AGTJ-100A), and second, a prototype plant (AGTJ-100B). The AGTJ-100A had been undergoing performance tests since 1984 at the Sodegaura Power Station of the Tokyo Electric Power Co., Inc. (TEPCO). The inlet gas temperature of the high-pressure turbine of the AGTJ-100A was 1573 K, while that of the AGTJ-100B was 100 K higher. Therefore, various advanced technologies like a ceramic coating had to be applied to the AGTJ-100B htp, while steam blade cooling was applied for the IGSC (Integrated Gas-Steam Cycle).

[225] See http://www.mhi.co.jp/en/news/story/1105261435.html Class designation letter 'I' was skipped and 'J' stands for 'Japan'.

[226] After 'Fukushima', this would allow the potential re-powering of several nuclear plants by 2×700 MW blocks.

5.3 GT Key Component: Combustor

It may be coincidental that the first document in our historical GT documentation address-es combustion problems in a (coal-burning) gas turbine in 1910, Figure 5-53, but the fact underlines the importance of the subject for a successful realisation of a 'gas turbine' from the very beginning.

It is requested expertise feedback from Oswald Richter, BBC Mannheim to Eric Brown[227], BBC Baden, who had received the proposal for such a gas turbine from the newly em-ployed Walter G. Noack. Richter first confirmed Noack's principle statement: '*Combustion under elevated pressure vs. ambient pressure certainly has the advantage that the same space contains more oxygen molecules. It is therefore also likely that poor-quality fuels burn better and more completely under pressure than otherwise.*' The exact configuration of Noack's suggested gas turbine is not known, but since Stolze's 'fire turbine' is mentioned, Sec-tion 3.2 and Figure 3-13, it is likely that an externally fired combustion chamber was under consideration. Richter expresses concerns that the combustion volume may not shrink un-der pressurised operation as much as predicted according to Noack's favourite idea, which he then put into practice in the Velox boiler 20 years later.

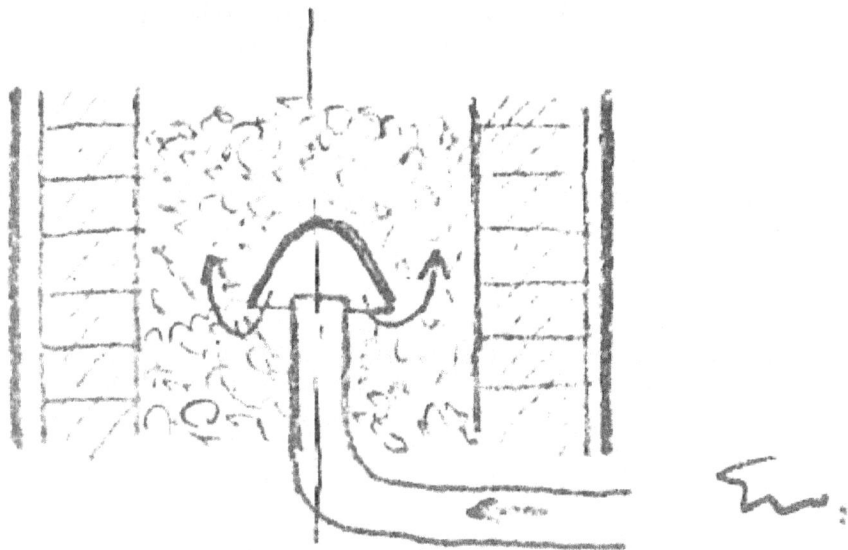

Figure 5-53   *BBC (coal) combustion technology: Sketch from a letter to Eric Brown (sig.), dated 14 June 1910* [228]

---

[227] Eric Brown, a cousin of company founder Charles E. L. Brown, was at that time Director of the BBC Turbine Factory at Baden. As orphans, Eric and Herbert Brown grew up with Charles E. L. Brown's parents at Win-terthur, and went also to school there. [Personal communication with Norbert Lang, see also Lang, Brown and Boveri]

[228] See BBC, Geschichte des Schweizer Turbomaschinenbaus

The other aspect of this early document is the inherent struggle with problematic fuels, which accompanied combustion development at BBC. Here the author suggests curing the threat of permanent clogging of the firing grate with a protected air feed line. Later combustion problems resulted from crude oil burning, and it was only recently that a preference for natural gas allowed a focus on clean, low-emission combustion.

5.3.1   Early Developments and Evolution of Combustor Arrangements

Astonishingly, the basic fluid dynamic phenomena in a cylindrical combustion chamber was clarified just in time for gas turbine applications[229] – and BBC immediately made use of these findings. As early as 1940 a combustion test facility with separate compressor for swirling burner testing was in place.[230] Burners of an optimised size were required for spraying fuel oil between 1 t/h to 10 t/h, with stable operation down to idle conditions.

The further development of the experimental combustion facilities is illustrated by Figure 5-54, which shows BBC's new combustion laboratory from 1964 with a few test rigs in use. A result of the combustor flow test rig is visualised in Figure 5-55, with measured combustor streamline pattern and the typical recirculation torus.

The silo combustor arrangement of Figure 5-55 is typical for a GT11B, Figure 5-23, from the 1970s, but it was very similar to established configurations dating back decades. It was mounted vertically beside the thermal block, as seen for the GT11B and GT13B, and horizontally for the GT9B. The basic elements were typical for all combustor arrangements up until then: one single burner, the cladding of the hot gas zone with metallic tiles and the general air/gas flow pattern. The compressor exit air enters from below and flows upwards through the outer annulus. The mass flow is split into 'primary air' which is used for fuel combustion, and 'secondary air' for cooling and mixing, thus limiting the flame length and equalising the temperature profile at the combustor outlet. In general, the combustor was designed for gas and oil firing, which allowed for simultaneous or separate burning of these fuels. The burner assembly was mounted at the centre of the top end of the combustion chamber and contained swirl vanes, an oil injection nozzle and control system, pilot and main gas burners, an ignition rod and the corresponding insertion and removal apparatus as well as two flame sensors. The lining of the combustor and the primary flame zone had five rows of metal tiles enclosed by a radiation shield. The tiles are unique features of this combustor design, Figure 5-56 (l); they are ribbed, rectangular plates which meet at the smooth inner gas path surface. The primary air passes between the gas path and radiation shield over the ribs and cools the gas path surfaces by convection and conduction, thus keeping metal temperatures below 850 °C. Originally the combustion side surfaces of the tiles had a welded deposition of chromium content to increase corrosion and oxidation resistance. Since the 1980s the tiles for GT8, GT11N and GT13E were manufactured as precision castings. The tiles had an inherently long life, but were individually replaceable if the need for repair developed. Access to the inner portions of the combustion chamber was gained by removing the burner assembly,

---

[229] See Meldau, Drallstroemung im Drehhohlraum 1935
[230] See BBM, Rueckblick... im Jahre 1940, 1941

so that a visual inspection and, if required, maintenance of the entire combustion system was possible, Figure 5-56 (r). Below the tile rows, in the secondary combustion area, was a formed sheet metal, double-walled liner with secondary air openings. Air therefore provided the dilution of the combustion gases to meet target design TIT values with sufficient uniformity prior to the circumferential flow distribution to the turbine itself.

Figure 5-54   *New combustion laboratory, 1964*[231] *1 – Conventional silo combustor with 2 – air pre-heater,*
3 – MHD high-temperature combustor with freely darting flame, 4 – combustor flow test rig
©ABB

---

[231]   Picture source: BBM, Forschung und Entwicklung im thermischen Maschinenbau, 1964, front page

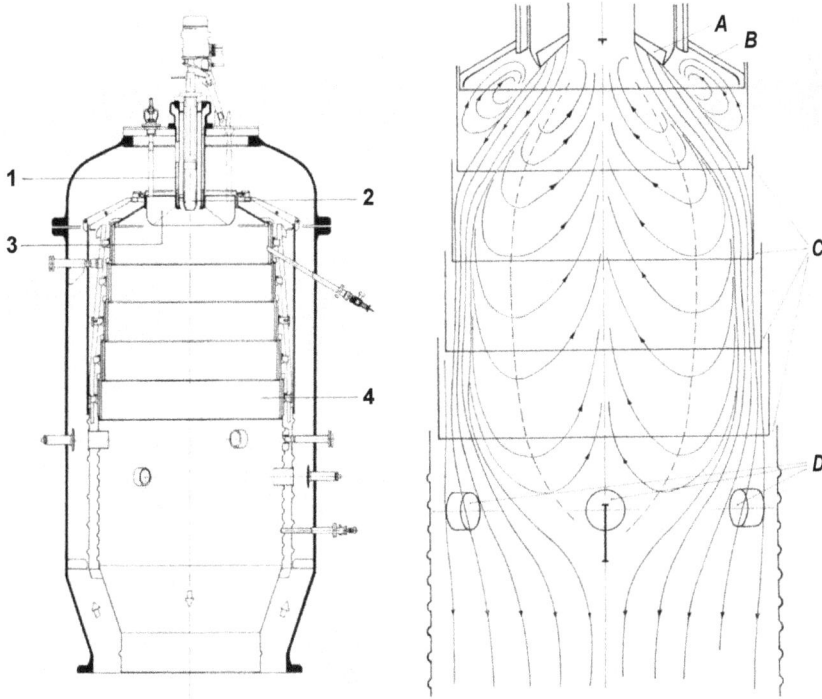

Figure 5-55   *Silo combustor arrangement and measured streamline pattern*[232] *1 – burner, 2 – fuel nozzle, 3 – main air swirler, 4 – ribbed combustor tiles, A – main air swirler entry, B – conus cooling air entry, C – combustor wall cooling, D – mixing air entry*   *©ABB (l)*

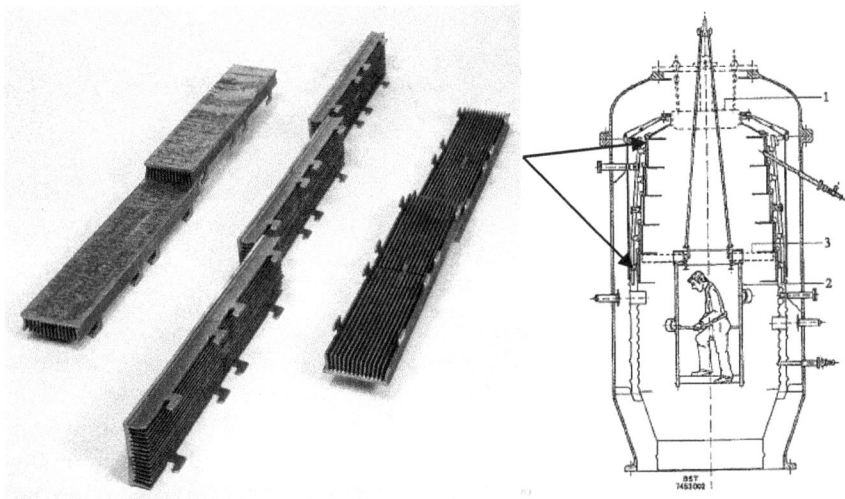

Figure 5-56   *Ribbed combustor tiles (l) and principle application of combustor inspection and maintenance basket (r)*[233]   *©ABB (l)*

---

[232] Figure source: BBM, Zur rechnerischen Erfassung der Stroemung in Gasturbinen-Brennkammern, 1964

Before any specifically targeted low NOx burner developments had been started, considerable progress in NOx reduction had already been demonstrated by a 2-stage reheat combustion process in a 290 MW CAES (Compressed Air Energy Storage) gas turbine plant, which BBC put into operation in November 1977 at Huntorf, Germany, see Section 5.5.3. A principle scheme of this sequential combustion concept is illustrated in Figure 5-57. There are several inherent advantages of this concept, which later led to its successful transfer to the GT24/GT26 engine family, here only the astonishing NOx reduction capacity of this process is outlined.

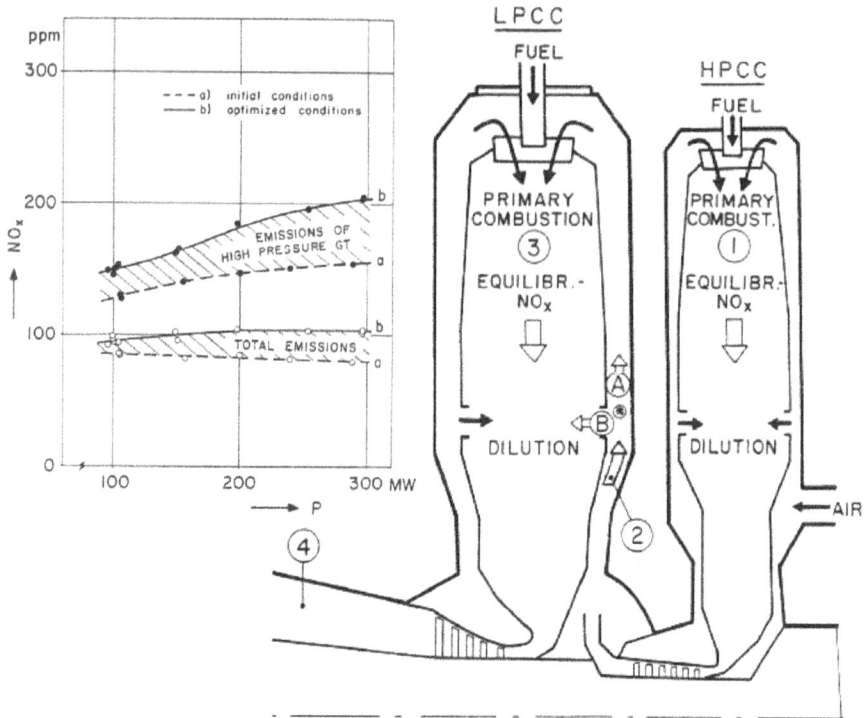

Figure 5-57   *Scheme of 2-stage combustion system and measured NOx values, volume fraction at 15 percent O₂ dry, in the exhaust gas of hp and lp turbine* [234]

With respect to early NOx reduction demands, water or steam injection into the flame was normal practice, which BBC's single combustion chamber with just one burner was ideally suited to. Depending on the required NOx values, the injected mass flow was in the order of the fuel flow with the extra benefit of increasing GT power output with higher water/steam addition. However, there were also severe disadvantages of this technique. Water use adversely affected the cycle efficiency and demineralised water and steam quality was not available in many cases. Therefore, the search for a satisfactory

---

[233] Source: (l) Endres, 40 Jahre Brown Boveri Gasturbinen, Figure 4 and (r) Hartmann, Inspection...

[234] Figure source: Koch, Application of the 2-stage combustion process

dry low NOx solution gained high priority. Though the combustion technology did not play the same role in GT development as it does today, the combustor design task was nevertheless equally complex. A major design target was to maintain stable operation with liquid and gaseous fuels over the whole load range and an acceptable hot gas profile. Secure ignition and sufficient burn-out were necessary to raise combustor loading and avoid both coking and overheating. Fuels ranged from gaseous (blast furnace gas, natural gas, LPG Liquid Petrol Gas) to liquid (diesel and fuel oil, heavy and crude oil). With regard to flue gas emissions the only concern was the abatement of visible smoke, Figure 5-14. In the 1970s the emissions of CO, UHC Unburned Hydro Carbons and especially the NOx Nitrogen Oxide emissions ('ozone hole') started to attract attention and triggered the first environmental regulations.

The 2-stage CAES combustion process takes place in the direction of flow after the previous, timely separated compression in a high-pressure HPCC and a LPCC (Low-Pressure Combustion Chamber), each of these followed by an axial expansion turbine, Figure 5-57. The air, compressed by cheap night energy into a 300,000 m^3 underground cavern, enters the first combustion chamber from there with a pressure of 40 bar and a temperature of 40 °C, prior to being heated up to about 550 °C. Following expansion in the hp turbine to 11 bar, the combustion gas at a temperature of 340 °C and an oxygen content of typically 17 percent enters the LPCC, where it is heated up to 825 °C. The total air mass flow is 420 kg/s and natural gas is used as heating fuel. As seen at point B of the LPCC, a split of the hpt exhaust gas takes place towards the LPCC primary zone and an immediate bypass to the secondary zone.

Two distinct combustion processes take place successively, each with its own fuel input. The generation of much less NOx (overall) is achieved in this case by extracting energy from the first stage combustion gases by allowing them to expand in the hpt. As a result, and due to the changed composition of the 'combustion air/gas', the flame temperature in the second stage and thus the amount of NOx produced is much lower. Furthermore, provided the second combustion zone is large enough, some of the NOx produced in the first stage is reduced to the lower level of the second stage.

Measurements in the Huntorf CAES plant and theoretical considerations[235] showed that the NOx formation in the second stage of combustion achieved only 36 percent of that in the first stage and that 25 percent of the first stage NOx was decomposed in the second stage. The inset graph of Figure 5-57 shows the values determined for the total NOx emissions, which were normalised to 15 percent O_2 in the dry exhaust gas of the lp turbine. In the course of the combustion optimisation, CO and UHC were reduced by a factor of about 2, at the expense of a slight rise in NOx. However, at full power the demonstrated overall NOx reduction potential is close to 50 percent, at that time a promising outlook for a process which in the future would be combined with effective NOx-reducing burners.

The first development strategy for dry (without water or steam injection) low NOx combustion was formulated at BBC in the 1970s. Right from the beginning it became clear that the past combustor technology with a single diffusional burner and a massive injection of secondary and mixing air (resulting in NOx values of 400 to 600 ppm, **parts NOx per mil-**

---

[235] See Koch, Application of the 2-stage combustion process

lion gas) had no dry, low NOx potential, excluding the possibility to achieve early target values of NOx < 75 ppm[236]. A 'dual column strategy' was formulated: one column was the development of new, lean premixed combustion and premix burner technology. The other column was to reduce cooling and mixing air consumption in the combustor in general, paving the way to annular combustor technology. The principal idea of this approach was to limit NOx generation in the combustor reaction zones by avoiding high flame temperatures. An immediate consequence was a generic change in burner/ combustor architecture and in the operation concept. The burners were now operating closer to the extinction limits, and stabilization of ignitable mixtures had to cope with new problem categories like mixing quality, flashback sensitivity, lean extinction limit and pulsation stability. The need to operate the combustor below the lean extinction temperature of a single premix burner led to distributed multi-burner systems with burner switching, staging and piloting concepts. The development of the combustor architecture focused on the other hand on improved liner cooling with impingement, film and convective cooling concepts; all of these technologies were first implemented in single-burner combustors.

The race towards higher gas turbine efficiency with increased firing temperatures led to a conflict with the simultaneous target to reduce NOx emissions. The consequence was to further reduce the amount of cooling air and to redirect the air as much as possible to the burner plane, in order to reduce the flame temperatures for a given hot gas temperature. The single combustor plus the hot gas casing, as the transition piece to the annulus at the turbine inlet, limited the cooling air reduction due to the relatively large hot gas surface. The solution to overcoming this drawback was the development of a compact annular combustor with low cooling air consumption and a distributed annular, multi-burner arrangement.

Figure 5-58 illustrates in an overview the decade-long development strategy for GT burners, sometimes with conflicting target settings, in power density from 20 up to 200 MW/m^2 and in emission perspective from 500 down to 10 ppm NOx.

The consequent application of the described fundamental development strategy culminated in several first time achievements:

–   In 1984, the world's first commercial lean premixed GT combustion with 32 ppm NOx in the GT13D at Lausward[237], Dusseldorf, D, Figure 5-24 (r)
–   in 1988, the world's first annular multi-burner combustor in a heavy gas turbine in the GT8B, Purmerend, NL, Figure 5-37 (l)
–   in 1995, the world's first reheat engine with a premixed, sequential annular combustor system in the GT24, Gilbert, NJ, USA, Figure 5-63.

---

[236] The US Clean Air Act of 1970, amended in 1977 and 1990, was the primary US federal environmental law applicable to power generation systems. The gas turbine New Source Performance Standards (NSPS) were issued in 1979. The NOx emissions of gas turbines in utility use with more than 100 BTU/h were limited at that time to 75 ppm at 15 percent O_2; however, more stringent regulations were already effective in California, for example.

[237] http://de.wikipedia.org/wiki/Kraftwerk_Lausward   See Marnet, NOx-Minderung

Single B.	Dry Low NOx Burners		
	1st generation	**2nd generation EV burners**	
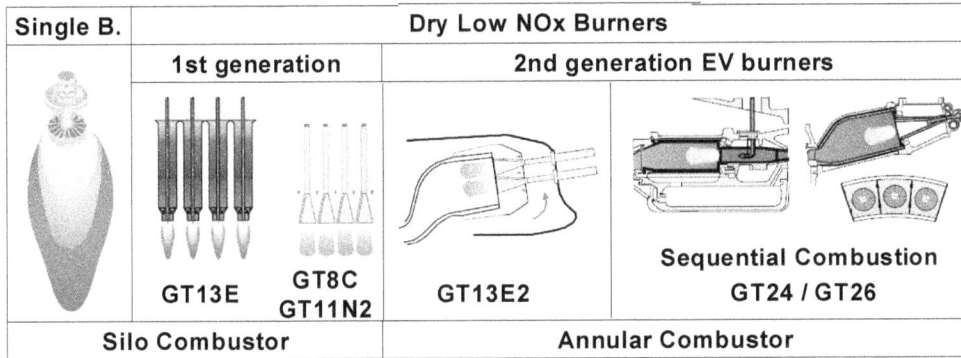	GT13E / GT8C GT11N2	GT13E2	Sequential Combustion GT24 / GT26
	Silo Combustor	**Annular Combustor**	

Figure 5-58   *Burner and combustor technology development strategy*[238]

5.3.2   First Generation Lean Dry, Low NOx Technology[239]

In 1978 a first BBC lean premix combustor concept was presented[240] based on theoretical insight that effective low NOx combustion required a) the separation of fuel/air mixing from the combustion process and b) combustion itself should take place under very lean conditions. The combustion chamber, Figure 5-59, was equipped with a great number of vertical, tube-type burner elements, each consisting first of a pre-mixing section in the top-down flow direction and at their lower end a swirler/ flame holder, achieving functional separation. A common combustion plenum guaranteed safe ignition and operation of the burners. In view of the strive for lowest NOx values, the burners were operated in a staged mode, allowing a load-variable combustion relatively close to the lean stability limit.

The cooling section of the original silo combustor flame tube was modified to reduce the amount of cooling air being adequate to the lower flame temperature, reflected by still only two rows of metallic tiles. Moreover, the secondary mixing air flow was reduced, which made up a significant portion in the former combustion arrangement. Most ambitious was the additional requirement to burn oil no. 2 in addition to standard natural gas fuel. The premixing section was therefore also the oil vapouriser and the corresponding tube size was primarily determined by the demand for a high degree of vapourisation. Under load, gas and oil were injected at the entrance of the premixing section and whenever necessary, a small amount of fuel was added in the centre of the swirler to improve the stability range. Full size tests with one burner element had to be carried out at engine conditions to obtain fundamental information about:

---

[238] Figure source: Doebbeling, 25 Years of BBC/ABB/Alstom lean premix combustion, Figure 2

[239] This and the structure of the following 'Combustion' section closely follows the historic ASME paper – see Doebbeling, 25 Years of BBC/ABB/Alstom lean premix combustion technologies.

[240] See Koch, The development of a dry low NOx combustion chamber. The concept, today also known as LPP Combustor (Lean Premixed Prevapourised combustion chamber) is an established possibility for low nitrogen oxide and soot emission operation. For idle and full-load operation with equivalence ratios, i.e. the actual fuel-air ratio referenced to the stoichiometric ratio, between 50 and 70 percent in the primary zone, the temperature level existing there is by about 600 °C lower than at stoichiometric combustion. The lean extinguishing limit lies at an equivalence ratio of < 40 percent. To guarantee a still stable combustion with lean mixtures, liquid fuels must already be vapourised and uniformly mixed with the air before entry into the combustion chamber.

Fuel supply

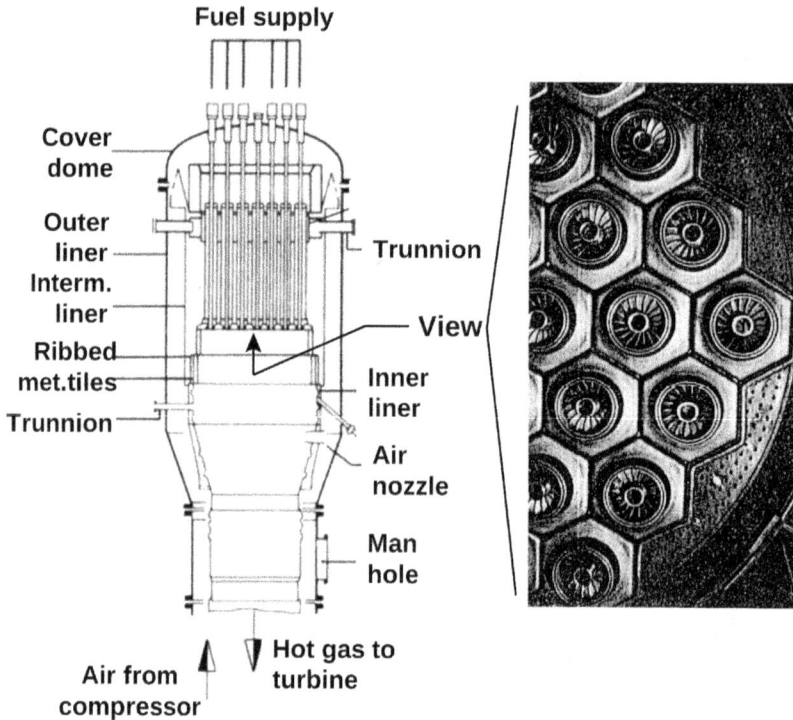

Figure 5-59   *First generation of dry, low NOx burners in a modified BBC silo combustor (l) and front view of hexagonal burner arrangement with swirlers and gas injection holes (r)* [241]   ©ABB (l)

− lean premix burning in general, stability limits and ignition behaviour
− NO, NO_2, UHC and CO emissions
− flashback safety for different types of flame holders
− autoignition, especially for oil burning
− oil vapourisation with different types of fuel injectors and under various load conditions, etc.

The majority of the tests were performed at GASL[242] (General Applied Science Laboratory, Inc.) in an excellent cooperation, so that the design of the engine combustor could be made in concurrence with the comprehensive test programme at the Baden offices. The first engine combustor was installed in an already existing GT13C in the municipal utility works

[241] Source: (l) Koch, The development ..., Figure 1 and (r) Jeffs, Asea Brown Boveri introducing...

[242] GASL is an American aerospace company, known as a pioneer of hypersonic propulsion. It was founded in 1956 by Antonio Ferri and became a developer and testing facility for advanced propulsion systems; Theodore von Kármán was another researcher there. Its expertise in hypersonic harsh environments allowed it to research and test materials and methods for extreme high temperatures as well as combustion systems relevant to current power generation and clean energy. Due to a limitation in pressurised air capacity, daily tests had to be accomplished within five minutes. Hans Koch led the testing either on the spot or by daily fax/phone communication; in parallel to the test activities the combustor design with 36 premix burners and all ancilliary systems was carried out. GASL, which was located at Westbury, NY, in the 1980s, is now based at Ronkonkoma, NY (Long Island), USA (75 km east of downtown Manhattan). http://en.wikipedia.org/wiki/General_Applied_Science_Laboratory

of Lausward, Dusseldorf, D in 1984, to become the mentioned < 32 ppm NOx record unit. The combustor consisted of a package of 36 burners. The injectors could be removed individually from outside. The fuel system, the control system and the air assist compressor were also new. Normally the burners were arranged in 5 groups for the staging schedule, with Group 1 in the centre always burning from the start and comprising the greatest number of burners, while the remaining groups were added step-by-step with increasing loads. Later on the first GT13E units were equipped with lean premix 'SBK (diffusion) burners' as well, Figure 5-30, before this type was modified by the introduction of a single-flame 'SBK Mark 2 (premix) burner'.

The commissioning of the new combustor concept went well; however, operation with fuel oil caused unexpected problems. After some time the swirlers stained with oil deposits from unvapourised oil of the premixing elements, leading to some swirler damage during subsequent gas operation. This problem could not be solved and therefore operation with oil no.2 was stopped.

Overall, this first generation of the DLN (Dry Low NOx) combustor was applied in seven GT units;[243] it proved the expected NOx reduction potential, but was considered as too complex and prone to deterioration after a while. Therefore, a second generation of lean premix burners was sought to overcome these disadvantages.

5.3.3   Multi-Injection Burner and GT8 Annular Combustor

Before discussing the further developments along the lean, premix combustion line, another technical approach for a low NOx burner shall be reviewed briefly due to its role in the realisation of the first compact, annular combustor at BBC. The idea behind the so called 'multi-injection burner' concept was to use the complete cross-section area of an annular combustor for a large number of small burners, thereby minimising flame length, residence time and consequently NOx formation. Short flames as well as a regular flow pattern allowed for the design of a relatively short combustor with reduced cooling air consumption (which again resulted in higher burner air flows). The burners were designed for normal diffusion flames.

There were 384 swirlers, each with a gas injector in the centre, Figure 5-37 (l), and with quenching air ports among the swirlers to reduce the high gas temperature immediately after burning. All burners were in operation over the whole load range of the gas turbine. The annular combustor was placed straight between the compressor and the turbine. The inner and outer annulus wall consisted of cooled tiles, arranged on a support carrier in 9 circumferential rows. In between, a cooling film was generated with a relatively small amount of cooling air.

Several GT8 gas turbines with an output of 45 MW were equipped with this type of combustor, the first one at Purmerend, NL in 1987. The NOx emission level at full load was approximately 70 ppm (at 15 percent O_2). This result showed that even with high burner air flow the combustion took place at near stoichiometric conditions and short residence time

---

[243] First generation dry low-NOx combustors ran (besides the Killingholme 13E) at Hemweg 13E (6 years), MK12 Pegus 13E (~11 yrs.), Lageweide 11D5 (~19 yrs.), Lausward A and B, 13C and 13B (up to now 25 yrs.), Korneuburg 13D (~17 yrs.). Information courtesy of H. Koch.

did not sufficiently compensate for the influence of high temperature on NOx formation. Consequently, it became clear that further improvements in lean premix combustion were required, for compact annular combustor applications as well.

5.3.4   Second Generation EV Premix Burner Technology

The first generation premix burners had to be excluded from further developments for future, more efficient gas turbines, as higher air inlet pressures and temperatures combined with a higher flame temperature would automatically increase the perceived autoignition and flashback sensitivity. The new, second generation of premix burners, so-called EV (EnVironmental) burners, diverged from established design philosophies for diffusional and premix burners, especially the dangerous mixing zones upstream of the swirlers were given up and fundamental principles of vortical flows applied instead, generating safe 'open premix zones'. The vortex breakdown of a strongly swirling core flow is used as an aerodynamic control for the inner recirculation zone, needed to stabilise a premix flame, without an auxiliary swirler or centrebody in a region with ignitable fuel mixture, Figure 5-60.

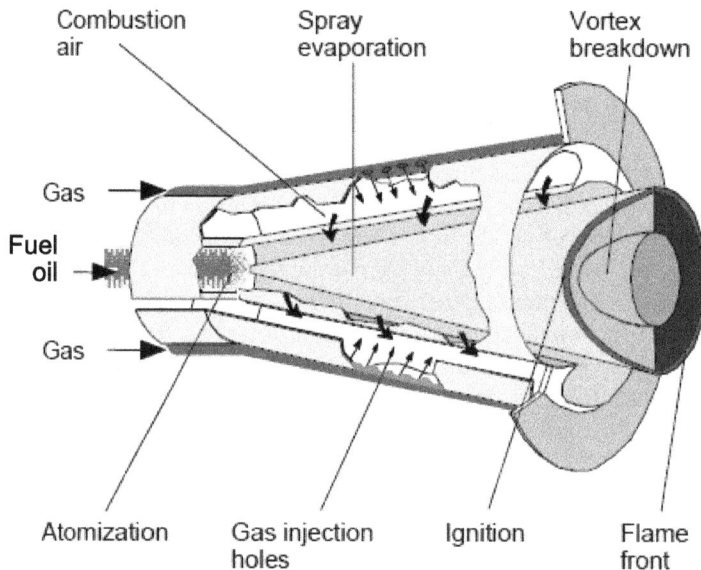

Figure 5-60   *EV burner principle* [244]

The EV burner consists of two half-cone shells, displaced parallel to the cone axis, so that two tangential slots are generated. The swirl strength of the flow entering through the tangential slots increases in the axial direction and is adjusted so that a vortex breakdown of the core flow occurs close to the burner exit. Since the EV burner requires no specific hardware for flame stabilisation, it has an inherent safety mechanism against autoignition and flame flashback.

---

[244]   Figure source: Doebbeling, 25 years of BBC/ABB/Alstom lean premix combustion, Figure 7

The EV burner and its inherent advantages represent a stroke of genius, comparable to the greatest inventions in the company's long history, combining fluid dynamic simplicity with reliable effectivity to achieve low emissions. Figure 5-61 illustrates the first patent ideas, filed in 1985[245] in the names of Jaan Hellat and Jakob J. Keller, which were simplified further in due course (as seen in comparison to Figure 5-60). As described in the patent, *'the object of the invention is fundamentally attained when slotted cones with appropriate semi-cone angles are provided. This provides an optimum possibility of combining the advantages of a free vortex tube with a swirler which is perfect from a fluid mechanics point of view. In this case, a vortex flow is obtained which has little swirl and an excess of axial velocity in the centre. Since the swirl speed of this burner increases strongly in the axial direction and reaches the breakdown or critical value at the end of the burner, a stable-positioned vortex return flow zone is produced.'* Expected difficulties that flow instabilities (due to compressor stalling, for example) may cause a critical flashback motivated this attempt to stabilise the flame fronts securely in the interior of a flow field, at a sufficient distance from the fuel injector. After comprehensive research it was concluded that the only element for this kind of free flame stabilisation was a bubble-type vortex breakdown structure or a vortex ring, as illustrated in Figure 5-61; Re_D denotes the Reynolds number with respect to the vortex tube diameter.[246]

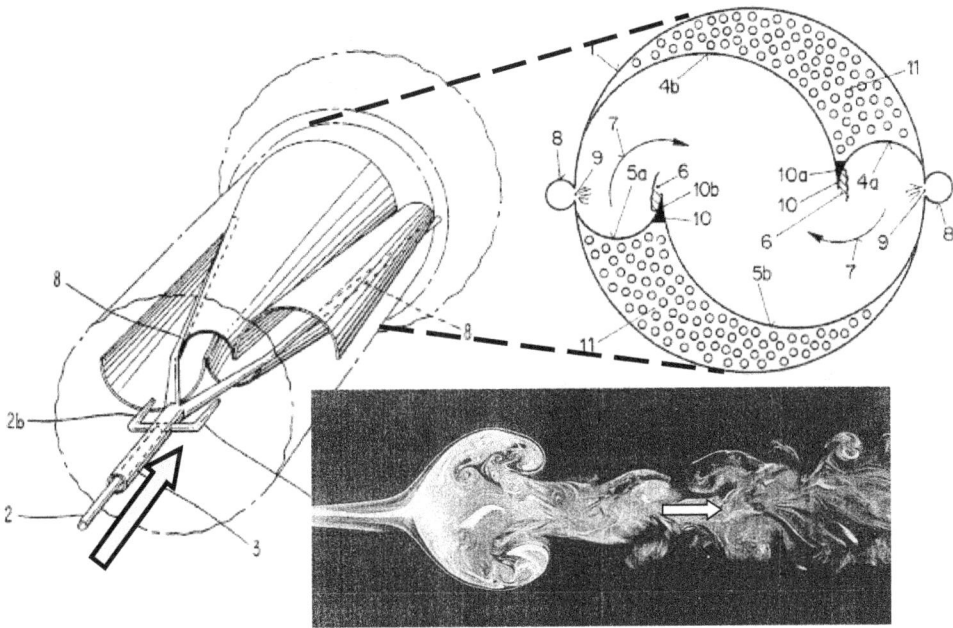

Figure 5-61   *Drawings of EV burner patent by J. Hellat and J.J. Keller, 1985, and vortex breakdown visualisation at Re_D ~3,300*

---

[245] The patent priority began on 30 July 1985, resulting in 'Dual Burner' patents CA1286886, EP0210462, IN167458, JP63038812, JP6103085 and US4781030. Inexplicably, the patent application was limited to the then urgent problem of 'liquid fuels' only. The first EV Burner in the known shape of Figure 5-60 was patented subsequently as EP0321809 etc. already in 15 countries/regions for Jakob J. Keller, Thomas Sattelmayer and Daniel Styner, with priority beginning on 21 Dec 1987 – again for liquid fuel only and it required 2 more years until finally in 1989 the wording of the patent EP0433790 (J. J. Keller) covered both liquid and gaseous fuels.

[246] See Keller, Double-Cone Burners..., Figure 1

As indicated in the calculated EV burner axial flow field of Figure 5-62, three aerodynamic features are important for stabilising the premix flow and the generation of flashback safety:

- Upstream of the vortex breakdown the core flow is strongly accelerated; a 'Jet-like Core Flow' forms a natural flashback barrier and prevents any flame inside the burner
- the 'vortex breakdown' generates a sudden transition flow pattern, thus fixing the flame root position
- downstream of the vortex breakdown a 'central recirculation zone' occurs, ensuring premix flame stabilisation near the burner exit.

The second important feature of the EV burner concerns the compact and effective generation of the fuel-air mixture. As shown in Figure 5-60 the alternative gaseous fuel is injected through a row of holes in a cross flow direction into the air, entering the tangential burner slots. Each gas injection jet has to penetrate a small portion of air and therefore only has to mix locally. Since the feeding holes are equally distributed along the slot length, the mixing boundary conditions are also equal and consequently, the gas-to-air mixing performance is equally spaced within the whole burner flow field. The gas mixing concept is the precondition for high mixing effectiveness which leads to a homogeneous air-to-fuel mixture, even in view of the compact mixing volume. This air-fuel mixing homogeneity within a relatively small cone volume is an essential precondition for the achievement of ultra-low NOx values.

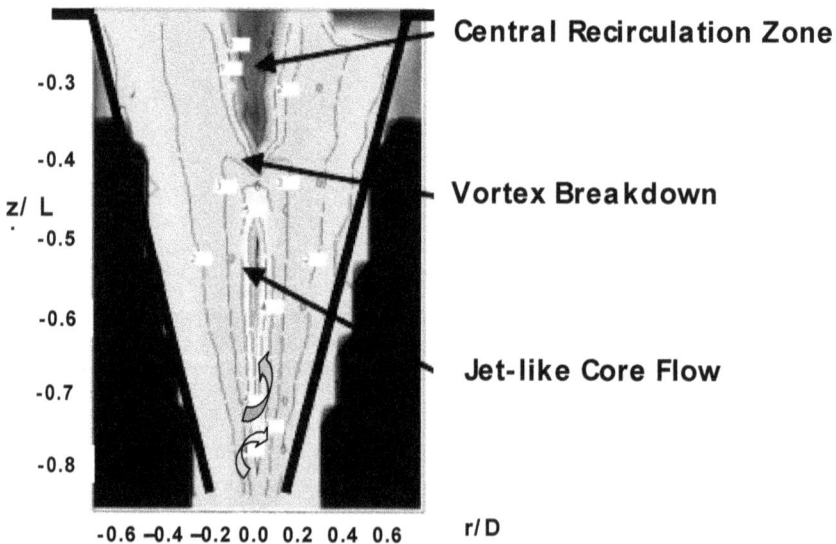

Figure 5-62   *Calculated EV burner axial flow field with prime flow features* [247]

In case, the liquid fuel is injected with a central plain jet nozzle in the cone head, Figure 5-60. The liquid plain jet disintegrates into small droplets within the burner leading cone section and the swirling air flow distributes the droplets throughout the whole flow field. Then the inner recirculation zone establishes the flame stabilisation near the burner outlet.

[247] Figure source: Doebbeling, 25 years of BBC/ABB/Alstom lean premix combustion, Figure 8

An additional pilot gas injection has been developed for annular combustor applications so that flame stabilisation even well below premix extinction limits is ensured. The pilot lance is in the centre of the burner cone and enriches the mixture of the core flow; in this pilot operation mode the flame is also stabilised by the inner recirculation zone.

The first time implementation of the EV burners was in the single combustor of a GT11N at Midland, Mi., USA in 1993, Figures 5-26 and 5-27, where 36 burners and an additional central ignition burner were placed in a hexagonal, three ring arrangement. The combustor was loaded by switching single burners on and off. The same principle of a hexagonal arrangement was later used for the single combustors of the GT13, GT9, GT8 and GT11N2 engines. The EV burner with central, inner pilot gas injection was introduced for the first time in the GT13E2 and GT10[248], and later in the GT8C2, GT24 and GT26. Today the whole Alstom GT fleet is available with a low NOx combustion system based on the unique EV burners.

During the comprehensive fleet implementation, the EV burner design was continuously improved for a higher reliability and longer lifetime, and at the same time the manufacturing process was considerably simplified from welded to a completely cast configuration at present. In the meantime the actual burner is a standard size cast body design, which is used for the annular and single EV combustors of all production engines. In addition, an effective and simple injection design has also extended EV burner applications for MBTU fuels.[249] Engine power is adapted by the number of installed burners; the standardisation has clear advantages with respect to an uniform fleet performance, but also from a logistics point of view.

5.3.5   SEV Burner Development for GT24/GT26 Applications

An important technology feature of Alstom's advanced GT24/GT26 engines, Section 6.1.3, is the sequential combustion system. The compressed air is heated in a first row of EV combustors, Figure 5-63. After the addition of about 50 percent fuel at full load, the combustion gas expands through the first, hp single turbine stage. The hpt lowers the pressure from approximately 30 to 15 bar.

The remaining fuel is added in a second, Sequential EV (SEV) combustor, where again the gas is heated to maximum turbine inlet temperature, Figure 5-64. The reheat cycle ends with expansion in the 4-stage lp turbine. Figure 2-21 already showed a comparison of the sequential reheat combustion process with a conventional GT cycle, clearly indicating that the sequential combustion cycle is characterised by a lower turbine inlet temperature for the same power output.

Sequential combustion is not new in the history of power generation, but BBC and other following companies had a strong preference to this superior GT cycle from the beginning.

---

[248] GT 10 combustor adaptation work was carried out at CTEC (Combustion Technology Centre), Burnley UK (former Lucas Industries Ltd.) in the winter of 1990/1 using a combustor segment with 4 burners (the GT 10 has 18) and the first production gas turbines with the EV burner system were commissioned at district heating plants in southern Sweden at the end of 1992, where NOx levels were observed in commercial operation of 15 ppm at 15 % oxygen with dry gas: an unprecedented achievement in a gas turbine of this size. See Turbomachinery International, 50 years of combustion research at CTEC

[249] See Doebbeling, 25 years of BBC/ABB/Alstom lean premix combustion technologies, p. 7, and – see Doebbeling, Low NOx premixed combustion of MBTU fuels

As outlined in Section 4.3.3.2, the great Aurel Stodola had already mentioned 'fractionalised combustion' as a promising measure to increase the thermodynamic efficiency of the GT cycle.[250] During the 1950s/1960s BBC was already delivering 24 plants with 2-stage combustion.

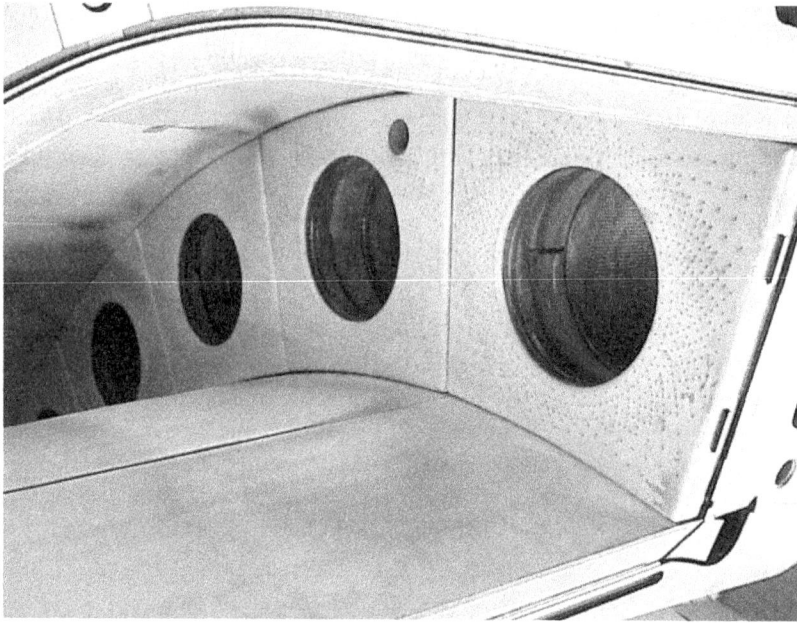

Figure 5-63   *GT24/GT26 EV combustor plenum* [251]

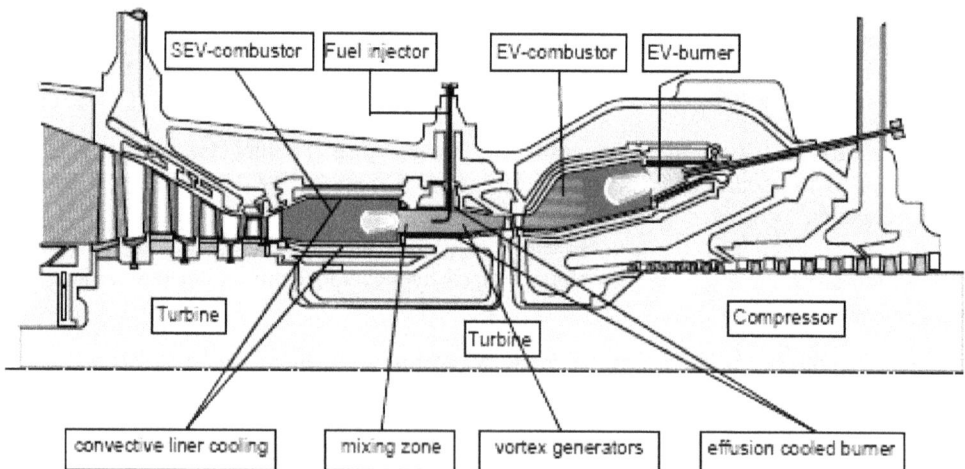

Figure 5-64   *Principle set up of GT24/GT26 sequential combustion* [252]

---

[250]   See Stodola, Load tests... Engineering. Stodola also first referred to the 2-stage combustion, reheat cycle by G. Mangold, VDI-Z, vol. 81, (1937) pp. 489–493.

[251]   Figure source: Doebbeling, 25 years..., Figure 23

In a conventional lean premix combustor, like the first EV combustor in the sequential combustion arrangement, spontaneous ignition must be avoided, as it could lead to overheating of the combustor components and to unacceptably high pollutant emissions. However, a reheat combustion system like the SEV combustion chamber can be designed to use the self-ignition effect deliberately for a simple and reliable construction.[253] The SEV annular combustor with self-ignition can be indirectly compared with the afterburner section of a military fighter aircraft, but in a low emission, premixed combustion mode. Combustor inlet temperatures of more than 1,000 °C were selected for the GT24/GT26 SEV combustor over the whole operation range in order to achieve reliable, spontaneous ignition with natural gas and to enlarge the stable operation range. Besides self-ignition, low emission values are the prime demand for a successful SEV operation; consequently, the second stage fuel and the hpt exhaust gas must be mixed completely before ignition. Otherwise, burning will occur in fuel-enriched zones only and locally high flame temperatures will cause high NOx generation rates. An optimum relationship between ignition delay and degree of premixing is therefore indispensable. In principle, the ignition delay has to be kept short to ensure self-ignition and to limit the general combustor size. Furthermore, this optimum should be maintained over widely varying fuel flow rates, which change under combustor load, and depending on fuel conditions, for example as a result of burning different blends of natural gas.

In the GT24/GT26 SEV combustor, the mentioned optimum is reached by using carrier air, i.e. secondary compressor air, which is injected into the SEV burner together with the fuel. The carrier air works both as a premixing enhancer by maintaining the fuel jet momentum, a factor critical to achieve high premixing quality, and as an ignition controller.

As in the EV burner, fuel distribution and mixing within the SEV burner is achieved with the aid of vortical flow. The flame is anchored at the vortex breakdown position. The vortices are generated by delta wings, formed like ramps and located oppositely on the four walls of the SEV burner duct, Figure 5-65.[254] The requirements of low drops in pressure and high safety against flames in the mixing zone led to these delta-wing type vortex generators, which can generate stream-wise vortices without any recirculation zones. A 'tetrahedral geometry' has been chosen to satisfy other practical considerations such as mechanical integrity and cooling.

---

[252] Figure source: Doebbeling, 25 years..., Figure 12

[253] See Joos, Development of the sequential combustion system. ASME Award 'Best Paper 1996'

[254] See Eroglu, Vortex generators in lean-premix combustion. After determining the principle SEV concept, the actual configuration selection only happened after some hardship. The basic idea of delta-wing shaped vortex generators (VG) was created through intensive R&D work at the ABB Research Labs. The specific geometry and the special arrangement with four VG's pointing into a central lance came about as solution #49 in Dec.1992 after numerous studies conducted with water tunnel, CFD, atmospheric and hp combustion tests. The SEV principle, fuel lances and operation are along with (at present, 2012) 50 patents part of the best IPR-covered areas. The first patent priority in this context dates back to 8 April 1993, US 5,518,311 by R. Althaus; A. Beeck; Y.-P. Chyou and A. Eroglu 'Mixing chamber with vortex generators for flowing gases' and US 5,513,982 by R. Althaus; A. Beeck; Y.-P. Chyou; A. Eroglu and B. Schulte-Werning 'Combustion chamber'.

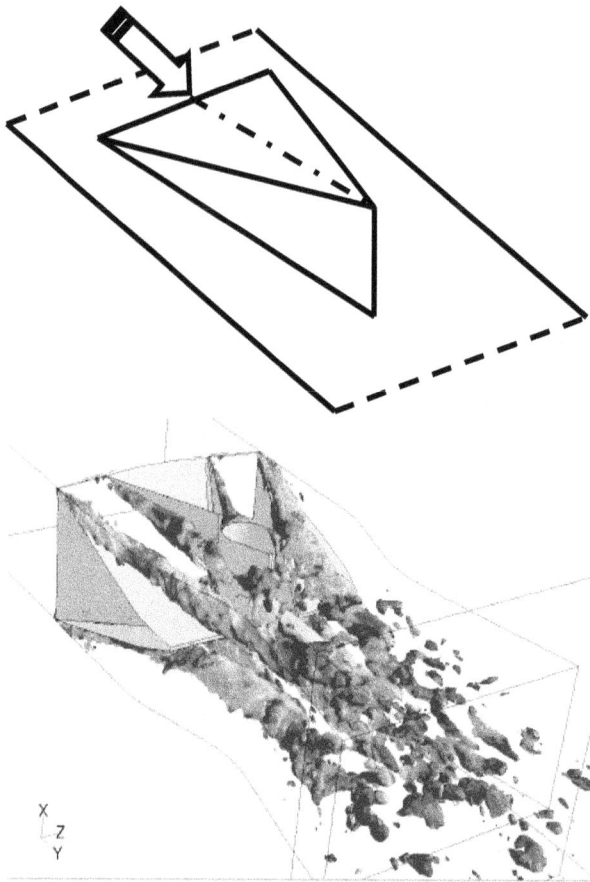

Figure 5-65   *Delta wing vortex generator and SEV mixing calculation* [255]

The principal arrangement of the second SEV combustor together with the fuel lance in the sequential combustion system was already illustrated in Figure 5-64. The fuel is injected through four holes, together with the carrier air from the fuel lance downstream of the delta wing vortex generators. Figure 5-66 shows an upstream view into a cast SEV burner duct with intense wall cooling and the correspondingly calculated vortical flow pattern in a normal plain downstream of the delta wing vortex generators. Two pairs of dominant vortices are generated so that each fuel jet is injected into one of the vortices.

After the assembly of a GT24 gas turbine for the first time in June 1995 on the premises of ABB Richmond, Va., USA, the initial implementation took place at Gilbert, NJ, USA at the end of 1995 and the corresponding GT26, 50 Hz version was launched in September 1997 at

---

[255] Picture source: courtesy of Peter Flohr. The used method LES, 'Large Eddy Simulation', is capable of reproducing the highly vortical flow much more accurately than standard time-averaged turbulence models (such as k-epsilon or Reynolds stress models). As a consequence, the location of stagnation points and recirculation zones, the fuel-air mixing process and flame front locations can be predicted.

EnBW Karlsruhe, D after a thorough test evaluation at the ABB/Alstom GT26 power plant at Birr, AG, CH. In the meantime, more than 80 GT24/GT26 engines are in commercial operation and achieved more than 4 million fired hours by the end of 2011. The GT24/GT26 engines are capable of running on natural gas and on oil with very low emissions, high availability, and especially attractive part-load operational parameters (efficiency, NOx < 10 ppm).

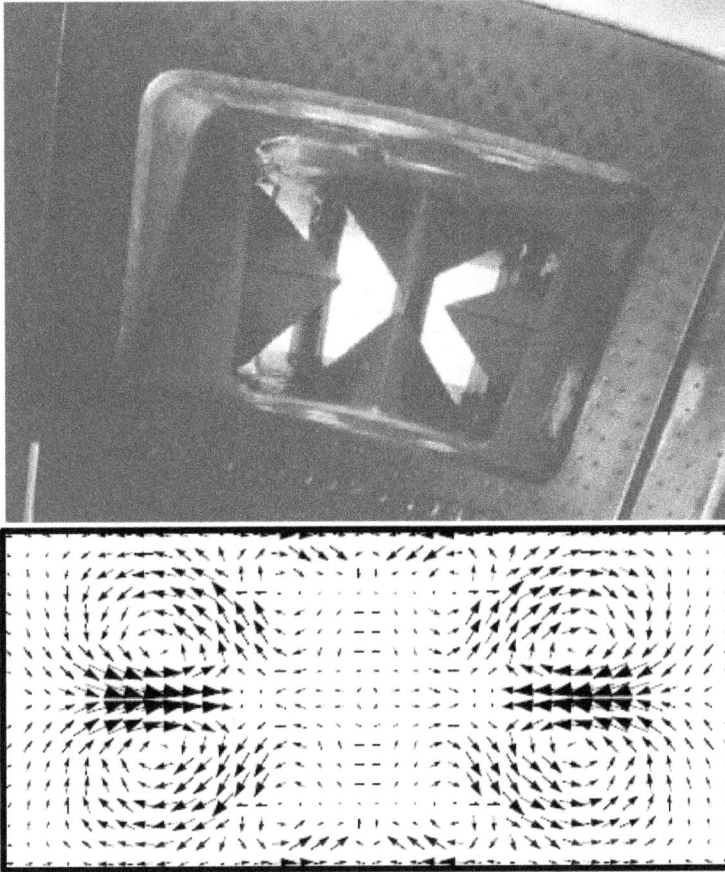

Figure 5-66   *Upstream view into an SEV burner duct (with fuel lance removed) and calculated vortical flow pattern downstream of the delta wings* [256]

5.3.6   Further Low Emission Combustion Development

An **AEV (Advanced EV) burner** was configured with special emphasis on liquid fuel application. The distribution of the radial inlet flow is made more even by using four inlet slots instead of two, so that the swirler becomes completely protected against wall impingement of fuel droplets. As illustrated in Figure 5-67, the addition of a mixing tube allows for a longer evaporation time and additional mixing in the gas phase.

---

[256] Source: Eroglu, Vortex generators...

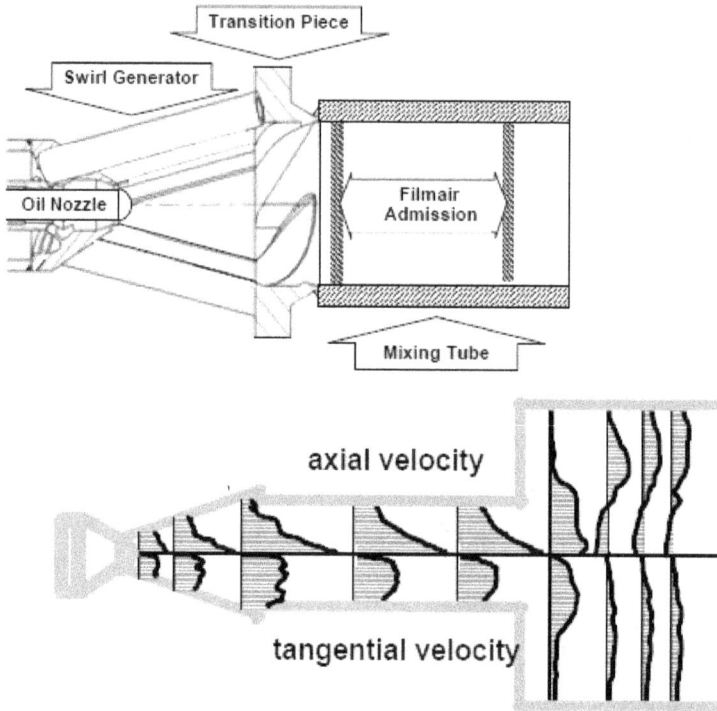

Figure 5-67   *Design features of the AEV burner with development of velocity profiles* [257]

The axial and tangential velocity distributions clearly indicate that a jet-like flow field with a distinctly small body vortex core is rapidly formed within the conical swirl generator. At the exit of the swirler, the maximum axial flow velocities on the centre line are twice as high as the bulk velocity. This high velocity jet flow is maintained throughout the mixing tube, providing high safety against flashback. High axial velocities can be achieved by the proper admission of small amounts of additional air (< 10 percent of total air flow). In combination with the dilution effect of this air addition, flashback along the mixing tube walls can be suppressed very effectively.

The AEV burner was developed at the ABB Research Centre in Daettwil, AG, Switzerland and it was extensively tested in both atmospheric and full pressure test rigs. At present the burner is available for the Swedish-developed GTX100 and also for the GT13E2; it meets the emission targets for both oil (35 ppmv) and natural gas (15 ppmv).

The **Annular Premix Combustor** development started at Alstom as a complementary activity to the premix burner development in general. The annular combustor concept was chosen in view of the reduced combustor surface exposed to hot gas and the resulting lower cooling demands. It implies the distribution of a multitude of individual burners across the annulus area. This requirement is in line with the generic premix combustor operation concept, based on a multi burner arrangement. The main advantages of this concept are:

---

[257] Figure source: Doebbeling, 25 years..., Figures 20 and 21

- Cross-ignition from burner to burner (without cross firing tubes)
- Part- and full-load operation with burner groups running at different flame temperatures or even with some burners switched off. Burners below extinction are stabilised
  by neighbours running above extinction, resulting in a complete burnout at the combustor exit and low NOx emissions
- The annular combustor generates a uniform hot gas temperature distribution at the
  turbine inlet.

The arrangement of the EV burners and their swirl direction must take into account the
induced tangential flows in the combustor annulus. Smart arrangements allow for substantial improvement of the cross ignition potential, the part-load stability, the mixing performance and the turbine inlet temperature profile. These were the main considerations for
the arrangement of 72 EV burners in the first annular premix combustor of the GT13E2,
Figure 5-33. Here, the EV burners are equally distributed over the circumference in a staggered two-row arrangement. Later, annular premix combustors were also realised for the
GT24, GT26 and GT8C2 with a simple, one-row arrangement, Figure 5-63.

The compactness of the annular combustor also enables the realisation of a closed-loop
combustor liner cooling system, where all liner cooling air is fed back to the compressor
main stream. The EV burners are separated from the combustor casing by an additional
hood, see Figures 2-18 and 5.64, to achieve this closed-loop liner concept. The precondition
is a small drop in burner pressure, as for the EV burner; the sum of cooling pressure loss
and burner pressure loss represents the total combustor pressure loss.

The development of advanced premixed combustion systems has been a key success factor
for the present generation of Alstom's GT products with very low emission values, and
high operational reliability and flexibility at the same time. The modular design of the annular premix combustor with EV and SEV burners has led to cost-competitive and reliable
combustor systems for highly efficient power generation gas turbines.

5.4 BBC Gas Turbine Mechanical Design and Manufacturing

5.4.1 General Gas Turbine Design Principles

Over the decades the mechanical design of the BBC/ABB/Alstom gas turbines developed an
independent, characteristic design philosophy, relying on:

- steady evolution of proven designs
- use of geometric scaling
- thorough pre-production testing.

In view of proven design elements there are numerous examples, the best known are the
unique welded rotor design, which is dealt with in the following section, EV/SEV combustors and combustor tiles which were described in the previous combustion section, special
patented blade root attachment techniques, Section 5.4.2, etc. The use of 50/60 Hz scaling
had already become a standard BBC design procedure in the late 1960s and has stayed so

since then, with the described inherent advantages of a GT family concept, Section 5.1.5.1 and 5.1.5.3.

As indicated in the individual GT component sections, thorough and comprehensive development testing was part of BBC's research and development activities from their beginnings in the 1920s, underlined in the following decades by the continuous adaptation of test facilities and special component test centres like the one illustrated in Section 5.3.1 for combustor development, and not least by the Corporate Research Centre, founded in Daettwil, AG, Switzerland by BBC in 1966 and maintained by ABB and Alstom since then. In 1998 a dedicated Gas Turbine power plant was inaugurated at Birr, AG, CH: first as an open cycle, single shaft GT26 test facility with attached generator and then modified in 2010 to simulate CCPP operation by means of an extension (with air preheater, once-through hp and lp coolers for cooling air cooling and dump condenser) to simulate CCPP performance for the newest, upgraded GT26, Section 6.3.2. Without a doubt Wilhelm Endres, Section 5.1.6, has contributed the most to developing GT mechanical design from an art to a science-based engineering discipline. For more than 20 years he wrote and published numerous seminal papers in this context with this purpose in mind, for external and internal use, not to be underestimated for continuity in engineering education in view of the long-term maintenance of proven design principles .[258] He typically deduces the system of GT design principles and rules from the requirements and problems for hot gas path components, i.e.

–   Gas temperature and temperature effects (*'The material temperature which the designer permits will naturally depend on the stresses present'.*):
    –   Expansion (*'Every part should be designed to expand freely under all operating conditions.'*)
    –   Strength/ductility (*'Interactions of temperature and stresses will cause creep and a reduction in time-rupture strength.'*)
    –   Fatigue strength (*'Both high and low cycle fatigue strength are affected by temperature; if the temperature is high enough to cause creep, then fatigue and creep cannot be considered separately.'*)

    'As a consequence of these phenomena, all high-temperature parts can only be designed for a limited operation time and a limited number of operating cycles.'

–   Forces and stresses :
    –   Gas forces, caused by pressure differences on the surface of a component
    –   Inertial forces by rotation and vibrations
    –   Impact of foreign objects
    –   Forces due to thermal expansion
    –   Stresses due to steady external or inertial forces
    –   Stresses from thermal gradient strains
    –   Stresses by transitory operational temperature gradients
    –   Forces and stresses in airfoil attachments

–   Corrosion and erosion

---

[258]   See Endres, Design principles of gas turbines, and – see Endres, Erfahrungen mit Gasturbinen

The counter-measures of a gas turbine designer to make hot gas components last under the described conditions according to Endres' experiences in 1973 were:

- Cooling (*'The rapid decrease in strength with increasing temperature ... is best defeated by cooling.'*)

- Choice of shape (*'This is the designer's main tool to achieve a successful design.'*):
    - Clear separation of functions
    - Free expansion of every part
    - No stress raisers and no notches
    - Minimal thermal stress, steady as well as transitory
    - No hot attachments
    - Ease of manufacture, preferably precision casting
    - Minimum machining of high-temperature alloys
    - Minimum cost

- In view of new materials, Endres then states:
    - Creep rupture strength (*'...Testing over 5,000 h before the alloy is seriously considered and over 10,000 h before it would go into a gas turbine.'*)
    - High corrosion resistance
    - High and low cycle fatigue well characterised
    - Reliable coatings available
    - Commercial availability without limitations

As an example of the mechanical design development of the mid-1960s, Figure 5-68 illustrates photoelastic stress studies on the root sections of a turbine blade, especially for BBC's typical fir-tree design.

5.4.2   The Welded Rotor Design and Manufacturing

All present power generation gas turbines follow the same basic design concepts: the compressor and turbine have one common rotor which is supported by two journal bearings. Thrust bearing and electric generator coupling are on the cold compressor end. The internal rotor design, however, varies widely among the different OEM manufacturers:

- rotor sections/discs welded together at the rim (BBC/ABB/Alstom)
- discs axially joined by one or several tie-rods and in parts, by additional (curvic) couplings (GE, Siemens, MHI).[259]

---

[259] The comparison of individual rotor designs corresponding to these concepts with respect to disc strength and lifetime showed no clear advantage for one specific design solution; advantages and disadvantages were observed for every design feature. See Endres, Rotor design for large industrial gas turbines, and – see Florjancic, Rotor design in industrial gas turbines

Figure 5-68   *Photoelastic stress study of the root section of a turbine blade, 1964* [260]   ©*ABB*

Rotor welding was applied by BBC as early as 1929, at the instigation of Adolf Meyer, the Technical Director at the time, Section 4.1.2, for steam turbines and 10 years later also for gas turbine rotors, Figure 5-69.[261] Before then, it was general practice that rotors were composed of keyed and/or shrunk-on-shaft, solid discs.

The cross-section of the Neuchâtel turbine reveals that the rotor is built from three discs without centre bores (with a centre bore, the core stress would double), of which the outer ones carry forged shaft stubs, while the middle disc was shaped as an 'equal-strength disc' and the three discs were welded together at the circumference. This design concept is especially suited for thermally and mechanically highly loaded gas turbines, as stated in the first comprehensive, published design concept assessment[262]:

–   *'The discs without bores can be formed as 'equal-strength discs' for highest mechanical loads*
–   *The single discs allow for best forging*
–   *There are no connections between discs and shaft by shrinking and keys, which might loosen under strong temperature variation*
–   *Small material amassment reduces the disc temperature reaction lag, especially during rapid operation load changes.'*

---

[260] Picture source: BBM, Rueckblick... in den Jahren 1963 und 1964, 1965, Figure 6

[261] There are indications that the previous forged solid shaft technique caused manufacturing and cost problems for turbo-generators, also underlined by pictures of an exploded rotor in the Baden spin pit test facility. The first welded steam turbine rotor in 1929 was delivered after thorough tests to Japan. See BBR, 50 Years of welded turbine rotors, 1981

[262] Remarkably, this positive expert statement came from Rudolf Friedrich in 1949, see Friedrich, Gasturbinen mit Gleichdruckverbrennung, p. 75. The Siemens GT concept is characterised by a substantially different tie-bolted rotor design with 'Hirth serration couplings' between the discs, a design approach not yet mentioned in Friedrich's book.

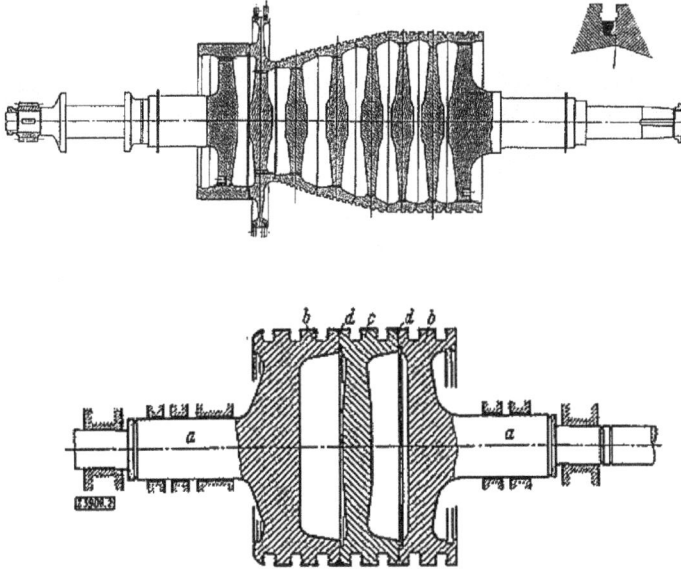

Figure 5-69   *BBC: First welded steam turbine rotor (Japan 1929, top ©ABB) and first welded GT turbine rotor (Neuchâtel 1939, bottom)* [263]

From a mechanical point of view the welded design concept has like a monolithic piece the major advantage of no mechanical joints which could slide as the rotor heats up and thus cause vibrations, wear and tear.

The development of rotor design philosophy at BBC can be reconstructed through the priority dates of a few selected, corresponding patent ideas, Figure 5-70. The first modifications of the established shrunk-discs technology can be deduced from patents with priorities as early as 1923, Figure 5-70 a), introducing differently shaped resilient elements for shaft-disc contact and damping, when the shrink force decreased at elevated temperatures. This approach (fruitless in the end) was continued for a surprisingly long time, until 1932, Figure 5-70 b). The welding of individual discs was applied in 1929, first for equal diameter forged disc configurations of an electric power generator shaft according to Figure 5-70 c) and in due course for the described steam turbine rotor, Figure 5-70 d) and 5-69. With this new rotor construction, the difficulties experienced in the production of solid forgings could be avoided.

The new rotor manufacturing concept consisted of a number of sections welded together around their periphery, which presented advantages not offered by any alternative form of construction. Since the welding zone is located some distance away from the rotor centre-line, fatigue due to alternating bending stress is low and vibration is avoided. In more than 80 years of widespread operational experiences in steam and gas turbines as well as turbo-generators in estimated more than 8,000 units, this design concept was found to be extraordinarily reliable; no catastrophic failures have been reported due to welding defects, in contrast to alternative design concepts.

---

[263] Figure source: Top – BBR, 50 years of welded turbine rotors, and Bottom – Stodola, Leistungsversuche p. 18

For the gas turbine development beginning during the 1930s, characterised by relatively small Velox boiler and Houdry refinery turbo-sets, the resulting stress level was sufficiently low, so a simplified built rotor manufacturing method was applied according to the sketch of Figure 5-70 e). Here, the attachment grooves for the axial compressor and turbine blading were machined in the cylindrical rotor tube with welded front and end pieces.[264] The bigger turbine dimensions of the first utility gas turbine at Neuchâtel, Figure 4-79, led to the first time application of rotor welding for this disc arrangement. In fact the early BBC gas turbines showed a mixture of both design concepts, the compressor blading inserted in a hollow rotor shaft as of Figure 5-70 e), while the hot turbine end requested the welded disc design of Figure 5-70 d).

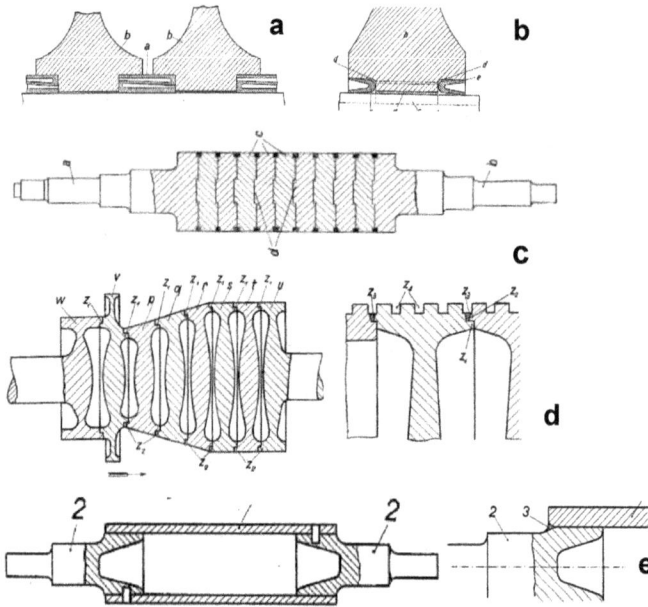

Figure 5-70   *Patent development* [265] *of BBC rotor manufacturing, with priority dates a) 1923, b) 1932,*
                    c) 1929, d) 1930, e) 1931

It was only in the 1950s that larger compressor dimensions demanded the welded disc approach throughout[266], all the more so when the concept of one single rotor running in two journal bearings, Figure 5-19, became a general design standard. This remains unchanged today, as illustrated in Figure 5-71 for the present advanced GT24/GT26 configuration.

---

[264] The documentation is somewhat fragmentary for these early compressor designs. It appears, however, as if the first BBC axial compressor delivery for ETH Zurich, Figure 4-22, and more clearly in Ackeret, High Speed Windtunnels, Fig. 43, still relied on the shrunk connection technology only.

[265] Expressed in GB patents for Brown Boveri and their corresponding priority dates, Figure 5-70 illustrates: a) GB214200 -11 April 1923, b) GB393555 – 5 March 1932, c) GB364027 – 2 November 1929, d) GB362813 – 2 May 1930, e) GB386931 – 5 November 1931.

[266] The decision was ultimately triggered by the GT15/GT17, Figure 5-43, with individual, highly-loaded disks for all compressor and turbine blade rows with up to 345 m/s rim speed. [Chr. Jacobi]

Welding enables machine components of enormous dimensions and extraordinary weight to be produced; the resulting mechanical properties can hardly be achieved with any other production process. Fully automated production processes have been developed for both technical and economic reasons. Figure 5-72 illustrates the welding machinery for gas turbine rotor manufacturing, both for preliminary Tungsten Inert Gas (TIG) root welding and subsequent multi-layer (powder) Submerged Arc Welding (SAW) process. Since the total reliability of the submerged arc process has been proven in service, all welding on the shafts of steam and gas turbines as well as compressors since the 1960s has been carried out according to this process; this fully-automated process produces perfect continuous welds. Both the tungsten/ inert gas root welds and the submerged-arc filler welds are examined with ultrasonic and other nondestructive testing equipment. Subsequent heat treatment eliminates shrinkage stresses set up during welding.

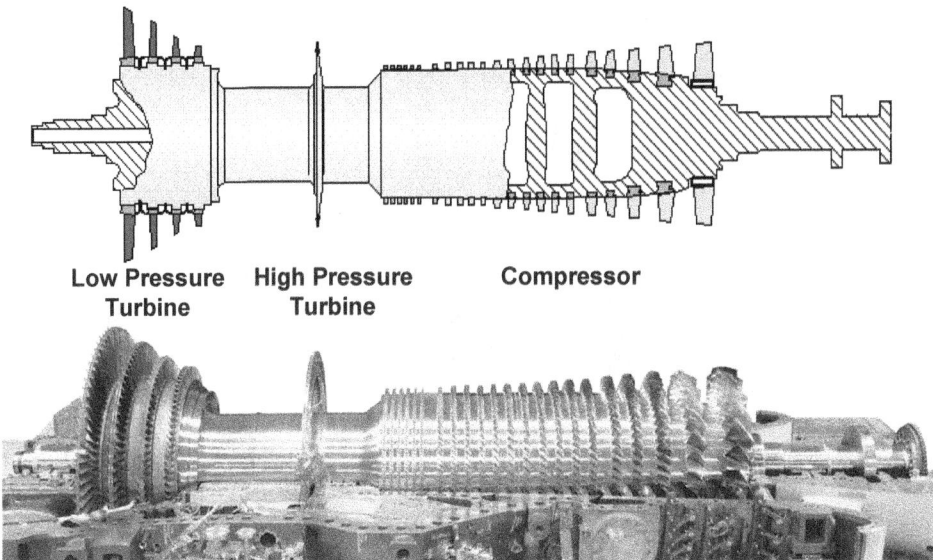

Low Pressure Turbine   **High Pressure Turbine**   **Compressor**

Figure 5-71   *GT24/GT26 welded rotor, overall length > 12 m*

Because of exceptionally high unit ratings there has been a corresponding rise in the mechanical strength requirements for the materials used in turbine shafts. While originally nickel alloy steels with elastic limits of around 300 N/mm^2 were used, it is now common practice to use CrNiMo or CrMoV tempering steels with elastic limits in the region of 600 to 800 N/mm^2.

The horizontal SAW facility of Figure 5-72 (r) is rather exceptional and was the largest of its kind in the world at the time of its erection in 1974. The facility can take workpieces of 22 m length, 4.2 m max. diameter and with weights up to 400 t. The facility is equipped with a turning device for the final finishing of rotors after the welding operation.

There are various stages in welding turbine or compressor shafts. The machined and inspected shaft components are stacked vertically, Figure 5-72 (l) in the correct order and accurately aligned to each other. Their temperature is raised by induction heating and the first fixation of the disc sections happens by an automatic root welding with the Argonarc process, where the electric arc burns in an argon (inert gas) atmosphere between the electrode and the workpiece. Subsequently, the rotor is moved on to the horizontally arranged SAW station, Figure 5-72 (r), where all disc joints are filled by multi-layer submerged-arc welding, which involves melting a bare wire in an arc under a layer of powder.

Alstom maintains overspeed testing installations for large turbines and generators at all its major manufacturing sites; in June 2010 the world's largest balancing facility was inaugurated at the factory in Chattanooga, Tn., USA, Figure 5-73. The overspeed test which serves as the final check on turbine and generator rotors, consists of briefly running the individual turboset at a speed 20 percent above the nominal value. Such test runs can only be performed in heavily reinforced pits or tunnels, so that in the highly improbable event of a rotor exploding, damage will be confined.[267]

Figure 5-72   *Vertical rotor stacking and preliminary TIG root welding (l), horizontal, automatic rotor manufacturing facility for final multi-layer submerged arc welding   (SAW, r ©ABB)* [268]

Before the overspeed test, any rotor unbalance resulting from asymmetries in manufacture must be removed. This balancing is also carried out in the overspeed test facility which is provided with the necessary measuring equipment. By increasing the speed in steps, meas-

---

[267] See BBC, Overspeed test facilities of the group, 1976
[268] Picture source: BBC, Welded turbine shafts, 1974

uring the imbalance and adding balancing counterweights as required, an optimum rotor balance can be attained up to nominal operating speed. The ensuing test, normally 2 to 3 minutes at 20 percent overspeed, will establish that the rotor runs sufficiently smoothly up to this overspeed limit. When subsequently operated in the power station, the regulation and protection equipment ensures that this speed limit is never reached again. The procedure of balancing up to operating speed followed by an overspeed test thereby guarantees that a turboset runs smoothly throughout its possible operation range.

The nature of the tests requires that overspeed installations be built to high safety standards without sacrificing accessibility. For large turbine rotors the enclosure pressure must be reduced due to the high windage losses of large blades; this reduces the necessary drive power and avoids unacceptably high blade temperatures. Consequently, pressures below 1 Torr (1 mm Hg or 1.333 mb) are necessary for very large rotor diameters. Figure 5-73 shows the newly opened rotor overspeed test facility and balancing preparations on a nuclear steam turbine at Alstom's Chattanooga plant[269]. The rotor balancing/ overspeed tunnel can handle rotors of 22 m length and 8 m diameter, weighing up to 350 metric tons. The tunnel can be hermetically closed to create the test vacuum and to drive the rotors up to 4,500 rpm.

5.4.3 Gas Turbine Manufacturing Sites and Network

Judging by the fast pace of construction at BBC in its first 25 years, this early period was certainly a stormy one. Scarcely a year went by without the addition of a new factory building. The first of them were built in 1897 to the east of Bruggerstrasse, which still divides the Alstom/ABB grounds at Baden, AG, CH today, Figure 4-2. At the beginning of the 20th century the first steam turbine production site was erected from the joint-stock capital of the specially founded 'AG fuer Dampfturbinen System Brown-Boveri-Parsons' (Public Stock Company for Steam Turbines Brown-Boveri-Parsons). The next notable addition was the 150 m long, 20 m high 'Halle 30', which BBC constructed in 1928 for large-scale assembly. Due to the economic upturn after the war, pressure to expand production areas mounted quickly after 1950. Turbogenerators were exceeding the 100 MW capacity and prospects for nuclear power machinery urgently required larger buildings. In 1956 the decision was made to build a large new factory on a greenfield site in Birr, AG, CH: 10 km as the crow flies on the west side of Baden. Large-scale industrial zones were planned for the region and excellent motorway connections were built nearby. Considerable investment was required for what is today a versatile manufacturing plant of 86,000 m^2, almost 50 percent of which represents actual shop floor production. Construction of the Birr plant began in April 1957. The first stage comprised a building of 270 m long and 78 m wide, divided lengthways into three halls, the characteristic saw-tooth shed roofs can be seen for miles, Figure 5-74, 1); by 1963 three more factory halls were added.

---

[269] The Chattanooga facility manufactures power generation equipment for the American market, amongst other the world's largest nuclear steam turbine, Arabelle. This is a versatile turbine that can be matched with any reactor, either pressurised or boiling water. The turbine weighs 120 t and, out of its 13 wheel rotor, has a yield capacity of 900–1,800 MW. Arabelle is approximately 2 percent more efficient than conventional steam turbines.

Figure 5-73 *Rotor overspeed test facility at Chattanooga, Tn., USA and rotor balancing of a large nuclear power steam turbine rotor*

From the beginning it was planned to build living quarters for the workforce nearby. In 1962 work began on a large housing estate called 'In den Wyden' (To the osiers), Figure 5-74, 6). The combination of factory and housing was a novelty at the time, which generated a lot of interest among planning experts. Originally, the estate was planned to become the heart of a new garden city with 40,000 inhabitants. This turned out to be over-optimistic[270], but nevertheless the factory contributed to the nearby village of Birr[271], Figure 5-74, 8): in 1960 the rural community had only 651 inhabitants, which had already increased to 2,600 by 1970 and by 2011 the population had reached 4,174, of which approximately 45 percent are of non-Swiss origin.

In 1976 the first spin pit test facility was attached and expanded in 1978, Figure 5-74, 3).

The transition from BBC to ABB at the end of the 1980s also led to a reorganisation in manufacturing; the group had to modify and rationalise its production capacity according to demand. Consequently, a production partnership with the factories at Mannheim was established, with a specialisation in stationary parts and assembly at Mannheim, while the newly focused 'rotor factory' at Birr concentrated on generator -, steam and gas turbine rotors and its blading.

---

[270] Needless to say, also inside the plant several features (like a 325 t crane) turned out to be overdone, especially when the development of nuclear industry did not materialise in Europe as originally expected.

[271] Historically, Birr belonged to the territory of the Count of Habsburg until 1415; the ruin of the family castle, the origin of the Austrian-Spanish Habsburg dynasty, is 3 km north-west of the village. The Swiss educator, school and social reformer Johann H. Pestalozzi (1746–1827) lived and died in Birr (Gut Neuhof); in 1792 the French national assembly awarded him the title 'citoyen d'honneur' (honorary citizen), the first person from Switzerland to ever receive the title. The remarkably close connection to France was continued in 1871, when the French Armée de l'Est (Bourbaki Army) crossed the border for internment to neutral Switzerland (after a failed relief attack on the occupied Belfort in the Franco-Prussian War of 1871). The 87,000 soldiers, who had to be supplied for several months, represented 3 percent of the entire Swiss population of that time, a considerable burden for the poor inhabitants. Since 1899 a memorial in Birr cemetery has commemorated this event, sculpted by Frédéric-Auguste Bartholdi (1834–1904), better known as the creator of the Statue of Liberty.

In 1996 the introduction of the new gas turbine family GT24/GT26 was accompanied by the opening of a GT26 test facility at Birr, which together with a link to the nearest gas-supply pipeline was erected in record time. In Figure 5-74, 2), only the exhaust stack is visible, which in reality measures 8 m in diameter.[272] Expansion works began for a facility adaptation to simulate KA26 combined cycle (steam and gas turbine) operation, which are accomplished in the meantime, Figure 6-40, providing a unique environment for product validation. The test power plant is equipped with a dual-fuel GT26 and a dual-fuel high temperature demonstration unit, dispatching directly to the Swiss electrical grid, like the previous single GT26 operation.

In 1999 the Birr manufacturing capabilities were considerably expanded by establishing the manufacturing of Hot Gas Path Parts (HGPP), mainly comprising all the finishing operations of cast combustor and turbine parts (Machining I, i.e. high-speed grinding, Coating – preferably LVPS, Low-Vacuum Thermal Plasma Spraying, Laser Cooling Hole Drilling and Machining II – e.g. brazing of vane sheet metal/ cooling inserts) – and by a GT reconditioning shop, again with a focus on HGPP reconditioning.

In 2006 the research and development laboratories were relocated from Baden-Daettwil to Birr. Twenty new laboratories were built for the researchers, mostly employed in developing gas turbine technology.

Finally, in 2007 the historic, first official utility gas turbine was relocated from its operational site at Neuchâtel (since 1940) to Birr and put on permanent exhibition there as an ASME Technical Landmark, as described already in Figure 4-81 in Section 4.3.3.3. Figure 5-74, 4), marks the location of the main entry gate to the Birr site and the nearby GT Neuchâtel Museum pavillon.

All the stages in the life of a power generation product are currently represented on site: research, development, production and service. In detail, the facility hosts hydro- and turbo-generator manufacturing, the rotor factory, hot gas path parts manufacturing, turbine reconditioning centre, test power plant, R&D laboratories, power generation training centre, field service and supply chain management, Figure 5-74, 5).[273]

The all-new GT24/GT26 represented in the 1990s also considerable challenges in production technology: Continuous Cost Cutting (CCC) and the shortening of throughput times were prime targets. In general in-house production was reduced and a widespread network of suppliers had to be qualified and continuously coordinated as part of the newly defined Supply Chain Management (SCM). The concept of manufacturing cells was established with machines grouped according to product requirements in a lean manufacturing set-up, distinct from the traditional functional manufacturing system in which all similar machines were grouped together. The use of manufacturing cells improved the material flow and is still especially suited for batch production, even in relatively low volumes as is typical for GT blading.

---

[272] In the beginning the stack and the resulting hot exhaust gas plume was of some concern to the operators of the Birr airstrip, Figure 5-74, 7), which is only 1 km east of the plant and where regular flight training takes place with light/composite aircraft. Comprehensive plume spreading calculations had to be carried out, but concerns were not justified in reality.

[273] In 2011 the total workforce at the Birr site comprised 1,750 Alstom employees, nearly 50 percent of these in Production, while the rest was split almost equally between Engineering, Hydro and Service.

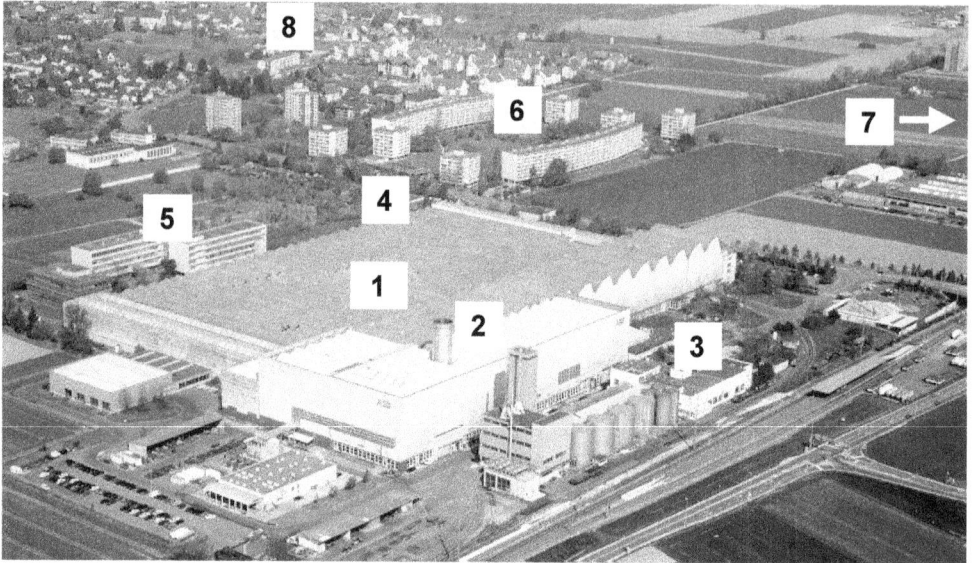

Figure 5-74   *Manufacturing site Birr, AG, CH ~1998 : 1 – Factory building, 2 – Exhaust stack of GT26 power plant, 3 – Rotor overspeed test facility, 4 – Main gate and Neuchâtel GT Museum, 5 – Supply chain management/ purchasing, 6 – Residential estate 'In den Wyden', 7 – Birrfeld airstrip, 8 – Birr village*

Supply Chain Management, developed from Just-In-Time (JIT) production, comprises engineering, costing, contracting, planning and management tasks from supplier selection and qualification, to procurement and parts flow coordination including logistics, both within a company and also beyond its boundaries. It is no longer the isolated supplier, but the optimised output of the complete, integrated supply chain which counts. Computer-supported parts and batch tracking tools are of utmost importance for supply chain steering (pull principle), to keep the stock inventory low and to limit the tied-up capital along the supply chain, which easily can reach several 100 million USD in a global supplier network.

One of the most effective tools for attacking costs and throughput times became the 'concurrent engineering' principle: GT design teams and production units, in-house or at external suppliers, working together from the very beginning of a new design. There is a co-ordinated, sequential process between engineering design and parallel manufacturing activities over a period of approximately 2.5 years (timewise slightly shifted). The savings in throughput time were significant. At the same time, closely cooperating concurrent engineering teams laid the ground for substantial cost reductions, especially necessary and attractive for newly emerging manufacturing techniques like single crystal or directionally solidified precision investment castings for GT turbine bladings, Figure 5-77. HGPP, Hot Gas Path Parts (turbine castings, corresponding machining operations, coating and laser cooling hole drilling) can make up one third of the total cost of a GT thermal block. In 1997, in a two-year-long, in-depth concurrent engineering team effort, the manufacturing cost of high-temperature turbine castings was reduced by up to 40 percent as part of a 500 million USD Long-Term Supply Management Agreement (LTSMA) with one of the leading US foundries, an essential prerequisite for the early market success of the newly designed, advanced GT generation.

The parts of a GT Thermal Block can be grouped in various commodities, which typically have considerably different lead times. Consequently, different strategies have to be applied for the acquisition of long-lead items which require project-independent orders, in contrary to activities which can be carried out during the actual 'Engine Project Phase' of 30–40 weeks.

Representing a key part of the casting materials commodity, Figure 5-75 illustrates the lower half of a horizontally split GT 26 casing, a complex and heavy (23.3 t) steel casting from Alstom's Elblag Foundry.[274] After first preparatory machining operations on the spot, these parts are transported by road to Mannheim, D, where the final machining operations take place in the Alstom plant.[275]

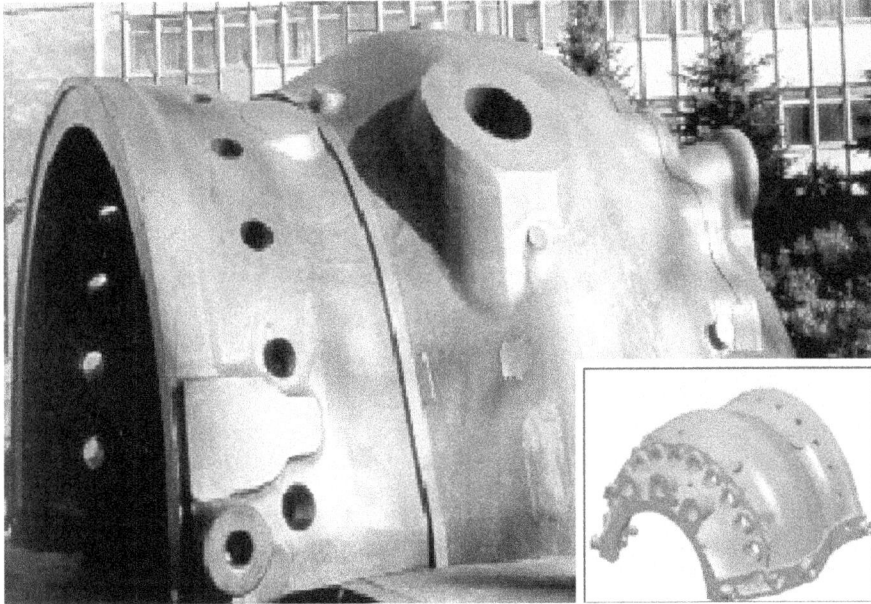

Figure 5-75   *GT26 casing, lower part – steel casting, Alstom foundry, Elblag, PL*

Large stationary parts with diameters of up to 6 m, lengths of up to 5 m and masses of up to 120 t are machined here with a multitude of exceptional machine tool equipment. For

---

[274] As outlined in the 'company tree' diagram Figure 2-1, the former Zamech plant in Elblag, PL, acquired by ABB in 1990, represents the eldest member, founded as F. Schichau Works at Elbing in 1837. The German entrepreneur Ferdinand Gottlob Schichau (1814–1896) founded a large shipyard in Gdansk/Danzig in 1890, as the water access to his hometown Elbing was too shallow. In 1950 this was merged with the Danziger Werft to become the new Lenin Shipyard, where a strike on 14 August 1980 led to 'Solidarność' and in due course to the 'wind of change' in 1990.

[275] In 2011 there were more than 620 employees, of which approximately one quarter worked in mechanical production.

example, Figure 5-76 illustrates the handling of a GT 26 casing on a Double Gantry I work centre.[276]

Figure 5-76   *GT26 casing machining at Alstom Mannheim, D – with Double Gantry I portal work centre, Waldrich Coburg, D*

Precision investment casting is used for the manufacturing of GT blading on the basis of a 'lost-wax method', one of the oldest metal-forming techniques, considerably improved in the 1970s with the introduction of especially designed, high-temperature alloys and the corresponding Bridgman[277] casting process of creating directionally solidified metal structures (SX Single Crystal and DS Directionally Solidified turbine blading).

Figure 5-77 illustrates the decisive process steps with the intended internal cooling flow pattern. Starting with a ceramic core which allows sophisticated internal cooling structures to be cast, this element is embedded in a net-shape wax model of the blade, which is then wrapped in a ceramic shell. After wax removal, the hollow ceramic mould, which comprises 3–5 turbine blade or vane shapes, is filled with liquid metal in a high-vacuum furnace. The metal then solidifies in a prescribed direction during an hour-long descent of the mould support platform from the hot upper to the cold lower zones of the furnace. Afterwards, the outer ceramic shell is removed mechanically and the inner ceramic core is etched out by concentrated acid (which becomes the hollow cooling airflow path). The casting process can produce complicated shapes that would be difficult or impossible to

---

[276] The Double Gantry production system is a renowned product of Waldrich Coburg, D. The total length of the facility is 48 m and the maximum portal operation height is 5.5 m at a width of 6.4 m. The spindle power for milling and drilling operations is 95 kW, at both ends there are installed additional 95 kW planing disks for turning etc. Each portal contains tool change mechanisms for up to 120 machining tools.

[277] Percy W. Bridgman (1882–1961) was an American physicist who won the 1946 Nobel Prize for Physics for his work on the physics of high pressures, (100,000 at) but unsuccessfully attempted the synthesis of diamonds many times. http://en.wikipedia.org/wiki/Percy_Williams_Bridgman

manufacture otherwise, it is precise down to the 0.1 mm range, it requires little surface finishing and in general, only final machining of the blade root and vane attachment sections.

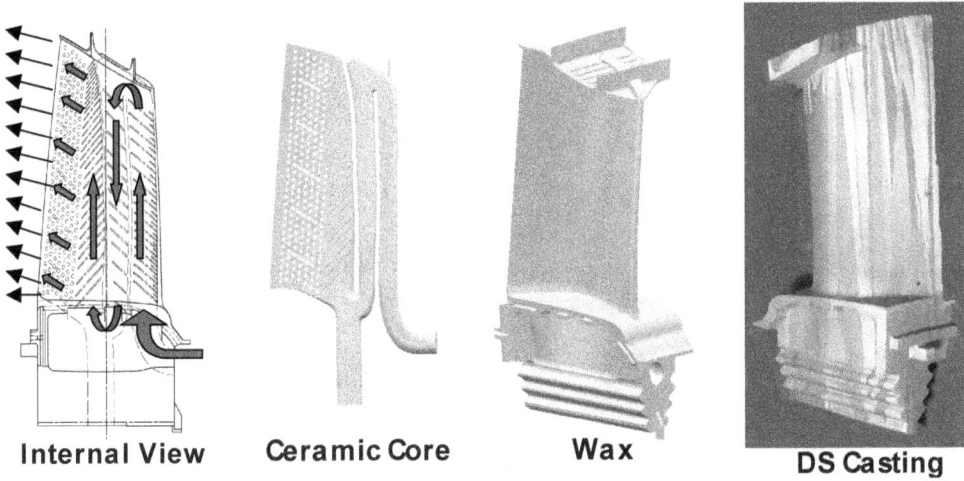

Internal View   **Ceramic Core**   **Wax**   **DS Casting**

Figure 5-77   *GT blading precision investment casting technology* [278]

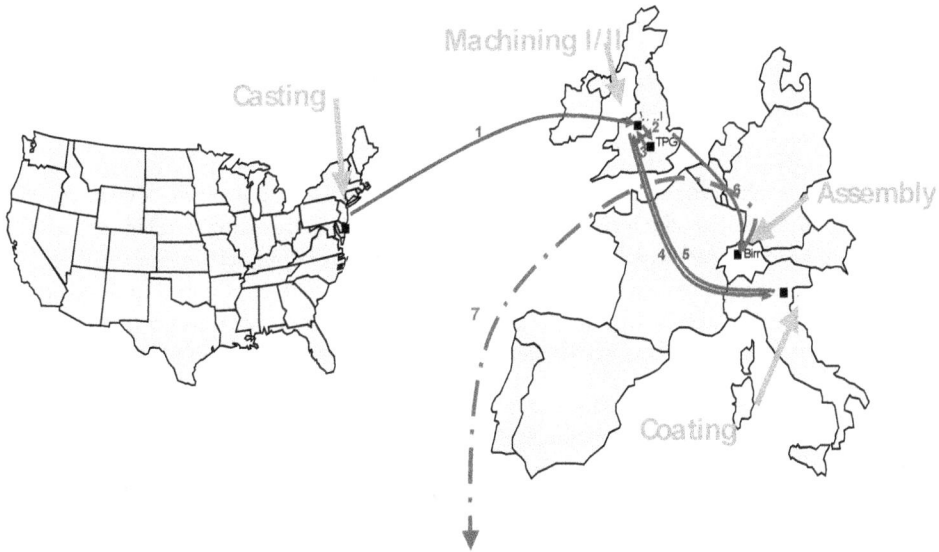

Figure 5-78   *GT blading sequential manufacturing & shipping flow* [279]

---

[278] See Eckardt, Advanced Gas Turbine Technology
[279] See Eckardt, Umweltaspekte bei Produktion und Betrieb

Figure 5-79   *GT blading – laser drilling of film cooling holes (l) and GT 26 thermal barrier coated turbine blade (r), with rows of cooling holes along the airfoil leading edge*

The application of such sophisticated manufacturing techniques may explain final parts prices of more than 5,000 USD per piece, e.g. for a 30 cm long, single crystal gas turbine tailor-made blade with the main strain-tolerant axis in line with the centrifugal load direction.

The trend towards economic globalisation supported the development of worldwide coordinated production networks which link highly specialised suppliers, as shown in Figure 5-78 for the example of high-temperature GT blading. A 8,000 km long supply chain can start here[280] on the US east coast at a qualified foundry, from where the castings are shipped to England (1) where Machining I operations are carried out, with intermediate local transports for special operations (2 & 3), before the parts are brought to and from a coating shop in Italy (4 & 5).

The following Machining II operations comprise parts for laser hole drilling as illustrated in Figure 5-79, by which (fan-shaped) cooling holes are drilled from/on the airfoil outer surface to the inner cooling flow channels, so that an isolating film of cooling air is wrapped around the turbine airfoil during GT operation. This has the consequence that the GT turbine inlet gas temperature is in fact several 100 degrees K above the melting temperature of the blade metal.

After accomplished Machining II operations the finished parts are forwarded to Birr, CH for the final rotor assembly, Figure 5-78. Afterwards, the rotor and the stationary parts are integrated at Alstom Mannheim, Figures 5-78 (6) and 5-80, from where the complete GT units are shipped to the power generation project site, as indicated in Figure 5-78 (7).

---

[280] The supply chain is obviously considerably simplified if the cast parts go from the casting house directly to the Birr HGPP facility for the remaining operations; however, especially strategic SCM considerations ask for a second sourcing line in general.

Figure 5-80   *GT final assembly floor – Alstom Mannheim, D*

5.5   Gas Turbine Special Projects

At the end of Part II of this historic power generation GT review it is worth addressing a few special BBC developments outside of the GT development mainstream.

5.5.1   Mobile and Packaged Energy Stations

Until the 1960s, movable power plants were an economical solution for areas where the increase in electrical energy demand had either outstripped the building capacity for new power plants, or as in the large 200 × 200 km Puebla-Veracruz area on the east coast of Mexico, where large seasonal variations caused problems. Though the basic power capacity of the state-owned energy provider CFE (Comisión Federal de Electricidad) was 108 MW and sufficiently high enough, the hydro share was close to 75 percent, meaning the actual power capacity fell to 62 MW during the summer months. GE had built a first mobile steam power plant there in 1948, followed by some smaller diesel units, until in 1955 CFE ordered the world's first three mobile GT plants to compensate for the energy gap during the annual drought period. Seven mobile GT plants were built by BBC in total, of which four alone were put into operation in Mexico.

The GT design was in principle similar to the stationary GT10* of that time, except that the 6.2 MW power plant was arranged to suit the mobile requirements in two railway waggons, Figure 5-81, of which the larger one (l) contained gas turbine and generator, while the electric equipment (main transformer, auxiliary diesel engine, etc.) was placed in the second switchboard car. Referred to the specific power kW/m train length, the diesel configuration achieved only 42 percent and the steam turbine 75 percent of the superior GT solution, while in reference to weight the kg/kW values of the diesel solution was 111 percent and the steam turbine 61 percent heavier. For BBC, previous experiences with compact GT locomotive designs, Figure 4-86, were a special advantage in the context of this project.

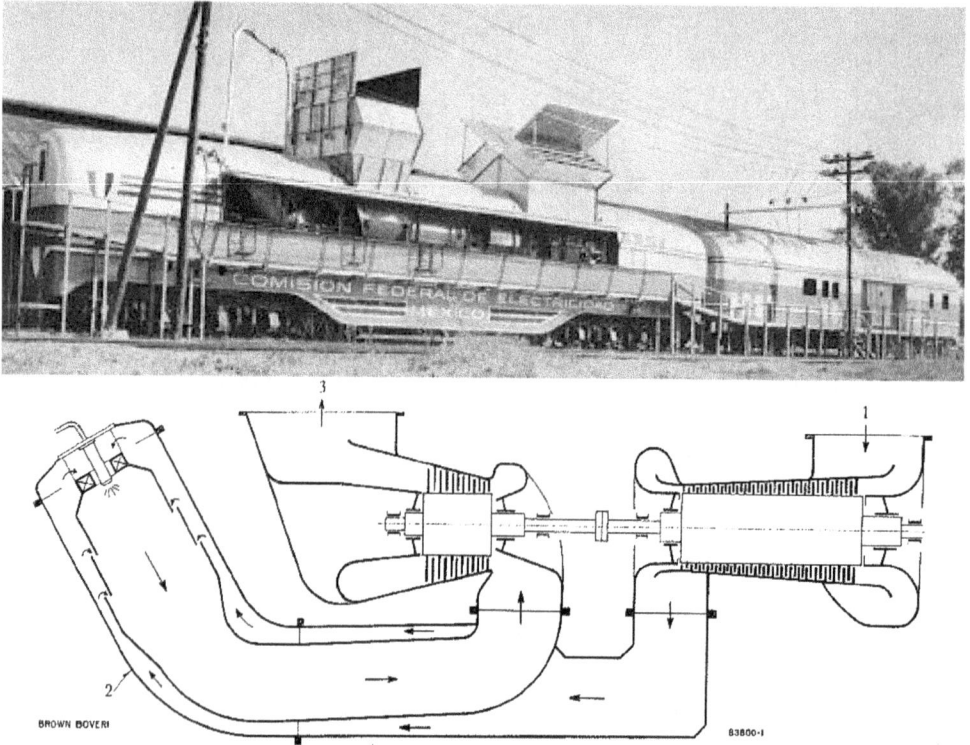

Figure 5-81   *6.2 MW mobile power station:* [281] *train cars attached to the CFE network Puebla-Veracruz, Mexico at 2,280 m altitude (top) and GT arrangement with compact, concentric ducting (bottom): 1 – compressor entry, 2 – combustor, 3 – turbine exhaust*

Although this mobile gas turbine had to close the power gap at an altitude of 2,280 m near Puebla, the measured power output surpassed the guarantee values of the first unit by more than 8 percent. Special construction problems had to be solved by the SLM (Swiss Locomotive and Machine) factory, Winterthur, CH, where the wagons were built with high frame stiffness in view of small gas turbine blade clearances: only 0.5 mm vertical bending was allowed for the longitudinal beams over the outer compressor-turbine bearing distance of 6.3 m. The total weight of the GT wagon was 150 t, as illustrated by the heavy crane construction of Figure 5-82.

---

[281] Source: BBC, Fahrbare Gasturbinenzentralen, 1958

Figure 5-82   *Bremen, D 1955: 250 t pontoon crane* [282] *lifts the two wagons of the first mobile 6.2 MW GT station on board SS 'Andyk'* [283] *on the way to Vera Cruz, MEX   ©ABB*

For a short while in 1955/1956 it appeared that the concept of shipping operationally-ready mobile power stations could overcome emerging difficulties due to steadily rising lead times. Once on site, the preparation up until power generation readiness lasted only 8 to 24 hours; but compared to the 7 mobile sets sold, the subsequent introduction of packaged

---

[282] Picture source: BBC, Fahrbare Gasturbinenzentralen, 1958, Figure 11. The pontoon crane, nicknamed 'Langer Heinrich' (Long Henry), officially 'Grosser Schwimmkran I' – was the biggest floating crane worldwide for ten years after its commissioning in 1915, with 250 t lift capacity. It served for BBC shipments several times, such as in 1937 by putting the 163,000 shp steam turbine propulsion unit from BBC Mannheim (the largest at the time) on board the heavy battle ship 'Tirpitz' at the Kriegsmarinewerft (navy yard) Wilhelmshaven. The crane had been built at DEMAG, Duisburg, D; it is 81 m high with 42 m reach. The pontoon had a displacement of 4,000 t with two propellers driven by two triple expansion compound steam engines of 1,000 hp combined, replaced by four 6-cylinder diesel engines in 1955; crane operation was by means of electric lifting/turning gears driven by two turbo-dynamos. In the meantime the crane is stationed, still in service and with the new name 'Maestrale', at Genova, I, harbour, Calata Gadda, expecting its 100th anniversary as a technical landmark in 2015. (Courtesy of Pietro Zunino, DIME, Università degli Studi di Genova)

[283] The SS Andyk (8,380 t), then named SS Groningen, was built in 1946 by the US ship yard Sun Shipbuilding, Pennsylvania, see Section 4.3.2.3. The order for the ship was originally placed by the Dutch Ministry of Shipping, since most Dutch shipyards had been destroyed during German retreat. This was a typical US-standardised 'C3-S-A5' class ship. C3 meant cargo ship with a length of 450 – 500 feet; in fact, SS Andyk had a length of 465 feet. The letter S stands for steam machinery, single screw and under 12 passengers. A5 is the design version (which started at A1). It was probably equipped with 2 Westinghouse turbines of 8,500 hp. The Holland America Line (HAL) took over the ship on 7 October 1946 and renamed it. On 5 September 1969 the ship was sold to Cyprus and renamed SS Aurora. The end came in Spain on 8 May 1971 when the ship was scrapped. Like all the other HAL ships, the ship was named after a Dutch village, Andijk, with some 6,000 citizens in the province of North Holland. The HAL passenger ships' names ended in 'dam' (e.g. Nieuw Amsterdam I-III, Noordam I-III) and the freighters in 'dijk'. http://eeuwen.home.xs4all.nl/andijk.htm and private communication with Ferry van Eeuwen on 23 January 2012.

GT units was considerably more successful, as outlined already in Sections 5.1.5.1 and 5.1.5.3 (Figure 5-19): 79 GT11L represented a kind of success story of the 1960s.

The package power plant concept had been introduced by GE in 1960 with the Frame 5 (MS 5001) engines. This milestone involved two successful features:

- completely self-contained packages capable of being installed with minimum time and cost, with all piping and wiring completed under closely controlled factory conditions prior to shipment
- manufacturing of 9 units in advance of orders for off-the-shelf delivery.

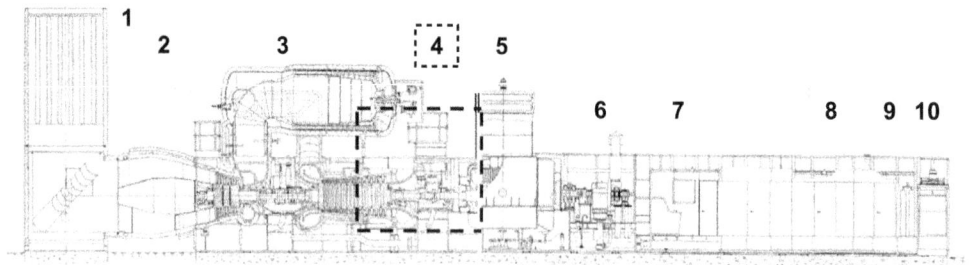

Figure 5-83   *Longitudinal section of the 15 MW GT11L power package* [284] *1 – Chimney with silencer, 2 – exhaust gas duct, 3 – gas turbine/ compressor block, combustion chamber, 4 – compressor air intake with filter/silencer, 5/6 – generator/starter motor block, 7 – control block, 8 – auxiliaries block, 9 – CO₂ fire-extinguishing equipment, 10 – ventilation block*   ©*ABB*

In due course BBC followed with the GT11L package, Figure 5-83, with the following basic blocks fully equipped in the factory and sent on site:

- Power block, i.e. turbine and compressor on common platform
- Combustion chamber
- Generator block
- Auxiliaries block
- Other items.

The packaging principle reduced the erection time on site to 4–6 weeks; the housing structure reduced the noise level to 60 dB within 300 m.

The first installation, again for CFE in Pajaritos, Veracruz, MEX, had been ordered in 1961, for a 60 Hz application and a rated power of 15 MW[285] with a total GT efficiency of 21.2 percent. The GT11L prototype was tested at Baden in 1964 up to idle, and afterwards up to full load on the Fundidora site; the plant is reported to still be in service (2010). In the mid-1960s the GT11L engine was slightly upgraded to 16 MW on the basis of an elevated turbine inlet temperature of 770 °C, which raised the thermal efficiency to 25.6 percent.

---

[284] Figure source: BBR, Development of a 15-MW Packaged Gas-Turbine Station, 1965, Figure 6
[285] See BBR, Development of a 15-MW Packaged Gas-Turbine Station, 1965 and – see BBC, The Brown Boveri Packaged Gas-Turbine Plant, 1968

The somewhat premature end of the GT11L packaged concept in comparison to the lasting success of GE's Frame 5 engines came in 1970, when the last GT11L was ordered and afterwards obviously intentionally replaced by the new GT9B and following derivatives[286], though documents about an official GT11L programme stop during the BST era are no longer available.

5.5.2   Aero Jet Expander

In the mid-1960s BBC was also at the forefront of development in terms of turbojet use for power generation. Figure 5-84 illustrates the first gas turbine plant on the European continent, equipped with 4 Rolls-Royce Avon turbojet gas generators and commissioned by BBC Mannheim in December 1966 for the Emden, D site of NWK (Nordwestdeutsche Kraftwerke AG).[287]

The decision for this special form of GT power plant relied on several advantages:

– the plant was constantly ready for peak load operation as well as sudden net stabilisation and emergency needs
– the traditional 'spinning reserve' of steam-power plants became obsolete
– the extraordinarily short starting period of only 2-3 minutes from standstill to full load
– low purchasing and operation cost (simple operation, low cooling-water requirement, remote control or one man per shift).

Technically, two Brown Boveri power turbines (1-stage impulse type) were directly coupled with the power generator; the driving gas stream was generated by 2 × 2 Avon[288] 1533A single-spool gas generators of 7,900 rpm design speed (17-stage axial compressor PR=9.8, 8 combustor cans, 3-stage un-cooled turbine TIT=865 °C).

The design of these turbo-expander sets again followed the GT packaging idea. Three blocks were made in the factory, assembled together with the gas generator engines and then shipped to site for final power plant integration. Each GT power plant had three blocks:

– Gas generator units (turbojet engines)
– Power units (expander turbines, generator)
– Control block.

The original impulse for this engine type came from the US as an immediate response to the 'Northeast Blackout' of 1965. On 9 November 1965 a significant disruption to electricity supplies occurred, affecting the north-eastern US states (New Jersey, New York, Rhode Island,

---

[286] The GT9 family accumulated sales figures of 106 units over nearly two decades.

[287] The electricity supplier NWK, Hamburg, D was founded in 1899/1900 by Siemens, and merged with Preussen Elektra, its former majority share holder, in 1985.

[288] The Rolls-Royce Avon was the first axial flow jet engine designed and produced by Rolls-Royce. Introduced in 1950, it went on to become one of their most successful post-World War II engine designs. It was used in a wide variety of aircraft, both military and civilian, as well as industrial applications, ending production after 24 years with 11,000 built units in 1974. The Avon axial compressor especially represents some early BBC and Swiss connections, both through the flow of ideas from the Metrovick F.2 to the Armstrong-Siddely Sapphire compressor, and in aircraft developments for the Swiss Airforce (FFA P-16, Hawker Hunter). http://en.wikipedia.org/wiki/ - cite_note-Gunston149-0

Massachusetts, Connecticut, New Hampshire, Vermont) up to Ontario, Canada. In total around 25 million people and 207,000 km^2 were left without electricity for up to twelve hours.

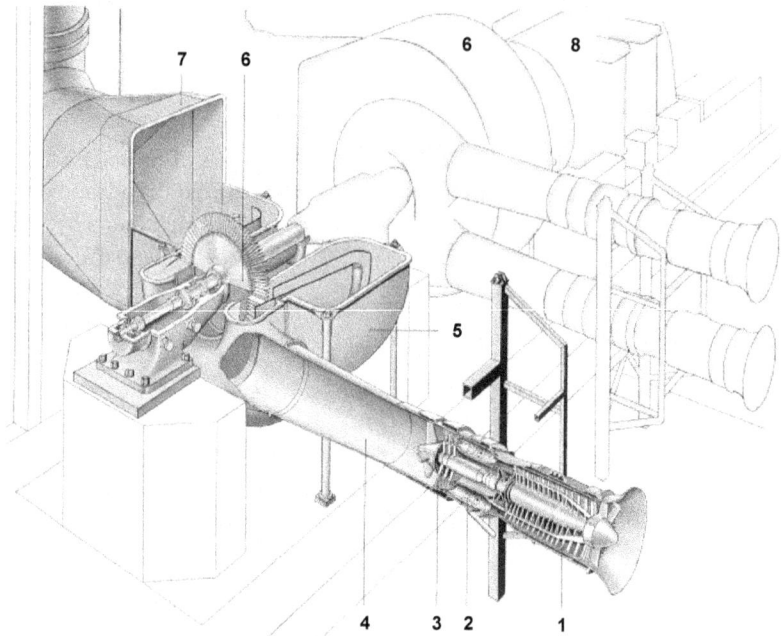

Figure 5-84   *52 MW aero jet expander, peak load power station at Emden, D* [289] *1 – RR Avon turbojet gas generator (GG) compressor, 2 – GG combustor, 3 – GG turbine, 4 – jet exhaust duct, 5 – power turbine casing, 6 – power turbine, 7 – exhaust duct, 8 – power generator*

In total BBC delivered 15 of these turbo-expander units between 1966 and 1969, with powers between 10 and 74 MW. The first four 10 MW units of this type were ordered by the Puerto Rico Water Resources Authority, Puerto Rico and put into practice on the basis of Pratt & Whitney JT3C-6 turbojets. The prime purpose was grid stabilisation and back-up during the hurricane season; consequently, the lightweight GT containers were designed to withstand wind velocities up to 250 km/h. The air-cooled 12,500 kVA turbo-generators, 60 Hz, were built with special heat-resistant winding to compensate for the strong heating gradient during the rapid acceleration of the gas generators. The units were operated with kerosene, typically for 1,500 h between overhaul. [290]

There were several versions of aero jet peaker plants from BBC Mannheim[291]:

—   one power turbine in combination with two turbojets was realised in 1965 at the Wiesmoor powerplant, located 30 km west of Wilhelmshaven, D. This unit (with approx. 25 MW power output) was erected to replace a unique, outdated peat steam plant (1909–1965) on the same site

[289]   Figure source: BBC, Peak load power station Emden, 1967

[290]   See BBM, Die Verwendung von Flugzeug-Duesentriebwerken zur Energieerzeugung, 1965

[291]   Information courtesy of Werner Steinbach, 23 February 2012.

- three turboexpanders charged by six RR Avon turbojets have been producing approx. 70 MW peak power at Itzehoe, D since 1971. The plant, refurbished in 2010, is operated via remote control from a E.ON power control station in Munich
- the BBC planning went up to a 100 MW version with four power turbines in combination with eight turbojets.

Technically extraordinarily challenging were the last two units in this series for ship propulsion. The activity started in 1961, when the German Ministry of Defense ordered a unit consisting of a Bristol-Siddeley Engines Ltd. (BSEL) marinised Olympus 200 series engine[292] as a gas generator, together with a Brown Boveri Mannheim A.G. two-stage long-life marine free power turbine. The responsible designer was Hermann Reuter, known for example as the war-time designer of the PGT 101–103 series of tank gas turbines and for several compressor designs for the BMW –003 turbojet series, whose work will be reviewed in detail in a book project on BBC's aero engine activities 1935–1955. In this context he had collected expertise which was also beneficial for integrating the first marine Olympus to this German Navy project. A test bed for extensive shore trials had already completed over 1,000 hours of test evaluation running, when the construction of the ship which was intended for gas-turbine power was abandoned. Test running of the next marine Olympus began in 1966. The power turbine was then single stage, operating at 5,600 rpm, utilising wide-chord blades.

Figure 5-85 illustrates the set up of the Olympus two-spool gas turbine and the early version of the BBC two-stage free power turbine.[293] A mechanical design highlight is the bearing structure of the free power turbine. The spokes of the shown rear disk shell are extraordinarily thin, so that they act as flexible springs and heat flow restrictors, but the thrust and buckling stiffness is high enough to absorb considerable bearing forces in case of a shock impact. In 1965 the Finnish Navy commenced a re-equipment programme with ships designed to suit the needs of protecting their shallow coastal waters in the Baltic Sea, especially for anti-submarine warfare and trade protection roles; internationally these vessels were labeled as corvettes.[294] The result was the order for the 700 t gunboats 'FNS (Finnish Navy Ship) Turunmaa' and 'FNS Karjala' from Wärtsilä's Hietalahti shipyard in Helsinki, both of which had been handed over to the Navy on schedule in autumn 1968, not least due to the fact that BBC Mannheim had accomplished the machinery work in time for ship integration.[295] The original requirement was for a compact vessel with 70 m length at

---

[292] The Olympus Mk 200 had a 16,000 lbf take-off thrust and was used for the first Avro Vulcan B2 bomber. The initial design of this second generation 'Olympus 6' began in 1952, a major redesign with five lp and seven hp compressor stages and a 'canullar combustor' with eight interconnected flame tubes. In spite of a much greater mass flow, the size and weight was little different to earlier models. Best known was the Rolls-Royce/Snecma Olympus 593, a reheated version of the Olympus which powered the supersonic airliner Concorde that was started in 1964. BSEL and Snecma Moteurs of France were to share the project. Acquiring BSEL in 1966, Rolls-Royce continued as the British partner. http://en.wikipedia.org/wiki/Rolls-Royce_Olympus http://en.wikipedia.org/wiki/ - cite_note-41

[293] BBN, Schiffs-Gasturbine mit Strahltriebwerk als Treibgaserzeuger (author H. Reuter). Especially the sophisticated bearing shock absorption structure indicates Reuter's familiarity with similar design tasks for tank GT suspension; modern tank GTs are specified for and undergo acceptance barge shock tests up to 50g accordingly.

[294] See Illingworth, Experience with the Olympus powered Brown Boveri gas turbine machinery of the Finnish Navy 700 t gunboat

[295] Beginning its sea trials in early 1968, the 'Turunmaa' was the first Olympus-powered warship to enter service, some six months before 'HMS Exmouth', the first British ship which had been refitted to trial the propulsion system for the Royal Navy and in due course, the first warship entirely propelled by gas turbines.

the water line, 7.8 m breadth and 2.7 m depth at the standard displacement, capable of carrying the new Bofors TAK 120 (Torn Automatisk Kanon, 120 mm) turret automatic gun as the main armament. In 1985–86 both ships were refitted, amongst other alterations the original Mercedes-Benz high-speed diesel engines were replaced by 3 x MTU diesel engines of 2,200 kW, each driving a propeller up to max. cruising speeds of 17 kn within a range of 5,000 nm.[296] The GT-powered speed was 37 kn, corresponding to 68.5 km/h.

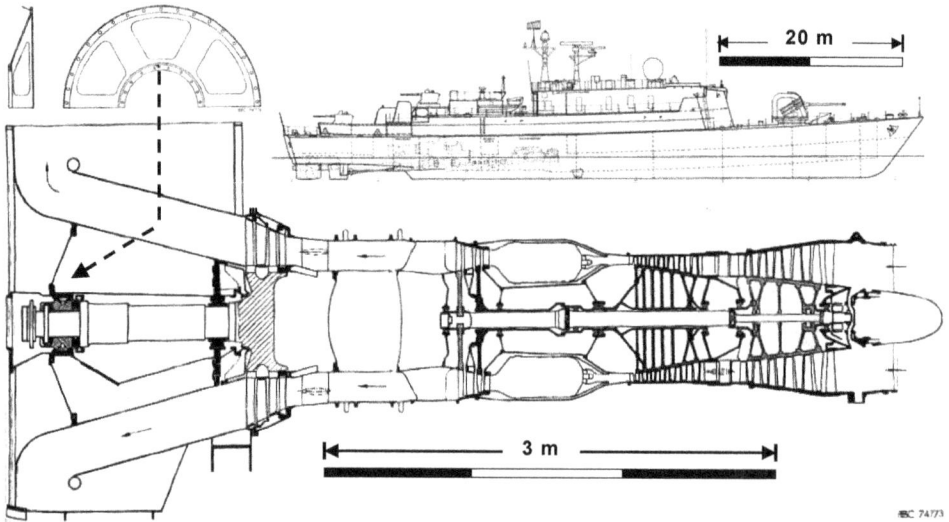

Figure 5-85   *Ship propulsion gas turbine Bristol Brown Boveri:* [297] *Rolls-Royce Olympus TM1a gas turbine with BBC free power turbine for driving a 16 MW pump-jet - Top inset left: conical rear, shock-absorbing and thermal elastic spoke-disk bearing support structure (upper part), – Top inset right: FNS 'Turunmaa' 700 t gunboat with 1×120 mm Bofors automatic naval gun on front deck (water-cooled barrel length 8.05 m), top speed 37 kn, 1969–2002   © Wärtsilä, Helsinki*

5.5.3   Compressed Air Energy Storage

Compressed Air Energy Storage (CAES) is a way to store energy generated at one time for use at another time. On a utility scale, energy generated during periods of low energy demand (off-peak) can be released to meet periods of higher demand (peak load). CAES stations correspond to hydraulic pumped-storage power plants with the inherent advantage that they can also be built in flat countryside. The first utility-scale compressed air energy storage project world-wide was the 290 MW Huntorf plant in Germany, realised by BBC Mannheim and in continuous operation since November 1978.[298] The initiative came then

---

[296] http://en.wikipedia.org/wiki/Turunmaa_class_gunboat

[297] Figure source: BBN, Schiffs-Gasturbine mit Strahltriebwerk, Figures 1 and 4

[298] The second was the 110 MW McIntosh plant of AEC (Alabama Electric Corporation, USA,1991). Both of these projects use salt domes for air caverns. Currently under development is the Iowa Stored Energy Park, a 270 MW project which will use aquifer-based air storage. A 300 MW project utilising depleted gas storage is being developed in California, and a 150 MW salt-based project is under development in upstate New York. The first adiabatic CAES project, a 200 MW facility called ADELE Adiabater Druckluftspeicher fuer die ELEktrizitätsversorgung (Adiabatic compressed-air storage for electricity supply), is planned for construction

from the NWK (Nord-Westdeutsche Kraftwerke AG), Hamburg, D, although Stal Laval in Sweden had propagated these ideas for decades on the basis of granite hard rock cavern storage.[299]

Brown Boveri had carried out the planning for the Huntorf plant and delivered the complete combustor/turbine group, as seen in Figure 5-86. The compressor group came from Escher Wyss GmbH, Ravensburg, D, belonging to Sulzer AG already at the time. The air cavern was built by Kavernen-Bau- und Betriebs-GmbH, Hannover, D.

Characteristic for this peak-load plant concept is the fact that compression of the air and expansion of the combustion gas are completely separate with respect to time, with the advantage that the full output of the turbine is available as useful power, whereas in a conventional gas turbine plant only about one third of the turbine output can be utilised. During the night two caverns in a salt stock on site at the Huntorf plant, with a total volume of 300,000 m^3, are charged with 50–70 bar compressor air during about eight hours, sufficient to operate the turbines with an output of 300 MW for approximately two hours.[300]

Before the project execution, BBC carried out thorough thermodynamic evaluations. As highlighted in Figure 5-86, the high-pressure (hp) turbine design basically followed intermediate-pressure steam turbine design principles, while the low-pressure (lp) turbine section reflects state-of-the-art gas turbine design practice:

- Consequently, though turbine inlet pressure should be maximised in order to reduce the storage volume, the inlet pressure of the air-storage gas turbine was limited to approximately 43 bar to stay in the range of steam turbine design experience
- a 2-stage expansion concept with intermediate reheat was selected for performance reasons
- the hp turbine inlet temperature was limited to 550 °C[301], i.e. without any need for novel cooling technologies for turbine rotor and blading, while after a hp expansion to approximately 11 bar, the lp section of the 50 Hz version at Huntorf could be loaded thermally higher up to TIT = 825 °C.[302]

The lp turbine expansion ratio, PR = 11, allowed this component standard GT13 turbine hardware to be used. The arrangement of both turbine bladings on one rotor necessitated a

---

in Germany in 2013 as a joint project from RWE Power, General Electric, Zueblin AG and DLR Deutsches Zentrum für Luft- und Raumfahrt (German Aerospace Centre), supported by Germany's Ministry of Economics and Technology (BMWi). Overall investments are in the order of 12 million €. Air will be compressed during periods when electricity supply exceeds the demand; the resulting heat will be buffered in a thermal energy store, and air will be pressed into underground caverns. When electricity demand increases later on, this compressed air can then be used to generate power in a turbine by simultaneously recovering the heat. The adiabatic approach will raise the efficiency level; furthermore, the compressed air reheating will no longer be based on natural gas, thus avoiding CO_2 emissions. The preferred site for this demonstration plant is Stassfurt, Saxony-Anhalt, D, an area of high wind energy usage. The planned output is 90 MW, and the air storage capacity of 360 MWh could thus substitute approximately 50 wind turbines of the local type for a period of four hours.

[299] Private communication with Sep van der Linden, 28 March 2011.

[300] See Koch, Application of the 2-stage combustion process

[301] This resulted in the given turbine pressure ratio in a hpt outlet temperature of 350 °C, equal to the compressor outlet temperature of the standard GT13B – and thus maintaining design commonality. Deviating from the standard, the hpt blading was made hollow to avoid excessive transient temperature stresses. [Chr. Jacobi]

[302] At that time a 60 Hz design version was released up to TIT = 900 °C.

relatively large balancing piston with movable sealing segments to minimise leakage losses; additionally, this leakage air is used for cooling the lp section. The entire air required in
the hp and lp sections for cooling rotor and blade carrier is preheated by exhaust gas in order to prevent excessive temperature differences in the machine. Both combustors have
been erected directly on top of the turbine casing. The low-pressure combustion chamber
(ahead of the lp turbine) is a standard design of the time, although it is supplied with exhaust gas from the hp combustion chamber and not with pure air. This gas still contains
sufficient oxygen for the current most advanced GT24/GT26 family combustion, Section
6.1.3, a general feature of sequential combustion and a frequent and successful design element in BBC/ABB/Alstom gas turbines dating from Beznau 1948, Section 5.1.5.2, via
Huntorf to the GT24/GT26 family, Section 6.1.3. The main characteristic of the high-
pressure combustion chamber is the high operation pressure of approximately 43 bar and
the low air temperature of ~50 °C at the combustor inlet, if the plant operates without air
preheating. Both combustors were designed according to the same principles, however, the
low air entry temperature of the first combustor necessitates a different distribution of wall
cooling and secondary air. A low cross-sectional load was selected for this combustion
chamber to eliminate any risks. The desired, favourable temperature distribution ahead of
the hp turbine blading was investigated in a model combustion chamber which was also
adapted to the potential use of inferior fuel quality.

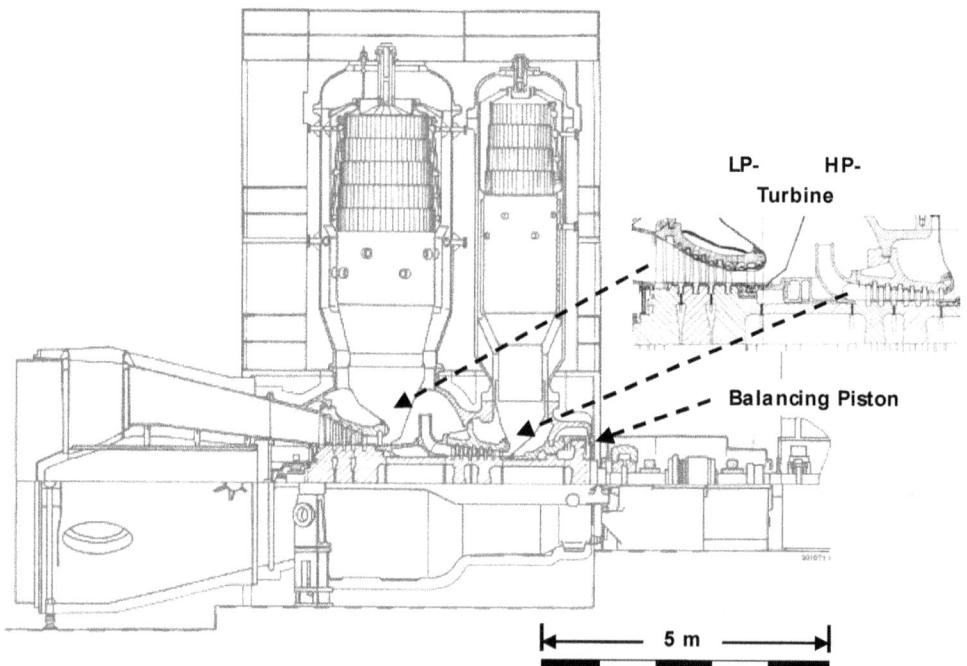

Figure 5-86   *Huntorf storage turbines and sequential combustion chambers* [303]

---

[303] Figure source: BBM, Brown Boveri air-storage gas turbines, 1977, Figure 3

The Huntorf CAES plant has now been operated successfully for more than 30 years. The number of starts has fluctuated widely during this operational period. This is attributable to:

- connecting the plant to a larger network in 1985, which added pumped hydro capacity
- the CAES plant's primary role as an emergency reserve in case of unplanned failure of other power plants
- the plant's role as an alternative option to purchasing expensive peak load from outside suppliers.

The power station is typically used today as a 'minute reserve'[304]: medium load (coal) power stations take 3–4 hours to generate full capacity before they can provide short-term power and the intervening time is preferably covered by the Huntorf CAES plant. Another typical use is for 'peak shaving' in the evening, when no more pumped hydro capacity is available. An additional application is associated with the strong increase in the number of wind power plants in Northern Germany in recent years: because the availability of this type of power cannot be reliably forecast, the Huntorf plant is able to quickly compensate any shortage in wind power.[305] Because of the intended shut down of large power generating capacities in Germany, the importance of 'minute reserves' is expected to grow in the near future. Another argument in favour of CAES is found in the steadily rising capacity of wind power, which creates less precise short-term predictability of necessary power production.

5.5.4 Low and Medium BTU Fuels (GT11N2-LBTU, GT13E2-MBTU)

The ability to burn a wide range of liquid and gaseous fuels, amongst its other well known advantages, has made the gas turbine indispensable across a wide range of economy branches and typical fuels. Consequently, much research over the years has gone into further improving and expanding the possible fuel spectrum. Market forecasts predict a significant shift in GT fuels over the next few decades. Shale gas has become an increasingly important source of natural gas in the United States over the past decade, and interest has spread to potential gas shales in the rest of the world. Globally, the first transition will be made from pipeline gas towards LNG (Liquefied Natural Gas), while high hydrogen synthetic fuels (syngases) may dominate later. Similarly, liquid fuels will shift from present 'No.2 distillate' to synthetics such as ethanol and methanol, or bio-diesel. As outlined already in Section 5.1.5.3, Sub-Section 'GT11 Development Line', Alstom's 115 MW GT11N2 is already well-positioned to deal with these expected changes. The flexible base concept and the choice of three combustors and four burner types allow it to burn a wide variety of 'exotic' fuels from blast furnace gases with a heating value as low as mere 2 MJ/kg, all the way up to 'synfuels' with a H_2 content of well beyond 50 percent. A specific feature of the GT11N2 is the large silo combustor, as illustrated for the GT11N in Figures 5-24 and 5-26, which enables operation with a wide range of fuels on the basis of three alternative combustor configurations.

---

[304] The term refers to power station output that can be made available within a few minutes.
[305] See Crotogino, Huntorf CAES

While the choice of combustors is important, the most essential feature in this context is in fact the flexible compressor, which due to its three rows of variable guide vanes allows significant airflow variations at high efficiency. This compressor technology, which the GT11N2 has in common with the GT24/GT26 product family, permits the significant airflow turndowns required to cope with fuels anywhere between 2–50 MJ/kg, i.e. fuel-to-mass flow ratio variations of 25, without hardware modification or excess bleed air.

The three alternative GT11N2 combustors are (in the rising order of heating values):

- the Low BTU combustor with a single burner for Low BTU gas
- the standard DLN Dry Low-NOx EV-combustor with dual-fuel EV burners configured in a multi-burner arrangement
- the SB Single Burner combustor, capable of burning natural gas as well as oil (distillate, heavy oil, crude oil) in diffusion mode with water injection to reduce NOx emissions.

LBTU Burner For Low BTU gases such as blast furnace gas, a dual burner with oil or natural gas as central injector is used. The design represents an evolution of technology from the 1950s and 1960s for the popular use of gas turbines in steel mills at that time,[306] but in the meantime has become a kind of standard, state-of-the art technology in a modified form. The air swirler basket provides compartments which are formed between neighbouring vanes.

Each of these compartments is sub-divided, one for air and another one for the LBTU gas, the relative size of which is determined by the heating value of the LBTU gas. Thus, the combustion air and LBTU gas leave the alternate compartments of the swirler basket to form a recirculation zone. Swirl intensity, swirl basket size and the resulting velocities determine the flame formation and stability. Although there is no premixing, the diffusion flame temperature is very low, resulting in very low NOx emissions down to single digit figures. Most requirements can therefore be met without dilution or other NOx abatement measures.

The corresponding burner options are illustrated in Figure 5-87.[307]

Modified Standard EV Burner and MBTU EV Burner The standard EV burner, Figure 5-60, has a proven capability with natural gas from 35–50 MJ/kg using standard hardware. A wider range of fuels may be covered with modifications of the gas hole size and injection pattern. In principle, the holes are made larger to reduce gas inertia and are moved towards the wide end of the cone to compensate for higher flame velocities. The burner's stable operation range with diluted natural gas fuels then goes down to 18 MJ/kg.

---

[306] See Zaba, On the Use of Gaseous Fuels of Low Calorific Value in Gas Turbines, and – see Zaba, Verwendung der BST-Gasturbinen fuer den Betrieb mit Brenngasen mit tiefem Heizwert (Low BTU-gas). A background report on the author of these papers, Tadeusz Zaba, (see Jenny, The BBC turbocharger p. 89), illustrates the sometimes surprising origin of the company's international workforce. During the early phases of WW II (1940), two Polish divisions fighting with the French Army were forced towards Switzerland where they were subsequently interned. Polish soldiers who had started studying at home were allowed to continue and finish their engineering studies at the ETH Zurich, where they were clearly identifiable during wartime by their brown uniforms. 'Tadek' Zaba managed to join BBC afterwards, where according to Jenny he became 'a pillar of the gas turbine department'.

[307] Figure source: Kenyon, Using non-standard fuels...; Figures 6-6, 6-4, 6-5 from left.

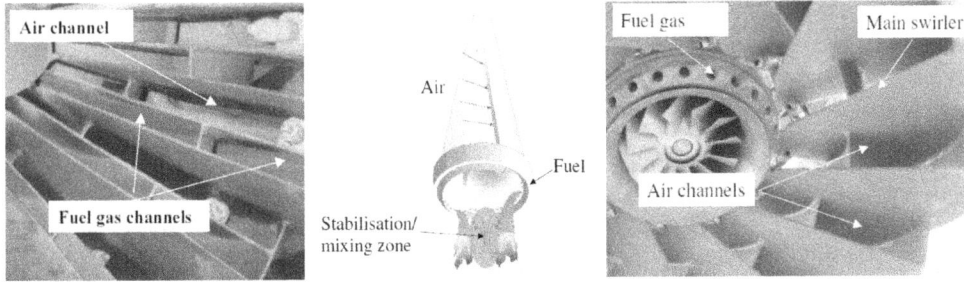

Figure 5-87   *GT11N2 burner options: LBTU burner (l), MBTU EV burner (m), SB diffusion burner (r)*

The MBTU EV burner represents a modified version of the standard EV double-cone burner, adapted for fast burning gases such as MBTU syngas or other gases with a high hydrogen content. Injection of these fuels along the air inlet slots is impossible due to the high flame velocity, which might lead to flashback into the burner cone. In addition, the higher fuel volume flow resulting from the lower heating values would lead to distortions in the incoming air profile if the fuel was to enter through the air slots. These constraints are overcome by extending the approach used on the standard EV burner to deal with lower heating values and faster combustion velocity, with larger fuel holes shifted to the wide cone end. For the MBTU burner the circular holes are located close to the burner end, injecting the fuel radially inward, a precondition for safe operation even with high hydrogen content fuels. Shortening of the standard EV burner to increase the air velocity towards the burner exit provides additional flashback protection. This high air speed together with thorough premix helps to minimise NOx values and further improve flashback resistance. Extensive laboratory testing and application of low NOx premixed combustion of MBTU fuels has already been demonstrated and reported for a 185 MW GT13E2 annular combustor application back in 1994.[308]

SB Diffusion Burner   In this burner the combustion of liquid and gaseous fuels is achieved by injecting the fuel in a combustion air circulation (vortex) zone created by a swirler. There are two concentric nozzles (injectors) for oil or natural gas in a dual-fuel single burner. The swirler basket consists of several vanes with air flowing through the channels formed between the two neighbouring vanes. The flame is anchored and stabilised in the recirculation zone and the swirler acts as a flame holder. Such a diffusion flame usually has a stochiometric air-to-fuel ratio $\lambda = 1$ with a relatively high flame temperature, leading to a high thermal NOx formation. Steam or water injection is used for NOx abatement in order to meet emission requirements. Such diffusion burners, if properly designed, can burn a wide range of fuels such as natural gas/LNG, heavy oil, crude oil, distillate and naphta. Due to their large size these burners proved to be very robust for crude and residual oil combustion, even with a high ash content, avoiding clogging and coking of injection holes.

The burner choice for process gases in the low and medium LHV (Lower Heating Value) range is solely LHV-dependent, while above LHV > 18 MJ/kg additional criteria like emission limits, fuel flame speed (flash back) and overall project economics have to be taken in-

---

[308]   See Doebbeling, Low NOx premixed combustion of MBTU fuels

to account. The SB diffusion burner can burn any gas in this range, but the high NOx emissions would require NOx to be controlled to acceptable levels. The EV-DLN burner is preferred for most natural gas applications, while the SB burner applies for process gases with high flame speeds.

In 1994 the first Low BTU application of the GT11N2 gas turbine was conceived together with Kawasaki Heavy Industries Ltd. from Kobe, Japan and supplied to Baoshan Steel Mill, Shanghai district, China, with another unit subsequently supplied to Mizushima, Japan, in 1998. Bao Steel Corporation installed a GT11N2 unit for operation with LBTU blast furnace gas, a by-product of the steel-making process. The combined-cycle facility aligns gas turbine, fuel gas compressor, steam turbine and turbogenerators on a single shaft. The gas turbine went into operation in the first quarter of 1997 and has accumulated more than 60,000 fired hours to date. The plant is designed to provide up to 150 MW of electrical power and up to 180 t/h of steam to the steel mill, while burning blast furnace gas with a heating value of only 2.2 MJ/kg. The GT11N2 gas turbine with the LBTU combustor confirmed the expected low emission values.[309] Since the beginning of commercial operation, the Baoshan plant has been operating continuously in base load mode, except for scheduled CCPP inspections and external events such as shutdowns of the blast furnace.[310]

---

[309] http://www.alstom.com/power/fossil/gas/gas-turbines/gt11n2/
[310] See Willfort, Gas turbine fueled by LBTU tailgas

6 Combined Cycle Power Plants – GT Technology Breakthroughs since the 1990s

The following section addresses the final Phase III of the BBC/ABB/Alstom GT development from the early 1990s under the newly founded ABB Asea Brown Boveri Ltd., up until today as part of Alstom's Thermal Power Sector. Although a comparison with former times is difficult, it appears that the past two decades were not only highly eventful and extraordinarily competitive, but also characterised by remarkable technological achievements, not least the stupendous breakthrough of the gas-powered combined cycle power plant.

6.1 Gas Turbine Development Survey

Industrial gas turbine technology received a pronounced boost at the start of the 1990s, resulting in a considerable increase in component performance and higher turbine inlet temperatures. Roughly estimated, turbine inlet temperatures increased by 200 K (degrees Kelvin) between 1990 and 2000, of which ¼ or approximately 50 K was owed to the introduction of new Ni-based high-temperature alloys, the development of durable thermal barrier coatings and corresponding (directionally solidified and single crystal) advanced casting and other advanced manufacturing technologies for turbine blading, while the rest largely resulted from considerably improved film and impingement cooling techniques, Section 6.2.2.2. These advancements were reflected by an increase in CCPP efficiency of nearly 10 points, close to the magic 60 percent total efficiency mark (which was finally achieved a decade later). All of these improvements were accompanied by a considerable reduction in CO_2 and NOx emissions due to DLN dry, low-NOx combustion techniques, but mainly due to the widespread use of natural gas fuel. Typically, the NOx emissions went down from 40 ppm (1990) to 15–25 ppm (2000) and in the meantime these have been further reduced to single-digit figures.

While turbine cooling technologies basically followed the established path of previous aero engine developments, the initial introduction of single crystal and directionally solidified materials for large power generation turbomachinery with heavier parts and considerably increased dimensions especially challenged the supply industries for precision cast parts and high temperature coatings. CFD Computational Fluid Dynamics expanded reliable flow prediction and analysis to the 3D viscous regime. Improved understanding of high-temperature turbine cooling technology in combination with extreme lifing demands open-ed up new technological frontiers.

Consequently, the introduction of the advanced GT24 (60 Hz) and GT26 (50 Hz) gas turbine technology, a lot of which was brand new, between 1995 and 2000 to an overheated power generation market, especially in the USA, led somewhat predictably to considerable engi-neering and management difficulties in the later years of ABB, Section 2.1.2. This has been steadily overcome in the meantime under Alstom ownership. The present Alstom GT port-folio [GT11N2, GT13E2, GT24, GT26 and derivatives] seamlessly covers a power range from 115 MW up to 326 MW for turnkey power plants including simple cycle, combined cycle, re-powering and retrofit. It supplies unique, historically evolved technology features to the power market which meet the requirements for extraordinary low emissions, high total efficiency and unique operational flexibility.

The opening chapter of this section provides a survey of gas turbine development, first covering the gas turbine technology and peculiarities of the business environment at the beginning of that period, followed by a short introduction to the ABB/Alstom GT develop-ment department. Afterwards the advanced GT24/GT26 development key project will be thoroughly reviewed and the contributions of a few prominent engineers highlighted briefly.

In line with the basic structure of the book, the description of this most recent third phase of the Companies' technical GT history also provides the basis to address the 'Turbine' as the last of the three GT key components in detail, followed by a chapter on 'Systems De-velopment' which here comprises a view of the comprehensive test facilities, high tempera-ture material and coating developments, as well as progress in advanced computational tools. Finally, the historical review will end with a presentation of the combined cycle power plants, both historically and in the context of its present overwhelming commercial and technical importance as the key carrier of gas turbine products.

6.1.1   The GT Technology and Business Environment in the 1990s

The restart of ABB in the power generation business coincided with a period of far-reaching changes in international politics: *'nineteen eighty-nine was the biggest year in world history since 1945*[1]. The end of the Cold War had an immediate and significant im-pact on the military-industrial complex, especially in the United States. It is said that ap-proximately 20,000 engineers were shifted from military to civil applications, mostly in the aero gas turbine section, but there were also substantial relocations from advanced high technology aero engine development to the neighbouring area of power generation gas turbines. In hindsight these circumstances contributed decisively to the breakthrough in

---

[1]   *'Nineteen eighty-nine was the biggest year in world history since 1945. In international politics, 1989 changed everything. It led to the end of communism in Europe, of the Soviet Union, the cold war and the short 20th cen-tury. It opened the door to German unification, a historically unprecedented European Union stretching from Lisbon to Tallinn, the enlargement of Nato, two decades of American supremacy, globalisation, and the rise of Asia.'* Timothy Garton Ash, Guardian, 5 Nov. 2009, http://www.guardian.co.uk/commentisfree/2009/nov/ 04/1989-changed-the-world-europe

combined cycle gas turbine technology. However, it also increased both the real and imagined risks of the ABB endeavour considerably.

When the huge advantages of combined cycle gas turbine power plants were becoming clear, the market leader General Electric introduced their F class, a new type of gas turbine based on newly available state-of-the-art GT technology, made larger and more powerful by the application of higher compression ratios and combustion temperatures. The first F-class machine, rated 147 MW and with a firing temperature Thg of 1,530 K (2300 F), was shipped to Virginia Electric & Power Co's (Vepco) Chesterfield Power Station, Va., USA in 1988 and started its commercial service as Chesterfield 7 on 6 June 1990. Its sister unit Chesterfield 8 began commercial operation in May 1992, the year the 159-MW GE 7FA was introduced with a 1,560 K (2350 F) firing temperature.[2] This technology had its origins in the field of aviation and posed a huge challenge for manufacturing/casting technology and in view of the considerably higher service life demands of power generation applications compared to aero gas turbines, also for engineering disciplines like mechanical integrity. GE was quickly followed by Westinghouse in 1989, who launched the 501F in collaboration with Mitsubishi, and in 1991 by Siemens with its V94.3. Starting from scratch and eager to catch up with its competitors, ABB decided for a 'leap frog' strategy[3] to position itself at the head of this promising market development. Consequently, ABB launched its own GT24/GT26 project, the first for the 60 Hz segment and the second, in a factor 1.2 geometrically scaled version for the 50 Hz market segment, in December 1991.[4]

With the first round of development completed in record time, details of the innovative new ABB design were announced as early as 1993[5], followed by a comprehensive ASME conference paper[6] in 1994. The first GT24 prototype was installed two years after the first announcement at the Gilbert, NJ, USA power generating station, operated by Jersey Central Power & Light. At the time of GT24's disclosure the strategic target of superior performance values was achieved with a nominal SC power output of 165 MW and an efficiency of 38.5 percent. Presented as a revolutionary solution, it was the most compact model available on the market and the only one to use sequential combustion with a particularly high compression ratio. The GT24 could also boast the cleanest operation and an official net plant combined cycle efficiency of more than 56 percent, 2–3 percent higher than its closest competitors.

However, the competitors quickly counter-attacked with their own innovations by the time the 50 Hz GT26 was brought to market with 265 MW power output in 1997. General Elec-

---

[2]   http://www.ccj-online.com/3q-2011/7f-users-group-history/

[3]   As referred to by Goeran Lundberg, then the head of ABB Power Generation, Honda's story as an archetype of a smaller car manufacturer entering a new market already occupied by highly dominant competitors was an admired example. The story of their market entry, and their subsequent huge success in the US and around the world, has been the subject of some academic controversy. Competing explanations have been advanced to explain Honda's strategy and the reasons for their success: Lundberg's conclusion was simply 'Speed is everything!' [Private information J.-E. Bertilsson, 4 May 2010]

[4]   See Nieto, MW and Km/h

[5]   The official press release for the all-new GT24 took place at the Waldorf-Astoria Hotel, New York on Thursday 9 Sept. 1993 with parallel announcements at London, Frankfurt and Zurich, followed by adequate customer programmes including fun regattas with traditional America's Cuppers during the following two days at Newport, RI, USA.

[6]   See Imwinkelried, The advanced cycle system gas turbines GT24/GT26

tric announced a 7FA version upgraded to 170 MW power output, in addition to the new steam-cooled H-class turbine concept. Westinghouse introduced the 501G series, still developed with its partner Mitsubishi, but with additional assistance from Rolls-Royce, while Siemens sought out the specialised expertise of Pratt&Whitney to develop the V94.3A/V84.3A, an impressive improvement in comparison to the previous status.

With some delay, the excitement that had surrounded the launch of these various advanced GT systems was overshadowed by disappointment, when repeated operational problems occurred and dissolved expectations of a straightforward marketing success. First GE was affected when the new F-class turbines suffered a series of severe rotor damages. In 1995, GE immediately mobilised all of its resources to resolve the early lifetime failures of its F-class gas turbines and recalled the defective rotors in an unparalleled airlift exchange programme.[7]

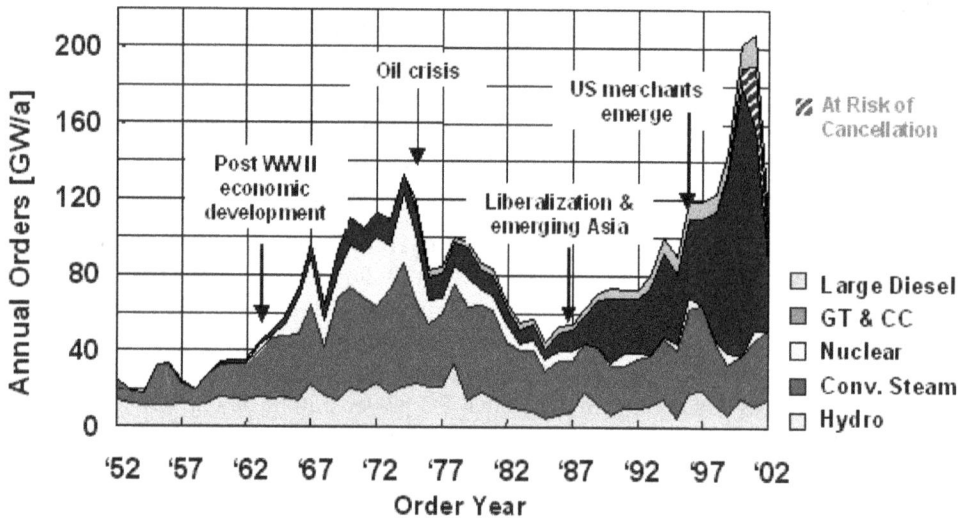

Figure 6-1   *Fifty years development of annual orders for worldwide power generation equipment, broken down into 5 PG categories*

GEC Alsthom also made necessary modifications to the 50 Hz F class turbines for which it possessed the manufacturing licence. Two years later these severe problems were solved and GE recovered its lead position in time to successfully participate in one of the strangest market phenomena so far, the US power generation 'bubble', Figure 6-1, for the power generation category 'GT&CC' (Gas Turbines & Combined-Cycle plants).

This figure illustrates the annual order distribution of power generation equipment from a global perspective over a period of fifty years from 1952 to 2002. In relation to a long-term average of approximately 40–60 MW/a over the previous decades, there are two pronounced elevations: a rather flat saddle from the late 1960s to the early 1980s, peaking in

---

[7]   In total 22 F-turbines had to be shut down and the rotors removed and repaired, while another batch of 28 F-engines had to be refurbished before delivery. See Catrina, ABB – Die verratene Vision, p. 114

the oil crisis of the mid-1970s, and a slowly accelerating, then pronounced peak, culminating in 1999–2000 with an sharp increase where the annual order volumes surpassed the long-time average by 300–400 percent for a short while. The main reason for this development was a transfer of power market deregulation mechanisms from a relatively small experimental field in the UK during the Thatcher era to the much bigger US playground, where a backlog demand for new power generation equipment had accumulated over the years (often demonstrated clearly by the increasing frequency of regional power 'blackouts'). This situation was accompanied by a fundamental change in the structure of the power generation industry, away from the classically established, long-term-oriented, technically qualified utilities to a new form: IPP, Independent Power Producers and Power Merchants, who followed strictly financial RoI (Return-on-Investment) optimisation criteria. Between 1990 and 2000 the ratio of utilities to IPPs in the USA changed from 3:1 (1990) to < 1:4 (2000).[8] This trend was technically supported by two striking advantages of the broadly emerging combined cycle GT power plants: a) their superiority in total efficiencies (close to 60 percent), i.e. 12–18 points more in comparison to large steam and nuclear plants, and b) the advantageous specific cost relation, where the typical 800 USD/kW CCPP value (= 100 percent) beat large steam and diesel plants by at least 50 percent, while nuclear power plants required 100–250 percent more. These facts explain a fundamental difference to the peaks of the mid-1970s that basically relied on conventional steam and nuclear investments, while the 'US power bubble' of the late 1990s was clearly related to the breakthrough in combined cycle GT technology.

Especially when it comes to a judgement about the responsible management team at the time, one aspect of this boom-bust cycle should not be overlooked. It seems this 'US power bubble' in power generation equipment orders with a culmination in 1999/2000 was the first in a while, followed by a series of similar phenomena over a short period: the dot.com or internet bubble of 2001, the stock market crash of 2002 (largely influenced by the events of 9/11/2001), and finally the more recent US housing bubble of 2007. Sensitivity towards potential, sudden collapses of overheating markets has considerably increased in the meantime. At that time, in the middle of an unprecedented, nevertheless explainable upturn of the US power market and under the historic circumstances of ABB's recovery in the power market, seizing opportunities in this rapidly expanding market for sellers was a logical and natural decision. The special situation of the booming US power market at the end of the 1990s and the described changes in the customer structure are two of several reasons in the context of difficulties which arose during the introduction of the new gas turbines; technical explanations will follow.

6.1.2   The ABB/Alstom Development Department

The implications of the BBC/ABB/Alstom company transitions for the workforce were discussed from a global perspective in Figure 2-2, with special focus on Switzerland in Figure 2-3. Though the corporate dominance of a Swiss parent company, as during the BBC era disappeared, Alstom's GT business still has its headquarters in Baden, Switzerland, unchanged for more than 80 years.

---

[8] This structural change played also a certain role when Alstom decided to put its US power activities on hold from 2000–2005 and several financial settlements had to be negotiated with these clients.

Throughout this book patenting records have been used as an indirect measure of innovation and creativity, especially of the development departments. The results of Figure 2-4 have been used before to illustrate principal changes, especially during the transition from BBC to ABB/Alstom. It becomes obvious that patenting per se received considerably more appreciation from the Companies during the recent ABB/Alstom period. In addition, Figure 2-5 and several corresponding detail analyses[9] appear to illustrate a certain trend towards a consistent broadening of the inventor base, especially in direct comparison with the historic Phase I until 1945 and the present Phase III from 1988. The large increase in patenting activities between 1990 and 2000 is rather unique, also comprising the early development period of the new gas turbines GT24/GT26.

In a more personalised form, Figure 6-2 illustrates the distribution of patent grants for the 10 most productive engineers[10] between 1991 and 2010. The quantitative difference up to the year 2005 (still based on filings during the peak time of GT24/GT26 development activities) in comparison to Figure 4-6 (1921–1955, based on 4 inventors only) and Figure 5-1 (1956–1990) is obvious.

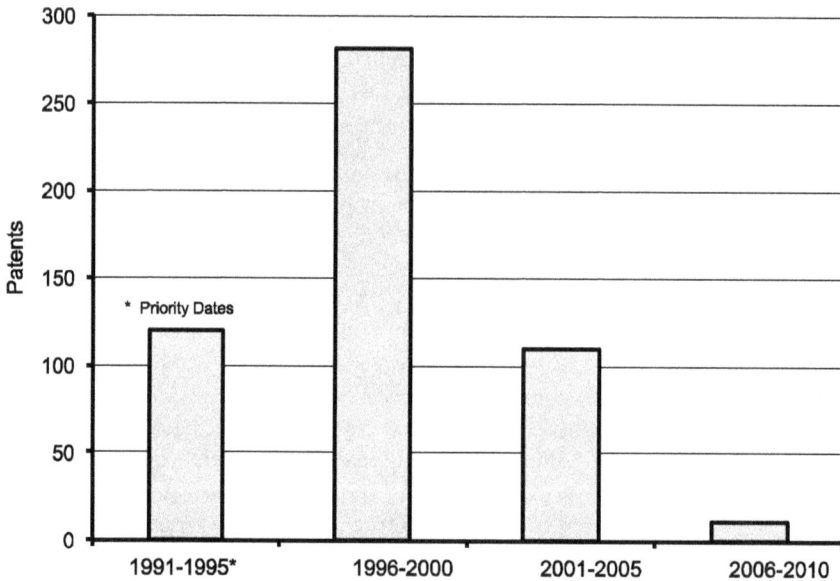

Figure 6-2   *GT patent development between 1991 and 2000 (ABB) and 2001–2010 (Alstom) from the 10 most productive engineers of that period*

Quantitatively, patenting activities over equal periods were roughly 5-times higher in Phase III than in Phase II during BBC ownership. This result can only be partially explained by BBC's late reservations about official IPR protection. This marked difference is

---

[9]   See also Figures 4-6 and 4-7 for Phase I, Figures 5-1 and 5-2 for Phase II and finally Figures 6-2 and 6-3 as representative for Phase III.

[10]   The total number of 537 patents in this 20 year period came, in descending order, from the following 10 engineers: H.U. Frutschi, R. Althaus, A. Beeck, H. Wettstein, B. Weigand, J.J. Keller, R. Fried, J. Hellat, F. Joos, F. Kreitmeier.

also clearly visible in Figure 6-3 in comparison to the reference Figure 5-2; the effect of considerably strengthened technology developments and simultaneously strong commitment in favour of patenting innovative ideas is especially pronounced for the engineering disciplines of 'combustor, turbine and turbine cooling and GT/CC systems'.

The impression of exhaustion in terms of technology in the GT product development, as described before by the low level of patent activities up until then, is also underlined by looking at the GT product portfolio from 1990 at the very beginning of ABB responsibility, Figure 6-4. At that time the ABB gas turbines under 30 MW were manufactured in Sweden by ABB STAL, while the Types 8, 9, 11 and 11N were built in Switzerland and the Types 13 and 13E in Germany.

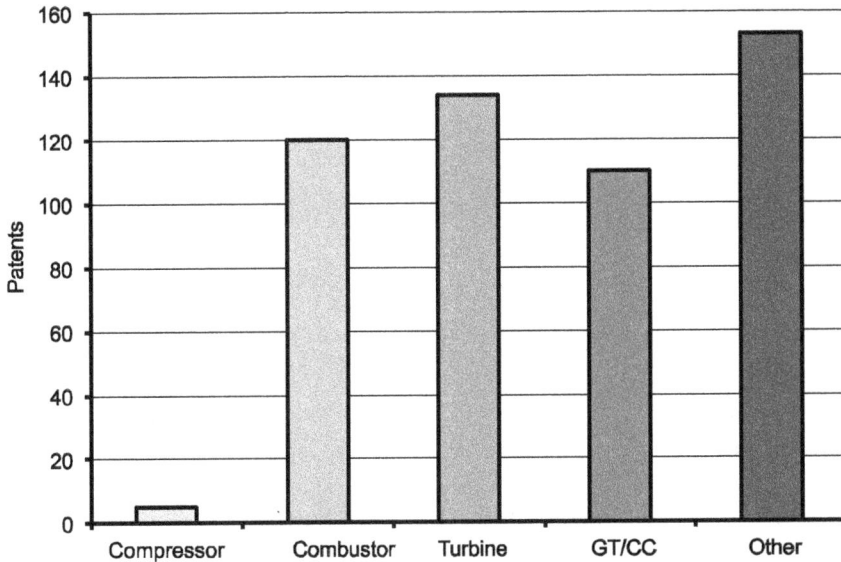

Figure 6-3   *GT patent contributions between 1991 and 2010 from the 10 most productive engineers – by engineering discipline*

The portfolio of 9 engines in total indeed covered a 'full range from 8 to 150 MW', but a more detailed look confirms just the quantitative impression given by the sheer number, and a considerable lack of up-to-date technology. The roots of the GT13 family dated back to 1972; except for the GT11N2 (introduced in 1994), the GT13E2 (1997) and the GT8C2 (2000) which were used for a partial portfolio renewal, Section 5.1.5.3, the rest were rather outdated. It was obvious that immediate action was needed, especially at the decisive, market-defining upper end of power.

A full range from 8 to 150 MW

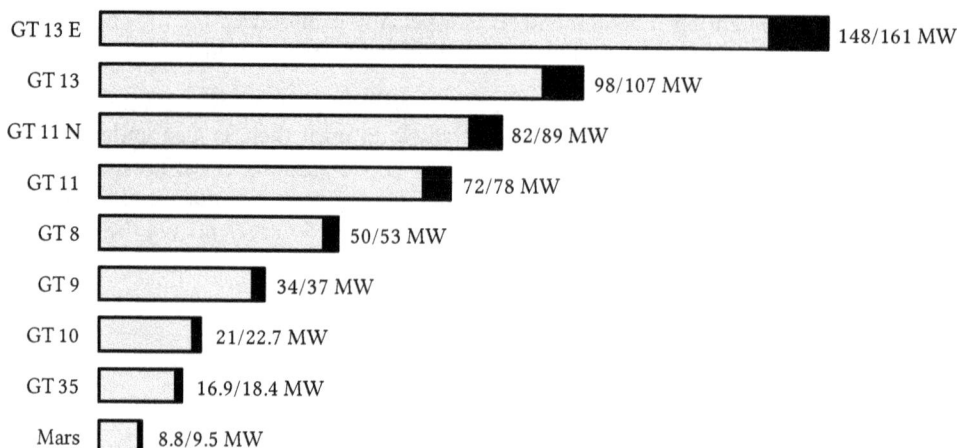

GT 13 E	148/161 MW
GT 13	98/107 MW
GT 11 N	82/89 MW
GT 11	72/78 MW
GT 8	50/53 MW
GT 9	34/37 MW
GT 10	21/22.7 MW
GT 35	16.9/18.4 MW
Mars	8.8/9.5 MW

Figure 6-4   *ABB GT product family, as announced in 1990:* [11] *Base-/peak-load ratings at generator terminals for natural gas*

6.1.3   The Key Development Project GT24/GT26

The initial considerations concerning a GT development restart at ABB tried to take emerging business trends into account:

- The power generation market was clearly moving to higher output and efficiency
- Environmental aspects, especially substantial NOx reductions, became increasingly important
- Internationally spreading ideas towards deregulation supported the design intentions to focus on operational flexibility from the very beginning.

The technical heritage clearly demanded reliance on a number of established success features:

- The 'welded rotor' had been one of the Companies' best proven design assets since 1929 and should be maintained for future designs as well
- 'Advanced axial turbomachinery' was traditionally a prominent design principle which had been underlined by the first time market introduction of the IGT transonic compressor in 1980; a tradition well worth being continued
- The successful introduction of 'lean premix combustion' as of 1982 in the GT13D with NOx < 25 ppm should be continued
- Since the realisation of the first utility 'combined cycle power plant' in 1961, a typical strive towards a breakthrough in combined cycle gas turbine technology characterised the Companies, emphasised by approximately 53 percent CCPP efficiency on the basis of in total 15 such GT13E installations in the 1992/3 period.

---

[11]   See ABB, Heavy-duty gas turbines ranging from 8 to 150 MW, 1990

On the other hand there were also recent experiences where a lesson was learned:

- The impact of mechanical integrity limits on the allowable machine size, as represented by previously cancelled projects in the 150–210 MW power range (GT15, GT17), based on GT8 test failures between 1982 and 1983, Section 5.1.5.4
- The clear correspondence between turbine inlet temperature and the achievable emission targets.

Given ABB's special situation at the 'time of change', i.e. the competition were highly capable and their own GT portfolio somewhat weak, it quickly became obvious that the intended 'leap frog' strategy of challenging the competition from a lead position required an innovative approach, already for the cycle selection. ABB decided to rely on historical experience, as the 'GT reheat cycle', Figure 2-21, had already contributed to the superior performance of plants such as the Beznau power plant, Section 5.1.5.2, and more recently the Huntorf CAES plant, Section 5.5.3. As mentioned before, the recommendation for performance improvement through 'fractionalised combustion' dated back to as early as 1940, when Aurel Stodola commented on the results of the commissioning tests of the first Neuchâtel utility gas turbine, Section 4.3.3.2.[12]

A comparison of idealised processes still at the time of the GT15N/GT17N projects in 1990/1991 revealed that on the basis of equal pressure ratios, the reheat process had somewhat lower efficiencies compared to the Joule process, but the specific work of the reheat process was nearly 30 percent higher. Under real operation conditions, for example by assuming a compressor efficiency of 86 percent and corresponding turbine efficiency/ efficiencies of 88 percent, the reheat operation principle even surpasses the Joule process[13]:

1. The reheat process reacts less sensitively to component efficiencies, which results with otherwise unchanged component parameters in a higher process efficiency.
2. The double expansion of the working medium preserves the aforementioned superiority in specific work for the reheat process; this allows a higher power density in comparison to a conventional gas turbine, i.e. smaller and cheaper frames and component sizes for the same power class.
3. Under the assumption of equal process efficiencies the reheat exhaust temperature is higher, which raises the reheat attractiveness, especially for combined cycle applications.
4. At 50 percent load the advantage of the reheat concept for CCPP part load efficiency was estimated to be 4-5 points better than a conventional GT approach.
5. The two combustors create additional operational flexibility, especially in terms of combustion stability, which can also be beneficially applied for an optimised CCPP part load operation at a superior performance level without increasing the turbine inlet temperature in general. Consequently, quick CCPP start-ups were predicted with hot

---

[12] In February 1939 in his seminal IMechE presentation, Adolf Meyer introduced an 'intermediate reheat stage' as a potential measure to improve cycle efficiency by 2.5 points. Wilhelm Endres later provided another hint towards the potential BBC/ABB preferences for 'sequential combustion': In 1947 recuperated 2-shaft-GT configurations achieved the performance level of steam turbines when inter-cooled. The steam turbines regained their performance superiority again between 1954 and 1965 with the introduction of intermediate reheat and higher temperatures in general; an experience which was certainly kept in mind within the ST-dominated BBC/ABB development department. See Endres, 40 Jahre Brown Boveri Gasturbinen, p. 4.

[13] See Lechner, Stationäre Gasturbinen, p. 36

starts below 60 minutes. The issue of operation flexibility also comprised different features such as power augmentation, ancillary services and upgrade packages.

6. As outlined in detail in Section 5.3.5, the sequential combustion concept contains a number of important advantages for achieving low NOx emission values, so that the NOx targets for dry fuel gas become as low as < 25 ppm and for wet oil combustion < 42 ppm, still with considerable development potential.

These thermodynamic and operational advantages must be measured against a number of drawbacks:

1. The exploitation of the higher thermal efficiency of the reheat process demands a higher compression ratio, consequently with more compressor and turbine stages.
2. The second 'SEV sequential combustion chamber' is highly loaded and as such an additional and costly gas turbine component.
3. The aforementioned flexibility of the sequential combustion simultaneously raises costs for process control and fuel supply of the gas turbine.
4. Power generation output and efficiency are somewhat reduced as not all of the intake air mass flow is participating in the reheat process.

In summary, the better performance of the reheat principle clearly demands more constructive expenditures.

6.1.3.1   Secret 'Skunk Works' 1991–1995

On 20 June 1991 the GT15N Programme Steering Committee stated:[14]

–   In retrospect, the decision to use a RR aeroderivative (compressor) was a mistake.
–   The turbine can not be completed by ABB without outside help.

In July 1991 the ABB Gas Turbine Department received the task of developing a new machine for the 60 Hz market at the top end power range of 150+ MW. This development had to be followed by the development of a correspondingly scaled gas turbine for 50 Hz applications.

As outlined already in Section 5.1.5.1, the naming convention can only be partially reconstructed. The new products were placed as GT14/GT16, in between the existing GT11/GT13 families with approximately 110 and 130 cm rotor diameter in the combustor region, and below the cancelled GT15/GT17 projects. In addition, there was a differentiation during the definition phase between single (1) and dual (2) stage combustion concepts, e.g. GT1.4 and GT2.4, so that after the decision to use sequential combustion, the new GT24/GT26 designations emerged. Finally, external partners were intentionally misled by using project designations like GT11N3, which was meant to insinuate a continuous development of the GT11 family instead of the radically new design approach actually followed.

Secrecy and disguise were part of the project set-up from the very beginning. A group of about 100 hand-picked engineers was isolated in a separate, inconspicuous office building at Sandstrasse 10, Gebenstorf, AG, Figure 6-5 – some 5 km west of the ABB office in Baden, AG, where several hundred more worked on detail solutions without knowing the overall

---

[14]   See ABB, GT15 Program Steering Committee Review, 1991

context. The reference to Lockheed's famous 'skunk works'[15] in this section's header applies to both the remote location and the enthusiastic team spirit. The Gebenstorf group, which returned to base when the newly built Konnex building was ready for occupancy, Figure 2-16, was led by Hans-Juergen Kiesow, Section 7.2, who succeeded Fredy Haeusermann in 1995, Section 6.1.4, as Director of the GT Development Department. The workload of this period can be barely overestimated, roughly 20,000 components and single parts had to be drafted, conceptually investigated, numerically assessed and finally drawn ready for production of this new engine family.

Figure 6-5   *GT24/GT26 secret design offices 1991–1995 at Gebenstorf, AG, CH*

In an incomparable document of 2,233 pages the design activities between 7 January 1992[16] and 17 June 1996 were recorded in detail in a collection of 'Weekly Reports'[17] by Pierre

---

[15] The designation 'skunk works' is widely used in business, engineering, and technical fields to describe a group within an organisation given a high degree of autonomy and freedom from bureaucracy, tasked with working on advanced or secret projects. It is an official alias for Lockheed Advanced Development Projects, responsible for a number of famous aircraft designs since the 1950s, including the U-2, the SR-71 Blackbird, the F-117 Nighthawk, and the F-22 Raptor. Its largest current project is the F-35 Lightning II. http://en.wikipedia.org/wiki/Skunk_Works

[16] Before that date there are two 'GT15/GT17 Project Meetings' documented, the first one on 29 October 1991 at Hotel Du Parc mentions 13 participants, in alphabetic order: Althaus, R; Farkas, F.; Fischer, M.; Frutschi, H.U.; Haeusermann, F.; Hellat, J.; Kiesow, H.-J.; Louis, Th.; Meylan, P.; Mueller, M.; Thoren, K.; Vogeler, K. and Wettstein, H. The second meeting on 5 November 1991 dealt with, amongst other topics, comprehensive information about the Japanese 'Moonlight Project' (1978–1987) of a 100 MW reheat gas turbine technology programme. This information may have additionally confirmed the already established trend towards the sequential combustion/reheat concept of what later became GT24/GT26. See ABB, GT15/GT17 project meetings

[17] See ABB, GT24/GT26 Weekly Reports. The collection of weekly reports reveals the mechanical design process from a long-term perspective, from early ideas to the final configuration with all the intermediate trials and

Meylan and Ernst Pauli, heads of the mechanical design groups. During that period in the early 1990s the mechanical design office underwent a fundamental change away from the drawing boards to CAD, Computer-Aided Design systems[18]. The General Arrangement I of Figure 6-7 presumably belongs to a final series of engine general arrangement drawings from a drawing board, after 100 years of such classic engineering activities at the former BBC premises. While hand drawn arrangements then disappeared definitively for general engine arrangements, they were and still are an indispensable tool for illustrating, developing and comparing conceptual mechanical design solutions. The 3D free-hand sketch in Figure 6-6 is a good example.

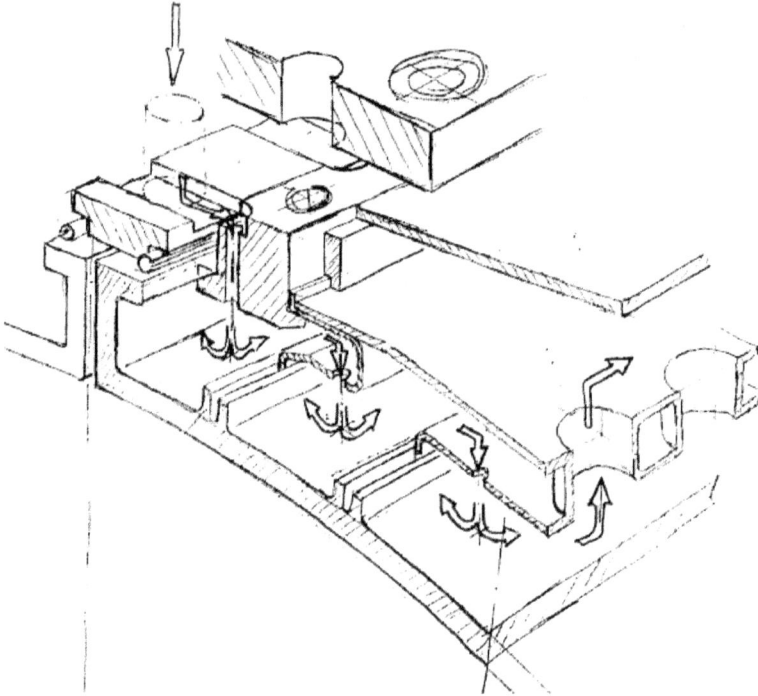

Figure 6-6   *Conceptual design: 3D free-hand sketch of 'cascaded impingement cooling, variant #2 of SEV combustor outer liner', Urs Benz, 15 January 1993* [19]

In creating this gas turbine, ABB drew upon the expertise of two aircraft engine manufacturers, the German MTU Motoren- und Turbinen-Union Munich GmbH[20] for compressor

---

18   errors. The reports are complemented by a comprehensive collection of 'general arrangement' drawings, see ABB's GT24/GT26 evolution from 1991 to 1994, Figure 6-7.

ABB/Alstom use CATIA, Computer Aided Three-dimensional Interactive Application, a multi-platform CAD/CAM/CAE commercial software developed by the French company Dassault Systèmes in the late 1970s for the development of Dassault's Mirage fighter jet. CATIA is written in the C++ programming language. http://en.wikipedia.org/wiki/CATIA

19   See ABB, GT24/GT26 Weekly Reports, p. 246 and p. 260 of 2'233

20   MTU Munich, today MTU Aero Engines Ltd., is Germany's leading aircraft engine manufacturer. Based on the former site of BMW Flugmotorenbau GmbH in Munich-Allach, the company started in the 1950s as BMW

and low-pressure turbine design and the Russian NPO Saturn[21] for the high-pressure turbine.

The first contacts between the ABB GT24/GT26 development team and the MTU APS Advanced Projects and Systems integration group[22] were initiated during 1992 by a phone call between Fredy Haeusermann and Hanns-Juergen Lichtfuss, head of MTU's development department. At first the cooperation was restricted to a few study exercises, mainly in the area of advanced axial compressor design, with and without intercooler. Mutual trust grew when MTU convincingly demonstrated its design capability of combining advanced CDA Controlled-Diffusion-Airfoil technology, Figure 4-52, with a clear-cut, rather unsophisticated annulus shape, Figure 6-13. This progress in aerodynamic compressor design technology reflected an important change in design methodology since the 1980s, away from the experimentally-based rig-after-rig compressor optimisation towards the reliable application of quasi-3D numerical flow predictions. The consequences were striking, by combining advanced aero-engine performance with a cost-saving, typical industrial gas turbine manufacturing approach. A design generation earlier the demand of a 100 percent scaled transfer of an experimentally optimised aero engine compressor ("...don't touch the V2500 annulus!") for the GT15N project, Figure 5-45, could not fulfil these expectations. However, more important for the joint ABB/MTU development effort was the confirmation that the high compression ratio of PR > 30, unheard of until then, could be realised with stable operation in 22 stages with only 3 rows of variable inlet guide vanes at an excellent polytropic efficiency level close to 90 percent. The achievement of this excellent performance level was confirmed in a 1/3 scale rig compressor ('Rig 250', Figure 4-53), built and tested at the MTU test department in 1993/1994.

Triebwerkbau GmbH, beginning in 1957 by producing GE J79 turbojets for the Lockheed F-104G Starfighter programme of the new German Luftwaffe. By 1965 it had been completely purchased by MAN AG to become MAN Turbo GmbH. In 1968/1969 MAN Turbo and Daimler-Benz formed a 50/50 joint venture for their aero engine development and manufacturing interests, considerably boosted by the Turbo-Union RB199 programme for the Panavia Tornado (MRCA Multi-Role Combat Aircraft). In 1985, Daimler-Benz bought MAN's 50% share in the company and made MTU part of its aerospace subsidiary DASA. In 2000, when DASA was merged with other companies to form the European Aeronautics and Defense Systems company EADS, MTU split from DASA and after intermediate private equity ownership became an independent joint-stock company in 2005, with at present nearly 8,000 employees and annual revenues of 2.9 billion Euros (2011).

[21] Immediately after the 'change' ABB started a design cooperation with NPO Saturn, Moscow which led to the foundation of an 80/20 joint venture/ engineering office in Moscow. Since 2004 it has belonged to Alstom's Gas Division as Alstom Power Uniturbo. NPO Saturn is a Russian aircraft engine manufacturer, formed from the mergers of Rybinsk and Lyul'ka-Saturn – after the designer A. M. Lyul'ka (1908–1984). Saturn's engines power many former Eastern Bloc aircraft. The 'lattice' cooling concept of the GT24/GT26 single-crystal hp turbine blade, Figure 6-23, is similar to that of the Saturn AL-31 turbofan engine (27–32 klb TO thrust, first developed in 1981), which in the meantime powers all derivatives of the renowned Russian air-superiority fighter Sukhoi Su-27 Flanker and the Chinese Chengdu J-10 multi-role jet fighter. Since 2000 Saturn has held a 50 percent stake in the PowerJet joint venture with Snecma to manage development of the new SaM146 engine for the Sukhoi Superjet 100 regional jet.

[22] The ABB-MTU contact for the GT24/GT26 project lasted from 1992–1997. At times the MTU personnel commitment was up to 50, mostly in compressor and turbine aerodynamics, mechanical integrity and mechanical design, with responsibility from preliminary design to manufacturing drawings for GT 24/GT26 16-stage lp compressor and 4-stage lp turbine. MTU Advanced Projects main activities of that time were in the areas of UHB Ultra-High-Bypass engines (CRISP Counter-Rotation Shrouded Integrated Propfan and corresponding technology development) and RTF 180, a Regional Turbo-Fan project with 18,000 lbs TO thrust. The power generation activities with ABB complemented MTU's own design and technology development between 1988 and 1992 for an SGT 12 project (6–16 MW intercooled/recuperated Ship Gas Turbine family), which was intended to address increased ship propulsion power demand above the 7 MW upper limit of high-speed diesel engines from MTU-Friedrichshafen at the time.

On the lp turbine design side MTU's technological strength had grown considerably in the 1980s thanks to the successful alliance with Pratt&Whitney on the PW2037 and PW4000 turbofans, where MTU specialised in the design, development and manufacturing of lpt modules. Repeated lp turbine rig testing at the altitude test facility of the Institute of Aircraft Propulsion Systems[23] at Stuttgart University had confirmed extraordinary high performance values of > 94 percent efficiency at sea level conditions for 5-stage lpt configurations, on the basis of newly developed laminar profiling, see also Section 6.2.1. In addition, as already mentioned, MTU carried substantial design responsibility for the civil V2500 turbofan (with typical applications in 150 seater aircraft), as one of the four European Turbo-Union partner companies in the military RB199 programme and as one of five IAE International Aero Engines members. In the turboshaft field, for example for helicopter applications, MTU had developed its own advanced gas generator GNT-1[24], which ran successfully in December 1984 at TIT levels up to 1,650 K. The demonstrator fully covered the size of the later 960 kW helicopter engine project MTR 390 in collaboration with Turboméca and Rolls-Royce.[25] Several new technologies had been tested in this context, including ceramic blade coatings, a film-cooled single-stage transonic hp turbine with single crystal blades and supercritical axial compressor blades.

In the beginning the connection to NPO Saturn implied for ABB's senior management not only presumed access to the somewhat mystified military technology of the former Soviet Union, but was also seen as a hardly beatable asset under production cost aspects, with huge potential in line with Percy Barnevik's overall growth strategy. This was an expectation which quickly evaporated with extended and more thorough system insight.[26] While Uniturbo's design offices at Moscow performed excellently both with respect to the quality of their engineering work and their adherence to schedules, the special requirements in a sequential combustion environment asked for certain compromises in the high pressure turbine design. While original performance tables of 1991 showed a hpt temperature loading some 100 K above that of the lp turbine, it was the GT24/GT26 lp turbine which actually had to cope with higher thermal loads in the long run.

After the official GT24/GT26 project kick-off in October 1991, the nucleus mechanical design team (of which P. Meylan's Weekly Reports remember) commenced its activities at

---

[23] ILA Stuttgart (Institut für Luftfahrtantriebe) was founded in 1948 by Professor Ulrich Senger (1900–1973), former Director and Chief Engineer of BBC Mannheim. During WW II he established the turbojet design department at BBC Mannheim under Hermann Reuter, pushed steam turbine propulsion for the German Navy to a record-high order of 200,000 shp for the aircraft carrier 'Graf Zeppelin', but most influential for aero developments were the unique high-altitude test facilities from BBC Mannheim as a single source. Consequently he also made high-altitude testing a trademark of his institute at the University of Stuttgart after the war. A more comprehensive study on BBC's aero turbomachinery and wind tunnel activities is planned to be published.

[24] GNT – Gas generator Neuer Technologie

[25] In 1976 the history of this engine commenced within the framework of a German-French anti-tank helicopter project PAH-2 (Panzer-Abwehr-Hubschrauber) which developed in the meantime to the tri-company MTR 390 engine for the Eurocopter 'Tiger', still with MTU responsibility for the core engine with its combustor and gas generator turbine, and a number of accessories.

[26] Actually, this position was revised somewhat earlier on the aero engine side in 1990, when e.g. Klimov's RD-33 engine for the Mig-29, with 24 such aircraft inherited by the Luftwaffe after the German unification, was immediately stripped and analyzed – with mixed results. Eastern materials' technology, as an example, was clearly assessed as inferior to comparable Western standards, but systems technology such as engine/airframe integration revealed numerous ingenious detail solutions.

Gebenstorf on 17 December 1991, recording the 'decisions' of the first team meeting[27] on 6 January 1992:

- Name of the design team is KWR15-D
- Regular weekly team meeting Thursday 10.00 to 11.30,
- Language inside project DE or EN, outside project EN at least, same as for summaries and pictures, (both) at least in EN,
- two concepts to be evaluated in phase 0:
  - Var 1: Single Shaft Engine.
  - Var 2: Twin Shaft Engine.[28,]

The following, deliberately picked list may illustrate the corresponding progress over time, with some insight into the team spirit and working atmosphere:

07 May 1992   Quitting of intercooler (twin-shaft) design configurations.

21 September 1992   Design freeze announced within the coming 7 months.

23 November 1992   First drawing of hp turbine from Saturn.[29]

21 December 1992   Draft of 'GT11N3' with 14-stage axial compressor ready to be sent to MTU for 'background information'.

11 January 1993   Announcement 'First engine out on 30.12.1994' and change to a flexible design with simple, sequential combustion 'Zone 2 variable'.

12 February 1993   MTU provides first GT24 (GT1.4) lpt annulus with profile shapes and blade count, in parallel Saturn work on hp turbine.

Two typical statements by Pierre Meylan (PM) about the early design phase follow, becoming less frequent towards the end:

19 February 1993   *'The ambiance during the Decision Meeting was very cooperative and the discussion objective. I wish to express my satisfaction and thanks to all participants. It was really a good Decision Meeting.'*

28 February 1993   *'Personal Remarks'* (on the sealing arrangement between the hp compressor and hp turbine) *'Such a Plethora* (bunch/Verhau) *of Pipework leads to the impression of a confuse design rather than a mature design. ...A much simpler solution has to be found to meet the Costs and Availability Targets.'*

03 March 1993   Order of IN 706 parts for first engine out, design team (9) complete.

---

[27] The small design group comprised besides P. Meylan: U. Benz, H. Brunner, P. Graf, E. Kreis, K. Matyscak, M. Rieder and R. Suter, see ABB, GT24/GT26 weekly reports, p. 545 of 2,233.

[28] 'Twin-shaft' stands here for intercooler integration between the lp and hp compressor, running at different speeds.

[29] The kick-off meeting for the GT24/GT26 turbines (hpt **and** lpt) took place in Moscow in May 1992. The GT24/GT26 hpt concept review was on 6 November 1992, presumably at that time the decision was made to continue with MTU on the lpt side. The head of Saturn's turbine department (aero, cooling, design) was Sergey Shchukin, since beginning of 2013 Chief Engineer Gas Turbines at Baden, CH.

30 April 1993	Design freeze, as announced on 21 September 1992, no further conceptual changes allowed, only small changes in 5 mm range.
09 September 1993	GT24 presentation at the Waldorf Astoria Hotel, NY, USA.
29 November 1993	*'Time schedule is more than critical'.* (PM)
06 December 1993	*'Time schedule is more than critical'.* (PM)
04 February 1994	Close cooperation with MTU team, strength calculations for GT26 lpt vanes 3 and 4 with improved MTU models.
22 April 1994	*'LPT GT24 (GT1.4) documents from MTU, first review performed by K. Vogeler and E. Kreis. General impression: a large part is OK, some changes and improvements are necessary.'* (PM)
06 May 1994	GT1.6 (GT26) is due to be scaled from GT1.4 drawings, final GT1.4 corrections are necessary before this.

In the broader context of the GT24/GT26 project the following milestones are noteworthy:

December 1994	GT26 engine project concept ready.
January 1995	GT26 start engine engineering.
May 1995	Start of construction works of first GT24 at Gilbert power station.
September 1995	GT24 first firing at Gilbert with subsequent opening; complete series of new seals had to be developed.
November 1995	GT24 first SEV ignition.
January 1996	GT24 first operational inspection.
January 1996	Start of GT26 manufacturing.
February 1996	Order intake for 8 GT24 units in Poryong, South Korea.
28 February 1996	ABB annual press conference at Warsaw, PL: *'Record growth in earnings, net income up 73 percent'* (ABB press release), market breakthrough in high efficiency gas turbines announced (Gilbert GT24 + 8 more in South Korea, EnBW Karlsruhe GT26 + 3 more ready to be signed in the UK and New Zealand), P. Barnevik becomes CEO and President of the Board.
16 March 1996	Zurich newspaper 'Tagesanzeiger' has a critical, insider-influenced story on the new engines: 'Spiel mit dem Feuer' (Playing with Fire), in due course final decision to build Birr GT Test Centre.
May 1996	GT24 base load operation on natural gas, in time for Summer Peak 1996.
October 1996	GT24 base load operation on oil, in time for Summer Peak 1997.

November 1996   First firing of GT26 at EnBW RDK 6S (Rheinhafen-Dampf-Kraftwerk) Karlsruhe, D, a 465 MW CCPP re-powering project, resulting in a 33 percent reduction of CO_2 emissions.

December 1996   Inauguration of GT26 Test Power Plant at Birr, AG, CH.

January 1997   First firing of GT26 SEV combustor.

February 1997   GT26 base load operation on natural gas and oil for Summer Peak 1997.

December 1997   GT24 customer acceptance.

Figure 6-7 illustrates the GT24 configuration development between July 1991 and September 1994 in a sequence of six 'General Arrangements GA I-VI':

GA I dates from July 1991 and therefore belongs still to the GT15/GT17 project phase. Unlike the other general arrangements, the overall flow direction is shown here from left to right. This co-axial two-shaft arrangement has a low-pressure (generator drive) shaft with 7-stage lp compressor and 4-stage lp turbine and a high-pressure shaft with 11-stage hp compressor and 2-stage hp turbine. The sequential combustion concept is indicated here by a first EV combustion chamber, while the second combustion stage was foreseen as simple interduct burning between hp and lp turbines. The split between lp and hp compressor allows for the integration of a performance-improving intercooler (IC)[30].

GA II from May 1992 is a single-shaft design with a 3-bearing arrangement, 14-stage lp compressor without intercooler but intermediate bearing, 5-stage ip (intermediate pressure) compressor and 9-stage hp compressor, sequential combustion, 2-stage hp turbine and 4-stage lp turbine, which already indicates the high overall pressure ratio demand of the sequential combustion cycle, but in a comparative stage count already illustrates the benefit of the later introduced MTU compressor design philosophy as visible in principle in GA IV from December 1992.

GA III represents a final drawing of an intercooler status from August 1992, several months after the decision to leave this design feature out. Here, the intercooler ducting is placed between the 5-stage lp and the 9-stage ip compressor. A third, intermediate bearing has been positioned again between ipc and 11-stage hpc. Two in-line, sequential combustors again frame a 2-stage hp turbine followed by a 4-stage lp turbine. This was by far the longest overall arrangement.

---

[30]   BBC had used GT intercooler arrangements (without additional exhaust gas energy recuperators) for the first time in 1959 in the 4×25 MW peaker plant at Port Mann, Vancouver, BC, CA, Section 5.1.5.2 and with a similar compressor configuration in the first combined-cycle power plant for utility applications with 75 MW net output at Korneuburg near Vienna, Austria in 1961. Interestingly, GE brought the LMS100[TM] to market in 2004, a new high efficiency simple cycle gas turbine in line with traditional BBC design considerations. This 3-spool GT configuration uses an intercooler between the lp and hp compressors and is specifically designed for peaking and mid-range dispatch applications, such as regenerative grid stabilisation with cyclic operation and increased efficiency demands along increased dispatch. The LMS100[TM] gas turbine intercooling technology provides outputs above 100 MW, reaching simple cycle efficiencies in excess of the claimed 46 percent. http://site.ge-energy.com/prod_serv/products/tech_docs/en/downloads/ger4222a.pdf

I. July 1991	
II. May 1992	
III. August 1992	
IV. Dec. 1992	
V. June 1993	
VI. Sept. 1994	

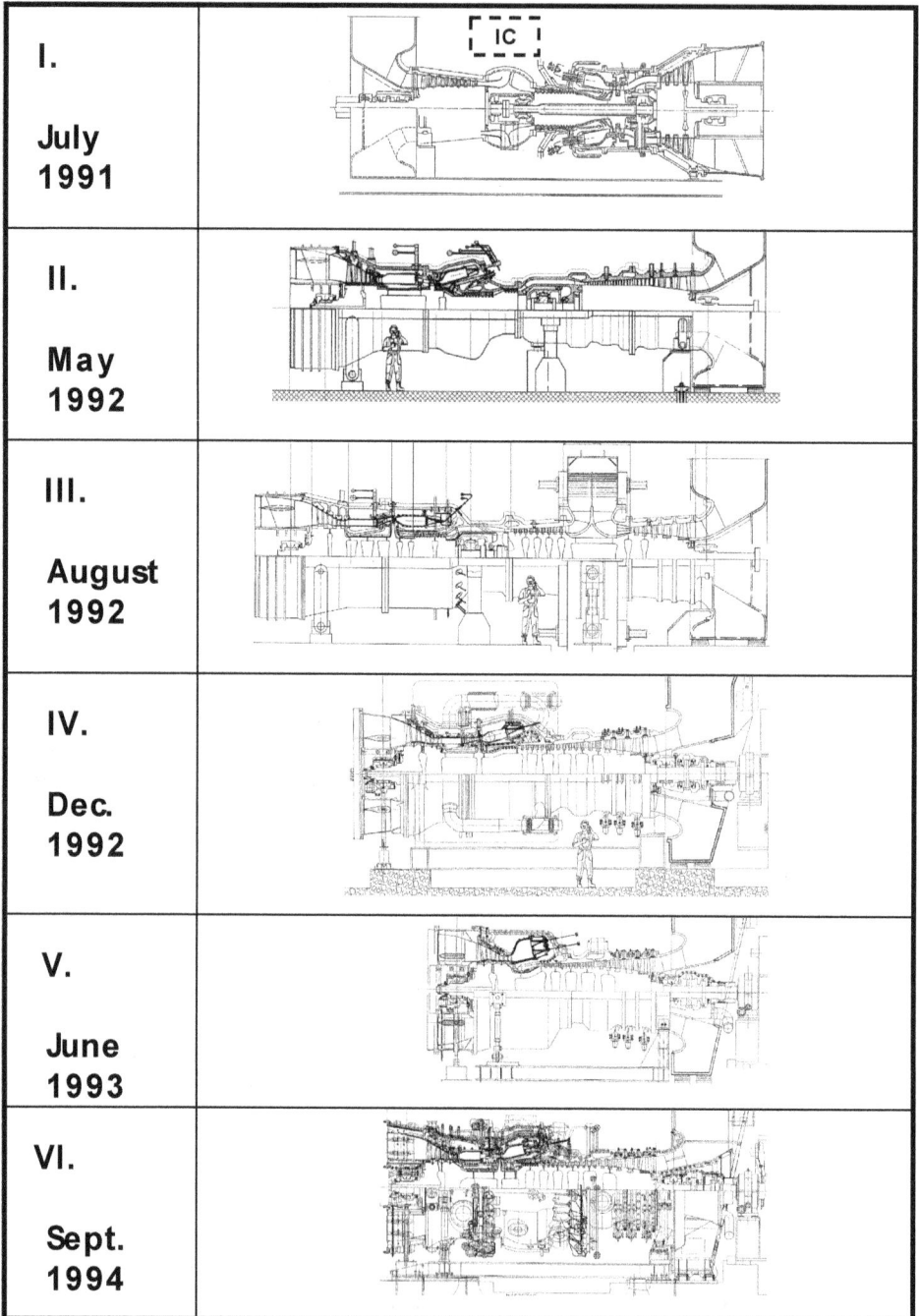

Figure 6-7   *GT24 development of general arrangements between July 1991 and September 1994* [31]

---

[31]   See ABB, GT24/GT26 weekly reports

GA IV dated December 1992, is very close to the final GT24 configuration selected some 10 months later. The design is now characterised by the typical welded rotor/two-bearing arrangement without intercooler. The MTU input shows up for the first time in a GA drawing. A 16-stage axial lp compressor with 3 VIGV rows (MTU), followed by 6 'repetitive stages' provides the required high PR > 30 for the sequential combustion chambers with 2×12 EV burners, 1 row of SEV combustors, with a first-time single-stage hp turbine from Saturn and the 4-stage lp turbine concept (established in the meantime) from MTU.

GA V surprisingly gave up the presumably definite sequential combustion concept in favour of a very simple, straightforward set-up. The drawing was officially released in August 1993, after the design freeze announcement at the end of April 1993. This 60 Hz GA (GT 14) combines a 15-stage compressor with PR ~16, an annular combustor with 2 rows of EV burners and the 4-stage turbine.

GA VI from September 1994 is the finally selected, well-known GT24 thermal block configuration, clearly optimised and axially compressed in comparison to the December 1993 preliminary reference case GA IV.

An equally scaled comparison of the 50 Hz GT26 configuration vs. the 60 Hz GT24 is depicted in Figure 6-8. This drawing shows the design status in May 1994, when a high degree of detailed design work had already been accomplished and the first round of the GT24 design work was practically finished. In principle the same configuration had already been under preliminary consideration as early as December 1992[32] (still as part of the GT15/GT17 investigations), and now a thorough comparative assessment pushed back all other alternate designs.

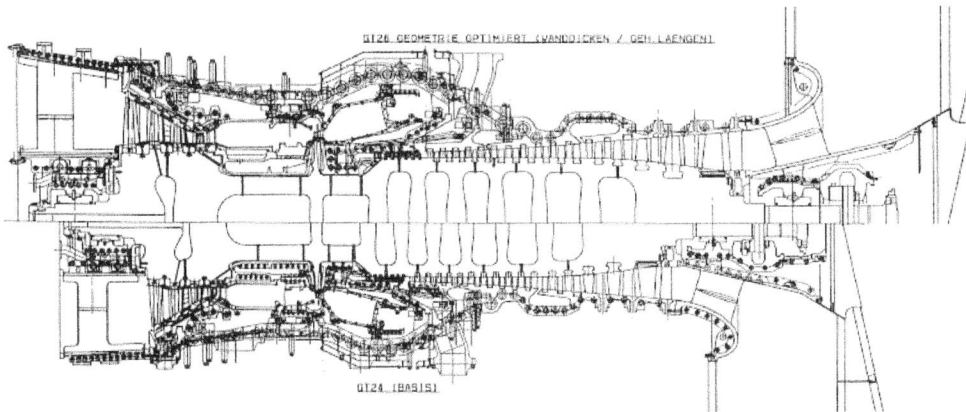

Figure 6-8   *Design status May 1994: GT26 (top) vs. GT24 (bottom), equal scale*

---

[32]   See ABB, GT24/GT26 weekly reports, p. 286

May 1994 was also the virtual branching point towards a decisive split, the initial realisation and firing of the GT24 at GenOn Energy's[33] Gilbert generating station in Holland Township, NJ, USA in September 1995, and at the same time for preparing scaled drawings for the first GT26 prototype manufacturing, which was to be put into operation in November 1996 at EnBW's[34] RDK 6S plant at Karlsruhe, D.

6.1.3.2   GT24/GT26 Design Stabilisation 1996–2002

After the two prototypes were installed, the normal development process would have been to establish and optimise the manufacturing processes, especially for parts from new high-temperature alloys: Single-Crystal (SX) and Directionally-Solidified (DS) turbine blade castings, and with the application of new manufacturing technologies (e.g. high speed grinding, TBC Thermal Barrier Coatings and the laser hole drilling of TBC-coated parts).

Some natural optimisation criteria would have been cost reductions by realising learning curve effects and reducing lead times. Practically, this optimisation/cost cutting phase started for the GT24 and GT26 engines in 1997/1998 only after considerable delays. Previously overriding interest was to get parts out of a slowly improving and accelerating production chain for the nine GT24 and four GT26, which had been sold more or less at once with the programme announcement on the basis of the existing design status A. At such a stage of development there are always remaining jobs to be accomplished, but in principle the headquarters had to wait and see how the new engines would perform in the field and be ready for corrective actions in the event of failures.

At first the optimum manufacturing concept had to be determined, and existing experiences from the influential turbojet business did not necessarily represent a guideline in the right direction, either for the casting processes or the subsequent manufacturing steps. In principle, there were two options, a) the integral 'engine-ready' approach, where the manufacturing was to be carried out in one shop, or at least within one supply company and b) a 'sequential supply chain', where the parts were shipped from one highly specialised shop/company to the next on the way to completion. The engine-ready concept had of course been propagated by the big manufacturers for precision investment castings looking for extra business, with the advantage that the ordering OEM got rid of the complex station-to-station order and logistics management within the supplier's premises. These attractive advantages were however soon compensated by the fact that a highly specialised foundry could not necessarily provide the best enabling technologies along the complete manufacturing chain. Therefore the sequential supply chain concept was often picked, especially for critical parts of quality, cost and/or throughput time considerations.

---

[33] GenOn Energy, Inc. is one of the largest competitive generators of wholesale electricity in the United States. With headquarters in Houston, Tx., GenOn is located in 12 US states with a generating capacity of nearly 24 GW, distributed to 47 owned, contracted or operated generating stations.

[34] With revenue in excess of € 18 billion in 2011 and some 20,000 employees, EnBW Energie Baden-Württemberg AG, headquarters in Karlsruhe, D, is one of the largest energy companies in Germany and Europe. EnBW was formed from the merger of two local utility companies (Badenwerk AG and EVS Energieversorgung Schwaben AG) on 1 January 1997, practically parallel to the operation launch of the first GT26 at the company's RDK Rheinhafen-Dampf-Kraftwerk, a 465 MW re-powered combined-cycle power plant.

This preference for the station-to-station, company-to-company transportation and supply chain is certainly not set in stone, but it was clearly the best way to overcome difficulties at times when even the foundries struggled with their own specific technologies. As previously indicated, most of the technology boost for power generation gas turbine in the early 1990s originally came from the aero engine side. However, as even enthusiastic foundries had to learn and accept, the SX manufacturing processes for a 2 inch hp turbine blade of a turbojet could only in principle be carried over to a 10 inch long, lp turbine blade[35] for a power generating gas turbine. Consequently, early cost estimates of typically 1.5 MUSD for a 200 MW 4-stage GT turbine (component) could easily be overrun by 200 percent after corresponding experiences had been collected in reality. Compared to the foregoing aero engine parts manufacture, the power generation parts first required huge investments within the foundries themselves, e.g. for heavy handling robots and 8 m high Bridgman furnaces for ultra-high vacuum casting processes. On the basis that a 'turbine commodity' (group of parts) amounted to roughly 35 percent of the cost of a GT thermal block, it became immediately clear that these uncertainties in costing could have over-shadowed the whole introduction process of the new generation of advanced power generation gas turbines.

Therefore, the ABB organisations responsible had to be adapted to these engineering peculiarities. The idea of an integrated, multi-disciplinary engineering team which had been successfully demonstrated in the 'GT24/GT26 skunk work activities' at Gebenstorf, Figure 6-5, was continued in the open-spaced office floors of the newly built Konnex building in Baden, Figure 2-16, from October 1995 onwards, now rolled out into the specialised TOY Turbine Optimum Yield[36] department.

Newly collected experiences in this context have been put into a 'Handbook of Lessons Learnt'[37]. It goes without saying that some of these lessons were the result of previous costly negative experiences.

Although the problems of 1996 were minor when measured against the later crisis of the year 2000, the reactions nevertheless revealed considerable strain below the surface of 'business-as-usual'. The local press was curiously used for communication by ABB's corporate management. On 14 March 1996 the 'Aargauer Zeitung' ran quotes such as *'The ABB gas turbines have some 'teething troubles' (Kinderkrankheiten), which have been cured in the meantime'*, followed by an article in the Zurich-based 'Tagesanzeiger' with the ambiguous title *'ABB spielt mit dem Feuer (ABB plays with fire)'* on 16 March 1996. Facts and details had been listed which clearly insinuated a high-ranking in-house whistle-blower. Here an indirect quote: costs of 100 million Swiss Francs needed to be spent to solve design flaws in

---

[35] It can be assumed that until the year 2000 the SX blading for the GT26 lp turbine blade 1 represented till the most challenging demands with respect to size, weight and design complexity in precision casting technology.

[36] The Yield Rate describes during the casting development process (and afterwards) the percentage of 'good' (further usable) cast parts in reference to the total number of castings; not usable are 'scrap parts' which have to be re-melted. Typically, casting yield rises during development from 40 to 80 percent and more with process improvement and design simplification, up to an optimum point after which benefits in manufacturing cost are compensated for by (costlier) GT performance drawbacks. This balanced process vs. design adaptation had to be carried out in lengthy, in-depth 'concurrent engineering' efforts between highly qualified representatives of (sometimes various) suppliers and the TOY design group.

[37] See ABB, Casting of GT24/GT26 turbine blades (A. Beeck), 1996 f.

the new turbines, caused largely by a lack of sufficient test experience which should have been rectified before market entry. Another quote: '*Panic spread at ABB*' as such otherwise strictly confidential business information had become public.[38]

At least, the concerns about the lack of testing for the new gas turbines may have accelerated the decision to build a unique GT test power plant within an empty hall of the Birr rotor manufacturing plant. This was a remarkable event with impressive dimensions, Section 6.2.4. The 'GT Power Plant', which allowed for complete engine testing up to the 265+ MW size of the GT26 including power generation and electric network supply, was erected in record time and became operational in December 1996. The validation tests in 1997 confirmed the intended GT26 performance with the expected output and efficiency. For the versatile sequential combustion, both EV and SEV systems could be demonstrated at target low emission values. The tests resulted in detailed adjustments of the operation concept (number of blow-offs, pilot/premix switch-over process, etc.).

As valuable as the new test gas turbine was (and will be in the future) it could not help to overcome any potential exposure against long-term warranties as part of any power plant sales contract. Traditionally, a common guarantee milestone for the power generation industry is at 24,000 EOH (Equivalent Operational Hours[39]) or the equivalent of nearly 3 years of uninterrupted operation. This describes a key figure of the thermal block sales contracts, followed by a considerable amount of 'small print', deviating from OEM to OEM.[40]

ABB's 'leap frog' approach to market and technology leadership also comprised a hidden cost aspect which obviously took the responsible management by surprise. It would not have materialised to that extent with a more moderate strategy, but from a rational perspective it did not come unexpectedly or in an excessive magnitude. A look at the comparative development of large turbojet engines in the upper thrust class would have immediately revealed a development cost scenario of 1-2 BUSD. It has to be understood that the load testing of aero and power generation gas turbines follows different load & failure modes which also imply different test times. The 'cyclic' aircraft/engine operation mode with frequent take-offs and landings can and must be simulated/tested before the first sales to customers for safety reasons. Typically, a new turbojet test programme in the upper thrust class[41] comprises eight test engines, which undergo an established, severe test programme of approximately two years endurance (if not partially shortened by AMT Accelerated Mission Tests). Unlike in aviation development, where public authorities make appropriate requirements for safety testing, such regulations and AMTs do not exist for power

---

[38] See Catrina, ABB – Die verratene Vision, p. 115

[39] Besides the number of actually fired Operational Hours (OH), EOH takes into account cyclic events such as starts and stops, rapid load gradients, protective load sheddings, trips, etc.

[40] In hindsight, after considerable amounts spent on warranty issues, one could speculate that ABB Power Generation might have committed to contractual lifing clauses too naively in this period. The small print was possibly not evasive enough. Especially for premature failures of SX/DS turbine blading in the observed 8,000–10,000 EOH range, typical of the early GT24/GT26, but also with nearly identical phenomena for competitive engines as well. This situation was severed by the described radical change in the customer structure, towards IPP dominance in 2000, Section 6.1.1. After the sudden collapse of the 'US bubble', rather profit-oriented IPP entrepreneurs, themselves financially highly exposed, often reacted completely differently in the context of guarantee settlements than what would have been expected from established utilities.

[41] New high-bypass turbofan engines > 90 klb take-off thrust such as GE 90, Trent 800 or PW 4000.

generation gas turbines due to TMF (Thermal Mechanical Fatigue) and a (material) creep-dominated development scenario. A 24,000 EOH guarantee value would have required a correspondingly long operational experience of at least three years, a proceeding which in the past had never been verified to the full extent. There were always certain trustworthy key customers, where the first engines were installed with a lead-time of say, one year, and the operational lessons learned were rolled out with refurbishments and upgrades to the remainder of the fleet. In the meantime, this thinking has been given up by all power generation OEMs, not least as a consequence of ABB/Alstom's negative experiences up until 2003. Today, an all new gas turbine development like the GT24/GT26 and a similar amount of new technologies that were not fully tested would certainly undergo an in-depth test and validation programme at the in-house test centre, accompanied by an accumulation of practical operation experiences at selected key customers.

This proceeding alone would easily drive the GT development budget beyond the 1 BUSD barrier, an order of magnitude which was apparently out of the question for ABB in 1997. Armin Meyer, responsible for all ABB Power Generation since 1994, with 40,000 employees worldwide and annual revenue of more than 9 BUSD, stated at the time in view of the GT24/GT26 development costs: '300 million Swiss Francs are enough, now we have to reap the harvest by intensifying sales !' It is very probable that the estimated 100 MSFr extra costs for the Birr test centre were already included in that amount. Ten years after the ABB Power Generation establishment, years filled with intense and sustained design and development efforts, the time appeared to be ripe to participate in the US power generation sales bonanza that was beginning. Alternatives such as a somewhat slower pace for introducing technology were presumably not taken seriously into account – 'sales' appeared to be the remedy for many, often somewhat heterogeneous expectations.

In Figure 6-9 the so-called 'US gas bubble' is put into perspective with the long-time annual gas turbine production value in BUSD. Clearly, the expanding US sales figures between 1998 and 2003 in gas-fired combined-cycle power plants overshot the long-time average, during the peak years 2000–2002 up to a factor of three! The other conclusion which can be deduced from this data is that the observation in the late 1990s that the global power generation business was growing healthily on the basis of a strong, global demand was justified. Between 1990 and 2015 the annual production rates grew nearly ten times. In hindsight, it was only the over-heating 'US gas bubble' with its negative crash-type consequences which overshadowed this bright perspective. During the steep upturn, it was difficult to hold back, as always during similar events in history; this was difficult for all the participants, but especially for ABB Power Generation under the described preconditions.

In addition to the global production numbers, Figure 6-9 also shows the sales development of ABB's GT24/GT26 fleet (since 2000, Alstom's). Up until 1998, with 12 sold units in total, the sales achievements and at the same time the warranty risks were rather limited and manageable. It was only then that a thorough market analysis revealed pronounced cost drawbacks for the new GT24/GT26 product family, mainly as a result of the expensive new technology content. Before a potential conflict between sales and development departments could develop, Rolf Kehlhofer, Managing Director of ABB Power Generation at that time and consequently responsible for both groups, came up with a supposedly wise solution: the net cost responsibility for the new engines was allocated exclusively to engineering, while the fixation of a market-attractive USD/kW price target remained with the sales organisation. Predictably, in view of the further US market developments, the tendency in engineering was not so much to tackle the existing cost barrier by a substantial, cost-cutting GT24/GT26 re-design, but rather by a strive for considerable price reductions on the supply side, on the basis of economy-of-scale effects. SCM, the newly founded Supply Chain Management organisation in engineering, took on the corresponding task of investigating possibilities on the supply side, first under an assumed potential sales volume of 30 units. 'Concurrent Engineering' was sought and found to be a suitable engineering tool for achieving substantial cost reductions for turbine castings as the most cost-critical parts' commodity – and in a fair, technically balanced way between the interests of the key suppliers and ABB. As an example the following, common press release conveyed the resulting message of this undertaking:

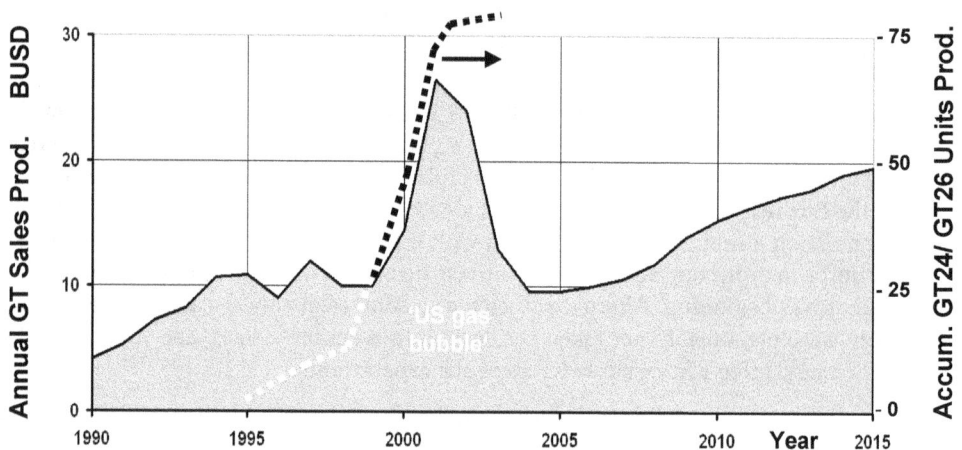

Figure 6-9   *Worldwide gas turbine production* [42] *1990–2015 and ABB/Alstom GT24/GT26 production 1995–2003 (units accumulated)*

'4 October 1999, Greenwich, CT (USA) – Howmet Corporation… today announced that ABB Power Generation Ltd, an ABB Alstom Power company, has extended its long-term agreement with Howmet to supply turbine castings for the advanced GT24 and GT26 gas turbines and other mature engines. The agreement begun in 1998 and now extended through the year 2003 with options to renew, represents approximately $500 million in future business and secures for Howmet a substantial share of castings for these power generation systems.

Dave Squier, President and CEO of Howmet, stated: "Current demand for industrial gas turbine systems is unprecedented. As production has ramped up to meet that demand, we have maintained quality and refined our manufacturing process to provide reliable, high-volume delivery of complex, advanced components. With this contract, ABB is locking in the capacity it needs to meet demand for these high-efficiency engines."

---

[42]   See Langston, Plowing new ground, quoted data from Bill Schmalzer, Forecast International, Newtown, Ct., USA. 'Production' is used in the sense of 'sales' at the time of delivery/commissioning. For actual sales/contracting dates the dotted curve can be assumed to be shifted to the left by 1–1.5 years in general.

John Gaskell, Director of ABB Alstom Power gas turbine segment said: "This agreement is an essential milestone in setting up and securing a high-volume supply chain output for our successful GT product lines. In close collaboration, both of our companies achieved significant technical and commercial progress. This is a basis for a long-term partnership, to the benefit of our customers and in favor of the rising demand of the deregulated markets."'

Not explicitly mentioned in the text was the fact that jointly agreed design improvements had to be implemented in the manufacturing process first, before they would actually become cost-effective. Therefore, the turbine casting prices decreased in several 5–10 percent steps from the starting level in 1998 until the contracted 2003 target values were achieved. However, it goes without saying that the additional boost for ABB's sales initiative was immediate, with sustainable success, sometimes regretted later, as illustrated by the accumulated sales record of Figure 6-9. Following the ABB slogan of the years 1993–2000, '*Go for growth*', and with the restrictive price barrier apparently removed, ABB sales figures for the GT24/GT26 soared up until 2000, the year of the complete takeover by Alstom, – rising to 46 engines in total[43], with the biggest increase with 24 sold units in 1999 alone.[44]

Another, immediate consequence for all OEMs was the need to increase the production capacity to consistently cope with the strong market demand all along the supply chain.[45] For ABB the new target was set in late 1999 to 80–100 engines annually, roughly six months before an abrupt sales stop and production reduction for the new engines was issued – interesting times, to say the least.[46]

'*A storm that had already been brewing made landfall in mid-July 2000, bearing down upon the recently created Alstom Power. The GT24 and GT26 heavy-duty turbines inherited from ABB, which had sold virtually all of the systems built at the time this lucrative business was acquired by Alstom, were not reliable.*' With this, somewhat dramatic wording, the presently official Alstom History account[47] describes the next round of events, and continues:

'*A new incident arose on the first "B" version of the GT24: the blades in the second row of the low-pressure turbine began to buckle after just 500 hours of service. At the end of 1999, on this*

---

[43] In the meantime, after the complete Alstom takeover of ABB's Power Generation business in March 2000, the GT24/GT26 fleet grew by another 52 units.

[44] Another explanation for the boosted GT order intake sees it in the context of the planned ABB-to-Alstom transition of ownership; the resulting volume of upfront payments is said to have been at 1+ BUSD cash.

[45] A considerable SCM negotiation and contracting challenge (pricing, delivery schedules, quality assurance, etc.). For ABB the situation was somewhat relieved due to a 40 MSFr investment in 1999 at Birr, where a Hot Gas Path Parts manufacturing & technology centre was opened for – intended – 50 percent of finishing operations (machining, coating, laser drilling, sheet metal brazing, etc.) of all precision cast parts.

[46] A strange phenomenon accompanied this period – engineers were drained in a bunch of business school' ideas nearly without time limitation, on large and small scale: Goeran Lindahl, since 1996 successor to Percy Barnevik as ABB CEO, ventilated in 1999 his favourite idea of KBE – the Knowledge Based Enterprise; Bart Huthwaite, a hired US consultant, entertained large groups of ABB GT designers 'always ahead of time' on fancy things like 'The Rule of the Evil Ings', leading to the *Universal Design Rule* ['Strategic Ilities (affordability, serviceability, performability) minus Evil Ings (complexity, precision, variability, ...) equals Lean Product Success'] – see Huthwaite, Lean Product Design, p. 272 –, and the GT development department's key personnel received from J.P. Womack and D.T. Jones (authors of must-read 'The machine that changed the world') their 2nd best bestseller on 'Lean Thinking – Banish waste and create wealth in your corporation', 396 p., for reading during Christmas vacation 1998/1999.

[47] See Nieto, MW and Km/h – a History of Alstom, p. 224; the following description in parentheses '...' relies partially on this source.

same turbine, which had just been installed at the Agawam combined-cycle power plant in Massachusetts (...) hair-line cracks had appeared in the first combustion chamber after only a few dozen hours in operation, due to an inefficient cooling mechanism. ... (These) two incidents in close succession on the first system installed raised concerns that design defects were to blame. ... Other turbines were checked and it soon became apparent that they all had flaws that prevented them from working correctly. ...In August (2000), modifications were made to a number of machines. The goal was to apply the same process gradually to the entire installed base. But this posed huge logistics problems. ... During this time, the teams began to develop solutions that would allow operations to be brought back to full load. The end of the crisis was still a long way off, however, since other problems would arise, underscoring the inherent deficiencies of the initial design, such as a fissure appearing inside the first row of blades on the low-pressure turbine. The conclusions were alarming: designed to run for three years before needing maintenance, the machines had not held up for longer than six to eight months! Some fifty turbines were already in operation and others with the same defects were in production, bringing the total to eighty engines. ... Given their increasingly tarnished reputation in the United States, further sales of the GT24 were suspended until 2003. Commercial efforts were focused instead on other gas turbines, known to deliver reliable performance that met specific market needs, and on steam turbines. Additional costs for the 2000/2001 financial year amounted to nearly 1.1 billion euros. ... In the first part of 2001, the slowdown in US economic growth that some had been predicting for months (the dot.com bubble had meanwhile burst) finally materialised. ... It came as no surprise either that the "gas bubble" created in the wake of electricity market deregulation and liberalisation would end up bursting one day. Driven by booming demand for electricity and the fact that purchasing agreements were tied to the price of crude oil, the average price for natural gas spiked in 2000. By January 2001, it had tripled ...' and together with the finally realised excess capacity, the market collapsed – leaving Alstom with a bunch of contractual 'open issues'.

Aside from the alarming tone, the journalist made valid points up until the end: technical problems naturally had to be expected in ABB/Alstom's gas turbine development, but the exposure to billable warranties on eighty sold engines caused financial strains. In hindsight it is therefore fruitless to discuss if and to what extent the full sales and delivery stop for the GT24/GT26 up until 2003 was justified – or additionally rather harmful.

Technically, the two addressed key issues were rather unimpressive, Figure 6-10, and after identification were cured rather quickly. For performance reasons, the second lpt blade had been designed with a tip shroud, see also Figures 5-77 and 6-23 – which required some extra-cooling for both the blade tip section and the opposing stator heat-shield.

The second problem appeared to be even more trivial, after the root causes for the failures were understood. Admittedly speculative guessing had previously increased the uncertainties. The lpt blade row 1 is from SX castings, with the intended design benefit that the single crystal alloy and manufacturing advantage alone counted for a 50 K higher temperature potential of that blade (which otherwise had to be cooled more). Due to limited information on the cracking and crack propagation behaviour of these new materials, the supposedly 'crazy alloy' was wrongly put into question intermediately. The final conclusion was rather simple as it can be reconstructed in reference to the internal cooling configuration, Figure 6-23. Heat transfer calculations had suggested some beneficial, local relief for the internal entry zone of the cooling flow, through a patch of 10 film cooling holes (which are circled in Figure 6-10 as 'critical cooling zone').

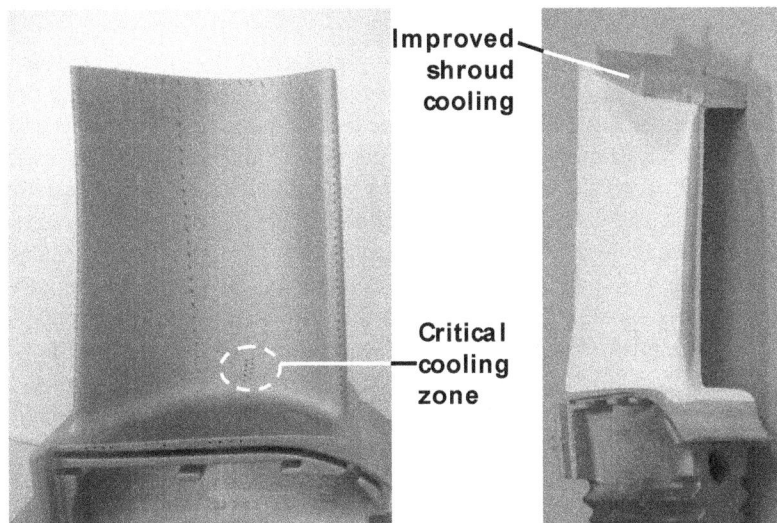

Figure 6-10 *GT24/GT26 lp turbine intermediate development problems – blade 1 local over-cooling (l), blade 2 tip shroud cooling adjustment (r)*

Though the principal design intent worked, a negative side effect had been overlooked, especially during engine shut-down, when the extra cooling measures were superfluous or even harmful. The blade was 'over-cooled' and correspondingly over-strained during engine run-down; the remedy, after the root cause had been isolated, was simple: the extra cooling holes were left out.

The short-term recovery actions to correct the identified flaws in the present hardware in field operation were started in 1998 and led finally in 2002 to the introduction of a further upgrade package with the following elements:

- a compressor upflow[48]
- optimized lp turbine blading for increased lifetime
- an optimised GT exhaust housing.

The new compressor design and the turbine design adaptations improved base-load performance, part-load efficiency and fuel flexibility. In September 2003 this 'lifetime & performance package' pushed the GT26B2.1 frontrunner engine at Ringsend, UK to a field-measured ISO-corrected gross efficiency of 38 percent and a gross power output > 270 MW. At the same time the combined-cycle KA26-1 performance guarantees grew to 408 MW, significantly above the 393 MW output guarantee of 1996, and the combined-cycle efficiency was recovered to 57.4 percent: back to and partially above the competitor engines.

---

[48] The compressor was redesigned for an increased mass flow of approx. 5 percent, with the goal of having a similar increase in the combined cycle power output. This was achieved both by optimised airfoil design and re-staggering of the compressor blades. The result was a design that required no change to the rotor and stator flow path contour. Compressor blade length and channel height, rotor and compressor vane fixation grooves, as well as blade and vane material all remained unchanged. The design was therefore able to be retrofitted.

Beginning in 2000 and intensified from 2002, thermal paint tests[49] on the GT26 Birr test engine made a considerable contribution towards regaining ground in the engine's thermal assessment. A detailed comparison of experimental vs. theoretical thermal tooling and the calibration process was initiated. Critical elements of the turbine blading, but also sections of the secondary air flow path (sealings, blade attachments and cooling air feeding stations) were covered with temperature-sensitive paints before a GT test run. These changed colour when the temperatures were above the colour transition level during a stable operation point locally. As a result the sealing system in critical areas could be partially improved or the cooling flows adapted accordingly. However, the fundamental possibility to calibrate the parts' life prediction tools was invaluable. Consequently, the thermal boundary conditions, an indispensable prerequisite for precise metal temperature and stress calculations, was no longer derived from relatively far-fetched correlations and questionable model tests, but for the first time validated in reality based on detailed engine operation conditions. Since then, this has been seen as a practical technological breakthrough with which Alstom's engineering set a new essential test standard in gas turbine development.

The actual proceedings have been illustrated in a simplified form in Figure 6-11. From the thermal paint covered hardware a) the analogue temperature information (thermal isolines) is transferred in the given example of a turbine vane surface into b) a digitalised metal surface temperature composite. The calibrated model with a validated, fully 3D temperature distribution c) is finally generated by taking into account the operating and boundary conditions of the specific test point, the measured data, information from corresponding detailed cold flow tests and, for example, the manufacturing tolerances of the analysed parts under consideration.

In February 2002 Alstom signed a new technology agreement with Rolls-Royce, which considerably helped to structure GT development recovery in the following years through the so-called 'gating' PDP Product Development Process[50], which allowed the initiation and roll-out of subsequent development steps only after certain pre-determined conditions (technolgy readiness demonstrations, test results, etc.) and approvals had been achieved. It should also not be underestimated that an increasingly reliable GT24/GT26 fleet in the field finally clocked up thousands of operating hours: from 2002 to 2006 the number of accumulated OHs rose from roughly 400,000 to > 2,000,000, and at the same time the number of new product support issues fell to only 10 percent of the 2002 level with a rapidly trained engineering team, notwithstanding the fact that an upgrade was then introduced.

The first genuine sign of regained market confidence came in January 2004, when Gas Natural Fenosa of Barcelona, E, placed a 430 million euro order with Alstom for a 1,200 MW turnkey combined-cycle power plant for its new site at Cartagena, near Murcia, E. This facility was to be equipped with three component groups, each consisting of a GT26, a waste heat boiler, a steam turbine, a generator and an instrumentation and control system. Gas

---

[49] This refers to the GT26 testing only, BBC actually used this technique as early as 1939, Figure 6-31, and continued these applications during the BST era in the 1970s.

[50] In the meantime part of the broader PDQ Product Development Quality process, which has mandatory 'gate reviews' between the development phases 'specification', 'concept', 'detail design', 'validation' and 'fleet introduction'. 'Validation' alone represents the most pronounced and rigid change in view of attitudes during ABB times.

Natural also signed a long-term O&M Operations and Maintenance agreement for the Cartagena plant, which entered service in 2006.

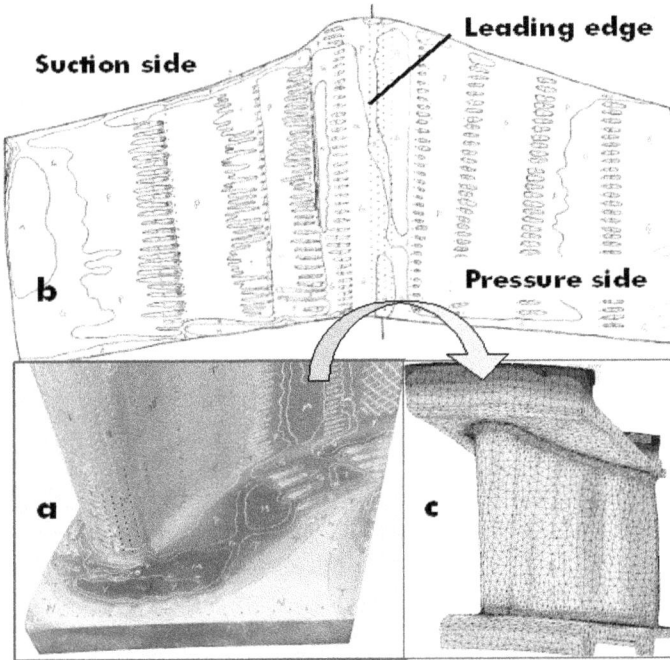

Figure 6-11   *GT26 engine thermal paint test: a – thermal painted hardware, b – 3D metal temperature validation, c – calibrated numerical 3D model for accurate temperature predictions*

For Alstom, which had already completed two other combined-cycle power plants using GT26 turbines for Gas Natural, this large new order confirmed the success of the improvement programme for these turbines and convincingly marked Alstom's return to the upper GT power generation segment with a competitive product.[51]

In review, beginning with the restart in 1990 after the ABB merger, nearly 15 years of comprehensive and intensive GT development efforts had come to an intermediate, successful end. It was a period with unforeseen technical difficulties and some imprudent management decisions, but first of all a huge multi-billion euro financial burden. This has only been brought under control since March 2003 by the new Alstom management under Chairman and Chief Executive Patrick Kron, initiating a successful turnaround accompanied by an intermediate 21 percent stake acquisition in Alstom's capital by the French State and the integration of financially strong partner companies.[52]

---

[51]   See Nieto, MW and Km/h - a History of Alstom, p. 244. Major GT sales also went to countries such as Italy, Germany and Thailand.

[52]   In June 2006, Bouygues group http://en.wikipedia.org/wiki/Bouygues_group acquired the French government's 21% capital holding for € 2 billion. Later in the year, Bouygues increased its Alstom shareholding to 24%, http://en.wikipedia.org/wiki/Alstom, and by March 2011 to 30 %.

The sometimes complex interdependency of several contributing factors has been outlined; a few of these merit being re-addressed, since they are normally overlooked in simplified explanation patterns of this dramatic engineering/management scenario:

1. The global change around 1990 spoiled the field of power generation gas turbines with aero capacities and technologies, except at ABB
2. The ABB initiated development speed race[53] in combination with a premature cut-back of resources later on, and an inadequate financial mind-set
3. ABB going public with the new projects too early
4. ABB pushing competitor-challenged upgrades without real need
5. The 'US gas bubble' with its change of customer structure
6. SCM and 'concurrent engineering' considerably reduced pricing barriers
7. Unrestricted sales campaign without sufficient development buffers
8. Warranty contracting
9. No established product support system
10. Owner and personnel changes increased instabilities.

6.1.3.3   GT24/GT26 Success Story since 2003

The situation steadily improved after 2003. Corrective action for a further upgrade package was launched as early as in 1999, ready for introduction in 2006. The main elements were:

- a compressor up-flow (again)
- the EV staged burner concept
- a SEV lance for lower NOx
- a slight turbine inlet temperature increase
- a general focus on fuel and operational flexibility.

The compressor was re-staggered in the front stages to increase the mass flow further. Additionally, for optimisation of efficiency and cooling air bleed conditions, some restaggering was carried out in the high-pressure part of the compressor. However, the actual flow path (as defined by the outer casing and the rotor profiles) remained unchanged, with the benefit that this compressor upgrade was also fully retrofittable to the earlier engines. Besides the restaggering, additional measures to optimise the blade clearances were introduced to raise the engine performance.[54]

The staged EV combustion was implemented in the GT26 fleet from 2006 and afterwards also appeared in GT24 commercial operation in Mexico from April 2011.[55] It further im-

---

[53] *"Ich sage es jetzt plakativ: Wir haben die ABB zügig aufgebaut, aber mit der dreifachen Geschwindigkeit hat man sie in den Boden reingefahren. Wir waren nämlich geübt darin, alles schnell zu machen. Wenn sie schnell das Falsche machen, geht es rasch bergab. (I'll say it boldly: we built up ABB in a speedy manner, but it was also wrecked with three times that speed. We had been used to doing things quickly. If you do the wrong things quickly, it all goes rapidly downhill.)"* Armin Meyer in December 2002, between 1995 and 2000 Exec. VP, ABB Ltd., responsible for Power Generation since 1994. See Catrina, ABB – Die verratene Vision, p. 256

[54] See Lanzenberger, A further retrofit upgrade

[55] Staged combustion is a method for reducing nitrogen oxide (NOx) emissions from combustion. The GT24/GT26 uses a fuel-staged supply a) by the conventional reheat combustion, where part of the fuel is burnt with less oxygen and b) within the new EV premix burners to expand the part load operation range towards a lower flame temperature.

proved performance and fuel flexibility, reduced part-load emissions, and enabled 'parking' of the engine at a low load level.

The cooling requirements in the SEV combustor were adjusted for an optimum trade-off between lifetime and efficiency, using the field experience and data from the GT26 test power plant at Birr, CH. Additionally, the lp inlet temperature was increased in order to improve the gas turbine engine and plant efficiency together with additional hardware modifications concerning Thermal Barrier Coating (TBC), cooling in the lp turbine and optimised leakage air consumption.

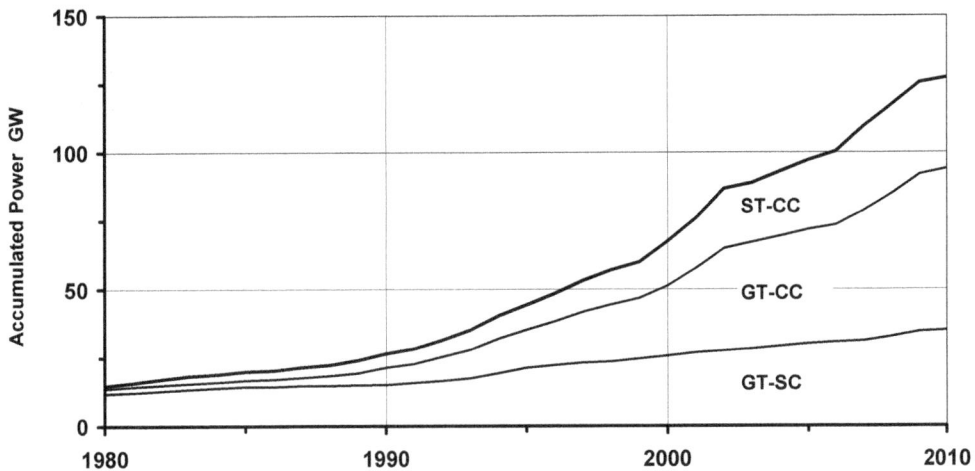

Figure 6-12   *BBC/ABB/Alstom accumulated installed power: GT-SC simple cycle share, GT-CC combined cycle share, ST-CC steam turbine combined cycle share*

The sustainable rise in Alstom's gas turbine business is illustrated by the plotted, installed power accumulation between 1980 and 2010, from alternate sources, gas vs. steam turbines and simple cycle vs. combined cycle operation mode, Figure 6-12. Clearly, the biggest contribution came from GT-CC, gas turbine installations as part of a combined-cycle power

---

Historically, staged combustion dates back to the 'hot bulb engine (Glühkopfmotor)' of the 1890s. The first stage of combustion occurred inside the hot bulb and the hot gases were then forced out into the cylinder, where they mixed with additional air and the second stage of combustion took place. At this time, staged combustion was used because it was a convenient method of ignition and it is unlikely that there would have been much concern about air pollution. http://en.wikipedia.org/wiki/Staged_combustion
The hot bulb engine became popular as a robust alternative to the diesel engine, for example in the Lanz 'Bulldog' tractor, manufactured by Heinrich Lanz AG in Mannheim, D between 1921 and 1960. The Lanz Bulldog was one of the most popular German tractors, with over 220,000 of them produced over its long production life. http://en.wikipedia.org/wiki/Lanz_Bulldog
The staged combustion cycle, also called topping cycle or pre-burner cycle, is also a thermodynamic cycle of bipropellant rocket engines, and was first proposed by Alexey Isaev in 1949. The first staged combustion engine was the S1.5400 (11D33) used in the Soviet planetary rocket, designed by Melnikov, Isaev's former assistant. At around the same time (1959), Nikolai Kuznetsov (1911–1995) began work on the closed cycle engine NK-9 for Korolev's orbital ICBM, GR-1. Kuznetsov later evolved that design into the NK-15 and NK-33 engines for the unsuccessful Lunar N1 rocket. In the West, the first laboratory staged-combustion test engine was built in Germany in 1963 by Ludwig Boelkow (1912–2003). http://en.wikipedia.org/wiki/Staged_combustion_cycle_%28rocket%29

plant. Instead of a visible plateau between 2003 and 2006, roughly ten percent annual growth was maintained during the past ten years, considerably more than the average four percent per year in the first decade observed. It appears that the (expensive) settling of the recent sales and development crisis also changed long-term stagnating market participation from the late BBC years into a healthy, sustainable Alstom growth period.

Another upgrade package was initiated in 2005 for introduction in 2011, which pushed the GT24 (GT26) output to 230 MW (324 MW), considerably above the original starting value of 165 MW (240 MW), and the simple cycle efficiency to an impressive 40 (39.6) percent. With respect to engine handling, this upgrade introduced the ability to switch between performance-optimised and maintenance-cost-optimised operation, depending on requirements. The next generation GT24, Figure 2-18, contains the well-proven EV burner from the 2006 GT24/GT26 upgrade. The hp turbine was also taken over from the previous ratings without design changes. Evolutionary modifications have been mainly limited to the following components:

- The modified compressor with enlarged flowpath allowed a further increase in mass flow for higher engine performance at again, further improved operational flexibility and high efficiency
- the SEV combustor was improved for increased fuel flexibility and even lower emissions
- the lp turbine gained efficiency with a unique, flexible operation capability at increased inspection intervals of + 30 percent
- enlarged turbine outlet duct for extra capacity and an improved exhaust duct
- improved fuel and operation flexibility in general.

The compressor annulus increase of Figure 6-13 resulted in a 25 percent increased inlet mass flow[56] for the upgraded version, on the basis of a highly efficient re-design for wide ambient and load range. The architecture is based on the 22-stage, well-established Controlled Diffusion Airfoils (CDA) design as already used in the GT24/GT26 engines. The outer annulus diameters were increased to match the requested mass flow increase. The number of VIGVs were increased from three to four, all GT24 upgrade design features are based on a scaling of the latest GT26 compressor upgrade.[57]

The SEV combustor architecture and structural parts remained unchanged in comparison to the previous GT24 version. Modifications were limited to the SEV burner and the SEV fuel lance as well as improved seals to reduce leakages. The burner modifications ensure better fuel mixing with the airflow, resulting in lower emissions over a wide operation range. Additionally, the robustness of the SEV combustion system against fuel gas composition changes could be enhanced: fuel gases with up to 18 vol.-percent of higher hydrocarbons can now be handled, without fuel preheating or other fuel conditioning measures.

The upgrade package includes an improved lp turbine which is scaled from the latest GT26 upgrade. The benefits are a) a higher lpt efficiency and b) switching on-line between two operation modes, thereby enabling the addressed extended operation intervals. All four lp turbine stages received airfoils with optimised profiles and cooling schemes, Figure 6-24.

---

[56] In reference to the original design value.

[57] See Hiddemann, The next generation Alstom GT26

The shroud design of blades 2 and 3 was improved to reduce over-tip leakages. In addition, the vane count per stage 1, 2 and 4 was reduced to minimise the cooled hot gas surfaces. The turbine outlet annulus was increased to accommodate the higher inlet mass flow delivered by the upgraded compressor.

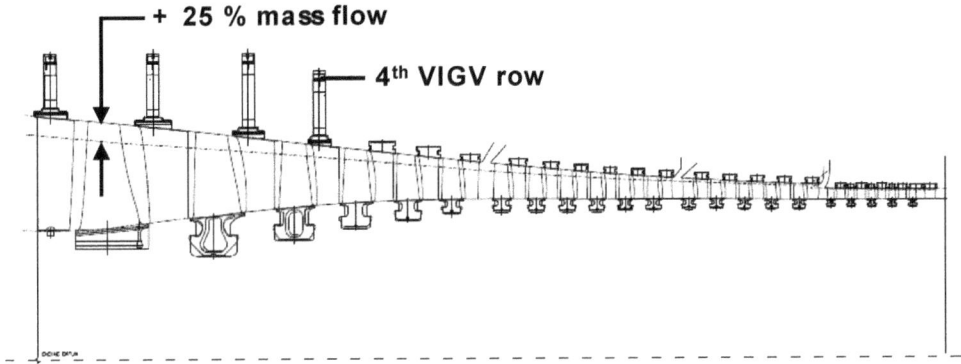

Figure 6-13   *GT24/GT26 compressor upgrade (rating 2011)*

SC Gas Turbines							
GT Type/ Year		Neuchâtel 1939	10/650 1953	11B 1968	13E 1983	13E2 1997	GT26 2011
Speed	rpm	3000	3600	3600	3000	3000	3000
Power	MW	4	7.2	33.3	140	165	326
Efficiency	-	0.174	0.192	0.255	0.330	0.360	0.396
Compressor							
Pressure Ratio	-	4.4	4.7	7.5	14.3	16.5	33.3
Mass Flow	kg/s	62.2	82	220	491	560	635
Entry Tip Speed	m/s	222	234	330	315	>370	>400
Entry Hub/Tip Ratio	-	0.76	0.72	0.67	0.59	0.57	0.54
Stages	-	23	17	15	21	21	22
Turbine							
TIT (Tmix)	°C	540	650	850	1070	1100	SEV>1250
Cooled Rows	-	-	-	-	6	6	6
Entry Tip Speed	m/s	196	207	280	268	319	>340
Exit Tip Speed	m/s	251	282	360	419	506	>525
Exit Hub/Tip Ratio	-	0.59	0.53	0.56	0.52	0.52	0.59
Stages	-	7	7	5	5	5	5

Figure 6-14   *Seven decades of BBC/ABB/Alstom GT development: Performance parameters of selected SC simple-cycle gas turbines and corresponding compressor and turbine components*

Despite its steadily upgraded performance, the new GT24 is still sold as F-class equipment after returning to the US market, where in the meantime competitors offer in addition to their F-class engines larger G-, H- and J-classes. Nevertheless, Alstom believes it has a competitive advantage because the GT24 is designed to provide the operational flexibility which generating companies need to be successful in the profitable ancillary services market, including spinning reserve, back-up of intermittent renewables, etc.

A further in-depth review of the new GT24 and GT26 design standard will follow in the context of the turbine and turbine cooling survey, Section 6.2, and in Section 6.3 on combined-cycle power plants.

In summary, Figure 6-14 illustrates the impressive historical development of simple-cycle GT main parameters over more than seven decades, from Brown Boveri's first utility gas turbine of 1939 to the present Alstom GT26: a stunning range from 4 to 326 MW output power, at total efficiencies from 17.4 to 39.6 percent.

6.1.4   Prominent Engineers

Unlike Phase II, which was over 40 years long, the present Phase III of ABB/Alstom's gas turbine development comprises only a quarter-century so far. However, this period was filled with a series of highlights, radical turning points and unforeseen events. The basic principle of this historical review: to emphasise the influence of a few important engineers in every phase under consideration, will be maintained in this section.

The first acknowledgement goes to Fredy Haeusermann, a prominent engineer who served the various Companies continuously from 1966 and was at the helm of hands-on gas turbine development activities during a decisive period between 1990 and 2002. Two other well-deserved acclaims go to Jakob J. Keller and Hans Koch, two engineers with sustained impact in promoting 'combustion' to an undisputed, lasting and stand-alone lead discipline within an excellent GT engineering environment. In addition, it is worth broadening the scope: The Gas Turbine Development Department created in 1995 and still maintain a 'Hall of Fame' today, where outstanding professional contributions of so far 21 engineers have been acknowledged over the years; this list of honoured members is attached with short descriptions in Section 7.3.

6.1.4.1   Fredy Haeusermann (*1942)

Born in Seengen AG, CH, Fredy Haeusermann completed his mechanical engineering studies at the ZTL Ingenieurschulen Zentralschweiz Luzern and the ETH Zurich from 1961–1966, finishing there as dipl. Ing. ETH. In 1966 he started his engineering career at BBC Baden, CH as a mechanical designer and development engineer for steam turbines. In 1970 he became responsible for the commissioning of steam turbines and plants, followed in 1975 by a further career step as head of industrial steam turbine design and becoming head of BBC's prestigious design department for large power generation steam turbines five years later. Between 1982 and 1984 he attended extra MBA studies in California. Finally, in 1985 he became President with technical and commercial responsibility for the section Steam Turbines and Heat Exchangers. Like many renowned predecessors in BBC's long technical history, he was picked in 1989 after considerable achievements in the steam turbine environment to the directorate for Gas Turbine Development, Section 7.2, with its inherent challenges in the 'dynamic' environment of the Asea Brown Boveri start-up phase.

Figure 6-15   *Dipl.-Ing. ETH Fredy Haeusermann (*1942, picture ~1995)*

The initiation, and to a large extent first realisation of design and development activities for an all-new power generation gas turbine in the upper power range, which became the GT24/GT26, occurred during the period of his responsibility for the Gebenstorf 'skunk works' offices. This happened together with the creation of a trustful and effective partnership network for engineering support from the MTU Munich and Uniturbo, Moscow. On 1 January 1995, Fredy Haeusermann handed the gas turbine development directorate over to Hans-Juergen Kiesow, while he himself was promoted to become new President and Business Area Unit Manager for Gas Turbines and Combined-Cycle Power Plants as successor to Rolf Kehlhofer, a post which he kept until the ABB to Alstom transition in 1999. In September 1999 he received till his retirement the global project responsibility for all gas turbine and combined-cycle power plants, especially on all repowering and GT13 projects. He is still connected to the power generation business on consulting basis.

Privately, Fredy Haeusermann served on his local council in Gebenstorf, AG, CH from 2002 to 2009 and in addition, from 1992 to 2011 he was a long-term member of the board (Schulrat) of the HWZ Hochschule für Wirtschaft, Zurich.

6.1.4.2   Jakob J. Keller (1950–1999)

For the gas turbine community the name of Jakob J. Keller will be lastingly connected with the invention of the EV burner, a stroke of genius in its simple and effective fluid mechanics set-up. He was born in 1950 in Oberaach, TG, Switzerland. He achieved his master's degree in physics from the ETH Zurich in 1973, and subsequently his Ph.D. too at Prof. J. Ackeret's Institute of Aerodynamics in 1975, with Prof. N. Rott as the doctoral advisor for Keller's thesis 'Resonant Oscillations in Closed Tubes: The Solution of Chester's Equation'. After two years on a kind of aerodynamicists' traditional 'grand tour' as a research fellow at Theodore von

Kármán's Guggenheim Aeronautical Laboratory, CalTech, Pasadena, Ca., USA, at the time the laboratory was led by H.W. Liepmann (1914–2009). Keller joined BBC's research centre at Baden-Daettwil, CH in 1977 as a scientist in the Fluid Mechanics department. In 1981 he received the Venia Legendi (the right to teach) Theoretical Physics at the ETH Zurich with a professorial dissertation on 'Nonlinear Self-Excited Acoustic Oscillations Within Fixed Boundaries'. Between 1979 and 1983, he was project leader for the aerodynamic design of Brown Boveri's new generation of governor valves for steam turbines. From 1981 he led several research projects on 'Combustion Instabilities', became leader of the research group 'Combustion and energy exchange', then head of the Aerodynamics department and finally in 1991, Vice President R&D at ABB Corporate Research.

Figure 6-16   *Prof. Dr. sc. nat. ETH Jakob J. Keller (1950–1999)*

Between 1988 and 1990 he had comprehensive responsibility for a development team of 40 scientists and specialised engineers to develop the new premix combustion technology for gas turbines, Section 5.3.4. This task was completed successfully with the introduction of the EV burner concept, which since then has had a major impact on ABB/Alstom's lead position in low-emission combustion for power generation applications. In 1991, Jakob Keller was the first recipient of ABB's Award of Merit for '*Outstanding contributions to the technology of the gas turbines, especially in combustion science and engineering over a sustained period*'. In 1992, he became a Member of the Swiss Academy of Technical Sciences under the presidency of Ambros P. Speiser (1922–2003), in appreciation of his scientific achievements in fluid mechanics and combustion, and '*his ability, to transfer research results in technical applications*'. In 1992, he left ABB for the Boeing Professorship at the Department of Aeronautics & Astronautics at the University of Washington, Seattle, Wa., USA, first as an Associate Professor, then was subsequently promoted to a full Professor in 1995. In May 1996 he returned to ABB Power Generation in Baden, CH as Vice President with various responsibilities in GT technology projects. After a brilliant academic and professional career, Jakob J. Keller passed away much too early in 1999, leaving his wife Marie Therese and three children aged between 15 and 19.

6.1.4.3   Hans Koch (*1929)

Hans Koch, born in 1929, finished his studies in mechanical engineering as dipl. Ing. at the ETH Zurich in 1952, after some specialisation in hydraulic machinery, combustion engines and unsteady flows with Prof. Eichelberg, and flight aerodynamics with Prof. Ackeret. After a first professional position in plant engineering at Luwa AG, Zurich from 1953 to 1955, he joined the gas turbine development department of Sulzer AG, Winterthur, CH for twelve years, from 1956 to 1968. His first activities comprised participation in the development of Sulzer's 3.5 MW GT1 engine under F. Zerlauth, GT performance design and calculations under A. Frieder and familiarisation with combustion technology, especially development activities for the multi-combustor GT7, including atmospheric combustor testing. An electronic control and servo system was also developed for the same engine in collaboration with Hawker-Siddeley, UK – the first time for a stationary gas turbine.

The BST period from 1969 to1974, saw Koch still at Winterthur, but now predominantly working for GT tasks from Baden as head of the combustion group. For the first time, low emission NOx targets were set for the GT9 engine test combustion chamber, with corresponding (and unsatisfactory) test activities at Muenchenstein, BL, CH.

Figure 6-17   *Dipl.-Ing. ETH Hans Koch (*1929, picture ~1988)*

For the next twenty years, until 1994, Koch worked in Baden as head of the combustion department since 1981. The top-mounted standard combustion chamber emerged for GT8 (new), GT11N and GT13E (in collaboration with BBC Mannheim). Primary zone water and steam injection technologies were extensively tested as preliminary NOx reduction methods for gas and oil burning, until the first dry low-NOx combustion chamber appeared. Experiences up until then had clearly showed that the only possibility to reduce 'thermal NOx'[58] was by means of 'lean combustion', with a high amount of excess air and premixed fuel to keep the flame temperature sufficiently low.

---

[58]   To differentiate from nitrogen-rich 'fuel NOx'.

A single-dome combustion chamber design with a large number of premix burners for part load operation emerged, Section 5.3.2, and was adapted for the GT13C operation with fuel oil EL and natural gas. Numerous operational problems had to be addressed and solved. The operation principle of staged, sequentially operated combustor groups was introduced, in combination with pilot burners to increase the LSL Lean Stability Limit. The corresponding GASL test activities as well as the first product introduction of the lean premixed combustion principle have already been described in Section 5.3.2, see also Figure 5-49.

Besides the trend towards more compact GT configurations, increasing GT turbine inlet temperatures accelerated the spread of annular combustors to improve the temperature distribution at turbine entry in general, but indirectly also ending the usage of low-quality fuels (with Na- and V-content) and as such, the peak of single combustion chambers.

The following development of a GT8 annular combustor with multi-injection burners has already been described in Section 5.3.3. The original proposal for an annular gas-fuelled combustor with 'Wabenbrenner'[59] came from the department TX (Theo Woringer and Gerassime Zouzoulas)[60]. After a lengthy GT8 test series at Beznau, the decision was made to switch to the multi-injection burner arrangement with 384 swirlers and quenching air ports, Figure 5-37(l). GT8 full scale engine tests, carried out by Andreas Vogel (1953–2003) and Philipp Brunner[61] confirmed the expected excellent homogeneity in the temperature distribution, but revealed also decisive limitations of the concept for further emission reduction. The path was cleared for the introduction of the second generation EV premix burner technology.

6.2   GT Key Component: Turbine

The description of gas turbine development has now reached the present, but for the historic review of the GT axial turbine component it is worthwhile to remember the early beginnings already addressed in Section 3.1.1, such as Leonardo da Vinci's chimney-integrated hot gas turbine.[62]

For BBC, axial turbine developments for Velox boilers in the early 1930s were determined almost exclusively by corresponding steam turbine experiences; consequently, the 7-stage

---

[59] See the hexagonal honeycomb burner structure in Figure 5-59.

[60] Theo Woringer was head of the BBC 'Feuerungslabor' (firing laboratory) until his retirement in 1984, followed by Jaan Hellat. The 'Wabenbrenner' for the GT8 RBK-W was Woringer's idea; thought of as a lean premix burner, the actual mixing of fuel and air behind the honeycomb grid was much too low for that purpose. Consequently, the emission results were unacceptable.

G. Zouzoulas was the inventor of the multi-injection 384 DK (DrallKoerper, swirler) burner. Here, excellent low emissions were achieved by low-volume, short flames with short residence times. This first successful annular combustor of BBC proved a high power density close to those of aero gas turbines, was easy to assemble and control by just one gas valve. A lasting problem was NO_2, characteristically visible from a distance as a yellowish fume cloud. [Information courtesy of Ph. Brunner on 11 April 2012]

[61] See ABB, GT-8 RBK-D(K) Purmerend, 1991

[62] As mentioned before, the word 'turbine' was introduced in 1822 by the French mining engineer Claude Burdin, derived from the Latin *turbo*, or vortex, in a memoir, 'Des turbines hydrauliques ou machines rotatoires à grande vitesse', which he submitted to the Académie royale des sciences in Paris. Benoît Fourneyron, one of his former students, built the first efficient water turbine according to consistent hydrodynamic principles, see Section 3.1.1.

uncooled axial reaction turbine of the first utility gas turbine at Neuchâtel, as illustrated in Section 4.3.3.3, also reflected the typical steam turbine technology design standard of the 1930s in detail. Claude Seippel[63] described the principle design considerations for the preceding Velox drive turbine developments:

> 'There was not much hesitation for a steam turbine manufacturer on how to build the expansion turbine to drive the compressor, which supercharged the Velox boiler. Due to the fact that no excess power was expected from the set ... a gas temperature of 500 °C (932 F) was sufficient. That was higher than current practice for steam turbines, but not too big a step. All that was required was to build the best possible turbine that available knowledge would permit. A five-stage reaction turbine was selected, not by tradition – the preceding turbine for the Holzwarth cycle had two impulse stages and the diesel superchargers had one – but because no better machine could be conceived.'

Turbine airfoil and the blade root section were milled as an integral piece of metal as for steam turbines. In comparison to steam turbines the airfoil length of gas turbine applications was relatively short, since the gas volume only expanded here at a maximum of 1:10. From the beginning the rear turbine stages carried damping wires to limit blade vibrations, as illustrated in Figure 4-79 for the last three blade rows of the Neuchâtel turbine.

While the evolution of GT turbine design at BBC appeared relatively straightforward as a deduction from steam turbines, the steam and gas turbine research still integrated at the time did not only take into account the results of the latest research in the field of aerodynamics, it also contributed in the short period of the late 1930s several essential 'firsts' which still have a lasting impact on turbine design as described in the following Sections 6.2.1 and 6.2.2:

- Flat water analogy of compressible flows was developed to become a practical industrial R&D tool for high-speed, two-dimensional transonic and supersonic steam and gas turbine cascade flow investigations
- secondary flows were visualised, analysed and described as a significant element for the performance in aerodynamically high-loaded turbine flow channels
- compressor and turbine cascade investigations to derive design rules for an optimum blade count, leading to the still widely used Zweifel loading coefficient

and finally, long-forgotten and rediscovered only recently, BBC used for the first time

- turbine blade film-cooling to open up the regime of elevated turbine operation temperatures beyond blade material restrictions.

6.2.1   Aerodynamic Turbine Development Milestones

The emerging science of lifting airfoil theory had been applied since the 1920s for axial compressor and turbine development at BBC. It appears that various flow visualisation techniques became a BBC speciality, soon to gain a comprehensive insight in complex flow patterns. A first example out of the newly established scientific test department is illustrated in Figure 6-18 by a water analogy streamline picture of a turbine cascade to evaluate the flow turning, from as early as 1926.

---

[63]   Seippel, The evolution of compressor and turbine bladings in gas turbine design

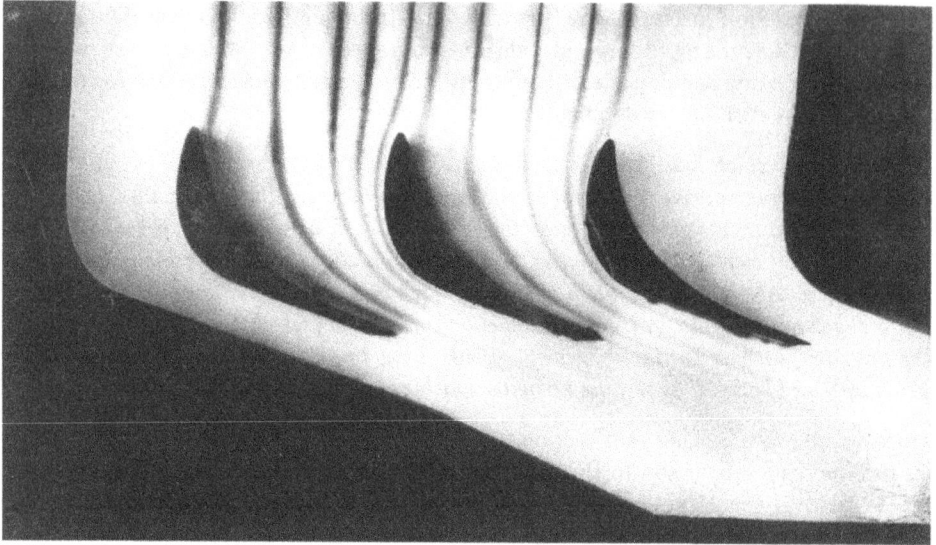

Figure 6-18   *BBC turbine flow visualisation of newly installed water tunnel, 1926* [64]   ©*ABB*

Ten years later the applied flow visualisation techniques became much more sophisticated. The newly emerging 'shallow water analogy' for compressible flows became a powerful tool to visualise and investigate transonic and supersonic turbomachinery cascade flows. The technique had been largely developed by Ernst Preiswerk as part of his doctoral thesis[65] at Professsor Ackeret's institute at the ETH Zurich between 1935 and 1938. The theoretical foundation of shallow water analogy states that under certain conditions a two-dimensional compressible gas flow and a flow of a liquid with a free surface are described by the same equations. The ratio of 'shooting water' velocity and ground wave velocity corresponds with the Mach number; well known phenomena of gas dynamics such as Mach lines or shock fronts which occur in both systems, can be visualised by a 'shallow or flat water channel'.[66] The method was afterwards used extensively by Preiswerk as a BBC employee[67] on the corresponding, newly built BBC test facilities; several in-house publications[68] documented the astonishingly appealing and revealing capabilities of this new technique.

---

[64] Picture source: BBM, Rueckblick... im Jahre 1927, 1928, p. 56

[65] See Preiswerk, Anwendung gasdynamischer Methoden auf Wasserstroemungen mit freier Oberflaeche

[66] The analogy requires neglect of vertical accelerations in the liquid, and has restrictions on the ratio of specific heats for the gas.

[67] Ernst Preiswerk (1911–1939) was born in Basel. After his studies in mechanical engineering at the ETH Zurich from 1930–1934, he received his Dr. sc. techn. degree ETH on 14 December 1938. From October 1937 to October 1939 he was employed at BBC Baden. He died on 21 November 1939 at the Sonnenrain Hospital in Basel after kidney surgery; his only child, daughter May, was born on 21 February 1940 in Chavornay, VD, CH. See Schopf-Preiswerk, Nachtraege zu dem Buche 'Die Basler Familie Preiswerk'

[68] See BBM, Rueckblick... im Jahre 1941, 1942, p. 77; in BBM, Forschung und Entwicklung im thermischen Maschinenbau, 1974: 'Untersuchung an rotierenden Schaufelgittern fuer transsonische Stroemung' (F. Rhomberg) pp. 762–773 and – see BBC, '75 Jahre Brown Boveri, 1891–1966, Festschrift 1966' p. 151. The fascination of these flow pictures, unique at that time, may also be illustrated by the fact that Seippel noted in his diary

After the publication of Preiswerk's thesis in German in 1938, it appeared as a NACA Technical Memorandum in 1940, indicating the international scientific attention and importance which this method publication gained in a short time. In 1946/1947 Preiswerk's hydraulic jump studies initiated systematic 'water table' experiments by Hans Albert Einstein, a Swiss-born US hydraulic engineer and first son of Albert Einstein, and Earl G. Baird, to test the validity of the analogy between the flow of a liquid with a free surface and the flow of a compressible gas.[69]

The literature list of Preiswerk's thesis has three references on the same subject from Dimitri P. Riabouchinsky, 1932 (DPR extended Jouguet's work to 3D motion), from himself, 1937 in the Schweizerische Bauzeitung and from Theodore von Kármán, 1938, which could be also interpreted as a kind of friendly competition for first publications between Ackeret and von Kármán. W. Albring, German rocket pioneer and renowned professor for fluid dynamics at the TU Dresden, D between 1952 and 1979, describes a meeting with D. P. Riabouchinsky (1882–1962) and a demonstration of his supersonic water analogy test facility in Paris in spring 1942, also with reference to Preiswerk's previous theoretical and experimental work.[70] This meeting presumably took place at DPR's L'Institut de Mécanique de Paris, now Supméca, at Saint-Ouen, 7 km north of Arc de Triomphe. In the meeting with Albring, DPR gave the credit for inventing the flat water analogy principles to N. Y. Zhukovsky (1847–1921), a Russian scientist and one of the founders of modern aero- and hydrodynamics (Joukovsky transformation, Joukovsky profile, etc.); young DPR was his scientific assistant.[71] In 1948 Albring brought the analytically improved technique based on Preiswerk's thesis back to Russia for ballistic rocket model test applications, where it was first received with scepticism, but then proudly accepted on the basis of the Zhukovsky background.[72]

Present day applications of this technique can be visualised for the tip sections of the longest blades of a steam turbine rotating at 2,000 km/h, i.e. the velocity of advanced supersonic jet aircraft. The steam expands and correspondingly accelerates in the rotating blade row from a subsonic to supersonic state. Figure 6-19 shows the flow pattern of such a turbine cascade on the surface of a flat water channel. 'Flowing water' at cascade entry and 'shooting water' at cascade exit represent the corresponding subsonic and supersonic regimes of

---

five numbers of a corresponding BBC picture series, which had just been archived on 5 September 1941 under 'Seippel/Zweifel/Meldahl' and were shown during his visit together with Ackeret et al. to Goettingen on 15/16 September 1941 to meet L. Prandtl, A. Betz, A. Busemann and O. Walchner, Section 4.2.4.3. Possibly the diary entry was also a kind of reassurance in case the subject of this delicate trip was questioned afterwards.

[69] See Krehl, History of shock waves, p. 512; this comprehensive, 1,300 page monograph confirms Preiswerk's prominent role in this context and names at the same time (1920) the Frenchman Émile Jouguet as well, who discussed the similarity between shooting channel flow and supersonic compressible flow and suggested that this analogy can be used to study 2D gas flows via experiments with a rectangular water channel.

[70] See Albring, Gorodomlia, pp. 130–134; special thanks to H.-J Lichtfuss for the hint.

[71] In 1904, the world's first institute for aerodynamic research was created, based on Zhukovsky's laboratory, practically on the family dacha site of the rich Riabouchinsky (banker) family at Koutchino, located 40 km SE of Moscow. In 1917 the Riabouchinskies had to flee Russia during the Revolution; in 1918 the laboratory became the nucleus of TsAGI Central AeroHydrodynamics Institute under the leadership of N.Y. Zhukovsky and in 1935 the place became Stakhanovo (after A. Stakhanov, Soviet record miner), renamed Zhukovsky in 1947.

[72] See Hildebrand, Praktikumsanleitung Flachwasseranalogie and – see TU Wien, Shallow water channel, for an introduction to this elegant scientific research tool.

an analogue steam flow. Oblique compression shocks in the steam flow are visible in water analogy by crossing 'water jumps' downstream of the test cascade.

Besides the described advanced flow visualisation techniques for detailed blade flow analysis, the first stand-alone multi-stage hot gas turbine component test rig of 500 kW and with airfoil blading was installed at BBC in 1943/4 to investigate the interaction of turbine blading at rotation and under realistic operation conditions.[73]

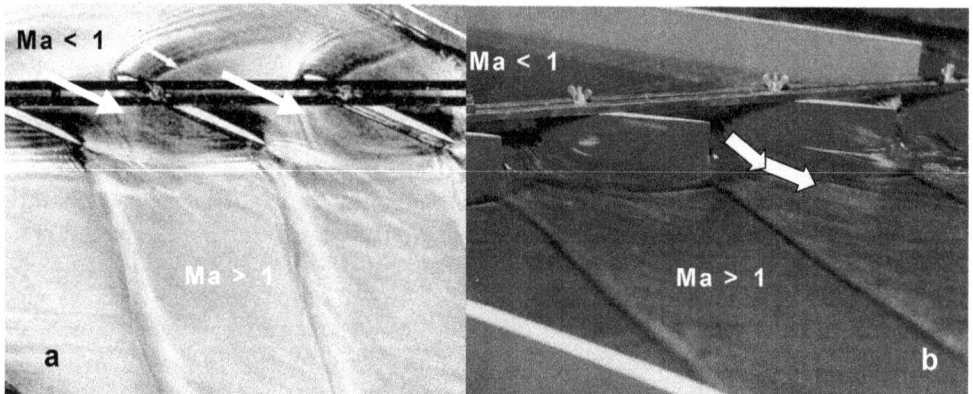

Figure 6-19	*BBC turbine flow visualisation 1941: flat water analogy, (steam) turbine flow channel, transonic flow shock pattern under different view angles: a – normal view, b – 'water fall' acceleration in perspective*

Another fundamental turbine flow phenomenon was discovered by Axel Meldahl[74] during intensive turbine research at BBC (loss measurements behind turbine airfoils/cascades and corresponding flow visualisation) from 1938 to 1940 and reported apparently for the first time in 1941[75], together with a first quantitative loss assessment: turbine secondary flows.[76]

---

[73]	See BBM, Rueckblick... im Jahre 1944, 1945, p. 10

[74]	Knud Axel Meldahl (1899–1972), a Danish citizen, was an assistant in BBC's Thermal Department, as far can be discovered from the BBC Hauszeitung (BBC's internal newspaper) between 1937 and 1948. He left BBC on 31 October 1948 to take a professorship in mechanical engineering at the Polytechnical College in Denmark, now DTU, Danmarks Tekniske Universitet, where he worked from 1948 to 1969. His major field of scientific research was power generation turbomachinery, and one of his specialities was calculations on rotating elements in turbine systems. (Information courtesy of Laila Zwisler, DTU, 2012)

[75]	See BBM, Ueber die Endverluste von Turbinenschaufeln (A. Meldahl), Nov. 1941 and – see Meldahl, End losses of turbine blades, Aug. 1942

[76]	In fluid dynamics, a secondary flow is a relatively minor flow superimposed on the primary flow, where the primary flow usually matches the flow pattern predicted using simple analytical techniques very closely, assuming the fluid is inviscid. In real flow situations, there are regions in the flow field where the flow is significantly different in both speed and direction to what is predicted for an inviscid fluid using simple analytical techniques. The flow in these regions is the secondary flow. These regions are usually in the vicinity of the boundary of the fluid, adjacent to solid surfaces where viscous forces are at work, such as in the boundary layer. Secondary flows are important in understanding the performance of turbines and other turbomachinery. Many types of secondary flows occur here, including inlet prerotation (intake vorticity), tip clearance flow (tip leakage), flows at off-design performance (e.g. flow separation), and secondary vorticity flows. Although secondary flows occur in all turbomachinery, it is a particular consideration in axial flow

Meldahl had observed that the turbine loss distribution some distance away from the endwalls was in agreement with the measured profile losses, while the losses increased considerably to the blade ends so that the overall turbine performance was strongly influenced by these effects. He concluded that in addition to

- the profile losses on the surfaces of the blades
- the skin friction on the annulus walls
- tip clearance losses

a new category of 'secondary flow losses' had to be introduced to explain the measurement results. Moreover, he stated that the turbine end losses were dependent on the form of the blade end, i.e. profile shape, including some edge sharpening, of blade pitch and tip clearance as well as Reynolds/Mach number and the degree of turbulence of the incoming stream. In a modern interpretation, *'the secondary losses are essentially those associated with the three-dimensional flows arising from the deflection of a stream of non-uniform velocity (the annulus wall boundary layers) through a row of blades. The losses in total pressure that arise from this motion may be viscous dissipation of the induced secondary velocities, or may arise from more complicated three-dimensional separation of the flow'.*[77]

Meldahl's combined theoretical/experimental approach consisted of

- the application of a formula from Betz on the induced drag of an airfoil with tip clearance flow at the free wing tip
- investigations on the impact of secondary flows at turbine blade ends without tip clearance
- flow visualisation tests in turbine cascades, Figure 6-20 and 6-21
- tests with an air turbine to determine the magnitude of end losses in a single stage reaction blading.

The result is a new formula for predicting the efficiency of turbine blading on the basis of similarity considerations. Without claiming here a comprehensive survey and review of the extensive turbomachinery literature on the subject, it appears that Meldahl clearly anticipated some fundamental thoughts on turbomachinery theory which only received a fully-fledged scientific analysis years later.[78]

---

compressors because of the thick boundary layers on the annulus walls. http://en.wikipedia.org/wiki/Secondary_flow

[77]   See Horlock, Axial flow compressors, p. 116

[78]   Our first other paper on this subject dates back to 1942 (at least one year after Meldahl's BBC publication), see Howell, The Present Basis of Axial Compressor Design, Part II, December 1942, where Howell states still somewhat ambiguously on p. 11: *'All miscellaneous losses such as bad incidences near the i/d (inner duct) and o/d (outer duct), secondary flows, tip clearance and wake interference are included in the secondary drag coefficient C_{Ds}.'* Howell only came up with his well-known IMechE compressor loss classifications in 1945 (see Howell, Fluid dynamics of axial compressors and – see Howell, Design of axial compressors, followed later by numerous publications on the subject, such as Squire&Winter (1951), Hawthorne (1951...1967), L.H. Smith (1955), etc.
Meldahl himself had already speculated whether he would be the first on this new subject. First he checked Stodola's Dampf- und Gasturbinen in vain; thereafter, he reviewed the proceedings of the IMechE Steam Nozzles Research Committee 1923–1930 in depth, especially their comprehensive works on short, low aspect ratio (steam turbine) nozzle segments, for which the end losses could add up to more than 50 percent, again without any hint towards an explanation of secondary flow mechanisms.

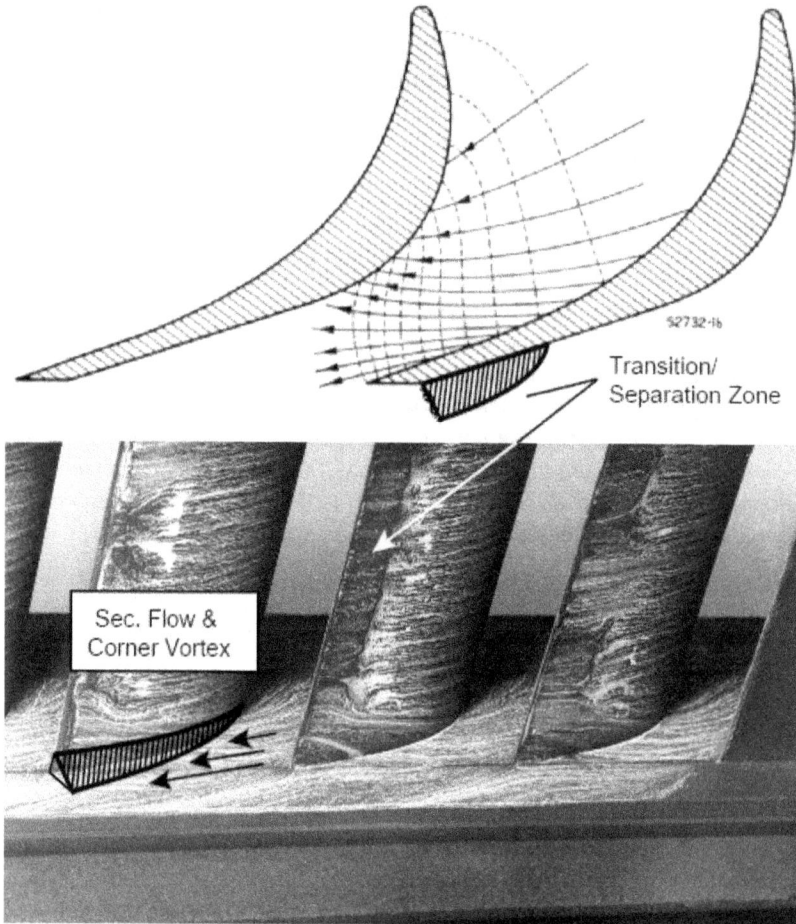

Figure 6-20   *BBC turbine flow visualisation 1941[79]: Turbine blade cascade, principle endwall secondary flow pattern, top (streamlines – solid, isobars – dashed) and experimentally determined endwall/airfoil streak lines, bottom   ©ABB*

After discussing the undisturbed flow conditions in the core of a turbine flow channel, Meldahl refers to the endwall flow pattern, as sketched in Figure 6-20, top. Here, the velocities within the wall boundary layer are considerably reduced, so that centrifugal forces are also diminished. This causes a flow pattern in the immediate vicinity of the wall which basically follows the superimposed pressure field, with resulting streamlines normal to the (dashed) pressure isobars. The effect of the secondary flows is an accumulation of boundary layer material in a suction side corner vortex, which as Meldahl speculates may even

---

Meldahl's US publication from August 1942 caused a commentary (not directly on secondary flows) which the editors of 'Naval Engineers' introduced with the following, unusual remark: '*Difficulties of present day communications make it impracticable to communicate to Switzerland in time to permit this discussion to be printed by the original publisher. Dr. Meldahl will be invited to make a reply, if he desires, as early as circumstances permit.*' (see Kreitner, Commentary).

[79] Source: BBM, Ueber die Endverluste der Turbinenschaufeln (A. Meldahl), Nov. 1941 – also Figure 6-21

induce additional suction side separation losses. The corresponding streakline pattern, Figure 6-20, bottom, was generated by wet oil paint, equally spread on the surface of the turbine flow channel before the blowing started. In this context Figure 6-21 illustrates the impact of a reduced blade count, i.e. an increasing cascade pitch/chord ratio t/l on the secondary flow effects. The development of the corner vortices on both ends of the blade suction sides is clearly visible, with increasing intensity towards higher t/l values, for t/l=1.55 with indications of flow recirculation and flow separation close to the blade trailing edge as well.

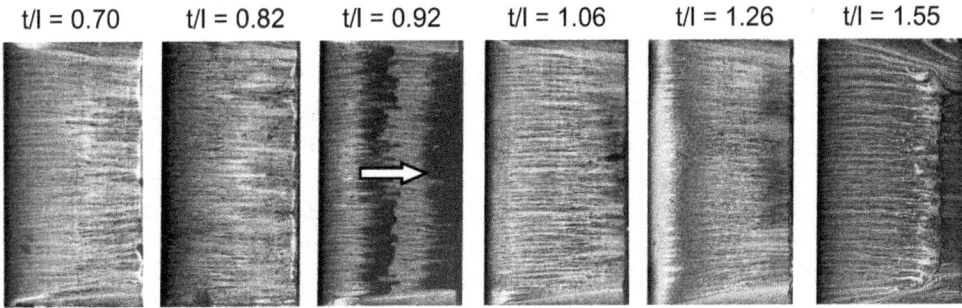

Figure 6-21   *BBC turbine flow visualisation 1941: increasing transition/ separation and endwall losses with pitch-chord-ratio t/l   ©ABB*

With all likelihood Meldahl's secondary flow experiments with varying blade count/ cascade loading stood in close context with Zweifel's search for an optimum loading coefficient (Zweifel number), which can be beneficially applied for both compressor and turbine design.[80] The choice of the blade spacing is a fundamental step within the design process of turbine blading. The pioneering publication by Zweifel (1945) and some years later by Ainley et al. (1951, 1957)[81] specified rules for the choice of optimal blade spacing and to realise minimal losses. In this context two conflicting phenomena have to be taken into account. On the one hand, a low blade spacing is associated with increasing profile losses due to high values of the blade wetted surface. On the other hand, high blade spacing is associated with higher aerodynamic loading which could lead to pronounced secondary flows, even flow separation, with corresponding losses. Zweifel introduced the aerodynamic loading coefficient, which is designated with his name in the literature, to quantify the maximal opposite pressure gradient realisable within the cascade. The Zweifel number is defined as the ratio between the peripheral force actually obtainable from the momentum theory, if the deflection diagram is given, and the force obtainable from an ideal pressure distribution, featuring the value of the cascade inlet total pressure on the pressure surface and the outlet static pressure on the suction surface.

---

[80] See a comprehensive review of Zweifel's work and the Zweifel number in Section 4.2.3. Contrary to Meldahl's advanced, nevertheless rather qualitative secondary flow analysis which was immediately released for broad international publication in 1941/2, Zweifel's valuable, immediately applicable design rule was rather hidden by the December 1945 publications, see BBM, Die Frage der optimalen Schaufelteilung and – see BBR, The spacing of turbo-machine blading, and became better known only in due course by frequent referencing in English publications in the 1950s, such as – see Horlock, Axial flow compressors.

[81] See Ainley, An examination, 1951 and – see Ainley, A Method, 1957

Such a profile pressure distribution would be represented by a rectangle and it would not feature a deceleration region. The optimum Zweifel number became a moving target, characterised by the developing state of the design art. While in the 1940s Zweifel himself saw values close to 0.8 as optimum, the evaluation of various measurements by Scholz[82] on turbine cascades in the 1960s indicated that for minimal drag-to-lift ratios this coefficient assumes a constant value between 0.9 and 1.0 for accelerating profiles featuring high deflections. This empirical loading limit has been widely used for the choice of the optimal spacing for turbine blades up to recent times, but obviously this experimental data (intermediately outdated) cannot be used for the design of innovative, advanced high-lift turbine blading where Zweifel numbers in the range of 1.0–1.1 have been achieved in the past 10 years. Nevertheless, Zweifel's loading coefficient is still widely used in scientific elaborations for both compressor and turbine investigations.[83]

The public introduction of the Zweifel loading coefficient in 1945[84] represents a preliminary end to BBC's creative peak of innovative features in turbine design; apparently the beginning of a stagnation phase which has already been observed in the context of Section 5.1.1. In 1960, not even the basic decision between the fundamentally different turbine types had been made (reaction vs. impulse/action) and finally some preference was hesitantly awarded to the reaction concept.[85] In this situation BBC could afford the luxury of offering two alternative thermal turbine options for their gas turbines, one for the challenging new temperature level of 750 °C and another one still for heavy crude oil operation at 620 °C, unchanged. The impression of development progress dragging on in parts is underlined by examples such as that of a turbine test series of 1959 on a new profile family, which was evaluated on a preliminary basis in 1964, and finally in numerical form in 1974(!).[86]

A revitalisation of turbine design in general based on new insights into fundamental fluid dynamic phenomena happened in the 1980s. The energy crises of the 1970s triggered developments in the aero engine industries to improve the fuel efficiency of high bypass ratio turbofans, with special focus on low pressure turbines at cruise conditions, the longest section of a flight cycle. This task was not easy, given the high turbine adiabatic expansion efficiency level of approximately 90 percent already achieved. Basic improvement potential was seen in the further reduction of profile (friction) losses by keeping the boundary layers (BL) on the high-velocity suction sides in a laminar state for as long as possible. This im-

---

[82]   See Scholz, Aerodynamik der Schaufelgitter and – see Scholz, Aerodynamics of cascades

[83]   For compressor application, see Lei, A criterion for axial compressor hub-corner stall, and for turbine, see Cardamone, Aerodynamic optimisation of highly loaded turbine cascade blades for heavy duty gas turbine applications: a dissertation, also with considerable Alstom (hpt) background.

[84]   The Zweifel design rule had presumably already been applied internally at BBC before 1940.

[85]   See BBC, Die Brown Boveri Gasturbinen mit einer Eintrittstemperatur von 750 °C, 1960. Impulse turbines change the direction of flow of a high velocity gas jet in the rotor. The resulting impulse spins the turbine and leaves the fluid flow with diminished kinetic energy. There is no pressure change of the gas in the moving blades, as in the case of a (reaction) steam or gas turbine, instead the pressure drop takes place entirely in the stationary blade rows. Credit for the invention of the reaction turbine component (and for the steam turbine concept) is given to the British engineer Sir Charles Parsons (1854–1931), and correspondingly to the Swedish engineer Gustaf de Laval (1845–1913) for inventing the impulse turbine. A generally named advantage of the impulse turbine is a minimum stage count (e.g. the first German turbojets Jumo 004B and BMW 003 had a single stage impulse turbine from AEG to save weight) and a correspondingly low cost level, while the inherent lower efficiency level of the impulse type influenced the lasting switch to the reaction concept for power generation and aero gas turbines. http://en.wikipedia.org/wiki/Turbine

[86]   See BST, Turbinenschaufelung N-Charakteristik, test turbine 'Brigitte'

plied the delicate task of delaying the laminar-turbulent BL transition to the rear profile section by maintaining sufficient flow acceleration, but limiting the losses of BL transition, pronounced transition 'bubbles' or even turbulent flow separation in the following, unavoidable flow deceleration section.

These developments started in secrecy in the late 1970s, at places such as MTU's OEM partner Pratt & Whitney[87] and the corresponding United Technologies Research Center at East Hartford, Ct, USA. This resulted in due course in independent research activities for lpt development programmes at MTU and partner institutes.[88] Typical was the newly developed, aft-loaded turbine profile family T106, Figure 6-22, which was thoroughly investigated in cascade wind tunnel tests, especially with respect to the low Reynolds number (Re<150,000) behaviour of the rear lpt stages at altitude. Complete 5-stage lp turbine tests at sea level condition confirmed the superiority of the new laminar-flow design by improvements in the turbine efficiency from 90 to 94+ percent. In the meantime, after the new design concept had become public around 1990, it developed into a comprehensive, thoroughly investigated new research field: UHL, Ultra-High-Lift turbine blading.[89]

The concept of low loss, laminar profiling became part of the GT24/GT26 configurations in the context of the already described design and development cooperation between ABB and MTU Aero Engines, Munich, Section 6.1.3.1. Typically, the change in design philosophy is illustrated by Figure 6-22, where a front-loaded BBC turbine cascade of the 1960s is compared to a 'laminar profile' of the 1990s. The isentropic Mach number distribution shows here a rather steady flow acceleration up to the first 60 percent of the axial profile chord length on the profile suction side. The following flow diffusion in the rear profile section has a certain tendency to develop laminar-turbulent transition bubbles. This phenomenon, first observed in steady cascade wind tunnel tests at comparably low turbulence levels, caused considerable concern as a potential source of loss. In order to avoid significant deteriorations in efficiency due to larger transition bubbles (or even non-reattached separation), the concept of controlled diffusion boundary layer design was introduced in industry. In addition, recent research results from cascades and test turbines revealed significant differences in the transition/separation behaviour between steady and unsteady, wake-induced flow conditions. In partiuclar, the loss of aft-loaded high lift profiles was de-

---

[87] See Gardener, Energy efficient engine low-pressure turbine boundary layer program (Om Sharma et al. carried out the principal investigations)

[88] See Hoheisel, Influence of freestream turbulence; Fottner, Anwendung neuer Entwurfskonzepte and – see Eckardt, PW2037 Entwicklung bei MTU. Corresponding turbine cascade tests were carried out at DLR Institute of Aerodynamics and Flow Technology, Brunswick and from 1985 at the Institute for Jet Propulsion (Prof. L. Fottner, today Prof. R. Niehuis) at the University of the German Armed Forces in Munich by means of the unique HGK Hochgeschwindigkeits-Gitterwind-Kanal (high-speed cascade wind tunnel), originally designed by Prof. N. Scholz. See Scholz, Der Hochgeschwindigkeits-Gitterwindkanal. The pressurized vessel construction allows for independent variance and thus separates the effects of compressibility (Mach number) and viscosity (Reynolds number). http://www.unibw.de/isa/forschung/versuchsanlagen/kva/hgk_ordner/hgk/index_html

[89] For one of the first comprehensive public reviews of 'turbine laminar profiling' and of 'controlled diffusion' BL control, see Hourmouziadis, Aerodynamic design of low pressure turbines. Robert E. Mayle, who contributed to early research on the topic, published a minimum acceleration parameter K as a kind of design guideline in 1991, required to maintain suction side laminar flow and prevent premature BL transition, see Mayle, The role of laminar-turbulent transition. The trailing edge diffusion was probably derived from the De Haller factor for compressor blading w2/w1 > 0.7 which is the square root of D = 0.49; typical lpt values were at D < 0.50–0.53 (courtesy of Jean Hourmouziadis).

termined considerably lower than that of conventional front-loaded profiles when wakes from foregoing blade rows were present.[90]

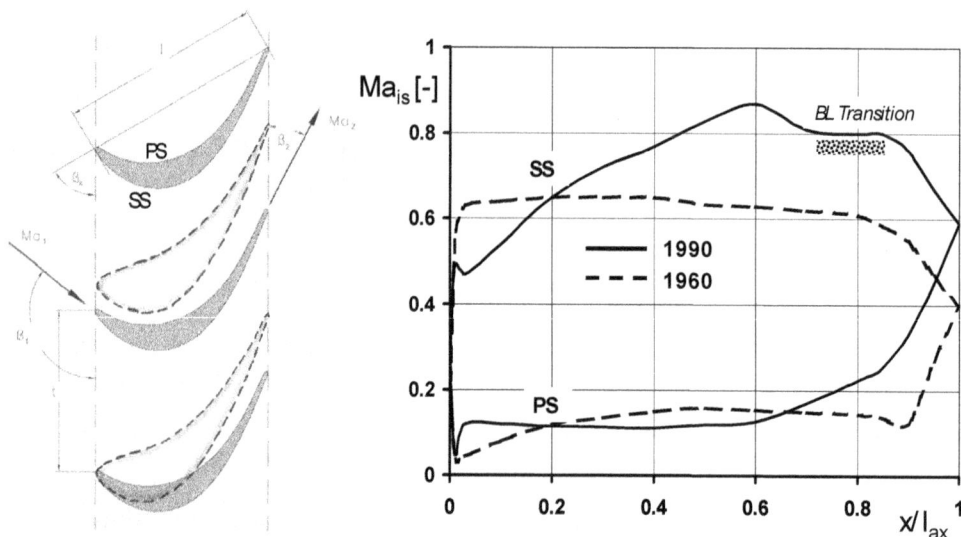

Figure 6-22   *Turbine profiling 1960 vs. 1990 [MTU T106]* [91] *and comparison of corresponding profile Mach number distributions* [92]

Interestingly, Jakob Ackeret was already very close to suggesting laminar profiling in 1938! In an article for the Schweizerische Bauzeitung[93] he speculated on the future of aircraft propulsion: *'Will it become possible to build compressors with 85 percent adiabatic efficiency and turbines with 90 percent?'* and then goes on to give the answer. By interpreting flat plate data for accelerated flow, he predicts that laminar flow might be maintained even at higher Reynolds numbers over a large part of the profile of sensibly curved turbine blading, meaning *'extraordinarily favourable (efficiency) values may be achievable'*.

The steady design progress during seven decades of BBC/ABB/Alstom turbine development has already been discussed in the context of Figure 6-14 for a few characteristic design parameters. The achieved advanced turbine design standard for the GT24/GT26 family is illustrated in Figure 6-23, where hp and lp blading are represented on the same scale as CAD surface models with a semi-transparent insight into the internal cooling structures.

---

[90]   See Hodson, Unsteady flow: Its role in the low pressure turbine

[91]   See Nagel, Numerische Optimierung dreidimensional parametrisierter Turbinenschaufeln, p. 68 http://ub.unibw-muenchen.de/dissertationen/ediss/nagel-marc/inhalt.pdf

[92]   1960: see BBM, Forschung und Entwicklung im thermischen Maschinenbau, 1964 (H.E. Imbach dissertation excerpt: Berechnung der kompressiblen, reibungsfreien Unterschallströmung durch ebene Schaufelgitter, pp. 752–761) vs. 1990: see Kožulović, Modellierung des Grenzschichtumschlags, p. 25

[93]   See Ackeret, Probleme des Flugzeugantriebes in Gegenwart und Zukunft, p. 4

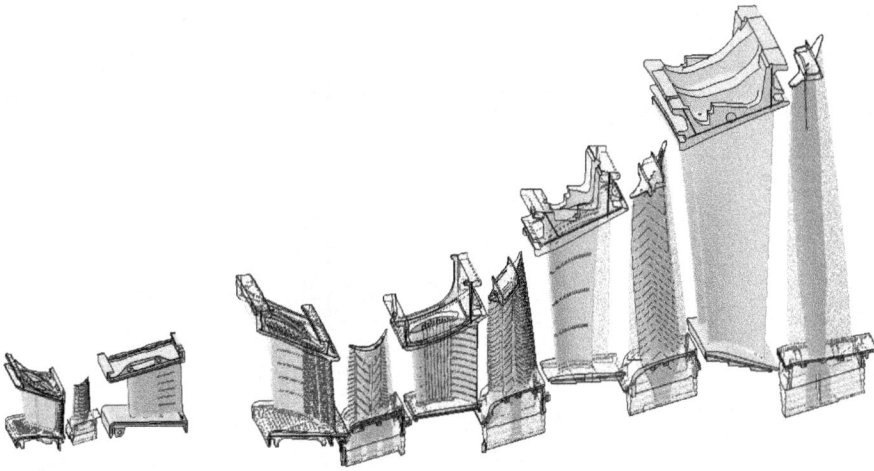

Figure 6-23   *GT24/GT26 turbine design standard 1995 with 2002 modifications, single stage hp turbine and 4-stage lp turbine, from left*

As previously addressed, optimum turbine design gets more complicated if cost and cooling aspects are taken into account. In fact, increasing the blade spacing and the deflection reduces the number of parts and is thus associated with reduced manufacturing and maintenance costs. This has particular benefits for high pressure turbine stages, where a reduction in the number of blades means a reduction in the components for cooling and therefore saves expensive compressed cooling air. Furthermore, the increase in stage loading is associated with higher drops in temperature and the cooling of the following stages can potentially be reduced/avoided. Higher pressure ratios across the stage increase tip leakage flows as well. Finally, higher pressure gradients and higher deflections lead to the development of more important secondary flow losses.

With more than 10 years of lessons learned, very effective remedies for some of these indicated turbine performance drawbacks were also integrated into the next GT24/GT26 design standard. As shown in Figure 6-24 for vanes and blades of the lp stages 1 and 3, three-dimensional airfoil profiling characterised by a circumferentially or axially, leaning and/or bowed airfoil shape has been applied for optimum loading distribution and to adapt secondary flow strength throughout all stages to achieve the best values of turbine efficiency.

The aerodynamic design focused on the optimisation of profiles for high lift levels[94] and the improvement of shroud aerodynamics[95], optimising the annulus flow path and the aerodynamic matching between the turbine stages. The flow path design was changed in or-

---

[94] See Cardamone, Aerodynamic optimisation of highly loaded turbine cascade blades; and – see Montis, Experimental and numerical investigation on the influence of trailing edge bleeding. It appears that film-cooled rows in particular have their optimum at less pronounced acceleration/laminarisation, an argument which accompanied early discussions within the design team (Prith Harasgama).

[95] See Porreca, Fluid dynamics and performance of partially and fully shrouded axial turbines; and – see Porreca, Optimised shroud design for axial turbine aerodynamic performance

der to accommodate a higher mass flow, with the additional benefit that the new turbine can be offered as a retrofit upgrade to existing customers.[96]

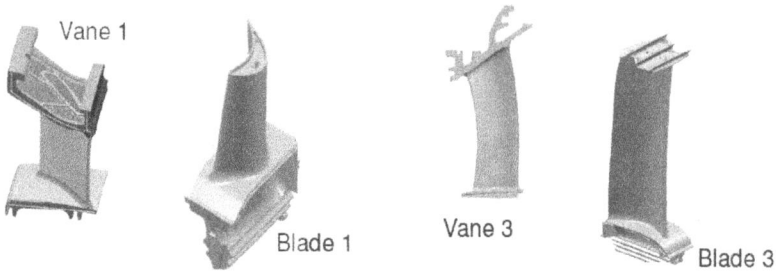

Figure 6-24   *GT24/GT26 turbine design standard 2011: lp turbine stages 1 and 3 aerodynamically improved*[96]

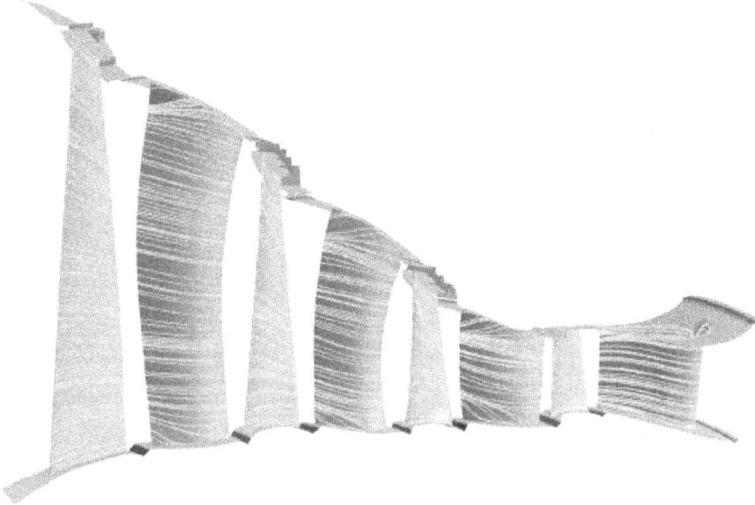

Figure 6-25   *Application of 3D computational fluid dynamics: GT24/GT26 multi-row turbine design, matching and optimisation*

Extensive use was made of 3D CFD; a fully featured multi-row CFD model was built up throughout the design phase, as shown in Figure 6-25, in order to confirm and optimise the stage matching and flow distribution.[97] The model also accounts for the main leakages and the addition of cooling and purge air. After 70 years this was an impressive repercussion of Meldahl's first hesitant attempts to find the right words to define his observations of secondary flows in a turbine cascade, Figures 6-20 and 6-21.

---

[96]   See Hiddemann, The next generation Alstom GT26
[97]   See Naik, Aero-thermal design and validation of an advanced turbine

6.2.2   Turbine Cooling Technology

6.2.2.1   Out of Time: Turbine Water and Film Cooling 1925–1945

To the best of our knowledge, the subject 'cooling of turbine blades' enjoyed its first comprehensive and specific review in the 1920 Eyermann-Schulz book on 'Die Gasturbinen'.[98]
In total eight different cooling techniques are described, mostly on the basis of water cooling, water and steam injection etc., which to a high degree are immediately independent
from the basic gas turbine arrangement under consideration. A few of these early trials in
advance of the finally, successfully realised constant-pressure GT concept was already described in Section 3.2. BBC was heavily involved in various stages of one of the concepts
from 1909 to 1938: the Holzwarth Gas Turbine, Section 3.3. Figure 3-18 illustrated the largest 5 MW version with participation from BBC in 1938. Development testing at BBC had
been carried out on a 2 MW prototype since 1930.

Figure 6-26   *Rotor set of 2 MW BBC-Holzwarth GT impulse turbines, generator at right, first blade row
water-cooled, 1930*[99]   *©ABB*

As shown in Figure 6-26, the 2 MW explosion turbine had two 2-stage impulse or Curtis
wheels. As the temperature of the gases at the inlet to the turbine was about 700 °C, the
blades of the first moving row were fitted with water-cooled ducts.

The corresponding hardware assembly is outlined in Figure 6-27, consisting of a water-cooled
core to absorb the stress and a heat-resistant shell which experienced hardly any stress. This
shell determines the turbine blade profile shape. These blades were threaded into the rotor of
the Holzwarth gas turbine and sealed off from the water ducts by a packing ring.

---

[98]   See Eyermann, Die Gasturbinen, ihre geschichtliche Entwicklung, Theorie und Bauart, pp. 248–256
[99]   Picture source: BBC, Geschichte des Schweizer. Turbomaschinenbaus, 1982

Figure 6-27   *Water-cooled turbine blade assembly for 2 MW BBC-Holzwarth gas turbine, 1929*[99]   ©ABB

Interestingly, the subject of 'turbine blade cooling' played a key role in the subsequent discussion following Adolf Meyer's comprehensive presentation on BBC's gas turbine development on 24 February 1939 at the IMechE 'Extra General Meeting' in London. The discussion was started by D.M. Smith[100]:

"It was stated in the paper that 1,000 deg. F. was the highest temperature which could be considered safe with uncooled blades. The author mentioned that numerous patents had been taken out for devices for cooling the blades." [An indication that Meyer's spoken presentation departed from its print version.[101]] ...'*Smith asked whether any device for cooling the blades could be regarded as commercially practicable, and to what extent the author would expect that cooling the blades would enable the temperature limit, and therefore the efficiency, to be raised.*

Adolf Meyer replied, that for the cooling of the blades some most ingenious and also very complicated devices had been invented, but experience indicated that those involving the use of cooling water must be discarded because of the danger of obstruction of the small passages by rust. Air cooling might however be found more interesting, and a cooling system of this kind, where the boundary layer of gas around the blade was replaced by fresh and comparatively cold air supplied to the hollow blade and emitted through a slit along the leading edge promised to give good results.

The prospects of the gas turbine, if it is possible – as the author believes it is – to raise the temperature in the near future to 1,200 deg. F., can be readily appreciated. This belief is based up-

---

[100] Dr. David M. Smith came from Metropolitan-Vickers (MV), Manchester, UK. MV's chief mechanical engineer Karl Baumann (1884–1971) was Swiss, part of Stodola's group of assistants at the ETH Zurich in 1906/7. He was obviously closely acquainted with A. Meyer, Section 4.1.2; both are assumed having initiated the contacts between RAE and BBC Baden before and during WW II, a fascinating part of the planned coverage of BBC's contribution to the early turbojet developments.

[101] See BBR, The combustion gas turbine, 1939

*on the experience obtained with a number of Diesel engine supercharging units which have been in operation for a considerable time at temperatures **approaching** this value.'* [102]

A few years later water as a cooling medium was still 'out' at BBC, but now the prime GT design philosophy was against turbine cooling at all, and instead, complex regenerative processes were favoured as a means for improving performance. This special development, which might have been partially encouraged by Ackeret (in publications and by his Closed Cycle GT example) had to be stopped in the 1960s after severe disappointments, not least due to the chemical incompatibility of crude oil combustion with the intercooling/regenerative approach.

In June 1945 the position on blade cooling at BBC Baden was documented by the group of US visitors (Th. von Kármán et al., for details see Section 4.2.4.6):

'Water cooling of turbine blades is not favored since the circulation of water is easily interrupted due to the collection of sediment near blade tips by the action of centrifugal force. They prefer to run at reasonable temperature where cooling is not required.' [103]

BBC was not alone with their scepticism about turbine cooling at that time. R. Friedrich, sometimes called the 'father of the Siemens gas turbine', wrote in 1949 after a thorough assessment of the state-of-the art of both, water and air-cooled turbines:

'There is the impression that the cooling of combustion gas turbines by water or air will not come for stationary (power generation) units and railway traffic in general. Potential applications may be limited to special constructions, exhaust gas turbochargers and high performance (aero) engines where other demands shift criteria such as cheap manufacturing, simple operation and longevity to a secondary role.' [104]

The early beginnings of turbine air-cooling had already been addressed for turbochargers in Section 3.1.2. The work of Christian Lorenzen gained particular attention, since it was brought into a direct development line with later German turbojet designs. [105]

E.W. Constant II knows about Lorenzen's background:

'Christian Lorenzen of Berlin began experiments on axial turbines with internally air-cooled blades shortly before 1930. His hollow-blade design apparently followed a Brown

---

[102] The mentioned 1,200 deg F corresponds to 650 °C. Eight weeks before the London meeting, on Wed 28 December 1938, Meyer's test department issued a test report TFVL 1234 on 'Air-cooled blade for gas turbines' with the conclusion '...*with the admissible cooling air flow [8 %] the blades are suited to be operated at gas temperatures of 800–1,000 °C.*'

[103] See Tsien, Technical Intelligence Supplement, p. 117. It goes without saying that the impressive achievements in turbine air cooling at BBC up until 1940, which will be discussed in the following Section, had not been mentioned in that meeting. Neither was Georges Darrieus addressed, an expert on turbine air-cooling and its applications in marine, land and aero gas turbines (who attended that meeting), nor was the corresponding BBC turbocharger hardware on show. A search for reasons must remain speculative, but the similarities to a failed meeting in Baden ten years before between Claude Seippel and Sanford Moss, GE on turbocharger issues is striking, see Section 3, Footnote 32.

[104] See Friedrich, Gasturbinen mit Gleichdruckverbrennung, pp. 73–74

[105] See Wilson, Turbomachinery – from Paddle Wheels to Turbojets. This author stated on p. 39: '*Christian Lorensen (sic) in Berlin began experiments on axial-flow turbines and hollow air-cooled blades, which were made by Brown Boveri, in 1929* (an isolated statement which could not be confirmed anywhere else). *His work led directly to German turbojet cooled blade designs.*'

Boveri patent used under license. Lorenzen was a pioneer in aeronautics and in auto-
mobiles. He had participated in early variable-pitch airscrew and turbocharger devel-
opment in Germany at the end of the First World War. In 1926, a turbocharger of his
design was tested at the German Aeronautics Test Laboratory (DVL Deutsche
Versuchsanstalt für Luftfahrt), Berlin-Adlershof. In 1928, he installed a turbocharger on
a Mercedes-Benz automobile engine in place of the Roots blower normally used. In his
experiments, Lorenzen encountered blade warping as a result of over-heating. He there-
fore began work on air-cooled blades, but had also designed a complete internal com-
bustion gas turbine around it. The cooling air was introduced into the hub of the tur-
bine and whirled outward through the blades, cooling the blades and compressing the
air at the same time. At its exit from the blades, the air was fed to a combustion cham-
ber, which exhausted through the turbine. The Lorenzen gas turbine proved impracti-
cable, probably because of internal flow losses, but his work on air-cooled blades proved
invaluable to the Germans during the Second World War.[106]

Confusion about Lorenzen's role grew from 1928 onwards when a positive review article
from DVL on the Lorenzen gas turbine in the prestigious VDI Zeitung[107] constructed an
inexplicably direct connection between Lorenzen and DRP 346 599. This patent on 'Gas
turbine cooling method and device' had been issued for **Brown Boveri** with German priority on 3 August 1920, deduced from the BBC patent CH 92250 from 21 July 1920, 20:00; a
corresponding patent illustration shows Figure 6-28, I). Eighteen months later, Lorenzen
received a patent (CH 101 035) for his 'gas turbine' with priority from 12 January 1922,
18:30. In hindsight, Adolf Meyer clarified the situation in 1947[108]:

'Brown Boveri ... obtained a patent some 20 years ago to use the hollow blades of an impulse
gas turbine to compress air by centrifugal action, so as to cool the disk and the blades and to
heat the air simultaneously... Such machines were built after the first world war on license
(from BBC?) by a Berlin firm (Lorenzen?) for some special purposes.'

At present, it is difficult to understand why DVL had referred to the BBC patent number without mentioning BBC in the context of the Lorenzen gas turbine. Either the patent numbers
were mixed up or the BBC licence was camouflaged intentionally. Nevertheless, the stated influence of these early gas turbine cooling activities on the German first turbojet configurations
is still clearly deducible. As an example, Figure 6-28 II) illustrates the turbine vane and III) the
turbine blade of the Junkers Jumo 004 B turbojet engine, which powered the Messerschmitt
Me 262. Clearly, there is a principal agreement between the patented air-cooled vane configuration (I-d) of 1920 and the realised engine vane of 1942, as well as the simple, straight air-
cooled blade of the B-4 engine version from the end of 1944, which has a high similarity to the
corresponding BBC patent (and the later Lorenzen patent/turbocharger).

Since AEG (the power generation 'Allgemeine Elektrizitäts-Gesellschaft') collaborated in
the design of the single-stage impulse-type turbine design of the Jumo 004 B, contemporary
steam turbine practice was followed here, similar to Brown Boveri's approach. A certain
amount of reaction was chosen for the blades to obtain the requisite amount of work from

---

[106] See Constant, E.W., The origins of the turbojet revolution, with reference to – see Lorenzen, The Lorenzen
gas turbine
[107] See Heller, Die Gasturbine von C. Lorenzen
[108] See Meyer, Recent developments in gas turbines

them with a relatively low turbine speed and without too much swirl in the exhaust gases, the intention of later using afterburning being kept in mind. '*The amount of reaction chosen was 20 percent, although Junkers wanted more and AEG wanted none at all*'.[109] This low reaction enabled cooling air to be drawn up over the blade roots. Although solid turbine blades were first used (Jumo 004 B-1 and B-2) for quick results, air-cooled, hollow blades were put under development from the beginning. Especially since it was realised that for a given limiting stress in a turbine wheel, a higher rotational speed and thrust was possible with the lighter, hollow blades at higher temperatures and thus lower fuel consumption. All versions of the engine used air-cooled, hollow turbine inlet nozzles.

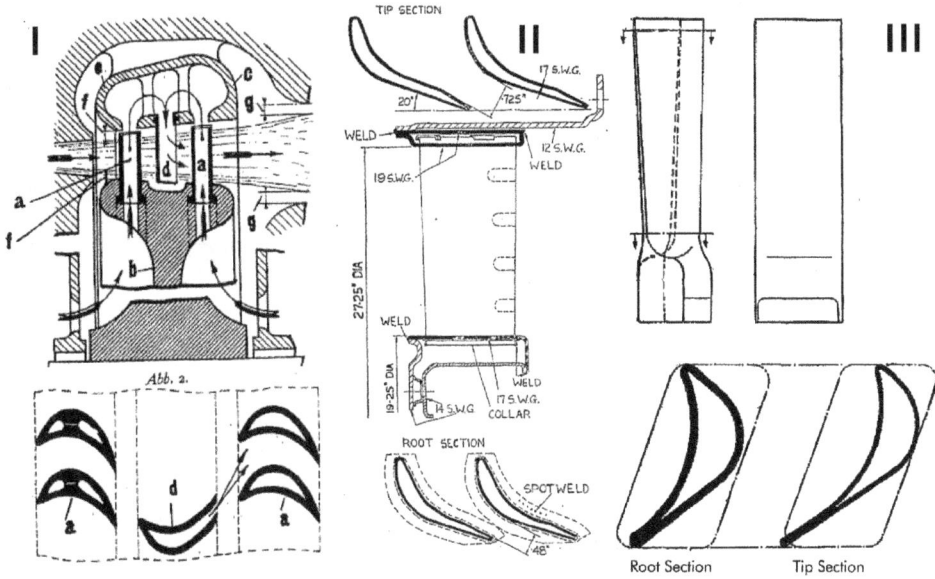

Figure 6-28   *BBC influence on Junkers Jumo 004 B turbine design I – BBC patents 1920, II – Jumo 004 B turbine nozzle assembly*[110]*, III – Jumo 004 B-4 hollow turbine rotor blade*

The need to develop hollow air-cooled turbine blading arose because of the German shortage of metals used in heat-resisting alloys. Compared with solid turbine blades, hollow blades needed less material and turned out to be easier and cheaper to manufacture, while their air-cooling permitted either a higher working temperature with increased efficiency and thrust, or the temperature as before increased engine life[111]. According to a post-war assessment, the Jumo 004 cooling air amounted to about 4 percent of the engine air throughput, permitting this engine to operate at a turbine inlet temperature of approximately 760 °C.[112]

---

[109] See Kay, German jet engine and gas turbine development 1930–1945, p. 71

[110] Picture source: Power Jets R&D, The Junkers Jumo 004 B jet engine, p. 346, II ©Emerald Group Publ. Ltd.

[111] At best, the engine life was 25–35 h.

[112] See Hafer, Gas turbine progress report, p. 130

The cooled turbine blade manufacturing process for the German turbojets had been developed during 1943 at the company W. Prym, Stolberg (12 km east of Aachen), based on a cold, deep-drawing process starting from 3 mm 'Tinidur' sheet metal.[113]

While the BBC patents of 1920 were applied and exploited 20 years later for German turbojet developments, BBC was already secretly carrying out the next innovative step at the same time: turbine blade film-cooling, Figure 6-29.

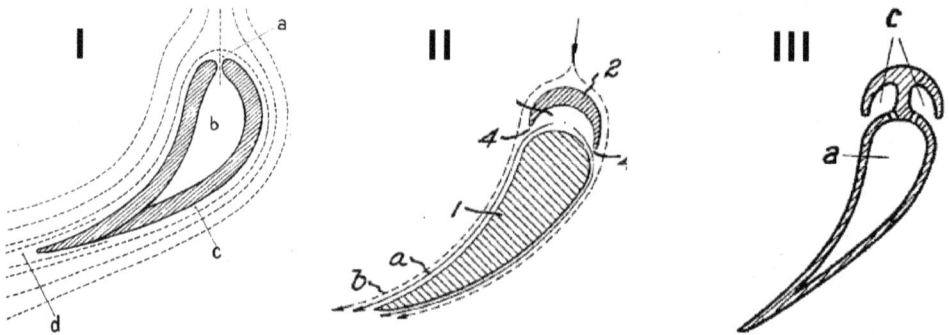

Figure 6-29   *Turbine film cooling – patent development I – Darrieus 1934, II – Faber 1938, III – Meyer 1938*

The foundations of this new advanced technology were laid by Georges Darrieus (Section 4.2.1, Figure 4-14) from BBC's French subsidiary C.E.M. (Compagnie Électro-Mécanique, Section 2.1.1) in 1934. A patent, US2149510 (also CA358651), was issued in his name for 'Method and means for preventing deterioration of turbo-machines', with French priority beginning on 29 January 1934. It consists (as illustrated in principle in Figure 6-29, I) of discharging a protective fluid/layer through one or more orifices/slits, located in the well-rounded leading edge of a turbine airfoil *'and spills smoothly and nearly equally on either side of this edge. Both faces of the blade ... are covered by a protective laminar and continuous layer of cooler or non-corrosive fluid which separates them from the hot or corroding working fluid'.*[114]

After a short while BBC contributed to improve its own field of innovation; Paul Faber from BBC, Baden CH filed a patent on a 'turbine blade', granted as US2236426 and based on a German priority from 27 July 1938. Introductory remarks of his patent text mentioned the drawbacks, also of the foregoing Darrieus patent: *'Hollow blades with outlet openings/slits along the*

---

[113] On 11 May 1943 a meeting to increase Jumo 004 production took place at Stolberg, besides RLM officials there were personnel from Junkers, Dessau; AEG, DVL, BMW, all Berlin; Krupp, Essen; Heinkel-Hirth and Daimler-Benz, Stuttgart and Brown Boveri, Mannheim in attendance. See Kay, German jet engine p. 75. For mid-1944 a total blade production target of 225,000, or 500 daily, was set-up. However, these targets were never reached, Prym's Stolberg production for the month of August 1944 was only 5,000 blades from 15–20 presses. An expansion of the facilities was needed and a new Prym plant was started for blade production in 1943 in Zweifall, 15 km SE of Stolberg. This plant was to be ready in October 1944 with 40–50 presses and a planned output of 300,000 per month. These plans collapsed sometimes between 21 October 1944, when Aachen was the first German city occupied by the Allies, and 5 February 1945 after the severe 'Battle of Hurtgen Forest' (Huertgenwald, 8 km east of Zweifall), impressively recounted by Ernest Hemingway in 'Across the River and into the Trees (Ueber den Fluss und in die Waelder)'.

[114] Excerpt from US2149510.

center of the leading edge of the blade have been used, but this location has the disadvantage that the slit may be clogged by dust, cinders or other foreign matter in the high temperature gas.'

Moreover, *'impurities in the gas may erode in a short time the blade leading edge section.'* Faber's patented blade shape is illustrated in Figure 6-29, II). Consequently, the improved patent idea was to form a 'crescent shape' with two emerging cooling films near the blade leading edge, basically in the same flow direction as the main flow, with this blade 'foresection' preferably formed of a particularly hard metal so that either a composite blade or vane is manufactured.

Again, this patent was almost immediately outdated by the next one, DE710289, again issued for BBC with first priority in Germany from 9 February 1938. The inventor was Adolf Meyer, Section 4.1.2, who turned negative experimental evidence from an investigated Darrieus-type cooling configuration, Figure 6-31, into a new idea of an improved blade shape, Figure 6-29, III), still known today as 'showerhead cooling'. Meyer's patent title indicates the contents: *'Blade with boundary layer protection against high temperatures and method for such blade production'.* Film-blowing directly out of a narrow radial channel in the blade's leading edge proved disadvantageous, as radially accumulated flow losses would result in a varying cooling mass flow density over radius. Separating the function of radial flow distribution in a wide main channel and only afterwards guiding the air into the film direction-turning head sections, led to several superior cooling blade shapes, such as the one shown as Figure 6-29, III).

The new turbine air-cooling philosophy was integrated by Brown Boveri for the first time in a 1938 prototype turbomachinery design. Figure 6-30 illustrates the VTF 201[115], a BBC turbocharger especially designed for what most likely became the Rolls-Royce Merlin[116] piston-engine. Figure 6-30a shows a VTF 201 cross-section, turned in a position so that the general turbine through-flow direction corresponds to that of the detailed measurements of the following Figure 6-31. Figure 6-30b illustrates an outside view and the corresponding detailed view of Figure 6-30c shows the turbocharger 'red-hot', operating under load on the test bench at Baden CH in 1938.

Interestingly, the VTF 201 base design (without special turbine blade cooling) already contains cooling features, which at best could be reconstructed from documented German wartime analysis.[117] The turbocharger shaft runs in two ball bearings: one in the cold compres-

---

[115] VTF 201 was renamed to VTF 280, when the compressor wheel diameter (280 mm) began to be used for reference, which again could be reduced to VTF 225 (225 mm wheel diam.) for the improved version with film-cooling.

[116] This speculation is likely to be true, though a final application clarification is still pending until the BBC aircraft involvement will be investigated as planned in the future. The Rolls-Royce Merlin is a British liquid-cooled, V-12, piston aero engine with 27-litre capacity, with 1,030 hp for Merlin II (1938). Rolls-Royce Ltd. designed and built the engine which was initially known as the PV-12, running first in 1933 and going into production in 1936; the engine is most closely associated with the Spitfire. In military use the Merlin was superseded by its larger capacity stablemate, the Rolls-Royce Griffon, another potential candidate for the 1938 turbocharger cooperation between BBC and RR. See http://en.wikipedia.org/wiki/Rolls-Royce_Merlin

[117] See Leist, Abgasturbinen mit Kuehlluftbeaufschlagung. The paper was presented to the Lilienthal-Gesellschaft fuer Luftfahrtforschung on 12 December 1940 in Munich, D. The author, then at Daimler-Benz Stuttgart, D, claims to know that this turbocharger from BBC Baden, CH was built under licence both at C.E.M. and Rolls-Royce. He quotes details from 50 hour altitude endurance test runs, which C.E.M. carried out in cooperation with Hispano-Suiza on their 12Y54(?) aero engine, which as indicated, he received from lead personnel involved in these tests. It is likely that the author mixed this up with the actual engine designation 12Y45 from 1938, with a Szydlowski-Planiol turbocharger. http://en.wikipedia.org/wiki/Hispano-Suiza_12Y The 'lead per-

sor entry section and one on the hot turbine exhaust side. This heat-exposed bearing is supported by four air-cooled (!) bearing struts. The air supply for this bearing is fed to the shaft centre by the curved tube at the extreme left of Figure 6-30a) and wraps the bearing struts with four cooling streams.

Figure 6-30   *VTF 201 BBC turbocharger for Rolls-Royce aircraft engine, 1938: a – cross section (radial compressor wheel diameter 280 mm), b – hardware, weight 32 kg, c – rear turbine view on test bench at red-hot operation* [118]   ©*ABB (b)*

The results of the turbocharger VTF 201 prototype testing with air-cooled turbine were taken together with the first time use of thermocolours for determining surface temperatures. Figure 6-31 shows the correspondingly evaluated surface temperatures at a turbine test point with 15,700 rpm speed, a gas temperature of 909 °C at turbine entry and a corresponding cooling air temperature of 150 °C:

– to the right (upstream) of the turbine rotor, the temperatures of the front rotor blade view and of the outer disk section are shown (with a variation between 670 °C at tip and 190 °C at the inner measuring diameter). In addition, the wall temperature distribution of the outer surface of the cooling air entry duct is shown, axially rising in the flow direction from approx. 140 °C to 190 °C at the blade disk front face,

– for the cooled turbine blade and the outer disk section, isotherms are drawn, which characterise a turbine blade leading edge variation from 290 °C at the hub to 655 °C at the tip, while the corresponding temperatures at the blade trailing edge vary between 390 °C at the inner duct and 680 °C at the tip,

---

sonnel' referred to could indicate a contact with Russian-Polish born Joseph Szydlowski (1896–1988), who stayed in Germany after WW I when he was a German prisoner (POW) and worked for companies such as Krupp, Daimler-Benz at Gaggenau, D, and Junkers until 1930. After founding Turbomeca (TM, now part of the SAFRAN group in France) in 1938, he was with a DB team (Fritz Nallinger) at the end of WW II in Bregenz on Lake Constance, where the German wartime project of a 'New York bomber' was transferred into a 120 PAX civil project 'Ultra Rapid' for the same target. As TM President he also hosted a small team of DB engineers until 1952 in Bordes, F, amongst these Bruno Eckert and Heinrich Kuehl. Karl Leist founded the Institut fuer Turbomaschinen at RWTH Aachen in 1948 and led it until his death in 1960.

[118]   All pictures, see Jenny, The BBC turbocharger; interestingly, Figure 6-30c had already been published separately in a kind of advertisement in 1943 as 'Aircraft turbocharger, 28,000 rpm at full load, exhaust temperature 900 °C'; see BBM, Rückblick... im Jahre 1942, 1943, pp. 52/53

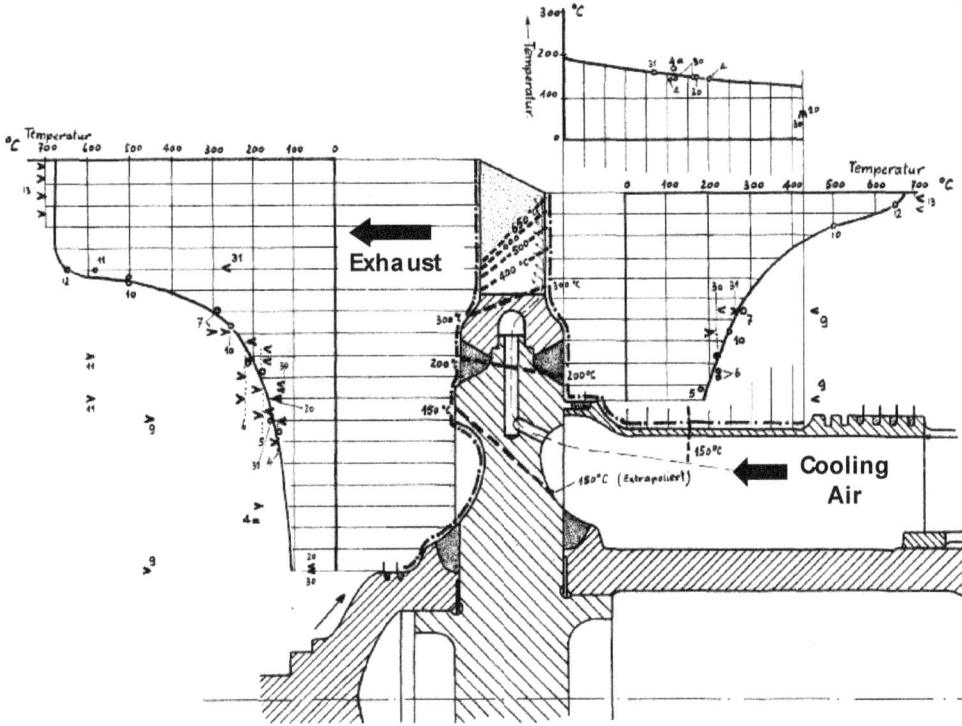

Figure 6-31   *First 'thermocolour' application for turbine surface temperature determination* [119], *original plot 1938 BBC air-cooled turbocharger turbine wheel VTF 201, n = 15,700 rpm, turbine entry temperature 909 °C* [120]

–   finally on the exhaust side the rear face temperatures of the blade row and of the outer disk section are depicted, with 2/3 of the blades' rear outboard section being constant at 680 °C, until the temperature curves down to 110 °C at the inner-most measured temperature of the turbine disk.

The general drawback of the cooling configuration after Darrieus, which has already been outlined during the discussion of step-by-step patent improvement, becomes clearly visible. Here a radially pronounced film-cooling effect can only be observed for the whole blade front zone, while the trailing edge situation confirms the known weakness, where a significant cooling effect is only visible for approx. one quarter of the inner trailing edge duct height.

---

[119]   First, according to BBC's own assessment in November 1939. The used thermocolours came from I.G. Farben A.G. (Ammoniak Laboratorium Oppau), Ludwigshafen, D.

[120]   See BBC, Flugzeug-Aufladegruppe VTF 225 mit Schaufelkühlung nach Darrieus, 1938/9. The complete, translated title of this BBC test report TFVL1204 from 1 December 1939 is '*Aircraft turbocharger VTF 225 with blade cooling according to Darrieus for 12 cyl. **Kestrel**, 4-stroke carburettor motor from Rolls-Royce, Derby.*' Since the original 'Kestrel' had only 700 hp and represented a design standard of the late 1920s, this reference in December 1939 must have been based on some disinformation. It was remarkable that the 'Kestrel' was used for Messerschmitt Bf109 V1 prototype testing in 1935 as planned German engines were not ready in time. The Reich Air Ministry (RLM) acquired four Kestrel VI engines by trading Rolls-Royce a Heinkel He 70 *Blitz* as an engine test-bed.

BBC also extended the test activities to a film-cooled blade shape in line with Adolf Meyer's patent, Figure 6-29, III). Five air-cooled turbine blades manufactured from 0.5 mm thick V2a sheet metal were arranged in a stationary wind tunnel cascade, and tested in a 1,000 °C gas stream (from gas oil combustion). The cooling flow was up to 8 percent of the main stream. The final test report TFVL 1234 from 28 December 1938 stated:

'As far as can be determined on the basis of this stationary test set-up, these turbine blades can be operated with acceptable cooling air quantities up to gas temperatures of 800–1,000 °C'.[121]

Darrrieus' outstanding contributions to BBC's advanced cooling (and other) technology developments were acknowledged in a touching 'Laudatio' from Claude Seippel in 1968, addressing Georges Darrieus in perfect French on his 80th birthday, as *"Vénéré maître, cher ami"* and continuing in a technical manner: *"...step-by-step, you accompanied the development of the gas turbine from its origins in the steam-generating 'Velox' to future applications at extreme temperatures. There are numerous memories to confirm this, here is just one example. By injecting a layer of cold air into the boundary layer, you perfectly managed to protect a blading which was flown through by a gas at **1,200 °C**. ..."*[122]

While the first time application of film-cooling for turbomachinery prototype testing has just been demonstrated for BBC, the honour of first time product introduction apparently remains with Georges Darrieus and his French affiliations. A future description of BBC's aero engine activities will address the TGAR-1008 axial turbojet with turbine film-cooling, where the first time usage in aero engines by the C.E.M daughter-company SOCEMA had been accomplished. In addition, the first industrial product application of film-cooling apparently goes to C.E.M. as well, as indicated by Hafer[123] in 1953:

' ...note that the first gas-turbine engine designed for marine service, by Compagnie Electro-Mecanique, made extensive use of boundary-layer cooling. The four-stage turbine had hollow stator and rotor blades with slits near the leading edges (Darrieus patent). Compressed air was fed into the blades and escaped through the slits to form a cool envelope around each blade.'

The author continues, obviously now with reference to TGAR-1008:

'This same arrangement was used in the nozzle blades of a French aircraft gas turbine engine, with the result that the blade temperature at the trailing edge was only 1,020 F (550 °C) in a gas stream having a temperature of 1,470 F (800 °C) with a 1 per cent cool-

---

[121] See BBC, Luftgekuehlte Schaufel fuer Gasturbinen (Blechschaufel Profil 581b), 1939. The first accordingly manufactured cooled blading had cross sections in the blade root and upstream of the exhaust slots that were too small, so that the cooling pressure had considerably to be increased for sufficient cooling. In due course, a phenomenon later known as 'ballooning' was observed during these first tests of air-cooled blading.

[122] See BBM, Hommage à Monsieur Georges Darrieus, Membre de l'Institut, Paris, 1968 « ... Pas à pas, vous avez suivi le développement de la turbine à gaz, depuis ses origines dans la chaudière 'Velox' jusque dans ses formes d'avenir utilisant des températures extrêmes. De nombreux mémoires l'attestent. Voici un exemple. A l'aide de l'injection d'une couche d'air froid dans la couche limite vous êtes arrivé à protéger parfaitement un aubage traversé par des gaz à 1200 °C. ... » It is likely that Seippel's reference to 1,200 °C had the military film-cooling application for TGAR-1008 in mind, as described in the following, but this required further investigation.

[123] See Hafer, Gas-turbine progress report, p. 130. At the time, the author calls 'film cooling' 'boundary-layer cooling'.

ing air flow (43)[124]. The turbine disk also was air-cooled in this plant (54)[125]. The first marine gas-turbine plant to go to sea, the Metropolitan-Vickers G-1, also used boundary-layer cooling for each side of the turbine disk (55)[126]. ...'

6.2.2.2   Heavy-Duty GT Turbine Cooling Development after 1965

The (re-)introduction of cooling to gas turbines was the most important technological breakthrough in gas turbine development since the end of World War II. Advancements in turbine cooling also helped gas turbines to penetrate today's power generation market. Like material advancements, which will be addressed in detail in the following section, cooling innovations allowed higher temperature inlet gases into the turbine blade path. Gas turbine operation at these higher temperatures granted higher efficiency and made these turbines more viable sources of electric power.

Figure 6-32 demonstrates how the introduction of cooling helped to increase turbine inlet temperatures (TIT) over time. For uncooled turbines the material development up until 1970 allowed for an annual temperature rise of 9 deg C on average; afterwards on the basis of remarkable improvements in cooling technology the annual growth value nearly doubled, and it is only after the mid-1990s that a certain flattening of firing temperatures becomes visible, which would have levelled off rather abruptly without the widespread usage of TBC (Thermal Barrier Coatings). The TIT increase helped to improve turbine efficiency while minimizing thermal damage, boosting the power output at the same time, a trend which strongly supported gas turbines in finding their way into the power generation market. As a kind of outlook, the figures mark an area of future 'Advanced Technologies' after 2010, which will be dealt with in detail in Section 6.4.2.

The breakthrough of cooling technology after the mid-1960s came rather unexpectedly. The 1958 Gas Turbine Progress Report of ASME still concluded:

'Turbine cooling has been very carefully studied during the past six years. Many tests have been made and certain results obtained. However, turbine cooling has not yet been used commercially except in a minor way. The desired breakthrough to materially increase turbine efficiency by using temperatures above 2000 F (1,100 °C) has not yet arrived commercially. ...The next six years should show a marked advance in turbine cooling and possibly turbine-inlet temperatures up to 3000 F (1,650 °C)[127].'

---

[124] See Hafer, Ref.43 'The Cooled Gas Turbine', by A.G. Smith and R.D. Pearson, Proceedings of the IMechE, 163, 1950, pp. 221–234

[125] See Hafer, Ref.54 'Gas Turbine Progress in France', The Oil Engine and Gas Turbine, 18 August 1950, pp. 139–144

[126] See Hafer, Ref.55 'Gas Turbine Propelled MGB 2009-No. 1', The Engineer, 184, 5 September 1947, pp. 218–220

[127] See ASME, 1958 gas turbine progress report. Here, the reference value is certainly referring to CET Combustor Exit Temperature, T_{mix} of Figure 6-32 is usually 150–200 deg C below CET; nevertheless, this forecast was clearly too optimistic in hindsight.

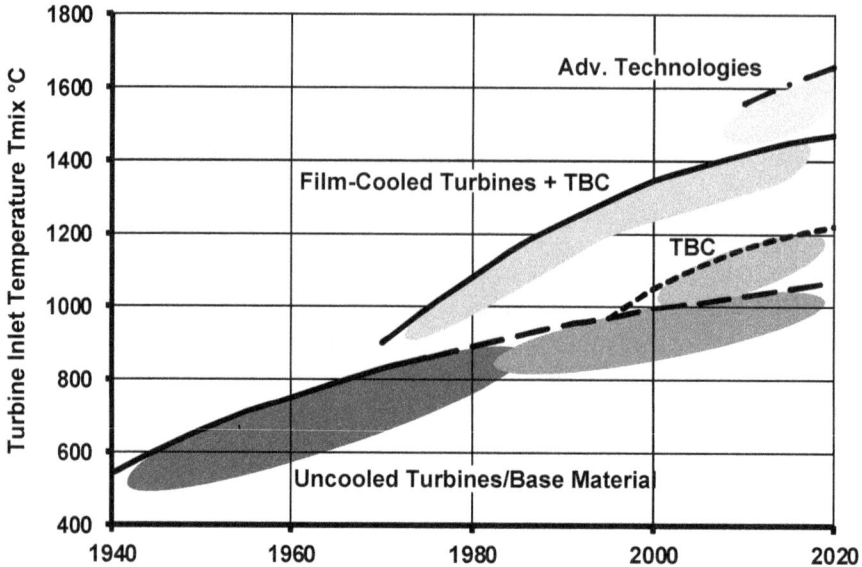

Figure 6-32   *Development of turbine entry temperature over time*

Traupel[128] reviewed the cooling transition period as follows:

> 'Turbine blades were originally uncooled, though air-cooling had already been proposed
> during very early gas turbine studies. This was mainly due to a lack of viable manufac-
> turing techniques for cooled blading. There was neither sufficient knowledge about heat
> transfer for a reliable blade design, nor an overall design base for gas turbines of that
> type. In addition, the relatively small increase in thermal efficiency appeared to be an
> argument against blade cooling. It was only the demand for a significant increase in
> specific power which promoted the introduction of blade cooling, first for aero engines,
> afterwards also in industrial gas turbines.'

However, up until 1960 leading aero engine manufacturers were also obviously hesitant
towards turbine blade cooling because of the associated costly design and unavoidable additional losses from cooling air. Only when the development of materials stagnated and the
heat-fatigue strength of high temperature materials reached its limits, did the comeback of
compressor-air-cooled turbine design succeed, mainly due to the initiative of Rolls-Royce
in Britain.[129] For power generation General Electric paved the way for the (cooled turbine)

---

[128] See Traupel, Thermische Turbomaschinen II, p. 358

[129] See Schubert, Progress in turbine blade cooling since 1945. Cooled blades entered commercial airline service
with the Conway engine in 1960, typically in the form of solid blade profiles with several radially straight
cooling holes. See Holland, Rotor blade cooling in high pressure turbines. The Rolls-Royce RB.80 Conway
(13–20 klbf TO) was the first bypass engine (or turbofan) in the world to enter service. Development started
at Rolls-Royce in the 1940s, but it was used only briefly in the late 1950s and early 1960s before other turbofan designs were introduced that replaced it.
Steady improvements in cooling have raised turbine entry temperature from below 1,000 °C to today's level of
over 1,500 °C for take-off conditions. Early blades relied upon convection cooling, but present-day blades are
cooled by mixed systems, employing cast multi-pass convection cooling and laser-drilled, film-cooled, fan-
shaped film cooling holes in sections with additional thermal barrier coating.

future in a 1966[130] ASME paper[131], simply by outlining the benefits of a rise in turbine inlet temperature. A TIT increase from 1,500 F to 1,700 F was determined to pay off with 6 percent improved GT thermal efficiency, but 25 percent in power output. Consequently, a cooled first stage turbine vane was introduced with a sheet metal core piece, so that the compressor discharge cooling air was fed from the outside diameter and distributed around the inner vane surface, cooling the leading edge and sidewalls. The air was then let out through holes on the profile pressure side near the trailing edge. The design was claimed to reduce the vane metal temperatures by more than 150 F (> 80 deg C). By the late 1960s, turbine baseload firing temperatures were near to 910 °C (1'670 F), and significant firing temperature increases depended on cooling first-stage vanes (or 'buckets' in GE terminology). With future increases in mind the MS7001 was designed to be readily adaptable.

A turbine cooling restart happened at BBC in the 1970s, completely independent of the accumulated film cooling experience of the Company up to 1945. First, the turbine inlet temperature was increased by introducing superalloy materials in general, followed by specially designed alloys for the corrosive IGT fuels, e.g. IN 738, later introducing step-by-step air cooling in the first turbine stages.[132] Even with these increased temperatures the simple cycle gas turbine did not reach the efficiency of the steam turbines, while at the same time the oil crisis of 1975 re-emphasised the importance of low fuel consumption.

Progress in turbine cooling expertise at BST/BBC can be measured by the following Company milestones:

- **1970** Preparation of analytical design tools for cooled blade design and life calculations[133]
- **1972** Eight patents[134] were filed with priorities between 18 January 1972 and 16 May 1975 for convectively cooled blades and vanes[135] as well as cooled rotor configurations, with the following inventors: J. Burket, W. Endres, C.J. Franklin, O. Frei, O. Iten, H. Melliger, B. Meloni, D. Mukherjee
- **1973–1981** Numerous experimental turbine tests[136] were undertaken to calibrate the spreading use of computer codes against measured reality; typical examples comprise flow and cooling tests in a cascade channel, measurements in a test turbine and for the

---

[130] Interestingly, first basic tests on turbine air cooling had been carried out in close cooperation between GE Schenectady and NGTE National Gas Turbine Establishment already before 1960, as outlined in detail in Kruschik, Die Gasturbine, pp. 251–256. The National Gas Turbine Establishment (NGTE Pyestock) in Fleet, part of the Royal Aircraft Establishment (RAE), was the prime site in the UK for design and development of gas turbine and jet engines. It was created by merging the design teams of Frank Whittle's Power Jets and the RAE turbine development team run by Hayne Constant. NGTE spent most of its lifetime as a testing and development centre, both for experimental development and to support commercial engine companies. NGTE Pyestock closed down in 2000 and was decommissioned to make way for a business park. http://en.wikipedia.org/wiki/National_Gas_Turbine_Establishment

[131] See Starkey, Long-life base load service at 1,600 F turbine inlet temperature

[132] As described already in Section 5.1.5.3, the first turbine vane internal cooling was realised after 1972, for four GT11C shipped to Blue Lake, Ca., USA.

[133] See BST, The determination of the life-time of cooled and uncooled gas turbine blades

[134] DE2202857; US3885609; CH547431; DE2442638; US3989412; US4021139; US4019831; US4056332; [761]

[135] Typically, the inventions refer to convectively cooled turbine blading, where the cooling air is split into two separate flows, of which each partial flow changes direction at least once.

[136] For example – see BBC, Development of the gas turbine blading

first time, enlarged, transparent blade model tests in a special water channel to determine the flow behaviour of the coolant directly in the rotating blades. In addition, this allowed checks of the internal flow pattern sufficiently in advance of actual blade precision castings

- **1983** A cascade wind tunnel for the testing of film-/impingement-cooled gas turbine blading[137] with automatic, computer-controlled instrumentation scanning was installed and applied.

A complete scheme of a BBC turbine cooling system from the mid-1970s is presented in Figure 6-33. Originating from the compressor discharge end, the cooling air is guided through bore (1) and the groove (2) to the outer rim of the rotor drum (3), to flow from here into the space between heat shields (4) and rotor (3) towards exit. Part of this stream is diverged to cool the first blade row (5). The blade root sections have openings so that the actual blade fixation zone (6) also remains at a relatively low temperature level. The residual cooling air continues to flow to the next blade rows for cooling (8) as long as there is sufficient momentum left in view of inevitable leaks. Another small amount of cooling air crosses the labyrinth seal (9), thus keeping the rotor front face typically below 500 °C.

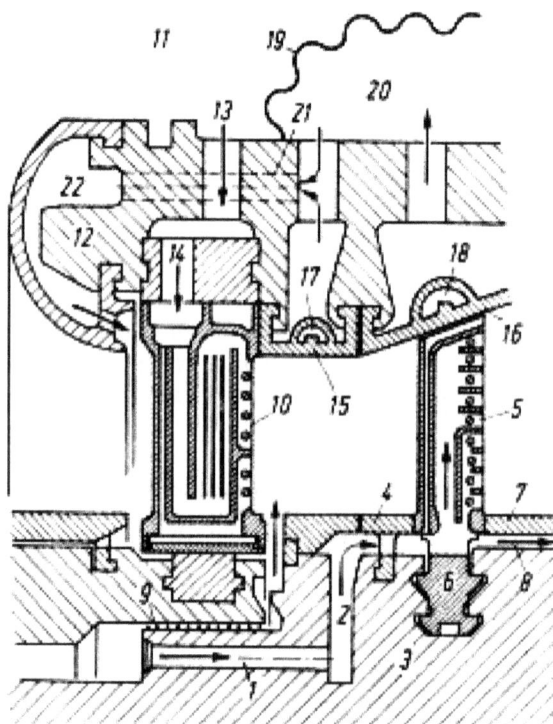

Figure 6-33   *Typical BBC turbine cooling flow scheme ~1975* [138]   *© 2001 Springer*

---

[137] See BBM, Gitterpruefstand fuer filmgekuehlte Gasturbinenschaufeln
[138] See Traupel, Thermische Turbomaschinen II, 4th. ed, p. 358 (figure source BBC)

The cooling air for the first row of turbine vanes (10) originates from the compressor ple-num (11), next to the turbine vane carrier (12). It flows through the bores (13) and (14) to the vanes, from where it escapes through the vane trailing edges to the hot gas stream. The turbine vane carrier itself is again protected by the heat shields (15) and (16). The latter car-ry circumferential channels (17) and (18) for further cooling air distribution. Space (20), en-closed by a flexible sheet metal wrapping (19), collects the escaping cooling air, so that it is guided through bores (21) to the space (22) upstream of turbine entry, to participate in the expansion process that follows.

Over time, cooling techniques were able to advance with improvements in computer codes and modelling. Engineering required models and tools for finite element analysis, heat transfer, and fluid dynamics to be able to design better and more complicated cooling sys-tems. Initially, engineers were only able to use simple two-dimensional modelling tech-niques for heat transfer calculations. In this sense, they were technologically limited in the same way as material engineers, who were unable to perform complex stress distributions in irregular shapes. To the benefit of both material stress and cooling efforts, finite element analysis codes improved during the following decades. General modelling applications such as Ansys were introduced, which was developed in great part by NASA.[139] Current versions of Ansys, Pro-Engineer (ProE) and CFD Computation Fluid Dynamics remain an integral part of modern turbine development and consequently, allow for much faster ad-vancements. As a result, cooling designs from BBC/ABB and Alstom have also become in-creasingly intricate and effective.

Though the use of cooling mass flows up to 20 percent has considerable drawbacks for GT performance, one essential advantage was already pointed out by BBC as early as 1974.[140] Based on the thin airfoil wall thickness, cooled blading has a superior material and thus more homogeneous temperature distribution, especially for transient operation conditions (start, shut-down). Therefore, the thermal stress of cooled blading is considerably lower for same operation conditions or the same variation of turbine entry conditions in comparison to uncooled ones, resulting in a beneficial, steep rise in the number of allowable cycles un-til first cracking (in the order of a factor x10).

A comprehensive survey on the turbine design tools at the time of the company transition from BBC to ABB is contained in a 1994 ASME paper[141] that describes the design of an up-rated 3-stage unit with two cooled stages for the 50 MW GT8C family. The entire aero-thermal performance of the original design was re-assessed, with the consequence of a 3+ percent improved GT efficiency. Turbine aerodynamics (annulus, airfoils) were improved by introducing a 3D lean/bow design philosophy with beneficial loss reduction. The tur-bine heat transfer and cooling was the subject of an intensive experimental study and the cooling design was radically improved. Consequently, the temperature distributions in the cooled blading became significantly more uniform. For the first stage a temperature reduc-tion of the vane was realised in the order of 75 deg C, while the corresponding blade tem-perature went down by 40 deg C. In addition, a material change raised the thermo-mechanical lifetime significantly above that of the preceding design.

---

[139] See CRIEPI, Comparative study on energy R&D performance
[140] See BBN, Entwicklungen fuer grosse Gasturbinen...1974, p. 417 (p. 8 in special print)
[141] See Harasgama, A new uprated turbine

In addition, this re-design effort was also the first opportunity ABB used to integrate the newly acquired design capacity in Moscow, Section 6.1.3.1. Consequently, the typical lattice or 'chevron rib' type cooling structure of GT8C blade 1 is very similar to the corresponding GT24/GT26 hpt configuration.

Further technology progress at ABB was represented by the cooling concept of the GT11N2, derived from GT8C and GT13E. The first stage vane has a combined impingement and film cooling system, comprising two inserts in the straight/cylindrical profile. Correspondingly, a simplified impingement cooling system with one insert is used on vane 2. The cooling configuration of rotor blades 1 and 2 is of the multi-pass, 'serpentine' type with a 'pin-fin' bank near the trailing edge.[142]

Over the years, a special relationship was built up with EPFL-LTT[143], the Turbomachinery Laboratory, Lausanne, CH on a variety of research topics, especially in the areas of turbine cooling and heat transfer. Between 1994 and 2012 more than 57 scientific theses and papers[144] were published with ABB/Alstom participation, a few of which are referred in detail:

- The Naphtalene Sublimation Technique (NST) has been applied for experimental heat transfer studies in comparison to the Liquid Crystal Technique (LCT) in compressible turbine cascade flows, also carving out NST limitations.[145] LCT was further accomplished by a sudden blade insertion mechanism, which became a kind of international test standard.[146]
- One of the first PIV Particle Image Velocimetry applications has been carried out for determining turbulence characteristics in an internal cooling passage; afterwards used as a kind of CFD reference case for further numerical studies.[147]
- Detailed measurements within a two-pass film cooling system of a gas turbine blade[148] took place, together with a pioneering application of the PSP Pressure Sensitive Paint method[149], which provides very high resolution contours of film cooling effectiveness, without being subject to the conduction error in high thermal gradient regions near the hole. Comprehensive experimental and numerical investigations/evaluations were carried out for showerhead cooling.[150]

Considerable progress in analytical cooling flow design and analysis has finally been demonstrated in the context of recent GT24 and GT26 engine upgrades.[151] Recent profile

---

[142] See Jacobi, Modernization and performance improved: ABB's GT11N2 gas turbine

[143] LTT Laboratoire de Thermique appliquée et de Turbomachines was led from 1983–2002 by Prof. Dr. Albin Boelcs (*1937), development engineer at Escher Wyss, Zurich between 1966 and 1971, and main collaborator of Prof. Dr. P. Suter at LTT from 1974–1983. After Prof. Boelcs' retirement, LTT is led by Dr. Peter Ott.

[144] For a summary/abstract list, see http://gtt.epfl.ch/op/edit/page-71678.html

[145] See Haering, An experimental study to compare the naphthalene sublimation

[146] See Hoffs, Transient heat transfer experiments

[147] See Schabacker, PIV investigations of the flow characteristics in an internal coolant passage

[148] See Chanteloup, Flow characteristics in 2-leg internal coolant passages http://lttwww.epfl.ch/publications/pdf/chanteloup_bolcs_piv_invest_a_2001.pdf

[149] See Wagner, Pressure Sensitive Paint (PSP) and Transient Liquid Crystal Technique (TLC)

[150] See Falcoz, Experimental investigations, and – see Falcoz, A comparative study on showerhead cooling

[151] See Savic, The next generation KA24/GT24 from Alstom; Hiddemann, Increased operational flexibility from the GT26 (2011) upgrade; Hiddemann, The next generation GT26; and – see Naik, Aero-thermal design and validation of an advanced turbine

design draws on the resources of numerous technology development programmes. As an integral part of the design process, selected profile sections were tested in high-speed linear cascade rigs for early feedback. Aerodynamic tests were carried out with and without showerhead, film and trailing edge cooling in order to verify the proper working of the profiles in such conditions and to confirm estimates of loss.[152] Figure 6-34 shows a direct comparison of the measured and predicted surface Mach numbers of an advanced hpt vane design[153] with and without film cooling air. Different film cooling arrangements were tested, allowing optimisation of the aero-thermal behaviour in terms of both losses and film cooling effectiveness. The agreement of numerical prediction in comparison to the measured data is very good.

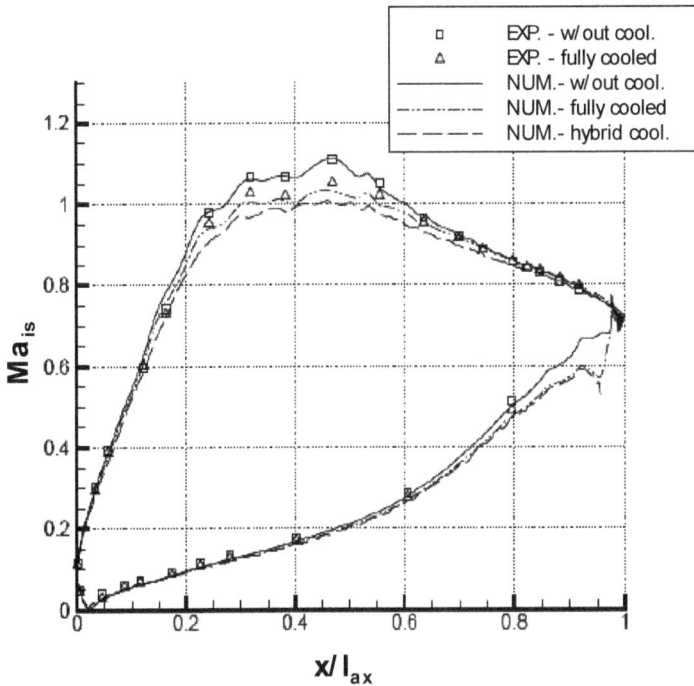

Figure 6-34   *Advanced turbine aerodynamics with cooling: measured and predicted surface Mach numbers* [153]

Finally, Figure 6-35 illustrates the impressive progress in heat transfer calculations. The laterally averaged film cooling effectiveness of the investigated turbine vane is compared against predictions. In general, the agreement between test and CFD data is convincingly good, particularly for the last film row on the suction side. As clearly visible, the trend in

---

[152] See Cardamone, Aerodynamic optimization of a hp turbine cascade blade; Montis, Experimental and numerical investigation on the influence of trailing edge bleeding; and Stephan, Investigation of aerodynamic losses and film cooling effectiveness.

[153] Figure source: Homeier, Impact of suction side cooling. In comparison to Figure 6-22 the changed design philosophy becomes obvious, with somewhat relieved laminar acceleration and corresponding, axially earlier turbulent transition/diffusion. Differences in peak Mach numbers can be traced back to hpt and lpt designs.

the midspan film effectiveness peaks; this and the film decay after the film rows is apparently very well captured by the predictions – on the basis of Alstom-internal film cooling correlations.[154]

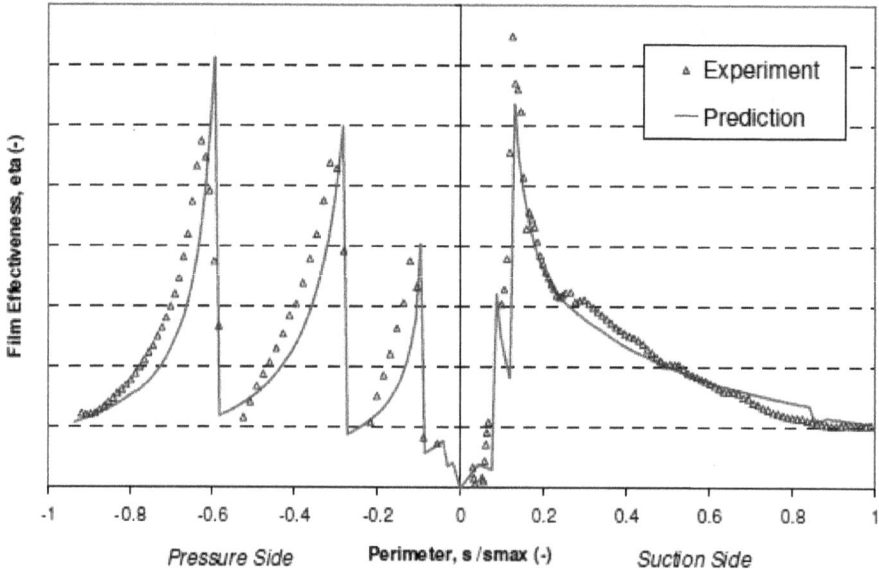

Figure 6-35   *Comparison of measured vs. predicted film cooling effectiveness on a turbine vane at mid-span*

6.2.3   High Temperature Materials and Coatings

Material development showed up already in the course of previous descriptions. An early BBC example is a 'rainbow set' of different materials of a (Holzwarth) test turbine blading from 1928, Figure 3-17.

As soon as the gas turbine developed from an auxiliary machine to a primary energy producer, the main problem was dealing with temperatures as high as safety would permit, which was of course sometimes also a matter of judgment, as illustrated by the following list of topics addressed by Seippel in 1966[155]:

–   development and selection of creep-resisting materials
–   chemical problems: corrosion of steel and deposits of fuel ashes
–   design of structures capable of important temperature expansion without causing undue stress or disappearance of running clearances
–   design of cooling channels to protect highly stressed and vital parts from excessive temperature.

BBC was far-sighted enough to invest in a new central laboratory in 1958, a 60 m long, 6 storey building on Haselstrasse in Baden, CH. This lab, which comprised chemical, physical

---

[154] See Naik, Multi-row film cooling performances of a high lift blade and vane (Figure 6-35)
[155] See Seippel, The evolution of compressor and turbine bladings in gas turbine design

and metallurgical-mechanical departments, grew out of the old materials testing departments. Originally the function was restricted to monitoring and inspecting of the properties of incoming materials and searching for defects. These tasks were soon expanded to include new material and process research in close cooperation with the design departments. In this respect the area developed into a fundamental pillar of technical progress.

The latter point of internally close cooperation was not always automatically granted. An example of this category was the development of turbine blading from ODS, Oxide Dispersion Strengthened alloys[156], an area where BBC claimed leadership in the 1980s. The technology was assessed to provide nearly a one percent rise in GT efficiency due to the resulting higher operation temperatures. ODS alloys can exhibit better stress-rupture properties than cast single-crystal alloys at temperatures above 800 °C. When manufactured by mechanical alloying, which has the ability to combine precipitation hardening with oxide dispersion strengthening, the alloys exhibit a range of properties which are attractive for blade applications at a wide range of temperatures. Due to the high intrinsic cost of mechanically alloyed ODS materials, considerable savings should have been achieved by forging or HIPping[157] cooled turbine blade configurations in halves and then brazing them together to be 'engine-ready' afterwards.[158]

In a thorough technology review workshop[159] with Dr. Anton Roeder, who had taken over development responsibility in 1987/1988, the BBC materials situation was also assessed. While the enthusiastic head of material development, after outlining the ODS and ceramics technology achievements culminated his presentation with the courageous conviction statement that *'after the metallic era now a non-metallic one has to follow!'*, the majority of attendants agreed on the conclusion that *'materials developments should follow the GT market in the future'* and that *'the existing supply base of available materials and processes should be taken more into account'*.

The indirectly indicated progress in material technology of Figure 6-32 with reference to the turbine base material can be roughly divided into four phases:

- the period of wrought alloys from 1940–1965
- the era of conventionally cast alloys 1955–1975
- the time between 1970–1990 can be characterised by DS directionally-solidified cast alloys
- from 1980 SX single crystal casting technology emerged and dominated GT advancements more and more, first for aero engine developments, then from 1990 fundamentally pushed by ABB's new family of power generation gas turbines GT24/GT26.

Soon the knowledge spread in GT development, that undesirable grain sizes, shapes, and transition areas were responsible for the premature cracking of turbine parts. This led to the development of the equiaxed casting process[160] which assured uniformity of the grain

---

[156] http://en.wikipedia.org/wiki/Oxide_dispersion_strengthened_alloy

[157] Hot Iso-static Pressing of ODS powder materials

[158] See Betz, High temperature alloys for gas turbines, p. 131

[159] See ABB, Seminar KWG Entwicklungsstrategie – on 20 October 1988

[160] For the 'equiaxed investment casting process' the molten metal is poured into a ceramic mould under vacuum, to prevent the highly reactive elements in the superalloys from reacting with the oxygen and nitrogen in

structure along all their axes. At elevated temperatures, component failure begins within and progresses through grain boundaries. Therefore, further study led to the conclusion that strength could be improved if grain boundaries were aligned in the direction normal to the applied force, i.e. for blades in the radial direction of the centrifugal force. This 'columnar grain' formation in a preferred direction, called DS (Directional Solidification), was introduced by Pratt&Whitney in 1965. As grain boundaries remain the weak element in turbine blades, numerous techniques have been evaluated to strengthen them since then. Even better than strengthening grain boundaries is the idea of eliminating them completely by producing parts consisting of a single crystal. In addition, the reduction in the amount of boundary-strengthened elements (at least in some alloys) has raised the incipient melting point by 50 deg C.[161] Nickel-based superalloys have come to dominate the high temperature stages of the gas turbine engine, from the high pressure compressor through the combustor and turbine stages to the exhaust outlet. Their success is due to their unique combination of mechanical strength and resistance to oxidation and corrosion at elevated temperatures.

As mentioned, a step change in temperature capability was realised through the introduction of directional solidification by the Bridgman process, Figure 6-37 (l), eliminating transverse grain boundaries, a source of weakness in a creep-dominated application. It was then a natural progression to the complete elimination of grain boundaries via single crystal casting. Consequently, the continued drive for more temperature capability has led to successive generations of alloys with ever more exotic alloying additions.

Creep is a common cause of failure in turbine blades and is in fact the main factor that limits service life. The single crystal (SX) material structure has the ability to withstand creep at higher temperatures than crystalline turbine blades due to the lack of grain boundaries. Grain boundaries are an area of the microstructure where many defects and failure mechanisms start. Nevertheless, creep will still occur in SX turbine blades, but due to different mechanisms at higher temperatures.

Relatively early in the GT24/GT26 development programme it was decided in agreement with the selected foundries to use CMSX-4 as a base material for the highest loaded turbine blades and vanes. CMSX-4 is an ultra high strength, single crystal alloy developed by Cannon Muskegon Corporation, Muskegon Mi., USA. This second generation nickel-based single crystal alloy containing rhenium is capable of higher peak temperature/stress operation of at least 1150 °C.

In close cooperation of both sides[162], in 2000 ABB was able to exclusively introduce a tailor-made alloy, CMSX-4 Mk4. Thorough test evaluations confirmed the high creep resistance of the new SX materials, but also revealed typical crystallographic defects in industrial SX parts at the so-called 'low-angle boundaries'. 'Freckles' occurred with a high concentration of eutectic materials and a tendency to recrystallisation after solution annealing (heat

---

the air. With proper control of metal and mould thermal conditions, the molten metal solidifies from the surface to the centre of the mould, creating an equiaxed structure. See Boyce, Gas turbine engineering handbook, p. 403

[161] See Giampaolo, Gas turbine handbook, p. 34

[162] CM was represented by their VP of Technology, Ken Harris, while for ABB/Alstom the material specialists and patent inventors Maxim Konter, Michael Newnham and Christoph Toennes participated in product development.

treatment). These negative features were suppressed by an alloy modification[163] with the addition of 0.02–0.04 weight-% C, 40–100 ppm B and a limited amount of Hf. All together B, C and Hf increased the strength of the low-angle grain boundaries. Thus, an acceptable angle between two sub-grains divided by the low-angle boundary increased from 5°–7° to 12+°, enabling economical production (with much higher yield) for massive SX parts.

During the past ten years the prices of several of the refractory metal[164] ingredients of the superalloys 'exploded', so that material strategies had to be constantly reviewed and adapted. At the time of the introduction of these second-generation alloys in 1997, for example, the cost of rhenium was 350 USD/kg; by August 2008 this reached 12',000 USD/kg. Consequently, rhenium's use in times of shortages became a struggle between the opposing forces of 'performance' and 'price', and the precious material applications had to be restricted to 'unsubstitutable' cases. Similarly, the price of tantalum, of which CMSX-4 has a relatively high share (approximately 7 percent), quadrupled during 2000 from a starting price of 200 USD/kg. In addition, the supply chain stability was considerably infringed by delivery guarantees decreasing to the next 3 months only. One of the consequences was to discuss the possibility of increased 'revert' (recycled) material usage under an extraordinary tight quality monitoring.

The manufacturing steps of DS/SX precision investment casting technology for turbine blades has already been outlined in Figure 5-79. In addition, Figure 6-37(l) illustrates schematically a corresponding Bridgman furnace: an cylindrical vessel up to 8 m high, evacuated to low vacuum. It works with three temperature zones. The upper zone has heater-stabilised temperatures above the melting point of the cast alloy. The lower zone has a temperature below melting point and an adiabatic zone as a baffle between the two. The ceramic mould, containing the top-down arrangement of e.g. five blades, filled with liquid metal, is positioned in the hot upper furnace zone on a 'chill plate' at the beginning of an approximately 2 hour long operation. Solidification starts by water cooling of the chill plate, which is slowly moved downward at the same time. Consequently, a plane solidification front progresses towards the upper, blade root section while the whole set-up passes steadily through the separating baffle zone.

---

[163] According to the ABB patents EP0914483 A1 and EP0914484 A1 'Nickel-Base-Superalloy' with priority 12 May 1999, the materials contain in weight percent: 6.0–6.8% Cr, 8.0–10.0% Co, 0.5–0.7% Mo, 6.2–6.6% W, 2.7–3.2% Re, 5.4–5.8% Al, 0.5–0.9% Ti, 7.2–7.8% Ta, 0.15–0.3% Hf, 0.02–0.04% C, 40–100 ppm B, 0–400 ppm Y, rest Ni.

[164] Refractory metals are extraordinarily resistant to heat and wear; the most common definition comprises the following five elements: Nb Niobium and Mo Molybdenum of the 5th period, and Ta Tantalum, W Tungsten and Re Rhenium of the 6th period. They all share some properties, including a melting point above 2000 °C, high hardness at room temperature, chemical inertness and a relatively high density. http://en.wikipedia.org/wiki/Refractory_metals

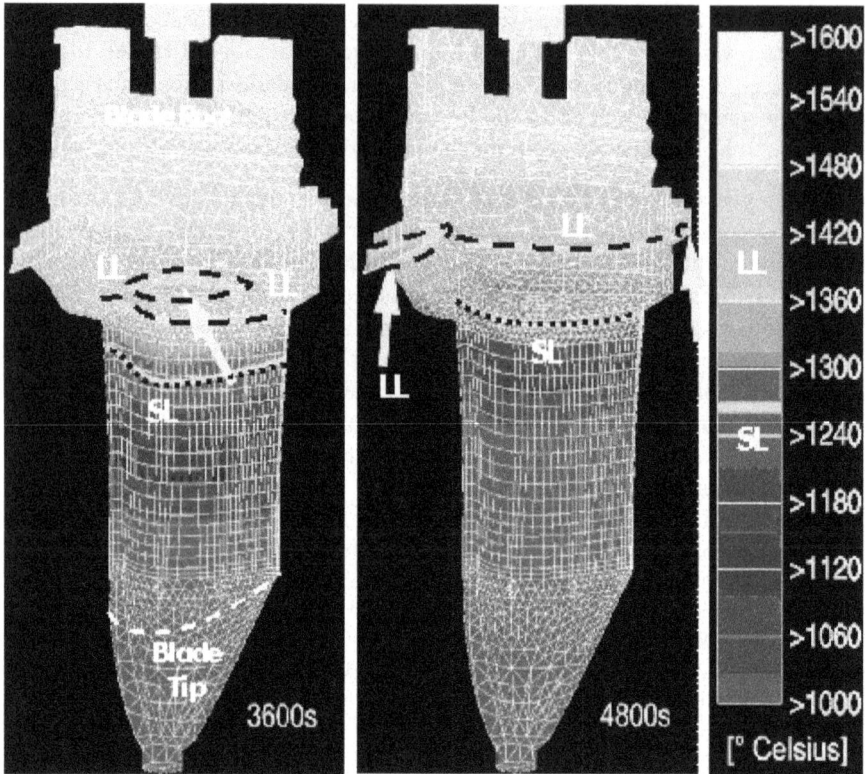

Figure 6-36   Turbine blade solidification modelling with CASTS®: Metal surface temperatures after
start: l) 1 hour, r) 1hour 20 min LL – Liquidus Line, SL – Solidus Line

As indicated before, the SX process development for large power generation blading represented a considerable extra challenge in comparison to the previous state-of-the-art experience with considerably smaller turbojet engine parts. ABB accompanied the concurrent
engineering design and product development effort for these parts by a comprehensive
numerical modelling initiative.

Traditionally BBC had already developed strong ties to the renowned Foundry Institute of
RWTH Aachen. In the 1990s the joint development of a 3D FEM software package was
launched which led to the successful programme system CASTS®, now independently marketed.[165] As a typical example of a CASTS application, Figure 6-36 illustrates the steadily pro-
gressing solidification front for a SX turbine blade casting, for two, deliberately selected mo-

---

[165] Prof. Dr. Peter R. Sahm (*1934), head of the Foundry Institute at Aachen, D between 1979 and 1999 started his
life-long dedication to DS superalloy developments in the 1960s at BBC Baden, CH, continuing in the 1970s at
the newly founded BBC Research Laboratory in Heidelberg, D. See Sahm, Gerichtet erstarrte eutektische
Superlegierungen für stationäre Gasturbinen. These early developments resulted in Access e.V., a registered
non-profit-making research association, arising in 1986 from a working group of the Foundry Institute of the
Technical University of Aachen (RWTH). Today represented by Robert Guntlin as Managing Director, Dr.
André Schievenbusch and Dr. Fredy Hediger, the latter with special responsibility for CASTS and thereof
improved software developments. Access specialised first in the fundamentals of solidification by using mi-
crogravity experiments and expanded in the meantime considerably, so that scientific expertise and technological skills are transferred into a wide range of best practice examples.

ments at 3,600 s (1 hour) on the left, and 4,800 s (1 hour 20 min) on the right. The Celsius temperature scale on the right illustrates the bandwidth between liquid state > 1,600 °C at the start of the process and solid state < 1,000 °C for blade sections below the baffle.

The relative baffle position is marked by the upper LL (Liquidus Line) and the lower SL (Solidus Line), limiting the intermediate 'mushy zone' of material solidification beginning. Without going into too much detail, the strength of the CASTS numerical predictions is indicated by the arrow-marked, isolated, undercooled and consequently defect-prone, potential freckles' zones. As a striking advantage of the numerical assessment, it is now up to the designer and/or casting specialist to either adapt the blade geometry or the mould set-up, for example by inserting 'grain continuators' to bridge these zones, or to control the furnace operation (solidification process) to prevent the indicated defects, for example by a drawing speed variation or extra mould insulation measures.

The intensity of the hot/cold transition i.e. a higher thermal gradient or a narrower LL-SL distance, improves the quality of DS/SX solidification processes as a decisive parameter. These simplified considerations laid the ground for several other ABB SX/DS process patents for GCC, Gas Cooled Castings[166], Figure 6-37 r). The principal idea is to place a number of small (supersonic exhaust) nozzles circumferentially in the plane immediately below the furnace baffle, so that the mould heat transfer is locally intensified by the impingement of inert gas jets. The main advantage of these inventions is the possibility of making the proposed additional nozzle equipment retrofittable to any existing Bridgman furnace without restricting the conventional standard operation mode.

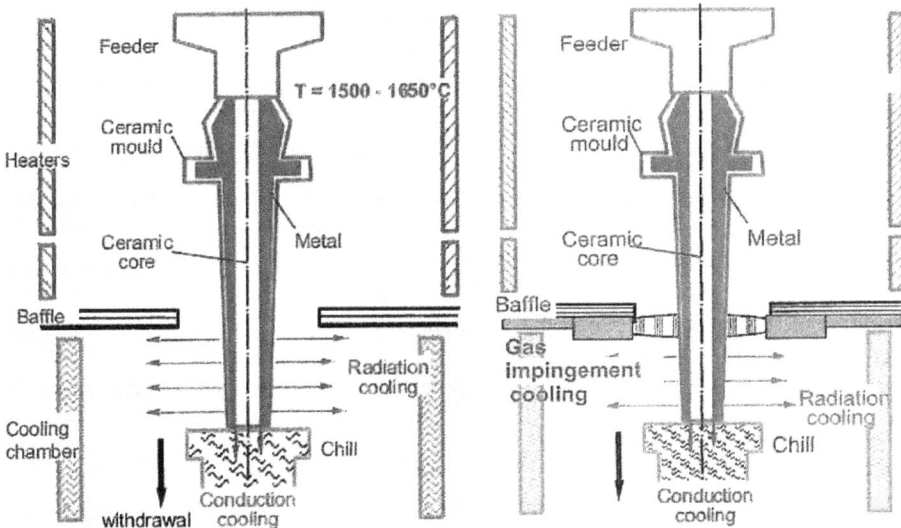

Figure 6-37   *Principle DS/SX casting processes: l) Bridgman furnace, r) GCC Gas Cooling Casting extension*

---

[166] US5921310 A, Process for producing a directionally solidified casting and apparatus for carrying out this process; priority date 20 June 1995, inventors E.L. Kats, M. Konter, J. Roesler and V.P. Lubenets. Also US20050103462 A1, Method for casting a directionally solidified article; priority date 06 November 2003, inventors M. Balliel, D. Eckardt, M. Konter and A. Weiland.

In the meantime the technique has been successfully applied to several large Bridgman units, especially for the production of large 1+ ft size SX and DS components. Inert gas feeding and control are automatically adapted to the casting process, so that only a 'GCC' button indicates this extra-quality choice to the furnace operator.

The net effect of all the advances in blade technology outlined above, coupled with alloy development, has been to increase metal temperatures by approximately 300 deg C over the last 50 years. With increasing blade operating temperatures, the intrinsic resistance of the metal to environmental attack is no longer sufficient. Protective coatings are therefore required to provide a thermal barrier and/or impart the necessary oxidation and corrosion resistance, Figure 5-79(r). The TBC Thermal Barrier Coating on combustor cans, nozzle guide vanes (NGV) and hp turbine blades provides a good example of a complex multi-step coating process. A 'bond coat' is applied by electroplating or thermal spray, followed by heat treatment and finally the application of a zirconia-based ceramic coating by LVPS Low-Vacuum Plasma Spraying or EB PVD Electro-Beam Physical Vapour Deposition.

While the turbine blade aerofoils are exposed to the highest temperatures from the gas stream, other parts of the blade may need to be protected from the effects of combustion products and wear and tear. Platinum aluminising and platinum chromising are two diffusion coating systems that provide such protection and are commonly applied to turbine blade shanks to improve sulphidation resistance. The effect of this surface modification is to increase the life of the blade system from hundreds to thousands of hours.

A variety of coating compositions and technologies are used in industrial gas turbines. Since the 1990s ABB/Alstom have consistently used Vacuum Plasma Sprayed (VPS/LVPS) MCrAlY coatings for the entry turbine stages, often in combination with Air Plasma Sprayed (APS) TBC as the best compromise for temperature capability, life and cost. Other systems used are PtAl protective coatings and EB PVD TBC. The last stages and structural parts are usually coated by oxidation or corrosive protective systems using lower cost processes (High Velocity Oxy-Fuel HVOF spraying, chromising and paint coat).

Despite high gas and metal temperatures in advanced turbines, corrosion must still be seriously considered in hot gas path material selection, especially in oil-operated and dual-fuel engines. Although fuel quality improved, memories of the first and severe GT design fuel crisis of the 1960s reappear as quick as a flash, Section 5.1.1. Significantly increased pressures on the first turbine stages enable the deposition of corrosion products at surface temperatures above 850 °C. Since alloying elements influence oxidation and corrosion resistance differently, it is important to have either a choice of MCrAlY compositions for various applications, or to use a balanced coating with a high Al and Cr content.

However, the most critical area in high temperature coating application currently seems to be not the coating chemistry, but the corresponding manufacturing process control. Out-of-spec coating thickness, porosity and interface defects result in much faster coating failure than an insufficient oxidation resistance. Alstom maintains a highly specialised manufacturing shop for Hot-Gas Path Parts (HGPP) within the Birr rotor factory, Section 5.4.3, where the post-casting processes (Machining I – root grinding operations, Coating and

Machining II – laser hole drilling, sheet metal finishing, brazing etc.) are kept stable at optimum specification standards in close cooperation with external suppliers.[167]

6.2.4   Tests at the Gas Turbine Power Plant Birr

The Alstom gas turbine power plant was already shown as part of the aerial view of the Birr rotor factory, Figure 5-74, #2. It lies 12 km west of the Gas Turbine R&D offices in Baden CH. The proximity of the facilities to design, validation and field service engineering is a considerable advantage for combining design activities with commissioning, testing and inspection experience. The GT power plant was inaugurated in 1996 with a 300 MW GT26 test unit for simple cycle operation, followed in 1999 by a 60 MW GT8C2. An extension of the GT26 facility towards simulated combined-cycle operation was launched in 2010, Figure 6-40.

The power produced at the GT centre is fed by means of a 360 MVA transformer into the Swiss national grid, so that there is a unique facility with the capacity to run a heavy-duty gas turbine, still in the development phase, under the commercial operation conditions of a simulated customer unit. A dedicated gas pipeline was installed to provide the facility with an ample supply of 100,000 m^3/h. Dual fuel scenarios are possible due to the integration of liquid fuel and demineralised water storage tanks (2 x 1,700 m^3), including a water treatment facility. In addition to performance verification over a wide range of operating conditions, start-up optimisation testing is just one example of how the Birr GT power plant can even be used for the investigation of specific needs of individual customers. Flexibility in test campaign planning and execution is a key factor for successful use. Rainbow test set-ups provide the opportunity to test and compare candidate design improvements under the same loading conditions. In addition to testing and the evaluation of operation regimes, boroscope and hardware inspections allow the monitoring of operation effects and the system's impact on seals and clearances.

Figure 6-38 provides an impression of the power plant at the opening and during the early commissioning phase in 1997 – with the assembly of the thermal block on the left hand side, a view of the impressive dimensions of the manufacturing hall environment on the right, and references to the most important building blocks.

The following list summarises the variety of testing capabilities demonstrated in different test campaigns :

- performance tests on fuel gas[168] and fuel oil
- fuel switch-over testing
- fuel flexibility testing with high hydrocarbon fuels
- advanced component testing and component matching
- secondary airflow sensitivity testing
- thermal paint tests
- component mapping including compressor surge tests
- compressor start-up and run-in procedures
- transient and protection tests

---

[167] See Thumann, Materials for power generation gas turbines
[168] It is also possible to validate different heating values and 'Wobbe indices' (used to measure the interchangeability of fuel gases) using different fuel gas mixtures.

- frequency response tests
- start-up optimisation.

Figure 6-38   *GT26 validation at newly erected GT power plant, Birr, 1997 1 – generator, 2 – GT26 rotor, 3 – GT exhaust casing, 4 – air intake, 5 – data acquisition centre, 6 – exhaust stack*

More than 3,000 instrumentation points are utilised to monitor and record key performance and detailed component data during engine operation. All standard operation and commissioning measurement points are instrumented and additional detailed measurements are taken for the validation of components and of special design features. The equipment is rounded off by advanced, special purpose instrumentation such as telemetry, pyrometry, tip clearance probing, active flame stability measurements and vibration monitoring, providing critical information on engine operation before a product improvement is commissioned in the GT fleet. The advanced telemetry system provides valuable data acquisition from sensors within the rotating system, such as temperatures, pressures, and vibration signals.

Figure 6-39 illustrates a spectacular view into the GT26 exhaust diffuser (which has become historic again in the meantime). Starting in 2010 the facility was newly built and upgraded by an adapted cooler configuration to simulate CCPP operation.

The corresponding sampling and analysis system is designed to handle frequent start-stop operation with minimal operator intervention. It features sample flow monitoring at instrument level and uses Profibus DP (Decentralized Peripherals) as a basis for signal exchange, thus ensuring the availability of all process parameters and status information of the process values on one side, and maximum flexibility for later upgrades on the other. The shelter based sampling system is designed for outside installation and equipped with 6 sample lines and 10 instruments for water/steam monitoring. An outside view of the upgraded plant after 2010 is shown in Figure 6-40.

Figure 6-39   *Birr GT power plant – exhaust diffuser, upstream view, 2000*

Figure 6-40   *Birr GT power plant – CCPP simulation extension, outside view, 2010*

6.3   The Combined Cycle Power Plants

A combined cycle power plant (CCPP) produces electricity from gas and steam turbines. The gas turbine drives an electrical generator and the exhaust gas energy from the gas turbine is used to generate steam in a heat recovery steam generator (HRSG) which then produces electricity from a steam turbine. The advent of the combined gas and steam cycle has enabled the gas turbine to leap to prominence as a primary power generator.

6.3.1   Historic Developments

In 1961, the Austrian utility Niederoesterreichische Elektrizitaets-Wirtschafts-Aktien-gesellschaft (NEWAG, today EVN AG), put the Korneuburg A power station into service, on the river Danube 15 km upstream of Vienna, Austria, Figure 6-41. This 75 MW plant from Brown Boveri can claim with good justification to be the world's first true combined cycle power station in the modern sense of that term.[169]

Figure 6-41   *Korneuburg A, construction of combined cycle plant 1960:*[170] *Two, twin-shaft 25 MW reheat gas turbines and two silo-type combustors are visible*   ©*ABB*

The idea of harnessing gas and steam turbines together is by no means new, as already demonstrated in Section 2.2.2 in the year 1825, when N.L. Sadi Carnot issued his ground-breaking essay on 'Reflections on the Motive Power of Heat'. Stodola[171] discussed the utilisation of waste heat in a steam circuit in 1922, while in 1936 Schroeder[172] drew up the main connections between boilers and gas turbines. Numerous patents[173] describe different vari-

---

[169] The following text relies to a great extent on an article which appeared in November 2011 in Modern Power Systems, see Eckardt, Celebrating 50 years of combined cycle; special thanks to James Varley, MPS editor, for upgrading the manuscript.

[170] Picture source: Eckardt, Celebrating 50 years of combined cycle

[171] See Stodola, Dampf- und Gasturbinen, pp. 984 and 995

[172] See Schroeder, Der innere Ausbau von Kraftwerken. In addition, Seippel and Bereuter quote eight related references between 1945 and 1960 in the BBR article, Footnote 174 of this Section.

[173] Seippel and Bereuter, see next Footnote, quote six patents (with inventors/owners and priority year): US1978837 (B. Forsling GE, 1933), CH246084 (BBC, 1945), CH245488 (MFO → BBC, 1945), CH275575 (MFO → BBC, 1949), CH287354 (BBC, 1951), CH336648 (W. Karrer MFO → BBC, 1956).

ants. It was therefore logical that Seippel and Bereuter started their 1960 review[174] with the question of why more combined plants had not actually been built. They concluded there was a principle dislike of complications in operation, especially if one had to fear new sources of trouble to be introduced into the installation. However, given the achieved experience with gas turbines in the meantime, the authors state that a combined GT/ST set-up would not represent more complications than the boiler fans which it replaces.

'Combined' or 'binary' cycle power generation installations had been constructed prior to Korneuburg A, mainly in the United States, notably GE's mercury/steam systems dating back to 1928.[175] These plants achieved high thermal efficiency, but they were expensive because of the substantial construction costs, associated particularly with the 'higher' mercury cycle plant.

There had also been conventional power plants using a gas turbine exhaust for preheating boiler feedwater. In addition, prior to Korneuburg A, there were already combined cycle cogeneration installations in industrial facilities, e.g. the BBC supplied facilities at the Dudelange steelworks in Luxembourg, dating back to 1956 and at Cornigliano, I, commissioned in 1961. A special highlight of the time (1959) was the first time realisation of a cogeneration/district heating plant by BBC Mannheim at Bremen-Vahr, D, for more than 30,000 inhabitants.[176]

However, Korneuburg A was the first power station in the world of any significant size to use the gas/steam combined cycle purely for electricity generation for a utility. This can be seen as ushering in the new era of the combined cycle plant as we know it today. In view of the key role which gas/steam combined cycle power generation has achieved in the intervening 50 years, and its continued and growing importance, it would seem worthwhile to shed some light on these early pivotal developments.

The idea at Korneuburg A was to produce electricity from natural gas, which had recently become available but for which no other consumer could be found. The 2 × 1 Austrian plant consisted of two 25 MW intercooled gas turbines and a 25 MW steam turbine from BBC, CH, and a HRSG with supplementary firing, designed as a forced-circulation boiler according to the Lamont process, which was supplied by Waagner-Biró, Graz, A. As illustrated by the principle scheme, Figure 6-42, the 26 percent-efficient GTs produced a relatively cool exhaust stream of only 310 °C (590F), resulting in low-quality steam. A fired superheater raised steam temperature to 440 °C (824F), boosting overall plant efficiency to more than 32 percent. Though the combined efficiency gain was relatively small, the power increase attributable to the added steam turbine was 43 percent. The plant construction work started in 1958 and the combined process operation was commissioned in January 1961.[176]

Though by present day standards, this first combined cycle plant was capable of only moderate efficiency, an average of 6,000 operation hours per year at full load conditions was achieved as a baseload plant from 1961 to 1974. From 1974 onwards the plant became uneconomical, mainly due to the rising fuel costs, and was therefore only used to cover peak load demands, replaced

---

[174] See BBR, The Theory of Combined Steam and Gas Turbine Installations, 1960
[175] The best known GE plant of that type being the 40 MW Schiller plant in Portsmouth, NH, from 1950. See Horlock, Combined power plants, p. 267
[176] See Auer, Practical Realisations of Gas Turbines Combined with Steam

eventually by Korneuburg B. This new plant, which went into operation in 1980, had a total power output of 125 MW and 47 percent combined cycle efficiency. There was no supplementary HRSG firing anymore and BBC's new GT13 gas turbine was employed with a capacity of 81 MW, an efficiency of 29.4 percent, a much higher turbine inlet temperature of around 1,000 °C, and no intercooling/reheat. The two plants together provide an instructive illustration of how combined cycle power plant technology developed in the 1960s and 1970s, in that both the relatively complex gas turbine configuration and supplementary HRSG firing were replaced by a gas turbine with a much higher turbine inlet temperature and a directly coupled dual-pressure HRSG. In fact, the Korneuburg site as a whole can be seen as encapsulating the history of the early search for the best combined cycle power plant concept.[177]

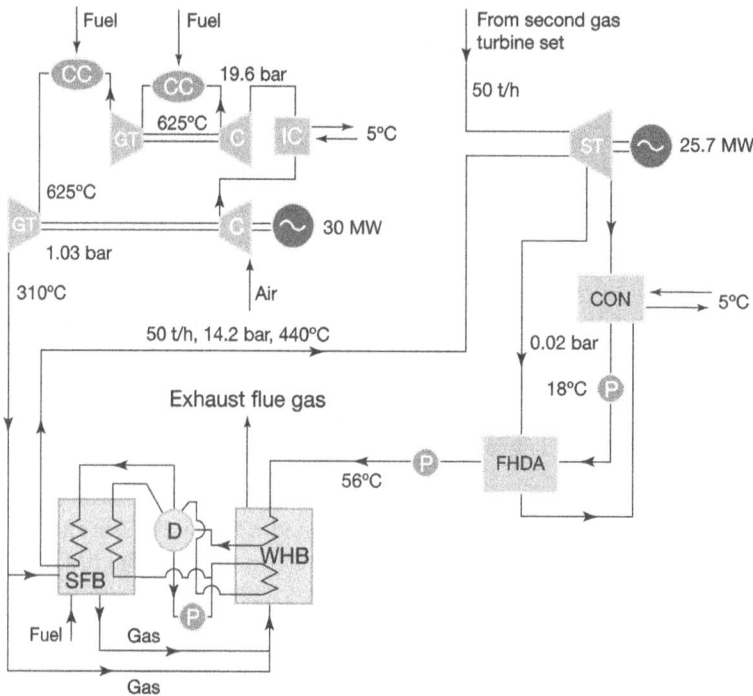

Figure 6-42   *Scheme of the Korneuburg A, combined cycle power plant* [178]

It is perhaps significant that BBC not only designed, manufactured and realised the first combined cycle plant for power generation, but also provided the theoretical compass for the new territory. In December 1960 Claude Seippel, Section 4.1.2, and R. Bereuther published a seminal paper[179] that reviewed many possible combinations of gas/steam turbine plants including the original concept of the pressurised Velox boiler, developed and produced by Brown Boveri from the 1930s to the 1950s. Particularly valuable are basic assessments of efficiencies that could be achieved in a combination plant both with and without

---

[177] Korneuburg is currently a 285 MW 'cold reserve' for the German power market (2012).

[178] Figure source: Eckardt, Celebrating 50 years of combined cycle

[179] See BBR, The theory of combined steam and gas turbine installations, 1960

supplementary heating, in comparison to those of a basic steam plant. The analysis is of historical importance as it gave considerable impetus to the development of combined cycle power plants. At the time of the final breakthrough of the CCPP technology, this tradition was maintained by several books and publications in close association to BBC/ABB, manifesting the special expertise of the Company and its lead personnel in this field.[180]

In the 1970s, the turbine inlet temperature of BBC gas turbines was increased, first by introducing superalloys and followed by specially designed alloys for the corrosive IGT fuels, e.g. IN 738, then later on by the introduction of air cooling in the first turbine stages (GT11B). Even with these increased temperatures the simple cycle gas turbine did not reach the efficiency of the steam turbine, while at the same time the oil crisis of 1975 re-emphasised the importance of low fuel consumption. A resolution of this dilemma was the further propagation of the combined-cycle power plant. When two thermal cycles are combined in a single power plant the efficiency that can be achieved is higher than that of one cycle alone, Section 2.3.3. The initial breakthrough of these cycles onto the commercial power generation market was possible due to the gas turbine, the key component of the combined-cycle plant, generating approximately 2/3 of the total output. Only since the late 1970s have turbine inlet temperatures, and hence exhaust-gas temperatures been sufficiently high for high-efficiency combined cycles. The result was a power plant with high performance, low installation costs and a fast delivery time.

The corresponding development of open circuit (gas turbine)/ closed cycle (steam turbine) power plants was extensively reviewed by Wunsch[181] in 1978. In addition, Horlock[182] tabulated for the following example of six BBC/ABB CCPPs with the associated changes in the main thermodynamic properties (e.g. gas turbine entry temperature and pressure ratio, steam turbine entry temperature and pressure, feed water temperature, etc.) the impressive historical development of this power generation system over the first thirty years:

- Korneuburg A[183], Austria, 1961, 75 MW power, 32.6 percent overall efficiency
- Geertruidenberg[184], NL, 1975, 120 MW, 46.1 percent
- Korneuburg B[185], A, 1980, 128 MW, 46.6 percent
- Roosecote[186], UK, 1991, 224 MW, 49.1 percent (with steam injection)
- Hemweg 7[187], NL, 1988, 600 MW, 45.9 percent
- Pegus 12[188], NL, 1989, 225 MW, 51.8 percent.

---

[180] See Kehlhofer, Gasturbinenkraftwerke, Kombikraftwerke...; Kehlhofer, Combined-Cycle Gas & Steam Turbine Power Plants, and – see Rufli, Systematische Berechnungen über kombinierte Gas-Dampf-Kraftwerke

[181] See BBR, Combined gas/steam turbine power plants, 1978

[182] See Horlock, Combined power plants, pp. 275–277. The book was published in 1992, with the dedication 'In memory of Dr Claude Seippel, Technical Director of Brown Boveri'.

[183] See BBR, Practical Examples of Utilising the Waste Heat, 1960 and – see BBR, Combined steam and gas turbine power stations, 1973

[184] See BBR, Combined steam and gas turbine power stations, 1973 and – see BBR, Combined gas/steam turbine power plants, 1978

[185] See Czermak, The 125 MW combined cycle power plant 'Korneuburg-B'

[186] See Jeffs, Britain's first combined cycle... (Roosecote) http://en.wikipedia.org/wiki/Roosecote_Power_Station

[187] See Pijpker, Amsterdam to have first 140 MWe gas turbine, and – see Keppel, Erste Betriebserfahrungen mit der 600-MW-Kombi-Anlage in Amsterdam. Hemweg 7 was erected in 1979, followed by the GT13E combined cycle extension in 1988. In 2012, there are plans for a more modern, gas-fired Nuon/Vattenfall replacement plant, Hemweg 9; the gas turbine shall remain in operation to cover peak load demands.

6.3.2   The Bright Presence of Combined-Cycle Power Generation

From a long-term perspective the high demand for additional power generation capacity is still unbroken; on a global basis some 120–160 GW are installed annually, corresponding to approximately 150 large, newly erected power stations. For years the strongest market growth rates have been observed in East Asia, especially China.

Figure 6-43 illustrates the rearrangement of the plant market shares by fuel category since the 1950s. Up until the mid-1980s steam plants clearly dominated with a share of > 50 percent. Since then, the **CCPP** combined cycle gas/steam turbine power plants emerged successfully as a dominating category – with approximately 1/3 of all new installations in the meantime, putting **Steam** into second place. The group **GT, ICE** of simple cycle gas turbines, internal combustion engines and fuel cells has also doubled its sales volume since the 1980s. **Renewables**, 3/4 of which are still represented by hydro power, remained rather stable over time and only increased in recent years, while the relative shrinkage of **Nuclear** power seemingly culminated around the year 2000. The important role of gas turbines in this development is further highlighted by the well-known fact that they contribute approximately 2/3 of the power output of an advanced CCPP, Section 2.2.2, so that in the meantime, some 40 percent of annual power investments overall are now GT-related.

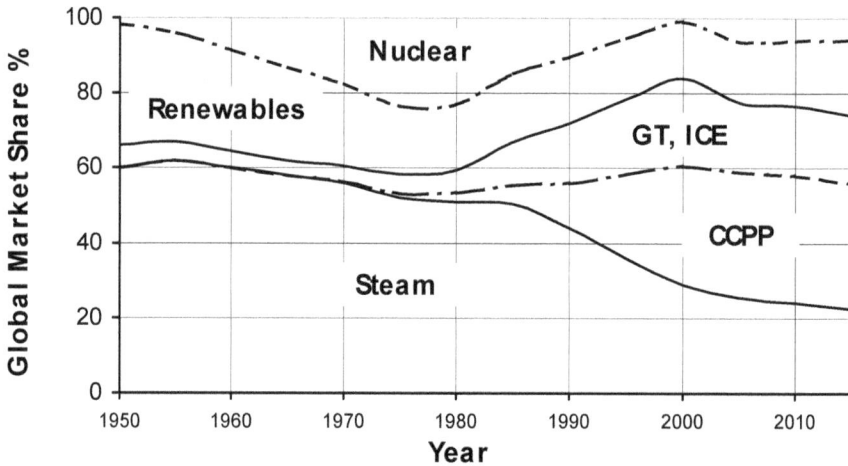

Figure 6-43   *Power plant market shares by category* [189]

Alstom's present CCPP portfolio is illustrated in Figure 6-44; with the majority of applications for the 50 Hz market, except for the 60 Hz KA24-2[190], and the KA11N2-2, which can

---

[188] See Ziegler, PEGUS 12, the world's most efficient power station. The UNA cogeneration plant comprises a GT13E, then the largest gas turbine in operation, and an unfired three-pressure heat recovery steam generator which feeds a triple-casing reheat steam turbine and a three-stage district heating system. The cycle is designed for a net base load of 223.5 MW, with a thermal efficiency of 51.78 % in straight condensing mode.

[189] Data source: BMWi, Leuchtturm COORETEC, p. 8

be ordered as 50 and 60 Hz versions. In addition, a KA11N2-1 LBTU with more than 150 MW power output is on offer for low calorific fuels. This, like the shown KA13E2-1 or the KA26-1 SS is a 1-on-1 configuration, where one gas turbine is coupled with one steam turbine and one common generator. All other versions are multi-shaft 2 GT-on-1 ST configurations, where each shaft has its own generator. The portfolio thus covers a net power range from > 150 MW to 935 MW with net efficiencies[191] between 51.3 and 60+ percent. Alstom has established a Plant Integrator™ approach in the way it designs and builds combined cycle power plants, bringing together the skills of a component supplier and an architect engineer. The focus lies here rather on overall plant optimisation and not so much on the individual component level.

Besides continuous technical advancement, the power generation market itself is undergoing a phase of principal change. The result of these changes and developments is that many of the advanced gas-fired combined cycle power plants that were installed in the late 1990s and 2000s are being called on today to operate under a wide ranging dispatch regime, including daily stop/starts and intermediate regimes. This was not foreseen at the outset, as due to their relative high efficiencies they were specified and designed based on base-load dispatch. Consequently, 'operational flexibility' is now becoming a central issue in the gas-fired power industry, and OEMs and operators alike have to re-define the way such power plants should be designed, for possible load regimes today and in the future.

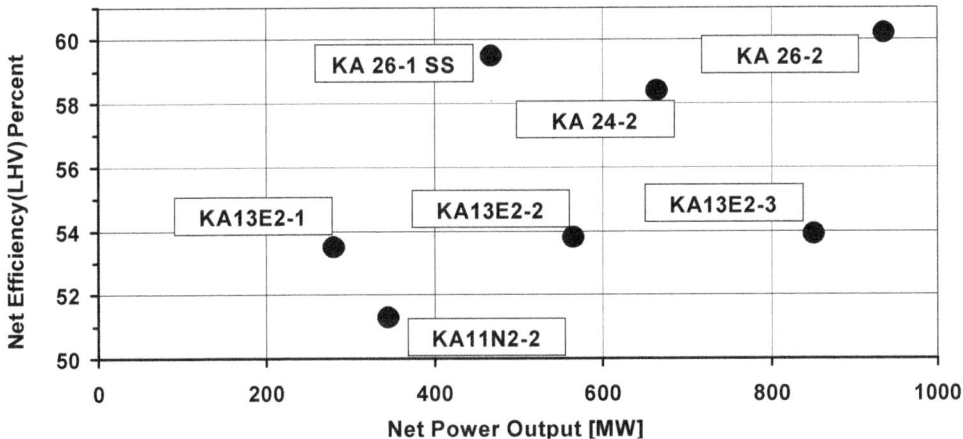

Figure 6-44   *Alstom's combined cycle portfolio, 50 and 60 Hz, ISO conditions*

The emergence and growth of renewable power, in particular wind farms, also brings new challenges and issues for power companies and grid operators, who have to balance power generation with load demand. Although the emerging renewables are expected to play an

---

[190]   In Alstom nomenclature KA24-2 stands for a GT24 'Kombi-Anlage' (CCPP) with 2-on-1 MS multi-shaft configuration, two GTs combined with one ST, – and a triple-pressure reheat drum- or once-through type HRSG. The exemption of this system, the KA26-1 SS, is a single-shaft arrangement with 1-on-1 configuration, where both GT and ST drive a hydrogen-cooled TOPGAS turbogenerator on a common shaft-line with a SSS Self Shifting Synchronous clutch.

[191]   Referred to LHV, the Lower Heating Value and ISO conditions

increasing role in the long term, fossil energy supply from fuel gas, oil and coal is likely to remain in use for decades. The increasing installation of power generation systems using renewable energies and their dependency on ambient conditions (like wind power) calls for a balance with the reliable and rapidly available power resources covering periods of sudden supply shortage, peak demands or simply following the automated generation control over a wide range of relative loads. It appears that the advantages of the GT24/GT26 sequential combustion system, now in combination with four rows of compressor Variable Inlet Guide Vanes (VIGV), have set a new industry standard in terms of part-load efficiency and turn-down capability. Given the fact that these two main contributors to operational flexibility were introduced in 1996[192] already, Alstom's present advanced GT engine family can be seen as a true pioneer of this important operational aspect.[193]

As an example from the middle of Alstom's CCPP portfolio, Figure 6-45 illustrates the set-up of a 1,000 MW GT13E2 power plant in Sohar, Oman. The plant was built by Alstom in 2009 based on a 476 MUSD turnkey Engineering, Procurement and Construction (EPC) and commissioning contract. The plant provides power to an aluminium smelter, so the reliability and robustness of the power supply are the main priorities. The plant consists of two blocks in a two-on-one configuration. Each block has two Alstom GT13E2 gas turbines with exhausts to a heat recovery steam generator with a common steam turbine. The gas turbines themselves can be operated either in a power or lifetime optimised mode. The Sohar Aluminium Power Plant (SAPP) normally operates them in the lifetime optimised mode, at high part load with a slightly reduced turbine inlet temperature, in order to permit an extension of the increased equivalent operating hours (EOH) to 36,000 EOH between hot gas path inspections. Additional power can be made available by operating the gas turbines in the power optimised mode at higher turbine inlet temperatures and, if necessary, at peak firing.

The steam turbines have a nominal site rating of 140 MW, at a seawater temperature of 35 °C, based on the steam delivered by two HRSGs without considering the input of supplementary firing. Evaporative cooling on the air inlet increases the gas turbine rating to a nominal 149 MW at 46 °C. Supplementary firing on the HRSGs can increase steam turbine power output to a maximum of 218 MW. Each CCGT block has a maximum net power output performance guarantee of 500 MW net at 46 °C when firing gas. The use of evaporative cooling and supplementary firing to provide this maximum level of power output is only foreseen in emergencies. As the gas turbines are dual-fuel fired a two-day supply (6,600 m^3) of distillate fuel oil is kept on-site for emergency back-up use. For this reason the gas turbines are equipped with an auto change-over facility from gas to oil in the event that the gas pressure falls below a minimum value.

As indicated before, both existing and newly built combined cycle power plants are changing their typical operation profiles, moving towards heavy cycling and intermediate dispatch regimes. These gas-fired power plants are experiencing increasing periods of operation at part- and minimum stable load, as well as frequent stop/starts up to daily cycling. Based on Alstom's experience of the KA24 fleet and its 50 Hz sister product, the KA26 fleet, with over 4 million fired hours, the base-load operation amounts to less than 25 percent of

---

[192]   At the time with 3 rows of VIGV still.

[193]   See Ladwig, KA26 combined cycle power plant

the overall operating profile. Plants also exist that mainly operate in the range of 60–95 percent plant load with some operation at part-loads as low as 40 percent. The impact of increased renewable power on grid networks is expected to see an even greater need for combined cycle to undertake a demand/supply balancing role, with increased periods operating at reduced or minimum load settings and/or actual shut-downs. All of this means that CCPPs will spend more and more time operating at part load, and part-load efficiency is therefore becoming a significant criterion in new CCPP projects, with Alstom's GT24/GT26 family especially representing excellent performance preconditions. Accurate modelling of thermal cycles is required to avoid premature fatigue failure of plant components for such requirements. Detailed temperature histories during transient events are required to determine the maximum stress ranges for life assessment. These details are obtained from a dynamic simulation of the thermal process in the component of interest. The simulation of the HRSG, for example, is used to tune the operation concept and then to determine the temperature history for the four main operation cycles: hot start, warm start, cold start, and trip. The heat-up, steady state, and cool-down phases of each cycle must be modelled. Thick walled components and those exposed to higher temperatures limit the life of the overall system.[194]

Figure 6-45   *SOHAR 2 × KA13E2-2 1,000 MW combined-cycle power plant 1 – air intake, 2 – GT13E2 (4), 3 – GT TOPAIR turbogenerator (4), 4 – main transformer, 5 – HRSG (4), 6 – steam turbine (2), 7 – ST TOPAIR turbogenerator (2), 8 – bypass stack, 9 – exhaust stack*

---

[194] See Stevens, Alstom's KA26 advanced combined cycle power plant

Solar power is the greatest of all renewable sources and at the same time has the best potential to be merged with combined cycle power generation applications. The energy that is received on earth is 60,000 times the annual global electricity demand and at the same time, only 0.1 percent of all suitable land area for solar thermal power generation could cover the global electricity demand. Figure 6-46 illustrates the concept of an advanced GT24-2 ISCC Integrated Solar Combined Cycle power plant project. The CCPP uses dry cooling in order to minimise water consumption. The ACC Air Cooled Condenser is visible at the front, the two GT24 are located at the back. Clearly visible is the central tower receiver, where the solar energy of the surrounding solar (mirror) field is concentrated and converted into steam; the steam from the solar plant can be used in the combined cycle plant to boost the steam turbine and to generate additional solar electricity, meaning that with this concept 15–25 percent of the CCPP output will be solar generated. The solar field of this GT24-2 reference plant comprises 36,600 heliostats, i.e. 555,000 m^2 reflecting area from the BrightSource partner company. The land requirement will be 3.4 km^2, based on a receiver tower height of 130–150 m. The Solar Receiver Steam Generation (SRSG) is 250 MW_{th} or 112 MW_e. In principle the concept is also adaptable to smaller solar field areas.

Figure 6-46   *The next step: KA24-2 ISCC (Integrated Solar Combined Cycle) solution with central receiver tower*

Depending on the plant configuration, two typical operation modes are possible:

–   The CCPP is operating at base load and the solar plant boosts the CCPP performance, which generates more power.
–   Alternatively, the GT load can be reduced by the amount of power that is generated from the solar field. The total load is unchanged, but fuel savings can be realised.

Interesting effects can be observed in the efficiency behaviour of this new process combination of a solar field and a large, independently operable CCPP. With two GTs in operation the maximum ISCC efficiency is around 70 percent and it **increases** with part-load > 70 percent. With one GT in operation, maximum ISCC efficiency increases from 80 to 90 percent, benefitting from the high part-load efficiency of the KA24 and the solar contribution. The corresponding CC efficiency during part-load operation is higher than the standard CC efficiency at part-load. For power output, the maximum solar contribution is 17 percent at base-load, but it is increasing to 45 percent when only one GT is operating at part-load. Correspondingly, the specific CO_2 emissions for CCPP are around 330–340 g CO_2 /kWh at base-load, which can be reduced in such an ISCC configuration to around 220 g CO_2 /kWh, further decreasing at a lower load.

Alstom has a first reference as equipment supplier for the Ain Beni Mathar ISCC power plant in eastern Morocco, 150 km south of the Mediterranean coast. However, this has a 1.66 km^2 'parabolic trough solar field' (without tower receiver) which provides 470 MW in total, 450 MW from the CCPP and 20 MW from the solar plant.[195]

The changed CCPP operational perspectives are highlighted if the characteristics required by power generators from their combined cycles 5–10 years ago are compared with those of today. In the past, the typical CCPP specification requirements focused on:

- highest base-load efficiency based on approximately 8,000 OH per year
- lowest specific sales price ($/kW)
- lowest overall operational costs
- shortest delivery schedule
- experience and reference cases
- low NOx/CO emissions at base load.

Due to the explained market changes, the present CCPP specification for operational flexibility must be based on:

- highest overall weighted efficiency based on expected OHs and load regime
- lowest cost of electricity based on base- and part-load efficiencies
- lowest overall operational costs based on the anticipated dispatch
- reliability, availability and starting reliability
- lowest minimum load during off-peak periods to minimise fuel consumption
- longer intervals between inspections to lower operation and maintenance costs
- start-up capabilities, particularly for hot/warm plant conditions
- loading/de-loading gradients and transient capabilities
- high cycling capability of all plant components
- low emissions (NOx, CO, CO_2, ...) over the widest possible load range.

These challenging targets represent the framework of future power generation developments, and consequently, lead directly to the final outlook section.

---

[195] Linear Fresnel or Parabolic Trough solar plant concepts mainly use oil that is heated up by the concentrated radiation, and are consequently temperature limited; the tower receiver works without oil. Alstom has supplied for the Ain Beni Mathar ISCC power plant already in 2007 2 GT13E2, 1 COMAX steam turbine and 3 TOPAIR generators.

6.4   Gas Turbine Outlook

6.4.1   Future Power Generation Market

There is apparently agreement on the development of the power generation market: at least in the industrialised OECD countries and for a near-term outlook over the next 20 years. Intermittent renewable energy (e.g. wind, solar) is becoming a greater part of the energy industry worldwide. Gas turbine plants must increasingly perform the role of backup capacity for those times when the wind is not blowing or the sun is not shining. The gas turbine is well-suited to play this role because of its ability to start/stop and change load relatively quickly and its relatively low initial cost. In areas where intermittent and cycling capacity is required, simple cycle and simplified (i.e. low cost) combined cycle plants will likely play a larger role. Large, high efficiency combined cycle plants will also be required in regions where natural gas provides base-load energy, but even then, the ability to start quickly and ramp up and down in load rapidly will be valuable.

As described, fossil-fired power plants will be required to meet peak demands and to compensate for output reductions and non-availability of the rapidly growing renewable energy sources. Medium-term power generation market scenarios, such as in Germany, assume that renewables will account on average for over 30 percent of all power generated. There will be phases during which renewables can meet the entire power demand, while in other phases back-up power generation from fossil power plants will still be necessary. The situation is complicated even more by the fact that the various phases frequently give way to each other within a short time, and above all the transitions are not entirely predictable. Consequently, fewer and fewer fossil power plants in the grid will have to cope with even steeper load ramps. If a power plant is seen as a system to be optimised as a whole, there are indeed solutions for meeting these future challenges: modern combined cycle power plants are particularly suitable. The reform of electricity and gas markets has led to major changes in the way decisions are taken on power sector investment. Opening the sector to competition has led to the internalisation of risk in investment decision-making. Investors now examine power generation options according to the different financial risks posed by the different technologies.

Given the long-term nature of electricity investments, investment decisions in base-load generating capacity are being made on the basis of long-term fundamentals, rather than looking at short-term behaviour in the spot or forward electricity markets. The current market preference for gas-fired, base-load power generation in many OECD countries can be explained mainly by the perceived lower cost. The characteristics of the CCGT, its low capital cost, and its flexibility have all added to its attractiveness. At the same time, the importance of CCGT means that gas markets assume a greater importance than ever for power generation development. For governments, this means moving forward on liberalisation and monitoring investment in both gas and electricity infrastructure.

The preference for gas-fired power generation does expose investors to the risk of increased fuel prices. The creation or development of electricity and natural gas markets has led to a system where the risks of price development can no longer be managed, but must be assessed with probable approaches. The adequacy of investment in peaking generation remains an even more sensitive issue. The low numbers of hours of operation of peaking plants have led to the question whether markets can bring forward adequate peaking capacity at all. Low capital and high flexibility in operation are particularly important attributes in an attempt to value peaking capacity. While investors have long been aware of the qualitative benefits of flexibility in a liberalised market, flexible peaking capacity may also become much more profitable than traditional approaches predicted.

Uncertainties in future price levels tend to favour flexible, short lead-time technologies. The current preference for gas-fired power generation in many countries can be explained in part by these new developments. The ability to add capacity quickly, to expand capacity at an existing power plant, or to switch fuels is becoming increasingly valued. By contrast, investments such as large hydro or nuclear plants bear a significantly larger investment risk.

6.4.2   Innovative Gas Turbine Technologies

Cutting emissions by 2020 to a level that could keep a global, 21st century, temperature rise under two degrees C is technologically and economically feasible, says a comprehensive new study released by the UN Environment Programme (UNEP) in 2011.[196] Accelerated uptake of renewable energy, fuel switching and energy efficiency improvements can deliver a large proportion of the necessary cuts. The deduced consequences for the development of future gas turbine technology[197] are:

- high load flexibility
- high component and cycle efficiencies
- fuel flexibility
- high exhaust gas temperatures
- thermally expanded cycle options (as a a rather long-term option).

High Load Flexibility The need for rapid load cycling capabilities in gas turbines in combination with alternatively generated, almost unpredictable power has been outlined before. These operation conditions will represent the future standard, with growing shares (if not already over-capacities in parts) in renewable energy supplies. Given the prime target of CO_2 prevention, the required fossil back-up power generation has to be ready on short notice.

Even more so than currently, the design of GT hot gas path parts has to pay attention to resulting TMF (Thermal Mechanical Fatigue) and LCF (Low Cycle Fatigue) problems. Increased demand for higher life cycles at uncompromised reliability triggers intensified research in materials and coatings, mechanical design principles and failure mode investigations.

High Component and Cycle Efficiencies Besides the parts' life, the further rise of gas turbine efficiency will be of prime interest. The efficiency of the key GT components (compressor and turbine) already achieve approximately 91 and 89 percent respectively, and di-

---

[196] See UNEP, Bridging the emissions gap
[197] See Vogeler, Zukunftsperspektiven fuer die Hochtemperatur-Gasturbine im Kraftwerksbau

rect improvements will remain predictably small and costly. A presently advanced gas turbine with an inlet temperature > 1,400 °C (Tmix, Figure 6-32) uses 18–20 percent of the compressor mass flow for cooling and leakages, which is consequently unavailable for its primary purpose. A reduction of cooling air mass flow by 50 percent has been set as a challenging, nevertheless realistic R&D target.[198] Consequently, the GT efficiency would rise by 2–3 percent, and the exhaust temperatures by 40–50 deg C (at unchanged TIT levels).

There are various promising technical possibilities for reducing coolant requirements and/or further increasing the gas temperature at turbine entry. They include the development of improved high-temperature materials and thermal barrier coatings. It is also hoped to make further progress by innovative component cooling and improved protection against high surface temperatures with the aid of thin cooling films. These innovative concepts are characterised by an increasing degree of design detail. The combination of impingement cooling and film cooling seems particularly suitable, and apart from transpiration cooling, which is more difficult to implement, it also seems to be the most efficient method. The 'Near Wall Cooling' concept uses innovative cooling systems with additional cooling air channels below the blade surface, with considerably improved cooling efficiency and as such, reduced cooling air consumption. The success of this development route depends on further progress in precision casting technology to a large degree, Section 5.4.3, especially on the manufacturing of complex ceramic cores. The principal manufacturing feasibility has already been demonstrated for advanced turbojet engines, but the cost-effective application for large power generation turbines is still pending. A characteristic feature of the required advanced casting technologies would be the handling of built core-assemblies to realise the intended complex cooling configurations. Figure 6-32 illustrates the potential TIT gains from the application of these 'Advanced Technologies' in turbine manufacturing.

One possible path for further cooling air savings comprises improved base material alloys and corresponding thermal barrier coatings. Considerable research activities are directed towards increasing the thermal load capacity of base materials, such as a switch from Ni- to Re-based superalloys, up to the possible use of Re-based high temperature composites with tungsten fibre reinforcements. Another development string tries to keep the present superalloy properties by circumventing extraordinary high-priced alloy ingredients, up to the option of combining several material options such as a separately cast single-crystal airfoil from a costly Ni-based alloy with blade fixations from conventional, cheaper materials. The long-established concepts of built and joined turbine blades may see a revival, with features such as cooled, internal load-carrying metallic core structures and CMC Ceramic Matrix Composite outer airfoil shapes, meaning a very thick TBC layer and thus contributing considerably to cooling air savings.

Improvements in the GT SAS Secondary Air System are an area for potential improvements, following successful application examples from advanced turbojet engines. It is imaginable that similar to the flying ACC Active Clearance Control concept (e.g. PW2037 lpt), an optimum clearance adaptation system for stationary gas turbines will be developed where the clearances increase during transient operation modes, and shrink by casing cooling during steady state operation. Both Siemens, with a complete rotor shift technique for

---

[198] See BMWi, COORETEC lighthouse concept

the new SGT5-8000H, as well as Mitsubishi[199], with a steam-heated casing in their G engine development, have already followed this approach.

Fuel Flexibility It is known that the CO_2 emissions are highly fuel-dependent: in relation to > 900 kg CO_2 for one MW of coal-generated power, the corresponding values for oil (600 kg CO_2/MW) and natural gas (450 g CO_2/MW) clearly indicate the preferred fuel strategy in favour of environmental impact. By coincidence, shale gas has become an increasingly important source of natural gas in the United States over the past decade, and interest has spread to potential gas shales in the rest of the world. Some analysts expect that shale gas will greatly expand worldwide energy supply, though both numbers and environmental impact of this 'just-in-time' solution are still under discussion.[200]

High Exhaust Gas Temperatures As outlined before, this feature is one of the inherent advantages of Alstom's GT24/GT26 sequential combustion concept. This technique represents a high CO_2 saving asset for future power generation in general, in combination with other mentioned technologies for rapid load adaptation.

At present, the ultimate targets in gas turbine power generation appear to be additionally addressed in new, thermally expanded cycle configurations :

– to increase the maximum temperature of the present 1,500 deg C gas turbines
– to achieve a new cycle to recover thermal energy at the 2,000 deg C level by other methods
– to develop a combined cycle 'cascade' to recover thermal energy, first of all without combustion, e.g. in a fuel cell system.

Thermally Expanded Cycle Options In a triple combined cycle power generation system, a SOFC (Solid Oxide Fuel Cell) power generation system is placed before the established CCPP combined cycle gas/steam turbine system. By generating power at three stages (the fuel cell[201], the gas turbine, and the steam turbine) the resulting Fuel Cell Combined Cycle (FCCC) system achieves outstanding efficiency in power generation from natural gas. The FCCC system is expected to achieve the world's highest power generation efficiency, exceeding 70 percent (LHV) for several hundred MW class power generation.

With an unusually high fuel-to-electrical efficiency, hybrid systems indicate a major paradigm shift for the future generation of power in a variety of applications. The first demonstrations of both high-pressure and atmospheric pressure hybrid systems verified the basic principles of the technology, described the component features that require technology advances, and confirmed the viability of the product for a short-term and long-term market for a broad variety of applications. Apparently, both the Molten Carbonate Fuel Cell (MCFC) and the Solid Oxide Fuel Cell (SOFC) are attractive candidates for hybridisation due to their high operating and effluent temperatures. Systems are emerging for distributed generation (15 kW to 50 MW) with combinations of High-Temperature Fuel Cells (HTFCs)

---

[199] See Fukuizumi, Large frame gas turbines

[200] http://en.wikipedia.org/wiki/Shale_gas

[201] A fuel cell generates electricity directly through electrochemical reactions and is more efficient than a heat engine because it eliminates the mechanical or rotating machinery. Because the performance of a fuel cell is not restricted by the Carnot Law constraints that limit heat engine efficiencies, the fuel cell will likely be the core of a high efficiency hybrid power cycle.

and Micro-Turbine Generators (MTGs). However, concepts are also evolving for central plant configurations (~300 MW) where ultra high-efficiency on both natural gas and coal are envisaged in combination with zero emissions of criteria pollutants, CO_2 sequestration[202], and hydrogen co-production.[203]

All projections for energy use predict a sound future for gas turbine plants, as long as the plant continues to use fewer resources by becoming more efficient and reducing emissions. This sets major challenges for gas turbine researchers and designers. Future gas turbines will need to reach nearly 50 percent efficiency in simple cycle and over 65 percent in combined cycle applications. There will be a trend towards Zero Emission Plant (ZEP), with – at best – no emissions into the atmosphere and any output (e.g. CO_2) disposed of in an environmentally friendly way. Even higher efficiencies will be achieved by combining gas turbines with fuel cells, once the latter have reached suitable scale, cost and reliability, and gas turbines will continue to contribute to the further, successful spread of renewable energies.

As over the past 100 years, none of these developments will happen by accident, and achieving them will set major challenges to gas turbine researchers and designers. But, as the gas turbine invention and development at BBC/ABB/Alstom made its way from an niche product to the core technology carrier for advanced power generation, there is confidence that a combination of entrepreneurial courage and engineering ingenuity will maintain this success story for decades to come.

---

[202] Carbon capture and sequestration/storage (CCS), refers to technology that attempts to prevent the release of large quantities of CO_2 into the atmosphere from fossil fuel use in power generation and other industries by capturing CO_2, transporting it and ultimately, pumping it into underground geologic formations to securely store it away from the atmosphere. Alstom is at the forefront of the versatile, power generation system-independent CCS technology. There are basically two development routes: oxy-combustion consists of burning the fuel in a mixture of oxygen and re-circulated flue gas, and post-combustion capture technology consists of separating CO_2 from exhaust gases using a solvent (amine or chilled ammonia). Alstom currently has approximately 10 CO_2 capture pilot plants operating around the world, with several others under construction and in planning. See http://en.wikipedia.org/wiki/Carbon_capture_and_storage, also for an extensive discussion of the potential risks and economic drawbacks of CCS. http://en.wikipedia.org/wiki/Carbon_capture_and_storage – cite_note-2

[203] See Samuelson, Fuel cell/ gas turbine hybrid systems

7 'Les Palmarès'[1]

7.1 BBC–ABB–Alstom Historical Firsts in Gas Turbine Technology[2] for Power Generation

1902 BBC starts steam turbine series production according to 'System Parsons'; first company on the European continent to do so

1905 BBC multistage centrifugal compressor (PR = 4.5), according to 'System Rateau' for the Armengaud-Lemâle first experimental gas turbine, Paris St. Denis, France

1910 First operative BBC gas turbine according to the Holzwarth principle at BBC Mannheim, Germany. This improved version followed immediately after Holzwarth completed a principal demonstration plant at Koerting Brothers AG, Hanover, Germany

1924 First two-stage centrifugal compressor VT402, turbocharging a two-stroke diesel engine, PR = 1.35, of SLM (Swiss Locomotive and Machine Works), Winterthur, Switzerland

1926 First all-axial aerodynamically profiled, 4-stage compressor test rig with PR = 1.2 at BBC Baden, Switzerland

1929 First welded steam turbine rotor delivered to Japan according to BBC's unique technology for welded gas turbine/steam turbine and generator rotors, patented e.g. as GB362813 with priority on 2 May 1930

1933 All-axial compressor (11 stages, PR = 3.3) and turbine set for first commercial Velox boiler with net power output (GT principle); two axial compressors for air and blast furnace gas applications in Mondeville, Normandy, France

1934 BBC equips the first closed-loop supersonic windtunnel at the ETH Zurich (Prof. J. Ackeret) with newly developed 13-stage axial compressor, PR = 3

1935 Patent in the name of Curt Keller for Ackeret-Keller closed cycle gas turbine, later via Escher-Wyss to BBC

1937 5 MW continuous air/electricity generating gas turbine installed in a chemical plant (Houdry oil refinery process) Sun Oil Co, Philadelphia, USA

1939 First proposal of a 2 x 18,000 hp CODAG (COmbined Diesel And Gas) turbine ship propulsion unit by Adolf Meyer, IMechE London

---

[1] In Switzerland this term is mostly used to describe a list of races a cyclist has won. In French it means a list of primary achievements or a list of winners.

[2] Without any further location reference, the place is meant to be Baden, Switzerland.

1939 First film-cooled axial turbine, film cooling patents and aero turbocharger application, presumably for Rolls Royce Griffon 37.V.12 motor

1939 4 MW first electricity generating, utility gas turbine, Neuchâtel, Switzerland

1940 First 3-stage axial experimental compressor with adjustable rotor blading at BBC Baden, Switzerland

1941 First gas turbine powered locomotive, 2,200 hp, put into operation for SFR Swiss Federal Railways, Winterthur, Switzerland

1943 First continuously operating coal-dust burning 1.6 MW gas turbine and coal-dust fired Velox boiler at BBC Baden, Switzerland

1944 After previously constructing several advanced altitude test facilities in Germany, BBC Mannheim inaugurated the first full-scale, fully climatised jet engine altitude test facility with a mass flow of 20 kg/s for simulated altitudes up to 13 km, Ma < 1 at BMW Flugmotoren, Munich-Milbertshofen, Germany (the so-called 'Herbitus Anlage')

1944 BBC Mannheim designs and delivers (part of) 8 subsequent axial and centrifugal compressor stages for 54 MW hydraulic drive power for first hypersonic 1×1 m Mach 10 windtunnel at WVA Kochel, Germany. Finally realised in the 1950s as 'Tunnel A' at AEDC Arnold Engineering Development Center, Tullahoma, Tn., USA

1944 Advanced gas turbine designs of BBC Mannheim's Dept. of Aero Engine Design TLUK/VE: first high-performance aero engine axial compressors with degree of reaction R = 0.5 for BMW 109–003 turbojet (Hermso I, II and III projects), first tank gas turbine, first long-range recuperated aero gas turbine for propeller drive with coal combustion

1945 World record 10 MW electricity generating, two-shaft reheat gas turbine with intercoolers and intermediate combustors. For the first time two axial compressors arranged in sequence from BBC Baden, Switzerland, for Bucharest, Romania

1948 40 MW (13 + 27 MW) world's biggest all gas turbine power station, with regenerator, intercoolers and sequential combustion in Beznau, Switzerland

1955 6 MW first mobile, railway packaged GT-powered generator sets for Mexico power network stabilisation

1960 4 × 25 MW world's biggest all gas turbine power station, Port Mann, BC, Canada and first high-power station remote operation and control between the Port Mann 100 MW GT plant and B.C. Electric Headquarters, Vancouver, Ca

1961 First gas/steam combined-cycle powerplant for 75 MW gas-fuelled utility application in Korneuburg, Austria

1965 Gas turbine powered sea-water distillation plant

1968 Finnish Navy's 'Turunmaa' with power turbine/installation from BBC Mannheim was the first large GT-powered ('RR/SN-Olympus') high-speed warship to enter service, top speed 37 kn

1977 300 MW CAES Compressed Air Energy Storage gas turbine power plant with sequential combustion, Huntorf, Germany; meanwhile a key reference project for the storage of renewable energies

1982 1,000 MW (20 x GT11-D5) world's largest crude oil GT power plant Riyadh 8, Riyadh, Saudi Arabia

1984 First commercial gas turbine operation with first generation lean DLN Dry, Low-NOx premix combustion in a modified GT13D at Lausward, Duesseldorf, Germany

1984 First transonic axial compressor within GT8 power generation gas turbine at Shell Pernis, Rotterdam, The Netherlands

1987 GT13E, world's biggest industrial GT > 140 MW and > 34 percent total efficiency, Hemweg 7 operational start, Hemweg, The Netherlands

1987 Second generation of 'multi-injection' lean premix burners introduced at GT8, Purmerend, The Netherlands

1990 First nuclear power plant repowered by 12 x GT11N sets, world's largest combined-cycle cogeneration plant, Midland, Mi., USA

1993 First vortex-breakdown Dry Low-NOx EV environmental dual-fuel burner as single combustor retrofit solution (Midland, Mi., USA) and in due course installed in GT13E2, GT8C2, GT24 and GT26 annular combustor configurations

1994 165 MW, 60 Hz high-efficient simple-cycle GT24 with the following for the first time:
- PR = 30 axial compressor on single shaft
- single crystal turbine blading for industrial GT application (GT26 lpt blade 1 ~25 cm casting length)
- low emission, sequential combustion system

1997 365 MW, 50 Hz combined-cycle GT26 with 58+ percent total efficiency and low NOx emissions

1999 World's biggest 140 MW turbo-gearbox from Renk AG, Augsburg, D; 60 → 50 Hz converter for GT11N2, with double helical gearing and directed lube oil injection: Brestanica, Slovenia (2), Fajr, Iran (5) and Tema, Ghana (2)

7.2 BBC–ABB–Alstom GT Development Directors[3] (Baden, CH)

Period	Name
1923–1946	Dr. h.c. Adolf MEYER (1880–1965)
1938–1965	Dr. h.c. Claude SEIPPEL (1900–1985)
1946–1948	Dipl.-Ing. ETH Kurt NIEHUS[4] (1900–1983)
1948–1968	Dr.-Ing. h.c. Hans PFENNINGER (1903–1989)
1968–1987	Dr.-Ing. Wilhelm ENDRES (1927–2007)
1981–1987	Dipl.-Ing. ETH Anton WICKI (*1944)

---

[3]   The term 'Director with responsibility for GT development' is somewhat ambiguous; at times these were shared responsibilities which explain a certain overlap in the periods given.

[4]   Confirmation pending

1988–1990	Dr. sc. techn. ETH Hans WETTSTEIN (*1941)
1990–1994	Dipl.-Ing. ETH Fredy HAEUSERMANN (*1942)
1995–1998	Dr.-Ing. Hans-Juergen KIESOW (*1949)
1999–2003	Dr.-Ing. Manfred THUMANN (*1954)
2003–	Dr. sc. techn. ETH Jürg SCHMIDLI (*1961)

7.3   The Members of the GT Development 'Hall of Fame'

Since 1995 the GT Development Department has awarded certificates of merit for dedicated and outstanding contributions to gas turbine development to:

Year	Name
1995	Franz Farkas (*1935)

joined the Companies in 1961 and held various positions until 2004. His major development contributions were the introduction of the transonic axial compressor to the gas turbine product line and the collection of design directives for all core competences. See also Section 5.1.6.

1995	Ferdinand Thueringer (*1936)

worked with the Companies from 1961 to 2001. He was test manager on the gas turbine test bed in Döttingen and initiated a field experience feedback system.

1996	Dilip Mukherjee (*1935)

worked within the Companies from 1970 until 2003. He re-introduced the air cooled turbine design at BBC. In his later years he managed the yearly cycle to define the Companies gas turbine product portfolio.

Thomas Meindl (*1960)
joined ABB Power Generation in 1991. He built up the compressor technology and supported the turbine design system.

1997	Werner Steinbach (*1938)

held various positions within BBC/ABB from 1954 to 1998. He was the chief designer for the GT13 product line. In this role he successfully managed the different discipline contributions towards the characteristically long life of the GT13 engine family.

1998	Jakob J. Keller (1950–1999)

As outlined in detail in Section 6.1.4, he began his career in the Research Centre in 1977. He contributed significantly to the advanced technology of the lean premix combustion. His EV burners have been implemented in all gas turbine products.

1998　　　　　Peter Rufli (*1953)
joined the Companies in 1992. He worked on the newly developed reheat gas turbines GT24/GT26 and became the first Chief Engineer in the gas turbine product line.

1999　　　　　Klaus Hollinger (*1956)
started at BBC in 1981 and has remained with the Companies since then. He put in place and maintained the field experience feedback process for gas turbines, closing the gap between customers and the R&D teams and driving further improvement to the satisfaction of both.

2000　　　　　Pavel Rihak (*1948)
worked in the Companies from 1980 until 2011. He built up the product support department, guided the teams to master field problems and worked on the implementation of the found solutions.

2001　　　　　Jaan Hellat (*1946)
joined BBC in 1979. He was project manager for combustion and drove the competition between the different design approaches forward. All current gas turbine combustors in Alstom's portfolio have been significantly influenced by him.

2002　　　　　Martin Scheu (*1962)
started in R&D in 1991. He created and maintained the gas turbine fleet lifetime consumption project and later was the initiator of the cost excellence programme.

2003　　　　　Peter Hartmann (*1938)
worked within the Companies from 1953 until 2003 and was rewarded for his 50 years engagement in the direction of continuous improvement of the products. He facilitated the interface with factories and initiated many improvement projects.

2004　　　　　Karl Reyser (*1953)
joined the Companies in 1979. He was Technical Director for the first integrated oil gasification combined cycle plant for API in Italy and later facilitated the transfer of customers' requirements into Alstom's product specifications.

2005　　　　　Peter Wollschlegel (*1947)
worked within the Companies from 1981 until 2008. He has successfully built up the gas turbine component sales business.

2006　　　　　Volker Eppler (*1963)
joined ABB Power Generation in 1995 and has been with the Companies since then. He successfully led the mechanical design of Alstom's 60Hz GT products and later facilitated the introduction of the adapted turbine design system.

2007　　　　　Fredy Haeusermann (*1942)
worked within the Companies between 1966 and 2002. He successfully rebuilt the gas turbine product portfolio and introduced the reheat concept into the large gas turbines, Section 6.1.4.

2008   Werner Balbach (*1946)
            joined BBC's material technology area in 1980. He introduced a struc-
            tured approach to the interpretation of material data and lifetime con-
            sumption of parts.

2009   Sandro Noro (*1952)
            joined the Companies in 1969. He successfully managed the procurement
            chain of Alstom's gas turbines towards on time product delivery.

2010   Stefan Florjancic (*1958)
            joined ABB in 1993 and has been with the Companies since then. He has
            been responsible for the development of turbines and the associated base
            technology. He later built up the service technology for the advanced gas
            turbine product portfolio.

2011   Martin Boller (*1966)
             began as commissioning engineer in 1989. As Vice-President of Sales and
            Tendering, he was a key contributor to the GT26 market re-introduction.

2012   Hermann Engesser (*1949)
            came to BBC in 1974 and stayed with the Companies till today. He spe-
            cialized first in the development of hydraulic control components, then in
            operational hazard identification in general and is since 2002 a recognized
            Expert in gas turbine safety.

8   Bibliography

Numbers in [ ] refer to the Alstom internal historic literature database.

ABB Asea Brown Boveri: 'ABB Gas Turbines for Power and Cogeneration Plants', ABB publication no. CH-KW 2006 88 E, status Jan. 1987, 11 p. [729]

ABB Asea Brown Boveri: 'Seminar KWG Entwicklungsstrategie. Sitzungsprotokoll vom 20. Nov. 1988', (E. Borinelli), Internal Note AN 88/52, 15 p. [997]

ABB Asea Brown Boveri: 'Development and Technology of a 145 MW Gas Turbine for the 50 Hz Market', ABB publication no. D KW 1089 89 E, status Feb. 1989, 12 p. [401]

ABB Kraftwerke AG: '20 Jahre ABB-Gasturbinen Typ 13: Von 55 bis 165 MW – Spiegelbild einer Evolution', (W. Keppel), VGB Kraftwerkstechnik 4/94, special print, 14 p. [361]

ABB Power Generation: 'Business Area PGT, Strategic Plan, Gas Turbines and Combined-Cycle Plant', (A. Roeder), internal document, 11 Aug. 1988, 28 p. [959]

ABB Power Generation: 'GT-Liste 9, 11, 13', internal document, 1990, 171 p. [469] and short version 15 p. [60]

ABB Power Generation: 'Heavy-duty Gas Turbines Ranging from 8 to 150 MW', ABB publication no. CH-KW 2022 90 E, status 1990, 8 p. [784]

ABB Power Generation: 'GT11N – The 82 MW Gas Turbine for Power and Cogeneration', ABB publication no. CH-KW 2025 90 E, status 1990, 6 p. [841]

ABB Power Generation: 'GT15 Program Steering Committee Review', internal document, 20. June 1991, 132 p. [1055]

ABB Power Generation: 'GT-8 RBK-D Purmerend: Bericht über den Schaden vom 91-03-10', (Ph. Brunner), internal Technical Note TN 91/225, 19 Sept. 1991, 34 p. [833]

ABB Power Generation: 'GT15(N) Turbine, Final Report', (A. Pfeffer), internal Technical Note TN 01-1544, 1992, 198 p., electronic issue prepared by H. Wettstein as Alstom TN 02/0016, 22 Jan. 2002 [953]

ABB Power Generation: 'GT13E2 – the most efficient low-emission gas turbine in the 165 MW class', ABB Review 6/7 1993, 8 p. [1101]

ABB Power Generation: 'The GT8C: An Improved, Low-Emission Gas Turbine', (F. Kreitmeier, M. Fischer, P. Harasgama, J. Krautzig), ABB Reprint, Power-Gen Americas 1993, Dallas Tx, USA, 17–19 Nov. 1993, 12 p. [967]

ABB Power Generation: 'New EV-Silo Combustor for the GT8C', (M. Aigner, M. Fischer, F. Magni), ABB Reprint, Power-Gen Europe 1994, Cologne, D, 8 p. [967]

ABB Power Generation: 'ABB's GT24/GT26 Evolution from 1991 to 1995', (P. Graf), internal collection of design drawings, 1995, 50 p. [1006]

ABB Power Generation: 'GT15/ GT17 Project Meetings on 29. Oct and 5. Nov. 1991', internal mail, 36 p. [1053]

ABB Power Generation: 'GT15 turbine cross section and WBZ 1979', internal document, 31. Dec. 1991, 39 p. [925]

ABB Power Generation: 'Casting of GT24/GT26 Turbine Blades. Handbook of Lessons Learnt', (A. Beeck), internal document, issued 18 May 1996 with upgrades in Aug. 1998 and Jan. 2000, 13 p. [926]

ABB Power Generation: 'GT24/GT26 Weekly Reports – from 07.01.1992 till 17.06.1996', (P. Meylan, E. Pauli), internal Technical Note TN 00/385, 4 July 2000, 2'233 p. [1045]

Ackeret, Jakob: 'High-Speed Windtunnels', paper at 5th convention of the Volta congress, Italy, Sept. 30 till Oct. 6, 1935, published as NACA TM 808, Washington Nov. 1936, p. 49 [3]

Ackeret, Jakob: 'Probleme des Flugzeugantriebes in Gegenwart und Zukunft', Schweizerische Bauzeitung, Vol. 112, No. 1, Zurich, 2 July 1938, pp. 1–4 [409]

Ackeret, Jakob: 'Auf dem Weg zur Gasturbine', Schriften der DAL Deutschen Akademie der Luftfahrtforschung, special print from Vol. 7b, Issue 3, 1943, pp. 72–75, presentation to the DAL Berlin on 4 Dec. 1942 [500]

Ackeret, Jakob: 'Das Institut für Aerodynamik an der Eidgenössischen Technischen Hochschule', Mitteilungen aus dem Institut für Aerodynamik, Leeman Zurich 1943 [801]

Ackeret, Jakob: 'Eulers Arbeiten ueber Turbinen und Pumpen', Vorrede zu Band II 15 von Eulers Werken, special print to 'Euleri Opera Omnia', Series II Vol. 15, Orell Fuessli Zurich, 1957, p. 61 [511]

Ackeret, Jakob and Keller, Curt: 'Eine Aerodynamische Wärmekraftanlage', preliminary announcement in Schweizerische Bauzeitung, special print, Vol. 113, No. 19, Zurich 1939, 3 p. [505]

Ackeret, Jakob and Keller, Curt: 'Aerodynamische Waermekraftmaschine mit geschlossenem Kreislauf', ZVDI, Vol. 85, No. 22, 1941 pp. 491–500 [505]

Affolter, Claudio and Siegenthaler, Urs : 'Architekturführer der Stadt Baden', Mueller, Baden 1994, 223 p.

Agricola, Georg: 'De Re Metallica Libri XII' (Basel, 1556, transl. from Latin to English by Herbert C. Hoover and his wife Lou Henry Hoover), The Mining Magazine, London 1912, 650 p. + drawings, reprint Dover, New York 1950

Agricola, Georg: 'De Re Metallica Libri XII', faksimile print, 5th ed., VDI Dusseldorf, 1978 and Springer Berlin, Heidelberg, 1987

Aicher, W. and Schnyder, S.: 'Umbau von Turbokompressoren', Techn. Rundschau Sulzer, Vol. 70, No. 3, pp. 17–24

Ainley, D.G. and Mathieson, G.C.R.: 'An Examination of the Flow and Pressure Losses in Blade Rows of Axial-Flow Turbines', Aeronautical Research Council, Report & Memoranda No. 2891, 1951

Ainley, D.G. and Mathieson, G.C.R.: 'A Method for Performance Estimation for Axial Flow Turbines', Aeronautical Research Council, Report & Memoranda No. 2974, 1957

Albring, Werner: 'Gorodomlia. Deutsche Raketenforscher in Russland', Luchterhand Vlg. Hamburg-Zurich 1991 255 p.

Alstom: 'New Gas Turbine Concepts Drafted by F. Zerlauth', (H. Wettstein) internal Technical Note TN 02/1520, 1985, 3+23 p. [47a, b]

Alstom Power: 'The Test Failures of the GT8A in the Years 1982/83', (H. Wettstein), internal Technical Note TN 02/0008, 10. Jan. 2002, 3 p. [951]

Alstom: 'The World's First Industrial Gas Turbine Set – GT Neuchâtel', Landmark documentation, 8 p., 2007 [205 a] http://files.asme.org/asmeorg/Communities/History/Landmarks/12281.pdf

Anderson, John D. jr.: 'Modern Compressible Flow with Historical Perspective', McGraw Hill, 2nd ed., 1990, 650 p.

Anderson, John D. jr.: 'A History of Aerodynamics', Cambridge University Press, New York 1997, 478 p.

ASME American Society of Mechanical Engineers: '1958 Gas Turbine Progress Report', (J.W. Sawyer) ASME, 1958, 141 p.

ASME American Society of Mechanical Engineers: '3500 kW Gas Turbine at the Schenectady Plant of the General Electric Company', special print on the occasion of the dedication as National Historic Mechanical Engineering Landmark, 8 Nov. 1984, 15 p. [223]

ASME American Society of Mechanical Engineers: 'The First 500-Kilowatt Curtis Vertical Steam Turbine', special print on the occasion of the dedication as International Historic Mechanical Engineering Landmark, 23 July 1990, 14 p. [298] http://files.asme.org/ASMEORG/Communities/History/Landmarks/5527.pdf

Auer, Werner P.: 'Practical Realizations of Gas Turbines Combined with Steam and Industrial Heat Cycles', ASME GT Power Conference Washington D.C., 5–9 March 1961, Paper 61-GTP-11

Auer, Werner P.: 'Entwicklung der Kraftwerksgasturbine', BWK Brennstoff-Waerme-Kraft Vol. 37, No. 5, May 1985, pp. 213–217

Bankoul, V. and Kean, J.H.B.: 'A New 30-MW Packaged Gas Turbine Power Plant', ASME paper 70-GT-11, 1970, 12 p. [271]

Barham, Kevin and Heimer, Claudia: 'ABB: The dancing giant – creating the globally connected corporation', Financial Times, London, 1998

Barth, Hermann: 'Zur Entwicklungsgeschichte der deutschen Flugtriebwerks-Hoehenpruefstaende', DVL special print to A. Baeumker's 75th birthday, 1966, 32 p. [586]

Bauersfeld, Walther: 'Die Grundlagen zur Berechnung schnelllaufender Kreiselräder', Z VDI, Vol. 66, No. 19, 13 May 1922, pp.461–465 and No. 21, 27 May 1922, pp. 514-517 [526]

Baumann, H.: ' Messung der Schaufelbeanspruchung für grosse Foedermengen', Schweizerische Bauzeitung 88, issue 24, 11. June 1970, pp. 541–543

BBC Brown, Boveri & Cie.: 'Ventilator für Generatorbelüftung (Geertruidenberg)', TF Test Report No. 762 and 762/I, 24. 10. 1929, 11 p. [304]

BBC Brown, Boveri & Cie.: 'Axialventilator im Doppelkonusgehäuse', TF Test Programmes and Reports No. 808 I-III, 16. May 1930 ff., 77 p. [305]

BBC Brown, Boveri & Cie.: 'Axialkompressor', TF Test Programme and Report No. 602, 7. Oct. 1926–23. July 1931, 58 p. [303]

BBC Brown, Boveri & Cie.: 'Vierstufiges Axialgebläse', TF Test Programme and Report No. 902, 30.7.1931–24.5.1932, 6 p. [851]

BBC Brown, Boveri & Cie.: '4-stage axial compressor test rig and blading, drawing set', 1926–1932, 10 p. [410]

BBC Brown, Boveri & Cie.: Offer for 'Axialgebläse für das Institut für Aerodynamik ETH', dated 24.12.1931, letter to Prof. Ackeret, dated 20.09.1933 and delivery contract with Direktion der Eidgenössischen Bauten Bern, dated 29.09.1933, 5 p. [412]

BBC Brown, Boveri & Cie.: 'Axialgebläse für Veloxkessel', TF Test Programme and Report No. 963/I, dated 22 Dec. 1932–8 May 1933, 11 p. [800]

BBC Brown Boveri & Cie.: 'Turbogebläsegruppe der Sun Oil Co. Philadelphia', TFVL Test Report No. 1113, dated 14 Aug. 1936, 8 p. [804]

BBC Brown Boveri & Cie.: 'Flugzeug-Aufladegruppe VTF 225 mit Schaufelkühlung nach Darrieus für 12 Zylinder-Kestrel, 4-Takt-Vergasermotor der Firma Rolls-Royce in Derby', TFVL Test Programme and Report No. 1204, dated 29 Nov. 1938–1 Dec. 1939, 17 p. [853]

BBC Brown Boveri & Cie.: 'Luftgekuehlte Schaufel fuer Gasturbinen (Blechschaufel Profil 581b)', TFVL Test Report No. 1234, dated 28 Dec. 1939, 6 p. [852]

BBC Brown Boveri & Cie.: 'Collection of GT Neuchâtel Original Photographs', 1939–1940, 26 p. [873]

BBC Brown, Boveri & Cie.: 'Gasturbine Neuchâtel', TFVL Test Report 1225, dated 1 April 1940, 7 p. [806]

BBC Brown, Boveri & Cie.: 'Gasturbinen-Elektrolokomotive', Test Programme and Report No. TFVL 1288, dated 04 Sept. 1940–21 Aug. 1942, 62 p. [754]

BBC Brown, Boveri & Cie.: 'Versuchs-Axialgebläse mit drehbaren Schaufeln (Laufschaufeln)', Test Programme and Report No. TFVL 1299, dated 14 Oct. 1940–18 July 1945, 12 p. [807]

BBC Brown, Boveri & Cie.: 'Gasturbine Filaret, Test Programme and Report No. TFVL 1504', dated March-Oct. 1946, 88 p. [315]

BBC Brown, Boveri & Cie.: 'Gasturbine Chimbote, Test Programme and Report No. TFVL 1515', dated 13 July-17 Dec. 1946, 63 p. [317]

BBC Brown, Boveri & Cie.: 'ND-GT Rotor und Regelung Beznau I, Test Reports No. TFVL 1530/I' in 2 parts – for Rotor dated 10 June 1948, 72 p. [309a], – for engine control (Regelung) dated 26 July 1948, 35 p. [309b]

BBC Brown, Boveri & Cie.: 'Gasturbinen-Anlage Beznau I and II, Test Reports No. TFVL 1530/II' in 2 parts – for Beznau I dated 22 June 1949, 28 p. [310], – for Beznau II dated 7 June 1949, 37 p. [311]

BBC Brown, Boveri & Cie.: 'Fahrbare Gasturbinenzentralen', Special print no. 2658 D, May 1958, 18 p. [11]

BBC Brown, Boveri & Cie.: 'Die Brown Boveri Gasturbinen mit einer Eintrittstemperatur von 750 °C', (H. Pfenninger) Special print 2795D, 1960, 31 p. [14]

BBC Brown, Boveri & Cie.: 'Brown Boveri Gasturbinen', Special print 2975 D, June 1963, 33 p. [17]

BBC Brown, Boveri & Cie.: '75 Jahre Brown Boveri, 1891–1966, Festschrift 1966', AG Brown, Boveri & Cie. Baden CH, 1966, 290 p. [849]

BBC Brown, Boveri & Cie.: 'Peak Load Power Station Emden with a 42-MW-Gas Turbo Group with Gas Generators', Special print 1967, 7 p. [1007]

BBC Brown, Boveri & Cie.: 'The Brown Boveri Packaged Gas-Turbine Plant', Special print 1968, 10 p. [268]

BBC Brown, Boveri & Cie.: 'Gasturbine Typ 13B, C und D', (W. Keppel), Technical Note TN 74/26, 27 Nov. 1974, 10 p. [467]

BBC Brown, Boveri & Cie.: 'Welded Turbine Shafts', (W. Dien), BBC Publication No. CH-T 060063E, 1974, 4 p. [680]

BBC Brown, Boveri & Cie.: 'Overspeed Test Facilities of the Group – Overspeed Testing and Balancing of Large Rotors', (W. Kellenberger, H. Weber, H. Meyer), BBC Publication No. CH-T 050083E, 1976, 12 p. [731]

BBC Brown, Boveri & Cie.: 'Aufgabe des Prototyps der Gasturbine 15 in Baden', (W. Endres), Aktennotiz AN 76/28, 5 April 1976, 18 p. [987]

BBC Brown, Boveri & Cie.: 'Studie GT-Typ 11E', (F. Farkas), Aktennotiz AN 76/53, 21 July 1976, 66 p. [447]

BBC Brown, Boveri & Cie.: 'Termine GT Prototyp 15 (Studie)', (F. Farkas), Aktennotiz AN 76/60, 26 Aug. 1976, 9 p. [457]

BBC Brown, Boveri & Cie.: 'BBC/TCT Kapazitätsbelastung', (A. Pesendorfer), Aktennotiz AN 77/8, 3 Feb. 1977, 4 p. [999]

BBC Brown, Boveri & Cie.: 'Pflichtenheft GT-Typ 11E/13E (Vorlaeufer)', (P. Zaugg), Aktennotiz AN 77/46, 22 July 1977, 16 p. [999]

BBC Brown, Boveri & Cie.: 'Besprechung mit Motor Columbus am 6. Okt. 1977', (W. Endres), Aktennotiz AN 77/61, 7 Oct. 1977, 3 p. [999]

BBC Brown, Boveri & Cie.: 'BBC Type 15, 165 MW 60 Hz Gas Turbine', draft of BBC PR leaflet HTCV-15901 E, ~1980, 9 p. [1002]

BBC Brown, Boveri & Cie.: 'Typ 15, einwellige 60 Hz Gasturbine für 150 MW', (W. Endres, F. Farkas), unpublished, prepared for BBC special GT issue 1981, 23 p. [377]

BBC Brown, Boveri & Cie.: 'Development of the Gas Turbine Blading', (D.K. Mukherjee, A. Wicki) BBC special publ. no. CH-T 113443 E, 1981, 12 p. [669]

BBC Brown, Boveri & Cie.: 'Geschichte des Schweizer. Turbomaschinenbaus, BBC Teil', unpublished, 1982, 133 p. [316]

BBC Brown, Boveri & Cie.: 'Projektauftrag GKR , Rationalisierung GT8', (A. von Rappard), Internal Note AN TCT 83/08, 3 Feb. 1983, 28 p. [989]

BBC Brown, Boveri & Cie.: 'Markt- und Wettbewerbssituation der Gasturbogruppe Typ GT13E', Internal Note, 30 Sept. 1983, 27 p. [755]

BBC Brown, Boveri & Cie.: 'Transonic Compressor Development. Testing – Test Rigs, Instrumentation, Results', (F. Farkas), internal document, 1984, 35 p. [414]

BBC Brown, Boveri & Cie.: 'Entwicklungs-Dauer fuer die Prototypen der Gasturbinen 11B, 9B, 11C, 13B', (P. Zaugg), BBC internal note, 11 Jan. 1985, p. 1 [952]

BBC Brown, Boveri & Cie.: 'GT15/ FPL, Sitzung vom 87-08-07', (H. Wettstein), Meeting Minutes SB KWG 87/05, 1 p. [924]

BBC Brown, Boveri & Cie.: 'GT11N early development – some strategic documents 1985–1990', memo collection, unpublished, 139 p. [991]

BBC Brown, Boveri & Cie.: 'Brown Boveri Gas Turbines', Special print CH-KW 1059 87E, 1987, 9 p. [402]

BBC Brown, Boveri & Cie.: 'Neuenburg – Hot Gas Path Inspection', (N. Sofia) Field Service Report No. 021111, 11 Nov. 2002, 23 p. [55]

BBM Brown Boveri Mitteilungen: 'Das BBC Flugzeug-Gebläse', (Ravoth) Vol. VI, No. 6, June 1919, pp. 119–126 [843]

BBM Brown Boveri Mitteilungen: 'Die Konstruktionen der Turbinenfabrik in 1926', (K. Sachs, P. Faber) Vol. XIV, No. 1, Jan. 1927, pp. 33–39 [189]

BBM Brown Boveri Mitteilungen: 'Rueckblick auf die Entwicklung der Brown Boveri-Konstruktionen im Jahre 1927', Vol. XV, No. 1, Jan. 1928, pp. 28–58 [191]

BBM Brown Boveri Mitteilungen: 'Die Konstruktionen der Turbinenfabrik in 1929', Vol. XVII, No. 1, Jan. 1930, pp. 29–43 [194]

BBM Brown Boveri Mitteilungen: 'Die Konstruktionen der Turbinenfabrik in 1931', Vol. XIX, No. 1, Jan. 1932, pp. 31–45 [87a]

BBM Brown Boveri Mitteilungen: 'Rueckblick auf die Entwicklung der Brown Boveri Konstruktionen im Jahre 1932', Vol. XX, No. 1, Jan. 1933, pp. 38–51 [88a]

BBM Brown Boveri Mitteilungen: 'Rueckblick auf die Entwicklung der Brown Boveri Konstruktionen im Jahre 1933', Vol. XXI, No. 1/2, Jan./Feb. 1934, pp. 25–39 [97a]

BBM Brown Boveri Mitteilungen: 'Rueckblick auf die Entwicklung der Brown Boveri Konstruktionen im Jahre 1935', Vol. XXIII, No. 1–2, Jan.-Feb. 1936, pp. 1–72 [89a]

BBM Brown Boveri Mitteilungen: 'Die Leistungssteigerung von Dieselmotoren und Flugmotoren nach dem Buechi-Verfahren durch das Mittel der Abgasturboaufladung. Ausblick auf andere Anwendungsmöglichkeiten der Abgasturboaufladung', (E. Klingelfuss) Vol. XXIV, No. 7, July 1937, pp. 175–190 [103]

BBM Brown Boveri Mitteilungen: ' Rueckblick auf die Entwicklung der Brown Boveri-Konstruktionen im Jahre 1939', Vol. XXVII, No. 1–3, Jan.-March 1940, p. 6 [81]

BBM Brown Boveri Mitteilungen: 'Leistungsversuche an einer Gleichdruck-Gasturbine der A.-G. Brown, Boveri & Cie. in Baden' (A. Stodola) Vol. XXVII, No. 4, April 1940, pp. 79–83 [124]

BBM Brown Boveri Mitteilungen: 'Rueckblick auf die Entwicklung der Brown Boveri-Konstruktionen im Jahre 1940', Vol. XXVIII, No. 1–3, Jan.-March 1941, p. 10 [108]

BBM Brown Boveri Mitteilungen: 'Rueckblick auf die Entwicklung der Brown Boveri-Konstruktionen im Jahre 1940', Vol. XXVIII, No. 8/9, Aug.-Sept. 1941, pp. 182–243 [106]

BBM Brown Boveri Mitteilungen: 'Ueber die Endverluste der Turbinenschaufeln', (A. Meldahl), Vol. XXVIII, No. 11, Nov. 1941, pp. 356-361 [109]

BBM Brown Boveri Mitteilungen: 'Rueckblick auf die Entwicklung der Brown Boveri Konstruktionen im Jahre 1941', Vol. XXIX, No. 1–3, Jan.-March 1942, 81 p. [116a and 116b]

BBM Brown Boveri Mitteilungen: 'Rückblick auf die Brown Boveri-Konstruktionen im Jahre 1942', Vol. XXX, No. 1–4, Jan.-April 1943, 71 p. [125b]

BBM Brown Boveri Mitteilungen: 'Rueckblick auf unsere Taetigkeit im Jahre 1943', Vol. XXXI, No. 1–2, Jan.-Feb.1944, 76 p. [148]

BBM Brown Boveri Mitteilungen: 'Rueckblick auf unsere Taetigkeit im Jahre 1944', Vol. XXXII, No. 1–2, Jan.-Feb. 1945, 57 p. [155b]

BBM Brown Boveri Mitteilungen: 'Zwei weitere Gasturbinen für Elektrizitätswerke bei Brown Boveri bestellt', (W.G. Noack), Vol. XXXII, No. 4, April 1945, 149 p. [161]

BBM Brown Boveri Mitteilungen: 'Die Brown Boveri Gasturbinen-Lokomotive', (H. Pfenninger), Vol. XXXII, No. 10–11, Oct./Nov. 1945, pp. 353–360 [162]

BBM Brown Boveri Mitteilungen: 'Die Frage der optimalen Schaufelteilung bei Beschaufelungen von Turbomaschinen, insbesondere bei grosser Umlenkung in den Schaufelreihen', (O. Zweifel), Vol. XXXII, No. 12, Dec. 1945, pp. 436-444 [159]

BBM Brown, Boveri Mitteilungen: 'Ein neuer Schritt in der Entwicklung der Verbrennungsturbine', (Cl. Seippel), Vol. XXXIII, No. 10, Oct. 1946, pp. 263–267 [143]

BBM Brown, Boveri Mitteilungen: 'Rueckblick auf unsere Tätigkeit im Jahre 1946', Vol. XXXIV, No. 1–3, Jan.-March 1947, 41 p. [149b]

BBM Brown Boveri Mitteilungen: 'Rueckblick auf die Entwicklung der Brown Boveri Konstruktionen im Jahre 1947', Vol. XXXV, No. 1/2, Jan./Feb. 1948, 4 p. [128b]

BBM Brown Boveri Mitteilungen: 'Aus unserer Strömungsforschung', Vol. XXXVII, No. 10, Oct.1950, pp. 357–367 [867]

BBM Brown Boveri Mitteilungen: 'Betriebserfahrungen mit Brown Boveri Gasturbinenanlagen', (H. Pfenninger), Vol. XL, No. 5/6, May/June 1953, pp. 144-165 [69]

BBM Brown Boveri Mitteilungen: 'Die Brown Boveri Gasturbinenlokomotive der British Railways', Vol. XL, No. 5/6, May/June 1953, pp. 166-177 [72]

BBM Brown Boveri Mitteilungen: 'Forschung im Dienste unseres Maschinenbaus', (Cl. Seippel), Vol. XL, No. 8, Aug. 1953, pp. 271–275 [742]

BBM Brown Boveri Mitteilungen: 'Die groesste Gasturbinen-Kraftzentrale der Welt "Port Mann" in Kanada', (R. Schmied), Vol. XXXXVII, No. 1/2, Jan./Feb. 1960, pp. 67–72 [244]

BBM Brown Boveri Mitteilungen: 'Rueckblick auf die Entwicklung der Brown Boveri Konstruktionen in den Jahren 1959 und 1960', Vol. XXXXVIII, No. 1/2, Jan./Feb. 1961, pp. 1–136 [563]

BBM Brown Boveri Mitteilungen: 'Rueckblick auf die Entwicklung der Brown Boveri Konstruktionen in den Jahren 1961 und 1962', Vol. L, No. 1/2, Jan./Feb. 1963, pp. 3–154 [257]

BBM Brown Boveri Mitteilungen: 'Versuchsarbeiten auf dem Gebiet der Axialverdichterbeschaufelung', Vol. L, No. 6/7, June/July 1963, pp. 32–39 [211]

BBM Brown Boveri Mitteilungen: 'Forschung und Entwicklung im thermischen Maschinenbau', Vol. LI, No. 12, Dec. 1964, pp. 727–824 [213a]

BBM Brown Boveri Mitteilungen: 'Zur rechnerischen Erfassung der Stroemung in Gasturbinen-Brennkammern', Vol. LI, No. 12, Dec. 1964, pp. 808–816 [213b]

BBM Brown Boveri Mitteilungen: 'Rueckblick auf die Entwicklung der Brown Boveri Konstruktionen in den Jahren 1963 und 1964', Vol. LII, No. 1/2, Jan./Feb. 1965 [247]

BBM Brown Boveri Mitteilungen: 'Die Verwendung von Flugzeug-Duesentriebwerken zur Energieerzeugung', (H.R. Bolliger), Vol. LII, No. 3, March 1965, pp. 239–241 [248]

BBM Brown Boveri Mitteilungen: 'Mr. Henry Thomas und die Geburt der Gasturbine', (Cl. Seippel), Vol. LIV, No. 2/3, Feb.-March 1967, p. 132 [706]

BBM Brown Boveri Mitteilungen: 'Hommage à Monsieur Georges Darrieus, Membre de l'Institut, Paris', (Cl. Seippel), Vol. LV, No. 8, Aug. 1968, pp. 404–410 [606]

BBM Brown Boveri Mitteilungen: 'Ueber den Reaktionsgrad der Turbinen', (Cl. Seippel), Vol. LXII, No. 3, March 1965, pp. 92–98 [67]

BBM Brown Boveri Mitteilungen: 'Die Entwicklung der Brown Boveri Gasturbinen-Kompressoren' (F. Farkas), Vol. LXIV, No. 1, Jan. 1977, pp. 52–59 [37]

BBM Brown Boveri Mitteilungen: 'Brown Boveri Luftspeicher-Gasturbinen (Brown Boveri Air-Storage Gas Turbines)' (P. Zaugg, H. Hoffeins), Vol. LXIV, No. 1, Jan. 1977, pp. 34–39 [41] and Publication No. CH-T 113 133 E in English [671]

BBM Brown Boveri Mitteilungen: 'GT8, eine neue BBC Gasturbine', (F. Farkas, A. von Rappard), Vol. LXX. No. 3/4, March-April 1983, pp. 125–129 [520]

BBM Brown Boveri Mitteilungen: 'Gitterprüfstand für filmgekühlte Gasturbinenschaufeln', (P. Iten, W. Baumann, D. Grob), Vol. LXX. No. 3/4, March-April 1983, pp. 151–157 [520]

BBM Brown Boveri Mitteilungen: 'Entwicklung und Erprobung einer 45-MW-Gasturbine' (A. Wicki, F. Farkas), Vol. LXXII, No. 1, Jan. 1985, pp. 4–11 [425]

BBN Brown Boveri Nachrichten: 'Die Betriebskennlinien mehrstufiger Verdichter', (Ulrich Senger) Vol. XXIII, No. 1–3, Jan.-March 1941, Mannheim, pp. 19–27 [830]

BBN Brown Boveri Nachrichten: 'Schiffs-Gasturbine mit Strahltriebwerk als Treibgaserzeuger', (H. Reuter) Vol. XXXXVII, No.12, Dec. 1965, Mannheim, pp. 599–603 [896]

BBN Brown Boveri Nachrichten: 'Entwicklungen für grosse Gasturbinen', (M. Simon) special print, Vol. LVI, No. 10, Oct. 1974, pp. 410–420 [32]

BBN Brown Boveri Nachrichten: 'Neue BBC-Gasturbinen grosser Leistung', (E.-O. Mueller, F. Poetz) Vol. LXII, No. 7, July 1980, Mannheim, pp. 247–255 and special print no. D GK 1314 80 D, 12 p. [750]

BBR Brown Boveri Review: 'The Combustion Gas Turbine: Its History, Development and Prospects', (A. Meyer), Vol. XXVI, No. 6, June 1939, pp. 127–140 [4]

BBR Brown Boveri Review: 'Load tests of a combustion gas-turbine built by Brown, Boveri & Company, Limited, Baden, Switzerland', (A. Stodola) Vol. XXVII, No. 4, April 1940, pp.79–83 [5]

BBR Brown Boveri Review: 'The Development of the Brown Boveri Axial Compressor', (Cl. Seippel), Vol. XXVII, No. 5, May 1940, pp. 108–113 [6]

BBR Brown Boveri Review: 'The Spacing of Turbo-Machine Blading, especially with Large Angular Deflections', (O. Zweifel), Vol. XXXII, No. 12, Dec. 1945, pp. 436-444 [159]

BBR Brown Boveri Review: 'Aerodynamic Methods Applied to Turbo-Machinery Research', (J. Lalive d'Epinay), Vol. XXXVII, No. 10, Oct. 1950, pp. 357–367 [172]

BBR Brown Boveri Review: 'Power Utilization in Trade and Industry, D4 Axial-Flow Blowers for Wind Tunnels', Vol. XXXIX, No. 1–3, Jan.-March 1952, p. 85 [179b]

BBR Brown Boveri Review: 'Progress and Work in 1952, IV Turbo-Compressors and Blowers', Vol. XL, No. 1–3, Jan.-March 1953, pp. 77–82 [181]

BBR Brown Boveri Review: 'The Determination of Optimum Compressor and Turbine Temperature Ratios for Two-Shaft Gas-Turbine Installations', (H. Baumann) Vol. XL, No. 5–6, May-June 1953, pp. 178–189 [180]

BBR Brown Boveri Review: 'The Axial-Flow Compressor in Industry', (A. Schramm) Vol. XXXXI, No. 11, Nov. 1954, pp. 395–404 [197]

BBR Brown Boveri Review: 'The Theory of Combined Steam and Gas Turbine Installations', (Cl. Seippel, R. Bereuter), Vol. XXXXVII, No. 12, Dec. 1960, pp. 783–799 [201]

BBR Brown Boveri Review: 'Practical Examples of Utilizing the Waste Heat of Gas Turbines in Combined Installations', (W.P. Auer) Vol. XXXXVII, No. 12, Dec. 1960, pp. 800–825 [196]

BBR Brown Boveri Review: 'Axial Compressor with Adjustable Stator Blades', (H. Baumann, P. Schmidt-Theuner) Vol. L, No. 6-7, June-July 1963, pp. 19–32 [210]

BBR Brown Boveri Review: 'Development of a 15-MW Packaged Gas-Turbine Station', (H. Pfenninger), Vol. LII, No. 3, March 1965, special print p. 19 [22]

BBR Brown Boveri Review: 'Combined Steam and Gas Turbine Power Stations', (H. Pfenninger) Vol. LX, No. 9, 1973 [715]

BBR Brown Boveri Review: 'Combined Gas/Steam Turbine Power Plants', (A. Wunsch) Vol. LXV, No. 10, Oct. 1978, pp. 646-655

BBR Brown Boveri Review: '50 Years of Welded Turbine Rotors', (J.-E. Bertilsson, G. Faber, G. Kuhnen), Vol. LXVIII, No. 12, Dec. 1981, pp. 467–473 [261]

BBR Brown Boveri Review: 'Computer program for verifying the operation states of gas turbines', (A. von Rappard, R. Dokkum) Vol. LXX, No. 3/4, March/April 1983, pp. 158–165 [520]

BBR Brown Boveri Review: 'Design and Testing of a 45 MW Gas Turbine', (A. Wicki, F. Farkas), Vol. LXXII, No. 1, Jan. 1985, pp. 4–11 [46]

BBR Brown Boveri Review: 'Type 13E – A Gas Turbine with High Rating and High Efficiency', (A. Eiermann, W. Keppel), Vol. LXXII, No. 3, March 1985, pp. 104–110 [466]

Becker, E.W. et al.: 'Present State and Development Potential of Separation Nozzle Process', Gesellschaft für Kernforschung m.b.H., Institut für Kernverfahrenstechnik, KFK 2067, Sept. 1974, 34 p. http://bibliothek.fzk.de/zb/kfk-berichte/KFK2067.pdf

Betz, W. (ed.) et al. : 'High Temperature Alloys for Gas Turbines and Other Applications 1986, Part I', D. Reidel Publishing Company, Dordrecht NL, distr. Kluwer Academic Publishers, 1986, 845 p.

Bidard, M.: 'Quelques considération sur l'aérodynamique des grilles d'aubes', Energie Aug. 1943, pp. 191–196 [562]

BMWi: 'COORETEC Lighthouse Concept. The Path to Fossil-Fired Power Plants for the Future', Federal Ministry of Economics and Technology (BMWi), Forschungsbericht Nr. 566, April 2008, 76 p. http://www.cooretec.de/publikationen

Bodemer, Alfred and Laugier, Robert. 'L'ATAR et tous les moteurs à réaction français', Ed. J.D. Reber, Riquewihr 1996, 336 p.

Boyce, Meherwan P.: 'Gas Turbine Engineering Handbook', 3rd edition, Gulf Professional Publishing, Boston 2006

Brecht, Bertolt: 'Wahrnehmung (1949)', in Werke, Große kommentierte Berliner und Frankfurter Ausgabe, Band 15 Gedichte 5; 1993

Briling, Nikolai: 'Verluste in den Schaufeln von Freistrahldampfturbinen', VDI-Forschungsheft 68, Berlin 1909, 75 p.

BST Brown Boveri-Sulzer Turbomaschinery Ltd.: 'Gas Turbines – Reference List', 1971, 33 p. [71]

BST Brown Boveri-Sulzer Turbomaschinery Ltd.: 'The Determination of the Life-Time of Cooled and Uncooled Gas Turbine Blades', (V. Beglinger, P. Suter) Turbo-Forum 1st ed., 1972, pp. 35–45

BST Brown Boveri-Sulzer Turbomaschinery Ltd.: 'BBC Patente betreffend aufgeladene Gaserzeuger für die Vergasung fester Brennstoffe', Technical Note TN 74/4, 19. 02. 1974, 17 p. [481]

BST Brown Boveri-Sulzer Turbomaschinery Ltd.: 'Leistungs- und Wirkungsgrad-Messungen an den Gasturbinen 11C, Blue Lake 1–4', (P. Zaugg) Technical Note TN 74/10, 16. 07. 1974, 25 p. [486]

BST Brown Boveri-Sulzer Turbomaschinery Ltd.: 'Turbinenschaufelung N-Charakteristik', Technical Note TN 74/12, 26. 07. 1974, 56 p. [488]

Buechi, Alfred: 'Ueber Verbrennungskraftmaschinen', in Zeitschrift f.d. gesamte Turbinenwesen, 1909, p. 313

Buechi, Alfred: 'Geschichtliches ueber den Ursprung der Idee, einige grundlegende Patente und die ersten kommerziellen Anwendungen der Buechi-Abgasturboaufladung an Brennstoffkraftmaschinen', in MTZ Vol. 18, No.6, June 1957, pp. 171–175

Cardamone, Pasquale: 'Aerodynamic Optimisation of Highly Loaded Turbine Cascade Blades for Heavy Duty Gas Turbine Applications', Dissertation Universitaet der

Bundeswehr Munich, 151 p., 2006 http://ub.unibw-muenchen.de/dissertationen/
   ediss/cardamone-pasquale/inhalt.pdf

Cardamone, Pasquale; Pfitzner, Michael and Loetzerich, Michael: 'Aerodynamic Optimization of a HP-Turbine Cascade Blade for Heavy Duty Gas Turbine Applications', AFT
   01/138, 6th European Conference on Turbomachinery , 7.-11.March 2005, Lille, F

Carnot, Sadi: 'Réflexions sur la puissance motrice à feu et sur les machines propres à
   développer cette puissance', Annales scientifiques de l'É.N.S. 2e série, tome 1 (1872),
   pp. 393–457 http://www.numdam.org/item?id=ASENS_1872_2_1__393_0 (French
   original) and

Carnot, Sadi and Thurston, Robert Henry (editor and translator) 'Reflections on the Motive
   Power of Fire and on Machines Fitted to Develop that Power', J. Wiley & Sons, New
   York 1890, 65 p.

Casey, M.V. and Hugentobler, O.: 'The prediction of the performance of an axial compressor stage with variable-stagger stator vane', VDI Report 706, 1988, pp. 213–227

Catrina, Werner: 'BBC Glanz-Krise-Fusion 1891–1991, Von Brown Boveri zu ABB', Orrell
   Fuessli, Zurich 1991 [631]

Catrina, Werner: 'ABB – Die verratene Vision', Orrell Fuessli, Zurich 2003 [654]

Chanteloup, D. and Boelcs, A.: 'Flow Characteristics in 2-Leg Internal Coolant Passages of
   Gas Turbine Airfoils with Film Cooling Ejection', Trans. ASME, J. of Turbomachinery, Vol. 124, No. 3, 2002, pp. 499–507

Chellini, Roberto: 'ABB introduces improved version of 50 MW class gas turbine', Diesel &
   Gas Turbine Worldwide, Vol. 24, No. 7, p. 12 and p. 14, Sept. 1992

Chellini, Roberto: 'Upgrade of the GT13E2 Gas Turbine', Diesel & Gas Turbine Worldwide,
   April 2012, pp. 40–43

Christiani, K.: 'Experimentelle Untersuchung eines Tragflügelprofils bei Gitteranordnung',
   Luftfahrtforschung Vol.2, 4, 27. Aug. 1928, pp. 91–110 [560]

C.I.O.S.: 'Gas Turbine and Wind Tunnel Activity Brown Boveri Cie.', Combined Intelligence Objectives Sub-Committee Item No. 5, File No. XXVII-22 1946, 8 p.

Conner, Margaret: 'Hans von Ohain – elegance in flight', AIAA 2001, 285 p.

Constant, Edward W., II: 'The Origins of the Turbojet Revolution', John Hopkins University Press, Baltimore, MD, 1980 [569]

Constant, Hayne: 'The Early History of the Axial Type of Gas Turbine Engine', Proceedings
   of the Institute of Mechanical Engineers 153, 1945, pp. 411–426 [570]

Cremona, C.: 'Investigations and tests in the towing basin at Guidonia', Lilienthal-
   Gesellschaft fuer Luftfahrtforschung Berlin 12.-15. Oct. 1938, published as NACA TM
   892, Washington, April 1939, 22 p., download at http://naca.central.cranfield.ac.uk/
   reports/1939/naca-tm-892.pdf

CRIEPI Central Research Institute of Electric Power Industry: 'Comparative Study on Energy R&D Performance: Gas Turbine Case Study', (D. Unger, H. Herzog) Aug. 1998,
   65 p. [308]

Crotogino, Fritz; Mohmeyer, Klaus-Uwe and Scharf, Roland: 'Huntorf CAES: More than 20 Years of Successful Operation', SMRI Spring 2001 Meeting, Orlando, FL, USA, 15.-18. April 2001, 7 p.

Czermak, H. and Wunsch, A.: 'The 125 MW Combined Cycle Power Plant "Korneuburg-B". Design Features, Plant Performance and Operating Experience', ASME paper 82-GT-323, 16 p. [650]

Darrieus, Georges: 'Hommage à Claude Seippel', in 'Dr. h.c. Claude Seippel zum 70. Geburtstag', Schweizerische Bauzeitung, Vol. 88, No. 24 and 30, 1970 [552]

Day, Lance and McNeill, Ian: 'Bibliographical Dictionary of the History of Technology', Routledge, London 1998

Dean, Robert C. jr.: 'The centrifugal compressor', ASME Gas Turbine Int'l, part I March-April 1973, pp.53–58 and part II May-June 1973, pp. 46-50

Destival, Paul: 'SOCEMA Aircraft Turbines. The Record of a French Development Programme', Flight Magazine, 18 Nov. 1948, p. 608 http://www.flightglobal.com/FlightPDFArchive/1948/1948%20-%201948.PDF

Diakunchak, Ihor S., Kiesow, Hans-Juergen and McQuiggan, Gerard: 'The History of the Siemens Gas Turbine', ASME Paper GT2008-50507, ASME Turbo Expo Berlin, Germany, 9–13 June 2008, 13 p. [349]

Dietler, Ursula and Lang, Norbert 'Tradition and Innovation', ABB Switzerland, Baden 1998 [395]

Doebbeling, K., Knoepfel, H.-P., Polifke, W., Winkler, D., Steinbach, C. and Sattelmayer, T.: 'Low NOx Premixed Combustion of MBTU Fuels Using the ABB Double Cone Burner (EV Burner)', ASME/IGTI Turbo Expo, The Hague, NL 13.-16. June 1994, Paper 94-GT-394, 11 p., [726] and ASME J. Eng. for Gas Turbines and Power, 118, (4), 1996, pp. 765–772

Doebbeling, Klaus, Hellat, Jaan and Koch, Hans: '25 Years of BBC/ABB/Alstom Lean Premix Combustion Technologies', Proceedings of the ASME Turbo Expo 2005, Vol. 2, 2005, pp. 201–213 [58]

Dunham, John: 'A.R. Howell – Father of the British Axial Compressor', ASME paper 200-GT-8, 9 p. [407]

Eck, Bruno and Kearton, W.J.: 'Turbo-Gebläse und Turbo-Kompressoren', Springer Berlin 1929, 294 p. [423]

Eck, Bruno: 'Ventilatoren', Springer Berlin, 1st ed. 1937 [533], 2nd ed. 1952 [534], 3rd ed. 1962

Eckardt, Dietrich: 'Untersuchung der Strahl/Totwasser-Stroemung hinter einem hochbelasteten Radialverdichterlaufrad', Doctoral thesis RWTH Aachen 1977 and DLR FB 77-32, 226 p. and 'Investigation of the Jet-Wake Flow of a Highly Loaded Centrifugal Compressor Impeller', NASA-TM-75232, 194 p., 1978

Eckardt, Dietrich and Trappmann, Klaus: 'PW2037 Entwicklung bei MTU, Auftrieb fuer die Boeing 757', MTU heute 1/82, 1982, pp. 10–13

Eckardt, Dietrich: 'Future Engine Design Trade-Offs', 10th ISABE International Symposium on Air Breathing Engines, Nottingham UK, 1991, 10 p.

Eckardt, Dietrich: 'Umweltaspekte bei Produktion und Betrieb fortschrittlicher Kraftwerk-Gasturbinen', X. International Colloquium on Production Technology, PTK 2001 proceedings, Berlin 2001, pp. 293–298

Eckardt, Dietrich and Rufli, Peter: 'Advanced gas turbine technology – ABB/BBC historical firsts' in ASME J. Eng. Gas Turb. Power, Vol. 124, 2002, pp. 542–549, [56]

Eckardt, Dietrich: 'Advanced Gas Turbine Technology – A Challenge for Science and Industry', International Symposium SYMKOM 2005, Technical University of Łódź, PL, Proceedings No. 128, 2005, Vol. 1, pp. 167–178

Eckardt, Dietrich and Ladwig, Michael: 'Celebrating 50 years of combined cycle', Modern Power Systems, Nov. 2011, pp. 23–24, http://www.modernpowersystems.com/digitaledition/ → Archive MPS Nov. 2011

Eckert, Bruno B.: 'Kuehlgeblaese für Verbrennungsmotoren', Motor-Techn.-Zeitschrift, Vol. 2, No. 10, 1940

Eckert, Bruno and Schnell, Erwin: 'Axial- und Radialkompressoren', Springer Berlin 1961, 2nd ed., 530 p. [514]

Eckert, Michael: 'The Dawn of Fluid Dynamics', Wiley-VCH Weinheim 2006, p. 286 [573]

Eichholtz, Dietrich: 'Geschichte der deutschen Kriegswirtschaft 1939–1945'. Saur, Muenchen 2003 [909]

Eiffel, A. Gustave: 'La résistance de l'air et l'aviation' (1910) and 'Les nouvelles recherches expérimentales sur la résistance de l'air et l'aviation' (1914)

Endres, Wilhelm: 'Three Gas Turbine Types for Outputs from 20–60 MW', BST Turboforum 1972, No. 1, pp. 27–34 [27]

Endres, Wilhelm: 'Operating experience with prototype gas turbines', BST Turboforum 1973, No. 3, pp. 107–113 [29]

Endres, Wilhelm: 'Design Principles of Gas Turbines', Reprint from 'High-Temperature Materials in Gas Turbines', Proc. Of a Symposium at BBC Research Labs, Baden, CH 12–13 March 1973, publ. Elsevier Amsterdam 1974, 14 p. [386]

Endres, Wilhelm: '40 Jahre Brown Boveri Gasturbinen', Special print 1979, 8 p. [45]

Endres, Wilhelm: 'The medium size gas turbine Type 8 from Brown Boveri' 83-Tokyo-IGTC-112, 1983

Endres, Wilhelm: 'Rotor Design for Large Industrial Gas Turbines', ASME Paper No. 92-GT-273, presented at the Intl. Gas Turbine and Aeroengine Congress and Exposition, Cologne, Germany, 1–4 June 1992, 9 p. [51]

Endres, Wilhelm: 'Erfahrungen mit Gasturbinen', ABB/Alstom internal design handbook, p. 176, 2000 [781]

Eroglu, Adnan; Doebbeling, Klaus; Joos, Franz and Brunner, Philipp: 'Vortex Generators in Lean Premix Combustion', ASME J. of Engineering for Gas Turbines and Power, Vol.123, Jan. 2001, pp. 41–49

Eyermann, Wilhelm and Schulz, Bruno: 'Die Gasturbinen, ihre geschichtliche Entwicklung, Theorie und Bauart', M. Krayn Berlin 1920, 310 p. [175]

Falcoz, Céline: 'A Comparative Study of Showerhead Cooling Performance', EPFL Thesis No. 2735, Lausanne CH, 2003

Falcoz, Céline; Weigand, Bernhard and Ott, Peter: 'A Comparative Study of Showerhead Cooling Performance', International Journal of Heat and Mass Transfer, Vol. 49, No. 07/08, 2006, pp. 1274-1286

Falcoz, Céline; Weigand, Bernhard and Ott, Peter: 'Experimental Investigations on Showerhead Cooling on a Blunt Body', International Journal of Heat and Mass Transfer, Vol. 49, No. 07/08, 2006, pp. 1287–1298

Farkas, Franz: 'Die Berechnung der transsonischen Beschaufelung "TS-Variante 1" für einen mehrstufigen Axialverdichter. Technical Note TN69/6, BST 1969 [312]

Airfoil geometries published in: Hirsch, C. and Denton, J.D (Ed's), Propulsion and Energetics Panel, Working Group 12 on Through Flow Calculations in Axial Turbomachines. Appendix II. AGARD-AR-175, Oct. 1981

Farkas, Franz: 'The Development of a Multi-Stage Heavy-Duty Transonic Compressor for Industrial Gas Turbines', ASME Paper 86-GT-91, 1986, p. 12 [141]

Feldhaus, Franz Marie: 'Ruhmesblätter der Technik von den Urerfindungen bis zur Gegenwart', Friedrich Brandstetter, Leipzig 1924 [559]

Ferri, Antonio: 'Investigations and experiments in the Guidonia supersonic wind tunnel', Lilienthal-Gesellschaft fuer Luftfahrtforschung Berlin 12–15 Oct. 1938, published as NACA TM 901, Washington, July 1939, p. 32, download at http://naca.central.cranfield.ac.uk/reports/1939/naca-tm-901.pdf

Ferri, Antonio: 'La Galleria Ultrasonora di Guidonia', Atti di Guidonia, Vol. I, Ufficio Editoriale Aeronautico, Rome 17 Feb. 1939, pp. 305–320 [1029]

Fetzer, G.: 'Windkanalanlagen für die Entwicklung der V-Waffen und anderer Raketenprojekte', Bestand RH 8 I Anh. III Wasserbauversuchsanstalt Kochel am See, Mitteilungen aus dem Bundesarchiv 1997, pp. 45–49

Flachowsky, Soeren: '"Das groesste Geheimnis der deutschen Technik", Die Entwicklung des Stratosphaerenflugzeugs Ju 49 im Spannungsfeld von Wissenschaft, Industrie und Militaer (1926-1936)' in TU Dresden, Dresdener Beiträge zur Geschichte der Technikwissenschaften, Heft 8, 2008, pp. 3–32

Flachowski, Soeren: 'Von der Notgemeinschaft zum Reichsforschungsrat', Studien zur Geschichte der Deutschen Forschungsgemeinschaft Vol. 3, Franz Steiner Vlg. Stuttgart 2008, 535 p. [768]

Flatt, R.: 'Entwicklung einer transsonischen Axialverdichter-Stufe. – Teil 1: Auslegungsprinzip, Vorversuche und Auslegung von zwei Laufrädern. Sulzer-Bericht Nr.1391, Abt.1504, Juli 1967

Florjancic, Stefan; Pross, Joerg and Eschbach, Urban: 'Rotor Design in Industrial Gas Turbines', ASME Paper No. 97-GT-75, presented at the Int'l. Gas Turbine & Aeroengine Congress & Exhibition, Orlando, Fla., USA, 2–5 June 1997, 7 p. [249]

Fottner, Leonhard and Lichtfuss, Hanns-Juergen: 'Anwendung neuer Entwurfskonzepte auf Profile fuer axiale Turbomaschinen. Teil B: Entwicklung und Erprobung optimaler

Profilformen fuer Verdichter- und Turbinengitter', MTU Munich, Techn. Report 77/045, 1977

Friedrich, Rudolf: 'Gasturbinen mit Gleichdruckverbrennung', G. Braun Karlsruhe, Germany 1949, 137 p. [593]

Friedrich, Rudolf: 'Dokumente zur Erfindung der heutigen Gasturbine vor 118 Jahren', VGB – B 100, Essen, Germany 1991, 100 p. [50]

Frutschi, Hans Ulrich: 'The closed cycle gas turbine: operating experience and future potential', ASME New York 2005, 294 p.

Fukuizumi, Yasushi; Shiozaki, Shigehiro; Muyama, Akimasa and Uchida, Sumiu: 'Large Frame Gas Turbines, the Leading Technology of Power Generation Industries', MHI Technical Review, Vol. 41, No. 5, Oct. 2004, pp. 1–5

Gardner, W.B.: 'Energy Efficient Engine Low-Pressure Turbine Boundary Layer Program', NTRS 05/1981, NASA CR-165338

Geidel, Helmut-Arndt and Eckardt, Dietrich: 'Gearless CRISP – The Logical Step to Economic Engines for High Thrust', 9th ISABE International Symposium on Air Breathing Engines, Athens GR 1989, ISABE Paper 89-7116

General Electric Co.: 'A Brief Critical Assessment of the Current BBC G.T. Line, Its Development, Experience and Problems', Dec. 1978, 45 p. (various documents) [449, 450]

Gersdorff, Kyrill von, Schubert, Helmut and Ebert, Stefan: 'Flugmotoren und Strahltriebwerke', Bernhard & Graefe Bonn, 4th ed. 2007

Giampaolo, Anthony: 'Gas Turbine Handbook. Principles and Practices', 3rd ed., The Fairmont Press Inc., Lilburn Ga., USA, 2006, 437 p.

Goede, E. and Casey, M.V.: 'Stage matching in multistage axial compressors with variable stagger vanes', VDI Report 706, 1988, pp. 229–243

Gorn, Michael H.: 'Harnessing the Genie: Science and Technology Forecasting for the Air Force, 1944–1986', Create Space Indep. Publ. Platform, June 2012, 238 p.

Green, J.J. et al.: 'Wartime Aeronautical Research & Development in Germany', The Engineering Journal Oct. 1948 pp. 531–538 and 545, Nov. 1948 pp. 584–589, Jan. 1949 pp. 19–25 [427]

Haering, M.; Hoffs, A.; Boelcs, A. and Weigand, B.: 'An Experimental Study to Compare the Naphtalene Sublimation with the Liquid Crystal Technique in Comprssible Flow', ASME Turbo Expo 1995, Houston Tx. USA, 5–8 June 1995, Paper 95-GT-016

Hafer, A.A.: 'Gas-Turbine Progress Report – Materials, Cooling, and Fuels', Trans. ASME, 75, 1953, pp. 127–136

Harasgama, S. Prith and Kreitmeier, Franz: 'A New Uprated Turbine for the GT8 and GT8C Gas Turbine Family', International Gas Turbine and Aeroengine Congress and Exposition, The Hague, NL, 13–16 June 1994, ASME Paper 94-GT-65, 9 p. [725]

Hartmann, M., Robben, R. and Hoppe, P.: 'Inspection, Maintenance and Field Repair of Heavy-Duty Industrial Gas Turbines', BST Turboforum 4/1974, pp. 213–216 [397]

Hawthorne, Sir William: 'The Early History of the Aircraft Gas Turbine in Britain', Revised Version in '50 Years of Jet-Powered Flight', DGLR Report 92–05, Vol. II, pp. 48–98 [602]

Haywood, R.W.: 'Analysis of engineering cycles: Power, refrigerating and gas liquefaction plant', Pergamon Press Elmsford NY 1986

Hearsay, R.M.: 'A revised computer program for axial compressor design', 2 volumes, ARL TR 75-0001, Jan. 1975

Heller, A.: 'Die Gasturbine von C. Lorenzen', ZVDI vol. 72, no. 51, 1928, pp. 1869–1872 [917]

Hendrickson, Robert L.: 'Gas Turbine Electric Powerplants', AIAA 13th Annual Meeting and Technical Display, Washington D.C., USA, 10-13 Jan. 1977, AIAA Paper 77–346, 8 p. [218]

Hiddemann, Matthias; Hummel, Frank; Schmidli, Juerg and Argüelles, Pablo: 'The Next Generation GT26, the Pioneer in Operational Flexibility', Paper presented at Power-Gen Europe, Milano, I, 7–9 June 2011

Hiddemann, Matthias; Stevens, Mark and Hummel, Frank: 'Increased Operational Flexibility from the Latest GT26 (2011) Upgrade', PowerGen Asia Paper, Bangkok, Thailand, 3–5 Oct. 2012, 18 p.

Hildebrand, Veit: 'Praktikumsanleitung Flachwasseranalogie', TU Dresden, Institut fuer Luft- und Raumfahrttechnik, HWK 2011, 17 p. http://tu-dresden.de/die_tu_dresden/ fakultaeten/fakultaet_maschinenwesen/ilr/nwk/dateien/ Anleitung_Flachwasser.pdf

Hirsch, Ch. And Denton, J.D. ed.s: 'AGARD PEP Propulsion and Energetics Panel, WG Working Group 12 on through-flow calculations in axial turbomachines, Attachment II by F. Farkas', AGARD-AR-175, Oct. 1981

Hirschel, Ernst Heinrich, Prem, Horst and Madelung, Gero: 'Aeronautical Research in Germany', Springer Berlin 2003, 694 p. and in German 'Luftfahrtforschung in Deutschland', Bernhard & Graefe Bonn 2001, 638 p.

Hodge, R.I.: 'Section 28. Industrial Process Applications' pp. 28–1 to 28–18 in Roxbee Cox, Harold (ed.): 'Gas Turbine Principles and Practice', Newnes London, 1955 [235]

Hodson, Howard P. and Howell, Robert J.: 'Unsteady Flow: Its Role in the Low Pressure Turbine', Whittle Laboratories, Cambridge UK, 2000, 33 p. http://www-g.eng.cam.ac.uk/whittle/publications/hph/2000-ISUAAAT-LPReview.pdf

Hoffs, Alexander; Boelcs, Albin and Harasgama, Prith: 'Transient Heat Transfer Experiments in a Linear Cascade via an Insertion Mechanism Using the Liquid Crystal Technique', Trans. ASME, Journal of Turbomachinery, 119, 1997, pp. 9–13

Hoheisel, Heinz; Kiock, Reinhard; Lichtfuss, Hanns-Juergen and Fottner, Leonhard: 'Influence of Freestream Turbulence and Blade Pressure Gradient on Loss Behaviour of Turbine Cascade', Trans. ASME, Journal of Turbomachinery, 109, April 1987, pp. 210–219

Holland, M.J. and Thake, T.F.: 'Rotor Blade Cooling in High Pressure Turbines', J. Aircraft, Vol. 17, No. 6, pp. 412–418

Holzwarth, Hans: 'Die Gasturbine', Oldenbourg, Munich and Berlin, 1911 [314]

Holzwarth, Hans: 'Die Holzwarth-Gasturbine', Holzwarth Gasturbinen GmbH, Mülheim/Ruhr, 1935 [576]

Homeier, Lars and Haselbach, Frank : 'Impact of Suction Side Cooling on the Losses of a Highly Loaded High Pressure Turbine Blade', Conference paper, 6th European Conference on Turbomachinery (ETC 6), 7–11 March 2005, Lille, F

Hoppe, P.J. and Keller, B.: 'Construction and Initial Operating Experience on Two Prototype Gas Turbine Power Plants', ASME Paper 74-GT-153, 1974, 8 p. [774]

Horlock, John H.: 'Axial Flow Compressors', Butterworths London 1958, 187 p. [513]

Horlock, John H.: 'Combined Power Plants, Including Combined Cycle Gas Turbine (CCGT) Plants', Pergamon Press Ltd. 1992, 228 p. [1026]

Hourmouziadis, Jean: 'Aerodynamic Design of Low Pressure Turbines', AGARD Lecture Series 167, 1989

Howell, A. Raymond: 'The Present Basis of Axial Compressor Design. Part I – Cascade Theory and Performance', R.A.E. Report No. E. 3946, June 1942, 6 p. [543] and 'The Present Basis... Part II – Compressor Theory and Performance', R.A.E. Report No. E. 3961, Dec. 1942, 25 p. [544]

Howell, A. Raymond: 'Fluid Dynamics of Axial Compressors', Proc. IMechE 1945, 153, pp. 441–452 [541]

Howell, A. Raymond: 'Design of Axial Compressors', Proc. IMechE 1945, 153, pp. 452–462 [828]

Hundertmark, Michael and Steinle, Holger: 'Phoenix aus der Asche: Die Deutsche Luftfahrtsammlung Berlin', Silberstreif Berlin, 1985, 120 p.

Hunt, Ronald J.: 'The History of the Industrial Gas Turbine', Part 1: The first fifty years 1940-1990, idgtE Power Engineer, Vol. 15, Issue 2, June 2011, 50 p. http://www.idgte.org/ IDGTE Paper 582 History of The Industrial Gas Turbine Part 1 v2 (revised 14-Jan-11).pdf

Huthwaite, Bart: 'The Lean Design Solution. A Practical Guide to Streamlining Product Design', www.innovationcube.com 2004, 284 p.

Illingworth, A. and Stanley Stys Z.: 'Experience With the Olympus Powered Brown Boveri Gas Turbine Machinery of the Finnish Navy 700 t Gunboat', ASME International Gas Turbine Conference & Products Show, Cleveland, USA, 10-13 March 1969, ASME Paper 69-GT-120, 8 p. [270]

Imwinkelried, Beat and Hauenschild, Rainer: 'The Advanced Cycle System Gas Turbines GT24/GT26, the Highly Efficient Gas Turbines for Power Generation', ASME Joint International Power Generation Conference, Paper 94-JPGC-GT-7, Phoenix, Az., USA, 2–6 Oct. 1994, 13 p. [691]

IPG International Power Generation: 'MCV-Midland 1'380 MW world's largest combined-cycle cogeneration plant', special print, 1989, 8 p. [995]

Jacobi, C., von Rappard, A. and van der Linden, S.: 'Modernization and performance improved: ABB's GT 11N2 gas turbine', ASME Paper 92-JPGC-GT-1, 8 p. [639]

Jeffs, Eric: 'Asea Brown Boveri Introducing a New Dual Fuel Dry Low-NOx Burner Design', Reprint from Gas Turbine World, May-June 1989, ABB publication no. CH-KW 2023 89 E, 1989, 8 p. [785]

Jeffs, Eric: 'Britain's First Combined Cycle in Service a Month Early (Roosecote)', Turbomachinery International (US), Vol. 33, No. 1, 2009, pp. 50-54

Jenny, Ernst: 'The BBC Turbocharger, a Swiss Success Story', Birkhaeuser Basel-Boston-Berlin 1993 [380]

Johann, Erik and Heinichen, Frank: 'Back to Back Comparison of a Casing Treatment in a High Speed Multi-Stage Compressor Rig Test', ISABE paper 2011–1223, 20th ISABE Conference, Gothenburg SE, 12–16 Sept. 2011, 11 p.

Johnsen, I.A. and Bullock, R.O.: 'Aerodynamic Design of Axial-Flow Compressors', NACA SP-36, 1965

Johnson, Dag and Mowill, R. Jan: 'Aegidius Elling – a Norwegian gas-turbine pioneer', Norw. Tech. Museum Oslo, March 1968, presented at the 1968 ASME GT Products Show, Washington DC and VGB Kraftwerkstechnik, Vol. 52, No. 2, 1972 [596]

Joos, Franz; Brunner, Philipp; Schulze-Werning, Burkhard; Syed, Khawar and Eroglu, Adnan: 'Development of the Sequential Combustion System for the ABB GT24/GT26 Gas Turbine Family', ASME paper 96-GT-315, ASME award 'Best Paper 1996', ASME 1996 International Gas Turbine and Aero Engine Congress & Exposition, Birmingham UK, 10-13 June 1996, 11 p. [704]

Kay, Antony L.: 'German Jet Engine and Gas Turbine Development 1930-1945', Airlife Publishing Ltd. UK, 2002, 296 p. [318]

Kay, Antony L.: 'Turbojet, History and Development 1930-1960, Vol. 1: Great Britain and Germany, 270 p. [597], Vol 2: USSR, USA, Japan, France, Canada, Sweden, Switzerland, Italy and Hungary', The Crowood Press, Ramsbury, Marlborough, Wiltshire SN8 2HR, UK 2007, 269 p. [598, 865]

Kehlhofer, Rolf: 'Gasturbinenkraftwerke, Kombikraftwerke, Heizkraftwerke und Industriekraftwerke', Resch, München 1984, 315 p.

Kehlhofer, Rolf; Bachmann, Rolf; Nielsen, Henrik and Warner, Judy: 'Combined Cycle Gas & Steam Turbine Power Plants', PennWell, Tulsa, Ok., USA 2nd ed. 1999, 300 p.

Keller, Curt: 'Axialgebläse – vom Standpunkt der Tragflügeltheorie', Dr. sc. techn. dissertation ETH Zurich, Leemann 1934, 190 p. [2] and 'The Theory and Performance of Axial-Flow Fans', McGraw-Hill Book Company, New York, 1937

Keller, J.J., Sattelmeyer, Th. and Thueringer, F.: 'Double-Cone Burners for Gas Turbine Type 9 Retrofit Application', CIMAC 19th Int'l Congress on Combustion Engines, Florence 1991, Paper G01, 15 p. [49]

Kenyon, Michael and Fluck, Markus: 'Using Non-Standard Fuels in the Alstom GT11N2 Gas Turbine', Paper presented at PowerGen International Conference at Las Vegas, Nv., USA, 6-8 Dec. 2005, 17 p. [647]

Keppel, Wolfgang E. and Janssen, J.B.: 'Erste Betriebserfahrungen mit der 600-MW-Kombi-Anlage in Amsterdam und ihrer Vorschaltgasturbine GT 13E', VGB Kraftwerkstechnik, 6/90, pp. 458–463 (special print D KW 6082 90 D) [940]

Keppel, Wolfgang and Meyer, J.: 'Entwicklung eines Axialverdichters fuer eine 150 MW-Gasturbine', VDI Conf. 'Therm. Stroemungsmaschinen', 15–16 Sept. 1992 Hannover, publ. ABB KW/GR Mannheim

Kirchhoff, Jochen: 'Wissenschaftsfoerderung und Forschungspolitische Prioritäten der Notgemeinschaft der deutschen Wissenschaft 1920–1932'. Dissertation Ludwig-Maximilians-Universitaet Muenchen 2003, 431 p., http://edoc.ub.uni-muenchen.de/13026/1/Kirchhoff_Jochen.pdf

Kingsbury, Nancy R. (ed.): 'World Directory of Aerospace Research and Development', GAO/NSIAD-80-71FS Foreign Test Facilities, 1990, 575 p.

Klapdor, Sebastian: 'Der Technologietransfer Deutschland – USA nach dem Zweiten Weltkrieg am Beispiel der Kochel Windkanalanlage', GRIN 2004, 176 p.

Koch, H. and Strittmatter, W.: 'Application of the 2-Stage Combustion Process in Connection with Gas Turbines and its Influence on NOx-Emissions', CIMAC Int'l. Congress on Combustion Engines, Helsinki, Finland 1981, Paper GT-12, 16 p. [472]

Koch, H., Bruehwiler, E., Strittmatter, W. and Sponholz, H.-J.: 'The Development of a Dry Low NOx Combustion Chamber and the Results Achieved', CIMAC Int'l. Congress on Combustion Engines, Oslo, Norway, June 1985, 18 p. [473]

Kožulović, Dragan: 'Modellierung des Grenzschichtumschlags bei Turbomaschinenstroemungen unter Berücksichtigung mehrerer Umschlagsarten', Dissertation, Ruhr-Universitaet Bochum 2007, 99 p. http://www-brs.ub.ruhr-uni-bochum.de/netahtml/HSS/Diss/Kozulovic Dragan/diss.pdf

Krehl, Peter O. K.: 'History of Shock Waves, Explosions and Impact', Springer Berlin Heidelberg 2009, 1288 p.

Kreitner, J.: 'Commentary on A. Meldahl's paper "End Losses of Turbine Blades", Journal of the American Society for Naval Engineers, Volume 54, Issue 4, Nov. 1942, pp. 574–576

Kruschik, J.: 'Die Gasturbine', Springer Wien 1960, 2nd ed. [508]

Ladwig, Michael and Stevens, Mark: 'KA26 Combined Cycle Power Plant as Ideal Solution to Balance Load Fluctuations', Hannover Messe – Power Plant Technology Forum 2011, Hannover D, 5–8 April 2011, 18 p. http://files.messe.de/001/media/02informationenfrbesucher/vortraege/2011_4/Power-Plant-Technology_Infrastruktur-Kombikraftwerk-als-ideale-Loesung-zum-Ausgleich-von-Lastschwankungen_Combined-cycle-power-plant-as-anideal-solution-for-load-balancing.pdf

Lakshminarayana, Budugur: 'Fluid Dynamics and Heat Transfer of Turbomachinery', John Wiley & Sons Inc. 1996, 809 p.

Lang, Norbert: 'Charles E.L. Brown 1863–1924, Walter Boveri 1865–1924, Gründer eines Weltunternehmens', Verein für wirtschaftshistorische Studien, Meilen, CH 2000 [767]

Lang, Norbert: 'Aurel Stodola (1859–1942) Wegbereiter der Dampf- und Gasturbine', Verein für wirtschaftshistorische Studien, Meilen, CH 2003 [320]

Langston, Lee S. and Opdyke jr., George: 'Introduction to Gas Turbines for Non-Engineers', Global Gas Turbine News, Vol. 37, No. 2, 1997 http://files.asme.org/IGTI/101/13001.pdf

Langston, Lee S.: 'Plowing New Ground. Gas Turbine Industry Overview 2008', Mechanical Engineering, May 2009, pp. 40-44 http://www.forecastinternational.com/notable/mechanicalengineering.pdf

Langston, Lee S.: 'Visiting the Museum of the World's First Gas Turbine Powerplant', Global Gas Turbine News – a Supplement to Mechanical Engineering Magazine, April 2010, p. 51 [886]

Lanzenberger, Kai; Daxer, Johann; Philipson, Stephen and Hoffs, Alexander: 'A Further Retrofit Upgrade for Alstom's Sequential Combustion GT24 Gas Turbine', Paper presented at the Power-Gen International 2007, New Orleans La., USA, 11–13 Dec. 2007

Lasby, Clarence G.: 'Project Paperclip: German Scientists and the Cold War', Atheneum, New York, 1971, 291 p.

Lecheler, Stefan; Schnell, Rainer and Stubert, Bertram: 'Experimental and Numerical Investigation of the Flow in a 5-Stage Transonic Compressor Rig', ASME paper 2001–GT-0344, ASME Turbo Expo 4–7 June 2001, New Orleans, USA

Lechner, Christof and Seume, Joerg (ed.s): 'Stationäre Gasturbinen', Springer Berlin 2003, 650 p.

Lei, V.-M.; Spakovszky, Z.S. and Greitzer, E.M.: 'A Criterion for Axial Compressor Hub-Corner Stall', Trans. of the ASME, Journal of Turbomachinery, Vol. 130, July 2008, ASME paper 031006, pp.1–10, http://www.ewp.rpi.edu/hartford/~ernesto/S2012/EP/MaterialsforStudents/Green/Lei2008.pdf

Leist, Karl: 'Der Laderantrieb durch Abgasturbinen', Luftfahrt-Forschung, Vol. 14, No. 4/5, 1937, pp. 238–243 [915]

Leist, Karl: 'Abgasturbinen mit Kühlluftbeaufschlagung', Jahrbuch 1941 der Deutschen Luftfahrtforschung, Vol. II, pp. 144–153

Leiste, Volker: 'Development of the Siemens Gas Turbine and Technology Highlights', Siemens internal document, Erlangen, Germany 2006, 8 p. [232]

Loft, Arne: 'From 3.5 MW in Oklahoma in 1949 to 60 000 MW Throughout the World in 1981', CIMAC 14th Int'l. Congress on Combustion Engines, Helsinki 1981, Paper GT22, 44 p. [747]

Lorenzen, Christian: 'The Lorenzen Gas Turbine and Supercharger for Gasoline and Diesel Engines', Mech. Engineering, Vol. 52, No. 7, 1930, pp. 665–672 [918]

Maier, Helmut: 'Forschung als Waffe. Ruestungsforschung in der Kaiser-Wilhelm-Gesellschaft und das Kaiser-Wilhelm-Institut für Metallforschung 1900-1945/48', Wallstein Goettingen 2007, 1210 p.

Marder, Maximilian: 'Motorkraftstoffe, Erster Band: Kraftstoffe aus Erdöl und Naturgas', Springer Berlin 1942, 596 p.

Marnet, Chrysanth, Kassebohm, Borchert, Koch, Hans and Sponholz, Hans-Juergen: 'NOx-Minderung im kombinierten Gas-/ Dampfkraftwerk Lausward', VGB Kraftwerkstechnik Vol. 65, No. 2, pp. 109–113 and Brown Boveri Technik, No. 3, 1985, pp. 120-124 [474]

Marriott, A.: 'Eine moderne umweltfreundliche Industriegasturbine', Technische Rundschau Sulzer, Vol. 69, No. 1, 1987, pp. 17–21 and Diesel & Gas Turbine Worldwide, Vol. 19, No. 6, July/Aug. 1987, pp. 47–49

Mayle, Robert E.: 'The Role of Laminar-Turbulent Transition in Gas Turbine Engines', Trans. Of the ASME, Journal of Turbomachinery, Vol. 113, No. 4, 1991 pp. 509–537

Meher-Homji, Cyrus B.: 'The historical evolution of turbomachinery', ASME Proceedings of the 29th turbomachinery symposium, Sept. 2000, pp. 281–321 [439]

Meier, Hans-Ulrich (ed.): 'Die Pfeilfluegelentwicklung in Deutschland bis 1945', Bernhard & Graefe Vlg. Bonn 2006, 476 p.

Meindl, Thomas; Farkas, Franz and Klussmann, Wilfried: 'The Development of a Multi-Stage Compressor for Heavy Duty Industrial Gas Turbines', ASME paper 95-GT-371, ASME International Gas Turbine and Aeroengine Congress and Exhibition, Houston Tx., USA, 5–8 June 1995, 8 p. [694]

Meldahl, Axel: 'End Losses of Turbine Blades', Journal of American Society of Naval Engineers, Vol. 54, August 1942

Meldau, Ernst: 'Drallstroemung im Drehhohlraum', Dissertation TH Hannover, Germany, 27 June 1935, 54 p.

Meyer, Adolf: 'The Velox Steam Generator, its Possibilities as Applied to Land and Sea', Mechanical Engineering, Aug. 1935, pp. 469–478 [79]

Meyer, Adolf: 'Recent Developments in Gas Turbines', Mech. Engineering, Vol. 69, 1947, pp. 273–277 [100]

Middlebrook, Martin and Everitt, Chris: 'The Bomber Command War Diaries: An Operational Reference Book 1939–1945', Ian Allen Ltd., April 2011, 804 p.

Montis, Marco; Niehuis, Reinhard; Guidi, Mattia; Salvadori, Simone; Martelli, Francesco and Stephan, Bruno: 'Experimental and Numerical Investigation on the Influence of Trailing Edge Bleeding on the Aerodynamics of a NGV Cascade', ASME Turbo Expo 2009, ASME Paper No. GT2009-59910, 8–12 June, 2009, Orlando, Fla., USA, pp. 1063–1073

Munk, Max M.: 'Elements of the wing section theory and of the wing theory', NACA TR No.191, 1924

Munzinger, Ernst: 'Düsentriebwerke – Entwicklungen und Beschaffungen für die schweizerische Militäraviatik', Baden-Verlag 1991, 226 p. [296]

Nagel, Marc G.: 'Numerische Optimierung dreidimensional parametrisierter Turbinenschaufeln mit umfangsunsymmetrischen Plattformen – Entwicklung, Anwendung und Validierung', Dissertation, Universitaet der Bundeswehr Muenchen 2004, 130 p. http://ub.unibw-muenchen.de/dissertationen/ediss/nagel-marc/inhalt.pdf

Naik, Shailendra; Krueckels, Joerg; Gritsch, Michael and Schnieder, Martin: 'Multi-Row Film Cooling Performance of a High Lift Blade and Vane', ASME Turbo Expo 2012, ASME Paper GT2012-68478, 11–15 June 2012, Copenhagen, DK, 9 p.

Naik, Shailendra; Sommer, Thomas P. and Schnieder, Martin: 'Aero-Thermal Design and Validation of an Advanced Turbine', Proc. of the ASME Turbo Expo 2012, ASME Paper GT2012-69761, 11–15 June 2012, Copenhagen, DK, pp. 647–656

Nauschuetz, Anneliese: 'Die Suche nach einem neuen Standort' in Kyrill von Gersdorff 'Ludwig Boelkow und sein Werk – Ottobrunner Innovationen' Bernhard & Graefe Bonn 2002, pp. 43–47

Neumann, Gerhard: 'Herman the German: Enemy Alien US Army Master Sergeant #10500000', Morrow New York, 1984, 1st US ed. and 'China, Jeep und Jetmotoren', Aviatic Vlg. P. Pletschacher Planegg 1989, 1st German ed., 279 p. [906]

Neumann, Herbert A.: 'Vom Ascaris zum Tumor. Leben und Werk des Biologen Theodor Boveri (1862–1915)'. Blackwell, Berlin 1998

Nieto, Françoise: 'MW and Km/h – a History of Alstom', Éditions Coop Breizh 2010, 304 p. [1036]

N.N.: 'BBC 120 and 210 MW gas turbines', Diesel & Gas Turbines Worldwide, Vol. 13, No. 2, March 1981, pp. 18–19

Noack, Walter G.: 'Flugzeuggebläse' in Rundschau ZVDI 1919, pp. 995–1026

Noack, Walter G.: 'Dr. h.c. Walter Noack – Lebenslauf und beruflicher Werdegang, von ihm selbst erzählt', own publication, autumn 1942, 56 p. [354]

Novak, R.A.: 'Streamline curvature computing procedures for fluid-flow problems.' ASME J. Eng. Power, Vol. 89, No. 4, Oct. 1967, pp. 478–490

Novaretti, Salvatore: 'Die Rolle sozialer Netzwerke in der BBC-Fruehphase', Badener Neujahrsblaetter, Jg. 87, 2012, pp. 14–23

Palucka, Tim: 'The Wizard of Octane: Eugene Houdry', Invention & Technology, Vol. 20, No. 3, Winter 2005, http://www.mindfully.org/Technology/2004/Eugene-Houdry-Octane1oct04.htm

Penzig, F.: 'Sichtbarmachen von Temperaturfeldern durch temperaturabhaengige Farbanstriche', Z.VDI, 83, issue 3, 21. Jan. 1939, pp. 69–74, translated as Penzig, F.: 'Temperature-Indicating Paints', NACA TM 905, Washington Aug. 1939, 19 p. http://ebookbrowse.com/naca-tm-905-pdf-d307241078

Pew, A.E.: 'Operating Report on Gas Turbine Use in Sun Oil Company's Refineries', The Oil and Gas Journal, 11 Aug. 1945

Pfenninger, Hans: 'Wirtschaftliche Betrachtungen zur Erzeugung von Energiespitzen durch Gasturbinen und Betriebserfahrungen mit dem Spitzenkraftwerk Beznau der NOK', BBC special print by Jean Frey Zurich, 1955, pp.1–12 [10]

Pfenninger, Hans: 'The Evolution of the Brown Boveri Gas Turbine', BBC special print 3217E-II.9 (3.66) 1965, 19 p. [18]

Pfenninger, Hans: 'Vergangenheit, Gegenwart und Zukunft der BROWN BOVERI Gasturbinen', MTZ 27/11, 1967, pp. 449–461 [369]

Pfenninger, Hans: 'Axialverdichterbeschaufelung für grosse Foerdermengen', Schweizerische Bauzeitung 88, Issue 24, 11 June 1970, pp. 535–541 [83]

Pfenninger, Hans: 'Die Gasturbinenabteilung bei BBC, Rückblick und heutiger Stand', Schweizerische Bauzeitung, Vol. 88, Issue 30, 23 July 1970, pp. 683–691 [23]

Pfenninger, Hans: 'Einige Gesichtspunkte zu den Verhandlungen mit Sulzer', BBC internal memo on tactical patent portfolio usage etc., 14 Dec. 1979, 95 p. [399]

Pfenninger, Hans: 'Zur historischen Entwicklung der Gasturbine', Internal BBC Memo TS-A, 3 Oct. 1981, 5 p. in [316]

Piantanida, Sandro and Baroni, Constantino: 'Leonardo da Vinci', Vollmer, Wiesbaden-Berlin 1977, 8th ed., 539 p.

Pijpker, B.B. and Keppel, W.E.: 'Amsterdam to Have First 140 MWe Gas Turbine', Modern Power Systems, Vol. 6, No. 5, 1986, pp. 28–33

Pohl, Manfred and Freitag, Sabine: 'Handbook on the history of European banks', European Association for Banking History, Edward Elgar Publ. Ltd., Aldershot UK 1994

Ponomareff, Alexander I.: 'Principles of the Axial-Flow Compressor', Westinghouse Engineer, March 1947, pp. 40-46 [416]

Porreca, L.; Behr, T.; Schlienger, J.; Kalfas, A.I.; Abhari, R.S.; Ehrhard, J. and Jahnke, E.: 'Fluid Dynamics and Performance of Partially and Fully Shrouded Axial Turbines', ASME J. of Turbomachinery, Vol. 127, Oct. 2005, pp. 668–678

Porreca, Luca; Kalfas, Anestis I. and Abhari, Rheza S.: ' Optimized Shroud Design for Axial Turbine Aerodynamic Performance', ASME J. of Turbomachinery, Vol. 130, July 2008, ASME Paper GT 031016, p. 12

Potter, J.H.: 'The Gas Turbine Cycle', ASME winter annual mtg. paper 1972, 8 p. [286]

Power Jets R & D: 'The Junkers Jumo 004 Jet Engine', Aircraft Engineering, Dec. 1945, pp. 342–347 [580]

Prandtl, Ludwig: 'Ergebnisse der Aerodynamischen Versuchsanstalt zu Goettingen', 1. Lieferung, R. Oldenbourg Muenchen 1921

Preiswerk, Ernst: 'Anwendung gasdynamischer Methoden auf Wasserstroemungen mit freier Oberflaeche', Dissertation, Mitteilung aus dem Institut für Aerodynamik (J. Ackeret), ETH Zurich, No. 7, Leemann Zurich 1938, 130 p. http://e-collection.library.ethz.ch/eserv/eth:21334/eth-21334-01.pdf and 'Application of the Methods of Gas Dynamics to Water Flows with Free Surface: Part 1 – Flows with No Energy Dissipation', Issue 934 of Technical Memorandums, NACA 1940, 69 p.

Rabl, Christian: 'Das KZ-Außenlager St. Aegyd am Neuwalde', Mauthausen-Studien Band 6, Bundesministerium für Inneres, Wien 2008, 165 p.

Reale Academia d'Italia (ed.): 'Convegno die Scienze Fisiche, Mathematiche e Naturali, 30 Settembre – 6 Ottobre 1935, Le Alte Velocita in Aviazione', 5th Volta Congress Rome, Proceedings Reale Academia d'Italia 1936

Riehm, W.: 'Leistungserhöhung der Diesel-Viertaktmotoren', in ZVDI 1923, pp. 763–766

Rott, Nikolaus: 'Jakob Ackeret und die Geschichte der Machschen Zahl', Schweizer Ingenieur und Architekt, Baar 1983, special print of issue 21/1983, 8 p. [499]

Ruch, Christian et al.: 'Geschäfte und Zwangsarbeit: Schweizer Industrieunternehmen im "Dritten Reich"', Chronos, Zurich 2001[601]

Ruehlmann, Moritz: 'Allgemeine Maschinenlehre: Ein Leitfaden für Vorträge, sowie zum Selbststudium des heutigen Maschinenwesens, mit besonderer Berücksichtigung seiner Entwicklung. Für angehende Techniker, Cameralisten, Landwirthe und Gebildete jeden Standes, C.A. Schwetschke und Sohn, 1865, 544 p.

Rufli, Peter: 'Systematische Berechnungen über kombinierte Gas-Dampf-Kraftwerke', Dissertation ETH Zurich No. 9178, 1990, 175 p.

Rummel, K.: 'Turbogebläse Bauart Brown-Boveri-Rateau von 750 PS, aufgestellt in Rote Erde bei Aachen', Springer Berlin, VDIZ publication No. 159, 1907, 8 p. [506]

Sahm, Peter R. and Hildebrandt, Uwe W.: 'Gerichtet erstarrte Superlegierungen für stationäre Gasturbinen', Publ. Fachinformationszentrum Energie, Physik, Mathematik, 1979, 186 p.

Samuelson, Scott: 'Fuel Cell/ Gas Turbine Hybrid Systems', ASME International Gas Turbine Institute 2004, 10 p. http://files.asme.org/IGTI/Knowledge/Articles/13043.pdf

Sané, Pranav A.: 'Wave Rotor Test Rig Design Procedure for Gas Turbine Enhancement', MSc. Thesis, Michigan State University, Dept. of Mechanical Engineering, 2008, 117 p. http://books.google.ch/books?id=fvI_DW5nmgIC&pg=PA4&lpg=PA4&dq=claude+sei ppel+bbc&source=bl&ots=edYb5oqh9i&sig=3jeeEBYEl9G818ZN_XfMjbsbm4k&hl=de - v=onepage&q=claude seippel bbc&f=false

Savic, Sasha; Lindvall, Karin; Papadopoulos, Tilemachos and Ladwig, Michael: 'The Next Generation KA24/GT24 from Alstom, the Pioneer in Operational Flexibility', Paper presented at PowerGen International, Las Vegas, Nv., USA, 13–15 Dec. 2011, 26 p. http://www.alstom.com/Global/Power/Resources/Documents/Others/110909 Alstom Power Industry Seminar - The Next Generation KA24GT24-USformat_v2.pdf

Sawyer, Robert Thomas (Tom): 'Sawyer's Gas Turbine Catalog 1967', Vol. 5, Gas Turbine Publications Inc., Stamford Ct., [75]

Sawyer, Robert Thomas (Tom): 'Sawyer's Gas Turbine Catalog 1969', Vol. 7, Gas Turbine Publications Inc., Stamford Ct., [278]

Sawyer, Robert Thomas (Tom): 'Sawyer's Gas Turbine Catalog 1974', Vol. 12, Gas Turbine Publications Inc., Stamford Ct., [77]

Sawyer, Robert Thomas (Tom): 'Sawyer's Gas Turbine Catalog 1976', Vol. 14, Gas Turbine Publications Inc., Stamford Ct., [78]

Schabacker, J.; Boelcs, A. and Johnsson, B.V.: PIV Investigation of the Flow Characteristics in an Internal Coolant Passage with 45 Deg. Rib Arrangement', ASME Paper 99-GT-120

Schiesser, Max and Bodmer, Leo: 'Richtlinien zur Behebung von Mängeln und Reibungen im Betriebe und zur Umsatzsteigerung, namentlich auf dem Gebiete der Normalprodukte', BBC Mitteilung A, Nr. 812, Baden 1939, 3 p. [398]

Schlaifer, Robert: 'Development of Aircraft Engines', Harvard University Press, Boston Ma., USA, 1949, 545 p. [583]

Schneeberger, Hans: 'Die elektrischen und Dieseltriebfahrzeuge der SBB', Band 1, Bj.1904–1955, Minirex AG, Lucerne, 18 p.

Scholz, N. and Hopkes, U.: 'Der Hochgeschwindigkeits-Gitterwindkanal der Deutschen Forschungsanstalt für Luftfahrt (DFL) Braunschweig', Forschung auf dem Gebiete des Ingenieurwesens, Vol. 25, No. 5, Duesseldorf, D 08/1959, pp. 133–147

Scholz, Norbert: 'Aerodynamik der Schaufelgitter', G. Braun, Karlsruhe, 1965, 545 p.

Scholz, Norbert and Klein, Armin: 'Aerodynamics of Cascades', AGARDograph-AG-220, 1977

Schopf-Preiswerk, Ernst: 'Nachträge zu dem Buche „Die Basler Familie Preiswerk"', 1961, 48 p. [ETH Bib.]

Schroeder, K.: 'Der innere Ausbau von Kraftwerken', Die Wärme, Vol. 59, No. 9, 1936, pp. 158–161

Schubert, Helmut: 'Turbine – The Hollow Metal Blade as Solution for Material Shortage', pp. 244–252 and 'Progress in Turbine Blade Cooling Since 1945', pp. 444–452, – in Hirschel, E.H.; Prem, H. and Madelung, G.: 'Aeronautical Research in Germany: From Lilienthal Until Today', Springer Berlin 2004, 694 p. [875]

Schwager, O.: 'Der Weg zum Höhenflugmotor', Luftwissen, Vol.5, No. 5, 1943, pp. 136-144

Schwarzenbach, Alfred: 'Meyer, Adolf' in Neue Deutsche Biographie, Vol. 17, 1994, pp. 325–326 http://www.deutsche-biographie.de/artikelNDB_pnd138343322.html

Seippel, Claude: 'Bemerkungen über Strömungen in Turbomaschinen', VDI Sonderheft (special publication) DW 472, 1933

Seippel, Claude: so-called 'Award Papers, a collection of CS assignments, awards, birthday recognitions, etc. between 1938–1980', 54 p. [818]

Seippel, Claude: 'Gas turbines in our century', Trans. ASME, Feb. 1953, pp.121–122

Seippel, Claude: 'Dampfturbinen von heute', Schweizerische Bauzeitung, Vol. 77, No. 20, 14 May 1959, pp. 305–316 [531]

Seippel, Claude: 'Entwicklungstendenzen im Dampf- und Gasturbinenbau', Oesterreichische Ingenieur-Zeitschrift, Vol. 9, Issue 7, 1966, pp. 233–241 [712]

Seippel, Claude: 'The evolution of compressor and turbine bladings in gas turbine design', Trans. ASME, J. Eng. for Power, Vol. 89, 1967, pp.199–206 [20]

Seippel, Claude: 'Die Entstehungsgeschichte des vielstufigen Axialverdichters bei Brown Boveri', Internal BBC Report TT 7509, 30 July 1974, 13 p. in [316]

Seippel, Claude: 'Die Zweifelsche Kennzahl und andere Kennzahlen im Turbinenbau (Reminiszenzen)', Schweizerische Bauzeitung, Vol. 94, No. 30, 22 July 1976, pp. 437–439 [415]

Seippel, Claude: 'Optimierung von HD- und MD-Turbinen: Rechenprogramm TRANCHE (Rep.83016, 15.06.1983), quoted in BBC TX Studies Survey 1971–1988, [808]

Seippel, Claude: 'Wie vollzog sich der Uebergang von der Verpuffungs- zur Gleichdruck-Gasturbine', Internal BBC Report T-SE 79/4, 7 Jan. 1980, 7 p. in [316]

Senger, Ulrich: 'Die Betriebskennlinien mehrstufiger Verdichter', BBC Nachrichten, Jan./March 1941, pp. 19–27 [830]

Smith, G. Geoffrey: 'Gas Turbines and Jet Propulsion for Aircraft', Flight Publ. Co. Ltd. London, 1st ed. Dec. 1942 [831] and 3rd ed. April 1944, 123 p. [770]

Smith jr., Leroy H.: 'Axial Compressor Aerodesign Evolution at General Electric', Trans. ASME, J. of Turbomachinery, Vol. 124, July 2002, pp. 321–330 [231]

Sobolew, Dimitri A.: 'Deutsche Spuren in der sowjetischen Luftfahrtgeschichte', Vlg. E.S. Mittler & Sohn GmbH Hamburg-Berlin-Bonn 2000, 311 p.

Speiser, Ambros P.: 'Episoden aus den Anfängen der Informatik an den ETH', in Informatik im ETH-Bereich: Vom Relaisrechner zum Grosscomputer, special print 1994, ETH-Rat, pp. 7–48

Starkey, N.E.: 'Long-Life Base Load Service at 1600 F Turbine Inlet Temperature', Trans. ASME J. of Engineering for Power, ASME Paper 66-GT-98, 6 p. [222]

Steigmeier, Andreas: 'Brown Boveri und Asea Brown Boveri: Schlaglichter auf hundert Jahre Unternehmensgeschichte', Asea Brown Boveri AG, Baden, CH 1991 [209]

Stephan, Bruno; Krueckels, Joerg and Gritsch, Michael: 'Investigation of Aerodynamic Losses and Film Cooling Effectiveness for a NGV Profile', ASME Turbo Expo 2010, ASME Paper no. GT2010-22810, 14–18 June 2010 , Glasgow, UK, pp. 1391–1400

Stevens, Mark; Hiddemann, Matthias; Ruchti, Christoph; Ruecker, Falk; Selby, Glenn; Perrin, Ian and Bauver, Wesley: 'Alstom's KA26 Advanced Combined Cycle Power Plant – Optimised for Base-Load and Cycling Duty', Power-Gen Europe paper, Amsterdam NL, 8–10 June 2010, 32 p. https://online.alstom.com/Businesses/Thermal-Power/Gas/PromotionTools/GasPowerPlants/Documents/02 KA26/04 Technical Papers/Alstom%27s KA26 Advanced Combined Cycle Power Plant Optimised for Base-Load and Cycling Duty_PowerGen Europe 2010.pdf

Stodola, Aurel: 'Dampf- und Gasturbinen', J. Springer Berlin, 5th ed. 1922, 1,111 p. [GT excerpt: 84]

Stodola, Aurel: 'Steam and Gas Turbine', McGraw-Hill, New York, 1927, 2 Vol., 1,356 p. [84, 177, 319, 346]

Stodola, Aurel: 'Load tests of a 4'000 kW combustion-turbine set', Engineering, London, Vol.149, 5 Jan. 1940, pp. 1–4

Stodola, Aurel: 'Leistungsversuche an der Verbrennungsturbine', VDI-Z, Vol. 84, No.1, 6 Jan. 1940, pp. 17–20 [822]

Stoff, Horst and Waelchli, René: 'Aerodynamische Entwurfsverfahren fuer transsonische Axialverdichter', Schweizer Ingenieur und Architekt, No. 33–34, 20 Aug. 1990, pp. 925–928

Stoff, Horst and Ebner, P.: 'Geeichte Wirkungsgrad-Nachrechnung fuer geometrisch aehnliche Axialverdichterbeschaufelungen', ABB KWR (internal) Techn. Note TN 91/293, 4 Nov. 1991, Baden [977]

Stoff, Horst: 'Warum die stationären Gasturbinen-Axialverdichter mit Überschall laufen', Ruhr University Bochum, In-Touch Magazine 2006, 4 p. [515] http://www.ruhr-uni-bochum.de/fem/pdf/in-touch-magazin2006.pdf

St. Peter, James: 'The history of aircraft gas turbine engine development in the United States: a tradition of excellence', IGTI International Gas Turbine Institute of ASME American Society of Mechanical Engineers 1999, 591 p.

Suter, Peter and Spaeti, H.: 'The surge limit on multi-stage axial compressors', BST Turboforum July 1972, No. 2, pp. 85–93 [74]

Taygun, Fikret: 'Die Gasturbine mit geschlossenem Kreislauf als Mehrzweck-Anlage', Schweizerische Bauzeitung, Vol. 84, Issue 10, 10 March 1966, pp. 175–190 [273]

Thad Allen, Michael: 'The Business of Genocide. The SS, Slave Labor, and the Concentration Camps', Chapel Hill and London, 2002, 375 p.

Thiel, Ernstfried: 'Von Oetztal nach Modane – aus der Geschichte des grossen Hochgeschwindigkeitswindkanals <Bauvorhaben 101> der Luftfahrtforschungsanstalt (LFM) Muenchen , spaeter Anlage S1 MA der Onera', DGLR-Jahrbuch 1986, Part II, pp. 773–795 [855]

Thoren, Kjell T.E.: 'Gas Turbine Development in Sweden after 1945 – A Historical Review', ASME Paper 98-GT-26, 1998, 9 p. [848]

Thumann, Manfred and Konter, Maxim: 'Materials for Power Generation Gas Turbines', Thermec 2000, Las Vegas, Nv. USA, Dec. 2000, Conf. Proceedings, pp. 378–385 [672]

Traupel, Walter: 'Die Entwicklung der Gasturbine in der Schweiz', Proceedings of the 5th World Power Conference, Vienna 1956, Section G3, Paper 203 G3/9 pp. 3765–3792 [347]

Traupel, Walter: 'Marksteine des schweizerischen Turbomaschinenbaues', NZZ Neue Zuercher Zeitung, 18 Feb. 1981, No. 40, pp. 50-52

Traupel, Walter: 'Thermische Turbomaschinen', Springer Berlin 1988, 3rd ed., Vol. I, 576 p. [516] and Vol. II, 530 p.

Trischler, Helmuth: 'Luft- und Raumfahrtforschung in Deutschland 1900-1970', Vol. 4 of Studies on the history of German Research Establishments, Campus Frankfurt/M. 1992

Trost, Wolfgang: 'Auf dem Weg ins All. Raketenforscher in Emmendingen, Riegel und Denzlingen 1946–1950' in 's Eige zeige', Jahrbuch des Landkreises Emmendingen für Kultur und Geschichte 23/2009, pp. 19–76

Tsien, Hsue-Shen et al.: 'Technical Intelligence Supplement', AAF Scientific Advisory Group, HQ Air Material Command Wright Field, Dayton Oh., USA, May 1946, 177 p.

Turbomachinery International: 'ABB launches GT13E2 with annular combustor at 164.3 MW', May/June 1992, pp. 41–44 [1100]

Turbomachinery International: '50 Years of Combustion Research at CTEC', May/June 1992, pp. 57–60 [1103]

TU Wien, Fluid Mechanics and Heat Transfer Laboratory: 'Shallow Water Channel', 5 p. http://www.fluid.tuwien.ac.at/319484?action=AttachFile&do=get&target=shallowwaterchannel.pdf

UNEP: 'Bridging the Emissions Gap', United Nations Environment Programme (UNEP) 2011, 56 p., http://www.unep.org/pdf/UNEP_bridging_gap.pdf

Viereck, Detlef, Wettstein, Hans E., Aigner, Manfred, Kiesow, Hans-Juergen: 'GT 13E2, the cleanest gas turbine for combined cycle and cogeneration application', IGTI Vol. 7, ASME COGEN-TURBO 1992, pp. 231–237

Vogeler, Konrad: 'Zukunftsperspektiven fuer die Hochtemperatur-Gasturbine im Kraftwerksbau (The Potential Future of High Temperature Gas Turbines in Power Plants)', VGB Power Tech, Vol. 10, 2011, pp. 29–33

Von Kármán, Theodore and Edson, Lee: 'The wind and beyond; Theodore von Kármán, pioneer in aviation and pathfinder in space' Little, Brown Boston 1967, 376 p. and 'Die Wirbelstrasse', Hoffmann & Campe Hamburg 1968, 434 p.

Von Rappard, Axel and Wicki, Anton: 'Modular Concept of a Gas Turbine Power Plant', ASME paper 85-IGT-1, ASME 1985 Beijing International Gas Turbine Symposium and Exposition, Beijing PRC, 1–7 Sept. 1985, 7 p. [643]

Wagner, G.; Vogel, G.; Chanteloup, D. and Boelcs, A.: 'Pressure Sensitive Paint (PSP) and Transient Liquid Crystal Techniques (TLC) for Measurements of Film Cooling Performances', Proceedings of the 16th Bi-Annual Symposium on Measuring Techniques in Transonic and Supersonic Flow in Cascades and Turbomachines, 2002

Wagner, Jens-Christian: 'Produktion des Todes. Das KZ Mittelbau-Dora', Wallstein, Göttingen 2001, 690 p.

Wegener, Peter P.: 'The Peenemuende Wind Tunnels', Yale University Press, New Haven and London, 1996, 187 p.

Welter-Thaler, Barbara: 'Flugpioniere in der Region Baden und Wettingen', Badener Neujahrsblaetter, Jg. 87, 2012, pp. 177–183

Wennerstroem, Arthur J.: 'Simplified design theory for highly loaded axial compressor rotors and experimental study of two transonic examples', Institute for Thermal Turbomaschinery, ETH Zurich, Note No. 12, Doctoral Thesis No. 3652, Juris Publ. Zurich, 1965

Wicki, Anton and Farkas, Franz: 'Entwicklung und Erprobung einer 45 MW Gasturbine', Brown Boveri Technik, Vol. 1, 1985, pp. 4–11

Wildi, Tobias: 'Organisation und Innovation bei BBC Brown Boveri AG 1970-1987, ETH Zurich, Preprints zur Kulturgeschichte der Technik, No. 6, 1998 [297]

Wildi, Tobias: 'Der Traum vom eigenen Reaktor. Die schweizerische Atomtechnologieentwicklung 1945–1969', Chronos Zurich 2003, 281 p. http://e-collection.library.ethz.ch/eserv/eth:30387/eth-30387-01.pdf

Willfort, Andreas and Tendick, Rex: 'Gas Turbine Fueled by LBTU Tailgas from the Syntroleum GTL Process', PowerGen International, Orlando, Fla., USA, 10-12 Dec. 2002, 8 p. [860]

Wilson, David Gordon: 'Turbomachinery – from paddle wheels to turbojets', Mechanical Engineering, Oct.1982, pp. 28–40 [426]

Wolf, Martin: 'Im Zwang für das Reich', excerpt from Ploechlinger, St.(ed.), Bauer, J., Wolf, M., Schroetter, B.: 'Verdraengt? Vergessen? Verarbeitet?' 3rd. ed. 2001 [859], download at http://goneu.tcs.ifi.lmu.de/facher/geschichte-sozialkunde/buch.pdf

Zaba, Tadeusz: 'Vergleich der BST-Gasturbinen mit denjenigen der Konkurrenz', BST TN 72/10, 7 Dec.1972, 12 p. [345]

Zaba, Tadeusz: ' Verwendung der BST-Gasturbinen fuer den Betrieb mit Brenngasen mit tiefem Heizwert (low BTU-gas)', BST TN 74/1, 25 Jan. 1975, 8 p. [478]

Zaba, Tadeusz: 'On the Use of Gaseous Fuels of Low Calorific Value in Gas Turbines', BBC Publication No. CH-T 113 153 E, ~1978, 8 p. [618]

Zaugg, Paul and Lang, Norbert: 'Ein doppeltes Gasturbinen-Jubilaeum und seine Bedeutung fuer die Region', Badener Neujahrsblaetter, Jg. 74, 1999, pp. 172–181 [360]

Ziegler, D.; Lercher, G. and Van Leeuwen, J.: 'Pegus 12, The World's Most Efficient Power Station', Joint ASME/IEEE Power Generation Conference , Boston Ma. USA, 21–25 Oct. 1990, Technical Paper 90-JPGC/GT-3

Zimmermann, Manfred W. 'Vom Pionier zum Leader', ABB Hauszeitung, 12/1988 pp. 2–10 [901]

Zwicky, Fritz: 'Report on certain phases of war research in Germany', HQ Air Material Command Dayton Oh. USA, Summary Report No. F-Su-Re, 1946, 186 p.

9 Index of Names

10 Subject Index

11 Nomenclature and Abbreviations

ABB	Asea Brown Boveri Ltd
AC	Allis – Chalmers, USA
AC	Alternating Current
ad	adiabatic
AEDC	Arnold Engineering Development Center, Tullahoma, Tn., USA
AEG	Allgemeine Elekricitaets-Gesellschaft, D
AEI	Associated Electrical Industries, UK
AEV	Advanced EnVironmental burner
AK	Ackeret – Keller process
AMT	Accelerated Mission Test
AS	Armstrong Siddeley, UK
ASEA	Allmänna Svenska Elektriska Aktiebolaget, S
ASME	American Society of Mechanical Engineers, USA
ASP	Air Storage Plant
AVA	Aerodynamische Versuchsanstalt Goettingen, D
BASF	Badische Anilin- und Soda-Fabrik, Ludwigshafen, D
BBC	A.-G. Brown Boveri & Cie., Baden, CH and Mannheim, D
BBM	Brown Boveri Mitteilungen, CH
BBN	Brown Boveri Nachrichten, D
BBR	Brown Boveri Review
BBT	Brown Boveri Turbomachinery Inc., St. Cloud, Mn., USA
BIOS	British Intelligence Objectives Sub-Committee
BL	Boundary Layer
BMW	Bayerische Motoren Werke, Munich, D
BPR	ByPass Ratio
BR	British Railways, UK
BST	Brown Boveri – Sulzer Turbomaschinen AG, Zurich, CH

BTU	British Thermal Unit
c	Absolute velocity
CAD	Computer-Aided Design
CAES	Compressed Air Energy Storage
CC	Combined Cycle
CCC	Catalytic Cracking Case (Houdry process)
CCC	Continuous Cost Cutting
CCGT	Combined Cycle Gas Turbine (in a CCPP)
CCPP	Combined Cycle Power Plant
CDA	Controlled Diffusion Airfoil
CE	Combustion Engineering, Windsor, Ct., USA
CE	Concurrent Engineering
C.E.M.	Compagnie Électro-Mécanique, Paris, F
CEO	Chief Executive Officer
CFD	Computational Fluid Dynamics
CFE	Comisión Federal de Electricidad, Mexico City, MEX
CGE	Compagnie Générale d'Électricité, Paris, F
CIOS	Combined Intelligence Objectives Sub-Committee
CM	Cannon – Muskegon, Muskegon, Mi., USA
D	Diameter
DB	Daimler Benz, Stuttgart, D
DC	Direct Current
DCA	Double Circular Arc
DFG	Deutsche ForschungsGemeinschaft, D
DFL	Deutsche Forschungsanstalt fuer Luftfahrt, Braunschweig, D
DFS	Deutsche Forschungsstelle fuer Segelflug, Ainring, D
DLN	Dry Low NOx combustion system
DLR	Deutsches Zentrum fuer Luft- und Raumfahrt (German Research Center for Aeronautics and Space), D
DM	Deutsche Mark
DS	Directionally Solidified casting
DVL	Deutsche Versuchsanstalt fuer Luftfahrt, D
EC	European Community
EDF	Électricité de France, Paris, F
EMM	Executive Management Meeting
EnBW	Energie Baden-Wuerttemberg, Karlsruhe, D

EOH	Equivalent Operating Hours
ETH	Eidgenoessische Technische Hochschule (Swiss Federal Institute of Technology), CH
ETHZ	Eidgenoessische Technische Hochschule (Swiss Federal Institute of Technology) Zurich, CH
EV	EnVironmental premix burner
FIAT	Field Information Agency Terminal
FKFS	Forschungsinstitut für Kraftfahrwesen und Fahrzeugmotoren (Research Institute of Automotive Engineering and Vehicle Engines) Stuttgart, D
FVV	ForschungsVereinigung Verbrennungskraftmaschinen e.V., Frankfurt/M., D
FWW	Forschungsinstitut fuer Wasserbau und Wasserkraft e.V., Obernach, D
GALCIT	Guggenheim Aeronautical Laboratory at the California Institute of Technology, since 1961 GAL – stands for Graduate Aeronautical Laboratories, Pasadena Ca. USA
GE	General Electric, Fairfield, Ct., USA
GEC	General Electric Company, Coventry, UK
GT	Gas Turbine
h	Total enthalpy
He	Ernst Heinkel Flugzeugwerke A.G., Rostock, D
HEX	Heat Exchanger
HGPP	Hot Gas Path Parts
hp	High pressure
hp	Horse power
HPC	High Pressure Compressor
HPT	High Pressure Turbine
HR	Human Resources
HRSG	Heat Recovery Steam Generator
HVA	Heeres-Versuchs-Anstalt, D
IC	InterCooler
IC	Internal Combustion
ICE	Internal Combustion Engine
IGT	Industrial Gas Turbine
IMechE	Institution of Mechanical Engineers, London, UK
ip	Intermediate pressure
IPP	Independent Power Producer
IPR	Intellectual Property Right
is	isentropic

Jumo	Junkers Motorenbau GmbH, Dessau, D
KA	Kombi-Anlage (CCPP)
KWU	Siemens KraftWerksUnion, Muelheim and Erlangen, D
LBTU	Low BTU
LFM	Luftfahrtforschungsanstalt Munich, D
LLOC	Low Load Operation Capability
lp	Low pressure
LPC	Low Pressure Compressor
LPT	Low Pressure Turbine
LRBA	Laboratoire de Recherches Balistiques et Aérodynamiques, Vernon, F
m	Mass flow
Ma	Mach number
MAN	Maschinenfabrik Augsburg-Nuernberg, D
MBTU	Medium BTU
Me	Messerschmitt AG, Augsburg, D
MFO	MaschinenFabrik Oerlikon, CH
MHD	Magneto-HydroDynamics
MHI	Mitsubishi Heavy Industries Ltd, Minato, Tokyo, J
MIT	Massachusetts Institute of Technology, Cambridge, MA, USA
MTU	Motoren- und Turbinen Union Munich G.m.b.H. (MTU Aero Engines Ltd), D
MV	Metropolitan – Vickers, Manchester, UK
NACA	National Advisory Committee of Aeronautics, USA
NASA	National Administration and Space Administration, Washington D.C., USA
NCR	Non-Conformance Report
NEBIS	NEtzwerk von Bibliotheken und Informationsstellen in der Schweiz (Swiss libraries network), CH
NOx	Nitrogen oxide
OEM	Original Equipment Manufacturer
OH	Operating Hours
ONERA	Office National d'Études et de Recherches Aérospatiales, Meudon, F
P	Power
pol	Polytropic
POW	Prisoner Of War
PP	Power Plant
PR	Pressure Ratio
q	Heat

R	Degree of Reaction
RAE	Royal Aircraft Establishment, Farnborough, UK
RAF	British Royal Air Force
R&D	Research and Development
RDK	Rheinhafen-Dampf-Kraftwerk (Rhine port steam power station), Ludwigshafen, D
Re	Reynolds number
RLM	ReichsLuftfahrtMinisterium (Reich air ministry), Berlin, D
RM	Reichs-Mark (German currency)
rpm	revolutions per minute
RR	Rolls – Royce plc, Derby, UK
RRD	Rolls – Royce Deutschland, Dahlewitz and Oberursel, D
s	Entropy
SAW	Submerged Arc Welding
SB	Single Burner combustor
SBK	Silo-BrennKammer (combustor)
SBZ	Schweizerische Bau-Zeitung (scientific newspaper)
SC	Simple Cycle
SCM	Supply Chain Management
SEV	Sequential/ Second EV combustor
SFr	Swiss Francs
SFR	Swiss Federal Railways
S.L.	Sea Level
SLM	Schweizerische Lokomotiv- und Maschinenfabrik (Swiss Locomotive and Machine works), Winterthur, CH
SNECMA	Société Nationale d'Études et Construction de Moteurs d'Aviation, Paris, F
SOCEMA	SOciété de Constructions et d'Équipements Mécaniques pour l'Aviation, Le Bourget, F
ST	Steam Turbine
SU	Soviet Union
SX	Single Crystal casting
T	Total temperature
TBC	Thermal Barrier Coating
TCh	Turbo Charging
TEL	Tetra-Ethyl Lead
TEPCO	Tokyo Electric Power Company Inc., Chiyoda, Tokyo, J
TH	Technical Highschool

THTR	Thorium High Temperature Reactor
TIG	Tungsten Inert Gas welding
TIT	Turbine Inlet Temperature
TMF	Thermal – Mechanical Fatigue
TO	Take – Off
TU	Technical University
u	Circumferential speed
UAH	University of Alabama, Huntsville, Al., USA
UHC	Unburned HydroCarbons
V	Volume flow
VDI	Verein Deutscher Ingenieure, Dusseldorf, D
VDIZ	VDI Zeitung (VDI newspaper)
VIGV	Variable Inlet Guide Vane
VKI	Von Kármán Institute for Fluid Dynamics, Rhode-St-Genèse, B
VP	Vice President
VR	VerwaltungsRat (supervisory board)
w	Relative velocity
W	Work
WVA	Wasserbau Versuchsanstalt GmbH, Kochel, D
WW	World War
ZF	(ZahnradFabrik) ZF Friedrichshafen, D
η	Efficiency
ρ	Density

Figures and Copyrights

This book – containing 290 figures – observes the Copyright principles :

12 The Author

Dietrich ECKARDT (*1944) has more than 40 years of professional experience in turbo-machinery research at DLR Cologne, in aero engine industry at MTU Aero Engines Munich and finally in the area of power generation gas turbines at Alstom/ABB in Baden/Switzerland, where this book project was realised between 2009–2013.

He had studied Mechanical/ Chemical Engineering at TU Berlin and finished a doctoral thesis on high-speed centrifugal compressor flow investigations under the supervision of Profs. W. Traupel, ETH Zurich and H.E. Gallus, RWTH Aachen in 1975. Since 1992 he has a honorary professorship for gas turbine development at TU Dresden.

He published approximately 50 technical papers, book articles and lectures on aero and industrial gas turbine engineering, turbomachinery R&D and corresponding history, touching also remote areas like strategic supply chain management, advanced project management (HSG-MBE) and intellectual property right protection.

Currently, he prepares a follow-on publication 'Jet Web' on BBC's role between 1935–1955 on international developments of early aero gas turbines.

eckardt@bluewin.ch

www.ingramcontent.com/pod-product-compliance
Lightning Source LLC
Chambersburg PA
CBHW081522190326
41458CB00015B/5434